Geoforming Mars

Robert Malcuit

Geoforming Mars

How could nature have made Mars more like Earth?

A treatise on some important factors involved in assessing the habitability potential of terrestrial planets that may aid us in our search for habitable terrestrial exoplanets

Springer

Robert Malcuit
Department of Geosciences
Denison University
Granville, OH, USA

ISBN 978-3-030-58875-5 ISBN 978-3-030-58876-2 (eBook)
https://doi.org/10.1007/978-3-030-58876-2

This Springer imprint is published by the registered company Springer Nature Switzerland AG
The registered company address is: Gewerbestrasse 11, 6330 Cham, Switzerland

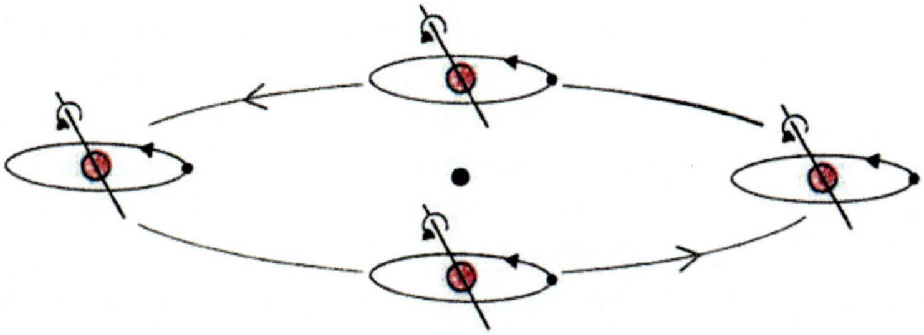

Pictogram illustrating a Mars-like planet in a prograde (counter clockwise) Mars-like heliocentric orbit with a moderately large satellite, about 1/100 the mass of Mars, in a prograde orbit around the Mars-like planet rotating in the prograde direction.

The general theme of the book is that regular tidal action (i.e., both rock and ocean tides) operating over geologic time (in this case marologic time) keeps a planet "alive" by aiding the processes that cycle and recycle planetary resources.

Note also that a satellite helps to stabilize the obliquity (tilt angle) of a planet over multi-thousand-year cycles. The obliquity for this illustration is 30 degrees. The initial obliquity could be more or less than this value but the main point is that it is stable over long periods of time. This stability of obliquity results in a seasonal cycle that is very consistent from year to year, decade to decade, and century to century. This stability of the seasonal cycle, along with other factors such as the presence of liquid water on the surface of a planet, is very conducive to the origin and evolution of life forms.

An extension of the tidal action concept is that the search for potentially habitable planets, in the author's view, should be refocused to the search for terrestrial planets with large satellites relative to the mass of the planet. As with planet Earth, the "Goldilocks" principle can be applied. A satellite in orbit around a Mars-mass planet must be large enough to do the tidal work associated with the recycling of critical planetary resources but not so large that it "despins" the planet to a synchronous spin-orbit condition over a short period of geologic (marologic) time.

This book is dedicated to my wife MARY ANN, for her patience, advice, consolation, and help over these past several decades. We certainly have enjoyed our time on planet Earth. Now, as elder inhabitants of the planet, we are beginning to understand the present environmental conditions on our planet as well as the importance of the operation of the Earth-Moon system.

Preface

Planet Earth has many unusual features. It has oceans of water, an oxygenated atmosphere, a complex biological system, an active magnetic field, a large satellite relative to the mass of the planet, and a prograde rotation rate of about 24 hr/day. Our "twin sister" planet, Venus, also has many unusual features. It has no oceans of water, no free oxygen in the atmosphere, no biological system, no active magnetic field, no satellite at present, and a very slow retrograde rotation rate. Planet Mercury also has a set of special features. It has a very slow prograde rotation rate because of tidal interaction with the Sun, that is, it is locked into a 3:2 spin-orbit resonance with the Sun. It may have an active magnetic field and it definitely has a remnant magnetic signature recorded in its ancient crust. Some of the crater-saturated crust of the planet Mercury probably dates back to the very early history of the planet. Its "sedimentary" rock record consists mainly of overlapping aprons of impact ejecta and it has lava flows of various ages on its surface. Planet Mars has some special features as a planet and, in my view, has the features of a normal terrestrial planet of its size and mass. It has a transparent atmosphere; it has ice caps of a combination of water ice, carbon dioxide ice, and methane ice that wax and wane with the seasons as well as with the long-term obliquity cycle of the planet, and it has evidence of flowing water at the surface. It has both water and wind deposited sediment units as well as a variety of sedimentary rock complexes. It has a great variety of impact craters that have accumulated over the history of the planet and it has some of the largest volcanic constructs in the terrestrial planet realm. It also has two small satellites, Phobos and Deimos, in prograde marocentric orbits. Mars has a prograde rotation rate of about 24.6 hr/day.

A major question is: What are the common features of a terrestrial planet – features to be on a checklist when searching for terrestrial exoplanets? A related question is: How common are terrestrial exoplanets? (About 10/4000 as of August 2019; a reasonable definition of a terrestrial planet is a rocky planet with about two times the mass of Earth or less.) Another unanswered question is: Will the common features of a terrestrial planet yield habitable conditions on a typical terrestrial exoplanet?

The only common feature of terrestrial planets in our Solar System is that they all have rocky surfaces. Mars and Mercury have some portions of their surfaces that are heavily cratered and those surfaces probably date back to their early history. In contrast, Venus and Earth have no portions of their surfaces that go beyond 1 billion years. So how can we attempt to create a story for terrestrial planets that may help us in our search for habitable terrestrial exoplanets?

My basic premise is that moderately massive satellites (1/100 of the mass of the planet, + or – a significant factor) are a necessary, but not sufficient, condition for the development of a habitable terrestrial planet. Details about the mass of the satellite and its direction of revolution are not as important as the presence of a sizeable satellite!

My preliminary analysis of our terrestrial planet group is the following: Mercury is too close to the Sun to yield any reasonable chance for habitability of biota. Our twin sister, Venus, in my view, captured a one-half moon-mass satellite in a sun-stabilized retrograde orbit early in its history and was potentially habitable for about 3 billion years during which the satellite gradually despun the planet and recycled most or all of the surface crustal features into the shallow mantle of the planet. The result is a "Hades-like" planet with a dense carbon-dioxide-rich atmosphere and a very slow retrograde rotation rate. Earth, in my view, captured a moon-mass satellite in a sun-stabilized prograde orbit, and after about 4.0 billion years, we end up with a lunar-mass satellite in a near circular prograde orbit at about 60 planet radii. The prograde rotation rate of the planet over these 4.0 billion years changed from a primordial rate of about 10 hr/day to the present 24 hr/day. The result is something like a "paradise" planet that we have grown to know and love!

So the question of this book is: How could nature make a mars-like planet (a normal terrestrial planet in a mars-like heliocentric orbit) into a planet that could harbor a biological system of some sort for a significant span of geological time? The answer, in my view, is to acquire a massive satellite relative to the mass of the planet and a few chapters of this book suggest different scenarios for accomplishing this task! In short, GEOFORMING MARS!

Chapter 1 introduces some special features of the terrestrial planets and the Solar System as well as suggests a procedure for solving problems in the natural sciences. Chapter 2 is about the early evolution of the Sun and the Solar System and why such a model is very important for an explanation and analysis of the condition of the terrestrial planets and asteroids. Chapter 3 discusses various models that have been suggested for the origin of the current satellites of Mars, Phobos and Deimos, and also discusses the origin of the only other extant natural satellite in the terrestrial planet realm, the Earth's Moon. Chapter 4 presents a prograde gravitational capture model for Mars in which a 0.1 moon-mass planetoid is captured from a heliocentric orbit into a prograde marocentric orbit early in the history of the Solar System. A crucial issue for any capture model is to identify potential Solar System sources for such a planetoid or family of planetoids. Chap. 5 presents a retrograde gravitational capture model in which a 0.1 moon-mass planetoid is captured from a mars-like heliocentric orbit into a retrograde marocentric orbit early in the history of the Solar System. Certain features of prograde and retrograde capture scenarios are similar (both are gravitational capture scenarios) but the long-term effects are as different as Dr. Jekyll and Mr. Hyde. Chapter 6 critiques the Giant Impact Model, which has been a "ruling paradigm" in the Earth and Planetary Sciences for the past three decades. Chapter 7 chronologically develops a Prograde Gravitational Capture Model that appears to offer solutions for many of the outstanding problems of the Earth-Moon system as well as for many other problems of Solar System Science. Chapter 8 ventures into Comparative Planetology under the guise of the GEOMETRY OF STABLE CAPTURE ZONES (SCZs). The discussion focuses on how the SCZ concept can be used to assess the probability of successful capture of satellites of specific masses by terrestrial (and non-terrestrial) planets. In the author's view, stable gravitational capture of sufficiently massive satellites is a major key process for the development of habitable terrestrial planets. Thus, the identification of large satellites should be a major factor in the search for potentially habitable exoplanets. Chapter 9 discusses the factors that are involved in making a planet habitable. There appears to be two categories for habitability: (1) short-term habitability (i.e., 1 or 2 billion years) and (2) long-term habitability (i.e., over 2 billion years). Mars, the normal planet, appears to be in the first category, and some of the features for a normal terrestrial planet are (1) that it has large portions of its primitive crust after 4.6 billion years and (2) that it has no large satellite at present or evidence of the existence of a large satellite in a stable orbit in its past. Planet Earth is in the second category and is a terrestrial planet that (1) has a large extant satellite in a stable orbit and (2) has very little rock, mineral, and chemical isotope evidence that could be considered as relicts of a primitive crust after 4.6 billion years of geological history. A major question emerging from this chapter is: WHAT WOULD PLANET EARTH BE LIKE WITHOUT ITS MOON? A reasonable response is that it would have a prograde rotation rate of about 12 hr/day and that it would have large sections of its primitive crust surviving at or near the surface of the planet after over 4 billion years of planetary evolution. The "normal" earth-like planet in an earth-like heliocentric orbit would have no deep oceans, no mobile continents, and no recycling system for volatiles (such as water, carbon dioxide, methane, and others). Chapter 10 discusses the importance of large satellites being associated with both our set of terrestrial planets as well as terrestrial exoplanets. The author suggests that a "LARGE SATELLITE" paradigm should be added to the "CONTI NENTAL DRIFT/PLATE TECTONICS" paradigm (i.e., the Wegener paradigm) and the "PLANETARY AND ORBITAL PARAMETER VARIATION" paradigm (i.e., the Milankovitch paradigm) for explaining the development of planet Earth. It is the view of the author that these three major concepts (i.e., paradigms or revolutions in thought) are all interrelated and are necessary for a reasonable scientific explanation of the special features of our very habitable "paradise" planet that has supported biological entities for at least 3.5 billion years.

Granville, OH, USA Robert Malcuit

Some Perspectives and Acknowledgments

This project is a progression of my long-term interests in the origin and evolution of the terrestrial planets of our Solar System and of the Solar System in general. Getting closer to home, we all know that the Earth-Moon system is very complex and its origin is still a very debatable issue in the Earth and Planetary Science community. Planet Venus, our sister planet, has a very stable, sun-centered orbit but apparently it has a very complex evolutionary history that is essentially shrouded in mystery. In my view, many scientists in the planetary science community are now only beginning to appreciate the apparent complexities of these two large terrestrial planets as well as the apparent long-term stabilities of their sun-centered orbits. On the inner side of Venus is Planet Mercury with its high specific gravity, elevated eccentricity relative to Earth and Venus, and many other unusual features. Mercury, however, is of very little interest to the biological community because of its proximity to the Sun. Planet Mars, on the other hand, also has many unusual characteristics but one outstanding feature is that its crustal complex dates back to the early history of the solar system. It has a decipherable rock and mineral record dating back some 4.5 billion years. Planet Mars has no serious complications like a large satellite that potentially could have caused the destruction of the surface rock complex and recycled much of it into the Martian upper mantle. Because of this lack of dynamical complexity and a rotation rate that could be very near to its primordial rotation rate, Mars is considered by some scientists, me among them, to be a "normal" terrestrial planet, perhaps a "prototype" of a terrestrial planet to be used as an aid in our search for terrestrial exoplanets.

The purpose of this book is to illustrate how an unusual process like acquiring a large satellite (1) could change the Martian planetary story from fairly simple to complex and (2) could theoretically convert a Mars-type planet into a habitable world.

I want to acknowledge the excellent geological and general education background I received in my undergraduate and graduate years at Kent State University and at Michigan State University. I am especially appreciative of the field work with Richard Hemlich (KSU) in the Bighorn Mountains, Wyoming, which eventually led to my acquaintance through the geological literature and later in person with Preston Cloud (UC, Santa Barbara). Many thanks to Tom Vogel (MSU) for his guidance through the experiences of graduate school and for the permission to do an unusual project in planetary science for a Ph.D. project. At MSU, I did get to meet Harold Urey in person, although very briefly, after a talk he presented on campus on the subject of the origin of the Moon. Harold thought that the body of the Moon was formed away from the Earth and was later captured (somehow) by the Earth. A "STELLAR" CONCEPT in my opinion. He also expressed on several occasions that the Moon may eventually turn out to be a "Rosetta Stone" for interpretation of the history of the inner solar system. I still think that his concept is very commendable. At AGU and GSA meetings while I was a graduate student, I did meet with such notables as Hannes Alfven, Fred Singer, Keith Runcorn, and Harold Masursky, all of whom at some time in their professional lives were proponents of the capture model for origin of the Moon. It is interesting to note that Al Cameron was friendly to the capture model in the early 1970s because it was a "default" model (i.e., fission and co-formation models did not seem to lead to a solution by Al Cameron's reasoning). Then Al went off on a 20-plus year excursion doing world-class computer simulations and other calculations in support of the Giant Impact Model only to find that it did not solve the lunar origin problem. OH WELL, SHUCKS!! Maybe the "default" model of gravitational capture is a reasonable solution for the lunar origin problem after all!

A recent pursuit of astronomers and planetary scientists is the search for exoplanets, especially habitable terrestrial exoplanets. The general theme in the search for potentially habitable planets is to "FOLLOW THE WATER." The general idea is that the presence of liquid water on a planet can lead to the development of life forms. My advice for exoplanet investigators is, yes, look for a planetary environment than can yield liquid water but of equal importance for long-term habitability is the presence of a large satellite relative to the mass of the planet and for a central star that is "not too hot" and "not too cold." We now know that there are many exoplanets out there but we also know that terrestrial exoplanets appear to be rare and that terrestrial exoplanets with both liquid water and a reasonably large satellite could be extremely rare. The bottom line, in my view, is that we humans and our plant and animal friends may be "alone" in a large region of space. We must realize that we live on a special planet and we as an "intelligent" species must develop a "special" attitude toward our planet and its limited resources.

In graduate school I did projects with Gary Byerly (LSU) and Graham Ryder (a long-term associate with the LPI before his premature death in 2002). In graduate school and over the years I had many fruitful interactions at national and international meetings with these two individuals.

Special thanks to Ron Winters (DU Physics and Astronomy) for collaboration on planetary science projects over the years and for his work with physics and computer science students, David Mehringer, Wentao Chen, and Albert Liau, in the late 1980s and early 1990s.

Many thanks to Harrison Ponce, Cheryl Johnson, Leslie Smith, and Madelin Myers (Denison ITS) for technical support and manuscript preparation. Thanks to

Adam Weinberg, President of DU, and the personnel in the Office of the Provost (Kim Coplin, Kim Specht, and Cathy Dollard and their very helpful assistant Jane Dougan) for office space, computer equipment, and other assistance and supplies since retirement. Thanks to Tom Evans (DU, Chemistry) for encouragement to pursue this project and for sharing his knowledge of the resources of the Harold Urey Collection in the archives of UC, San Diego Library. Thanks to my active departmental colleagues and generations of geoscience students for the many questions and comments over the years. Thanks to Ron Doering, Clement Wilson, Aaron Schiller, and others on the Springer team.

On a very personal level, special thanks to my wife, Mary Ann, for her patience during the preparation of this book project and for her perusal of each chapter and for her special advice and constructive criticism. The manuscript is much improved because of her careful attention to detail.

Contents

Author Biography

Robert Malcuit is professor emeritus at Denison University. Though he retired from teaching in 1999, he still pursues research interests in Earth and planetary sciences. His current research pursuits are: (1) gravitational capture potential for planets, (2) explanation of several major features of the Earth and Moon, (3) explanation of the major features of planet Venus, (4) explanation of many features of the Neptune–Triton system, and (5) explanation of several unique features of the Earth in terms of a capture origin of the Earth–Moon system and the subsequent evolution of the lunar orbit to the present condition.

Chapter 1
Introduction

> Mars is believed to be lifeless, but it may be possible to transform it into a planet suitable for habitation by plants, and conceivably humans. The success of such an enterprise would depend on the abundance, distribution and form of materials on the planet that could provide carbon dioxide, water and nitrogen. —From McKay et al. (1991, p. 489).

> Mars is in some ways the type example of a terrestrial planet. It is neither geologically stillborn, like Mercury or the Moon, nor so active that most of the geological record has been destroyed, like Venus or the Earth. —From Nimmo and Tanaka (2005 p. 133).

> We revisit the idea of 'terraforming' Mars – changing its environment to be more Earth-like in a way that would allow terrestrial life (possibly including humans) to survive without the need for life-support systems – in the context of what we know about Mars today.

> Recent observations have been made of the loss of Mars's atmosphere to space by the Mars Atmosphere and Volatile Evolution Mission probe and the Mars Express spacecraft, along with analyses of the abundance of carbon-bearing minerals and the occurrence of CO_2 in polar ice from the Mars Reconnaissance Orbiter and the Mars Odyssey spacecraft. These results suggest that there is not enough CO_2 remaining on Mars to provide significant greenhouse warming were the gas to be emplaced into the atmosphere; in addition, most of the CO_2 gas in these reservoirs is not accessible and thus cannot be readily mobilized. As a result, we conclude that terraforming Mars is not possible using present-day technology.— From Jakosky and Edwards (2018, p. 634).

Planet Mars appears to be a normal terrestrial planet. It still has much of its original (primitive) crust. Its prograde rotation rate, about 24.6 hour/day, is what is expected for a planet of that mass and density (MacDonald 1963, 1964). It has a rarefied atmosphere and a low mass hydrosphere mainly because the mantle of the planet has not been degassed to any significant extent over the last few billions of years. Figure 1.1 shows a typical view of planet Mars.

At this point I need to define some new terms to be used in this book. The traditional term for "engineering" a planet like Mars is TERRAFORMING. By traditional use the term "terraforming" is a project of engineering by humans to make a planet like Mars somewhat habitable for biological forms. After consideration of the information in Jakosky and Edwards (2018) it appears that the term "terraforming" could also be used for the construction of semi-sealed geodesic

R. Malcuit, *Geoforming Mars*, https://doi.org/10.1007/978-3-030-58876-2_1

Fig. 1.1 Photo of Mars showing many of the surface features of the planet such as the north polar ice cap, the three major volcanoes of the Tharsis plateau, and Olympus Mons (to the west of the plateau), and Valles Marineris. The thin clouds of Mars are water vapor and the transparent atmosphere is composed mainly of carbon dioxide. (Photo is courtesy of NASA/JPL)

dome communities or perhaps larger projects on Mars for human occupation. Jakosky and Edwards (2018) conclude that planet-wide TERRAFORMING of planet Mars may not be physically possible using current technology and resources on Mars but, in my view, smaller scale projects as described above could be constructed and operated using mainly local resources.

Here I am introducing the term GEOFORMING as a theoretical planetary development process resulting in conditions on a planet or exoplanet that are similar to those on Earth at the present time or at some time in the past or future evolution of planet Earth. Other suggested theoretical planetological development terms are Venoforming, Maroforming and Mercoforming. A VENOFORMED planet would have conditions similar to those on Venus today or at some time in the past or future evolution of planet Venus. A MAROFORMED planet would have conditions similar to those on Mars at the present time or at some time in the past or future evolution of planet Mars. A MERCOFORMED planet would have conditions similar to those on planet Mercury today or at some time in the past or future evolution of planet Mercury.

Some terrestrial planets like Venus and Earth can evolve considerably over geologic time. For example, conditions on planet Earth were significantly different one billion years or two billion years ago than they are at the present time. The same can be said for planet Venus. Other terrestrial planets evolve very little after the initial stages of planet formation: examples of this category, in my view, are planets Mercury and Mars.

The purpose of this book is to summarize the current state of conditions on Mars and then to present some theoretical planetological development processes that could have resulted in conditions similar to those of planet Earth today. In short, we want to GEOFORM MARS.

Planet Earth, in contrast to Mars, has a large satellite that has had major effects on the rotation rate and the degassing history of the planet. The primordial rotation rate of about 10 hour/day (MacDonald 1963, 1964) has gradually changed by rock and ocean tidal interaction with the Moon over geologic time to our present rate of 24 hour/day.

Planet Venus, in contrast to Mars, has undergone a global resurfacing event in the past 0.5–1.0 Ga (billion years) (Herrick 1994; Herrick and Parmentier 1994), the mantle has been degassed significantly [i. e., about 1.6 times that of Earth (Cloud 1972, 1974)], and the rotation rate has been changed, over geologic time, from the expected primordial prograde rate of about 13.5 hour/day (MacDonald 1964) to the present very slow retrograde rotation rate of about 243 earth days/year.

A major question is: Are there any major events in the history of planet Mars that could have resulted in a more complex planetological evolution, an evolution that may have been more favorable for the development of at least a relatively simple biosphere?

In this book we will explore the theoretical ramifications of the gravitational capture of a sizeable satellite of 1/100 the mass of Mars. This mass ratio is very similar to that of the Earth-Moon system which is 1/81. We will also consider the positive and negative effects of capturing an even larger satellite of about 0.02 of the mass of Mars. Along the way we will discover:

- that both prograde and retrograde gravitational capture scenarios for these sizeable planetoids are physically possible for planet Mars,
- that any of these capture scenarios could have had profound effects on the lithospheric, hydrospheric, and atmospheric evolution of planet Mars,
- that a major long-term effect for any stable capture scenario is a very significant change in the rotation rate of planet Mars by way of the rock and ocean tidal history of the planet,
- that any of these capture scenarios would have greatly increased the chances of development of primitive life forms on Mars over geologic time.

1.1 The Real Evolutionary History Versus "What Could Have Been" for Planet Mars

This book on geoforming Mars is partly a work of theoretical planetology but it starts with a set of facts (and some interpretations) about planet Mars at the present time in the scientific investigation of the planet. The early history of Mars (what I call in this book, the pre-capture era) is the same for all five models discussed below because any capture episode would probably occur within a few 100 Ma after the formation of the planet. In general, there are five evolutionary scenarios to consider.

Scenario One: This model begins with a primitive Mars in a mars-like heliocentric orbit with a primitive rotation rate of ~24.6 hr/day and no sizeable satellite (Phobos and Deimos are much too small to make a difference here). This scenario is essentially the Mars that we have today. If planet Venus or planet Earth would undergo an evolution scenario like this it would be a Maroforming process.

Scenario Two: This model begins like Scenario One with a Mars-like planet in a Mars-like heliocentric orbit. Then within the first few 100 Ma of its existence Mars captures a 0.1 moon-mass planetoid from a mars-like heliocentric orbit. The post-capture marocentric orbit circularizes in a few billion years during which the prograde rotation rate of Mars changes from 24.6 hr/day to ~88 hr/day. The rock tidal amplitudes at the time of capture are up to 9 km and rock tidal amplitudes up to 6 km are common for the first few thousand years after capture. But during the orbit circularization of the satellite, the prograde rotation rate of Mars is forever decreasing because of angular momentum exchange with the satellite in prograde orbit. The chances for life forms to evolve in the low to mid-latitude zones are favored by the high rock and ocean tides and associated volcanic outgassing soon after capture but the gradual decrease in rotation rate is not all that favorable for life forms because of the increasingly long, cold nights on the planet. This scenario if an attempt at Geoforming.

Scenario Three: This model is similar to the previous model in that it begins with a Mars-like planet in a Mars-like heliocentric orbit. Then within the first few 100 Ma Mars captures a 0.2 moon-mass planetoid in a prograde marocentric orbit from a Mars-like heliocentric orbit. The rock tidal amplitudes for this capture scenario are about 18 km and the timescale for orbit circularization is considerably shorter. The planet, however, loses much of its prograde rotation rate via tangential tidal action over a much shorter period of time than in Scenario two. Thus the time-frame for evolution of life forms is very favorable over that brief period of marologic time. As in scenario one, the "long-night" syndrome would probably make living condition intolerable within a few billion years of time. This is another attempt at Geoforming.

Scenario Four: This model has a similar beginning to Scenarios one, two, and three with a Mars-like planet in a Mars-like heliocentric orbit. However, in this case a 0.1 moon-mass planetoid is captured into a stable but highly elliptical retrograde marocentric orbit. The timescale for retrograde marocentric orbit circularization is very short relative to that for a prograde orbit and within a few 100 Ma the orbit is circularized to about 30 mars radii. The high rock tides and intense rock and ocean tidal activity associated with the capture event and subsequent orbit circularization would probably result in a large portion of the crust in the equatorial zone of Mars being recycled into the martian mantle. All this rock tide activity would result in massive volcanism and outgassing of mantle volatiles. During this retrograde capture scenario, the rotation rate of Mars decreases rapidly and any biological system on the planet finds itself in the "long night syndrome" fairly rapidly but the rock and ocean tides are forever increasing in amplitude and frequency as the orbit evolution proceeds. Eventually, after about 8 billion years of time (about 4 Ga in the future for Earthians), the satellite breaks up in orbit at the Roche limit for a solid body. The final result is a mars-like body with the environmental conditions of planet Venus today (see Malcuit 2015, Chap. 6). This scenario is an attempt at Venoforming.

Scenario Five: This model is similar to that of Scenario Four but the captured retrograde satellite is twice as massive. Thus the tidal amplitudes are twice as high and the post-capture orbit circularization is somewhat shorter. The planet is despun over a shorter period of time and there is a lengthy period of the "long night syndrome". Eventually the prograde rotation of the mars-like planet decreases to zero and the planet begins to rotate in the retrograde direction. There are high rock and ocean tidal amplitudes as well as a progressive increase in both tidal amplitude and frequency. The increasing retrograde rotation rate of the planet prolongs the lifetime of the retrograde marocentric orbit. THIS IS AN UNUSUAL ORBITAL SCENARIO BUT IT IS THE MOST FAVORABLE FOR THE DEVELOPMENT AND SUBSEQUENT EVOLUTION OF LIFE FORMS ON A MARS-LIKE PLANET! In this scenario the satellite orbit eventually evolves to the Roche limit for a solid body, breaks up in orbit, and the particles, large and small, fall to the surface of Mars. The final rotation rate of Mars, in this scenario, is about 11 hr/day retrograde and the environmental conditions on Mars are similar to those on planet Venus today. This is another attempt at Venoforming.

1.2 Discussion of the Primordial Planetary Rotation Rate of Solar System Bodies

There appears to be two schools of thought to explain the rotation rates of both terrestrial and gaseous planets. School One proposes that there is an "orderly component of planetary rotation" and that this orderly component resulted from the accumulation of small bodies from orbits of low eccentricity by a favored planetary embryo on a circular orbit. Calculations suggest that planetesimals on elliptical orbits from the outer regions of the accretion zone of the planet generally induce a relatively rapid prograde rotation (i.e., prograde rotation rates like we observe today for planet Mars) as well as prograde rotation rates that are in

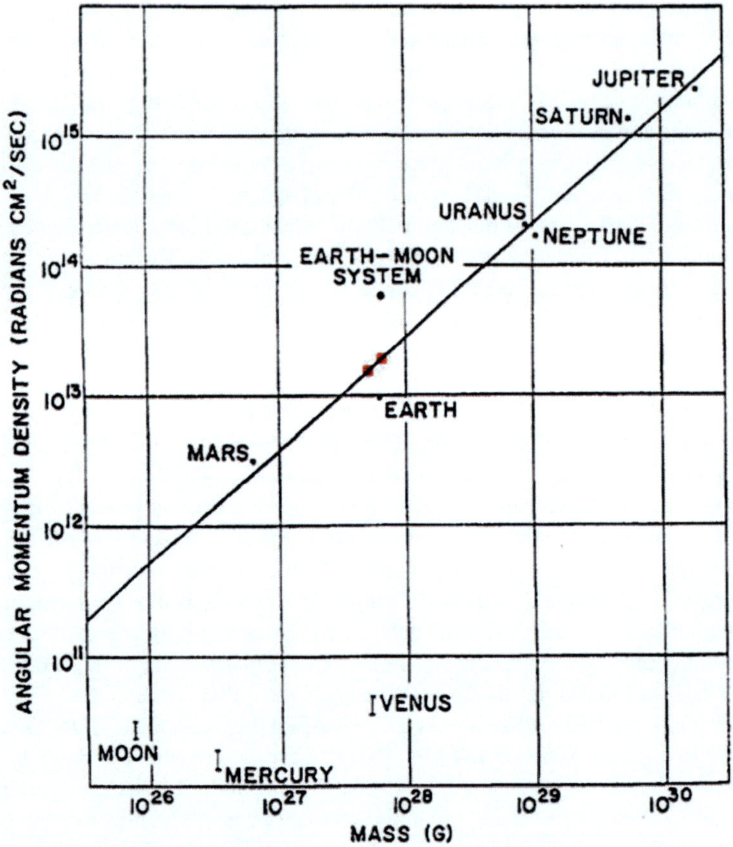

Fig. 1.2 Plot of Angular Momentum Density vs. Mass for the planets of the Solar System as well as for the Earth's Moon and the Earth-Moon system. (Diagram adapted from MacDonald (1963, Fig. 38, with permission from Springer.) The line was placed for the best fit of the information for the rotation rates for the outer planets and Mars. The assumption, which is suggested by the plot, is that Mars has a rotation rate very close to its primordial rotation rate (~25 h/day). If all the angular momentum of the Earth-Moon system is placed in the Earth, the rotation rate would be ~4.5 h/day. If the Earth is rotating ~10 h/day, the Earth would plot on the line between the position of the Earth-Moon system and the Earth (the *red square* symbol on the *right*). This information suggests that the original rotation rate for Earth was ~10 h/day. If the angular momentum of a lunar-mass body in a 30 earth radii circular orbit in a prograde direction is added to the prograde angular momentum of an Earth rotating at 10 hr/day, then that combination plots in the position of the Earth-Moon system on the plot. Likewise, if Venus is elevated to the line vertically above its position on the graph (the *red square* symbol on the *left*), then the original rotation rate would be ~13.5 h/day. The primordial rotation rates of the Moon as an independent planet and that of planet Mercury can be estimated using the same procedure. (Diagram from MacDonald (1963, Figure 38) courtesy of Springer)

agreement with the observational plot in MacDonald (1963, 1964) (see Fig. 1.2). These accretion schemes apply to both terrestrial planets and gaseous planets. Major sets of calculations for this model are in Lissauer and Kary (1991) and Lissauer et al. (1997).

The MacDonald (1963, 1964) plot suggests a primordial prograde rotation rate of about 10 hr/day for Earth and about 13.5 hr/day for Venus. These prograde rotation rates are reasonable initial conditions for explaining a prograde gravitational capture model for the earth-moon system and a retrograde gravitational capture model for a Venus-satellite system. Thus, the history of the four terrestrial planets can be explained in some detail using the "orderly component of planetary rotation" (Lissauer and Kary 1991; Lissauer et al. 1997) in conjunction with the plots of MacDonald (1963, 1964). Johansen and Lambrechts (2017) present a review of the process of the formation of planets by way of the accretion of pebbles; in this review they present calculations demonstrating how prograde rotation of a planetoid results from the accretion of pebbles in a gaseous environment.

Additional support for this "orderly component of planetary rotation" school comes from detailed calculations on the history of the Earth-Moon system when starting with the present conditions of the system. Hansen (1982) did a traceback calculation of the Earth-Moon system which included both the action of the ocean tides and the earth tides. He presented a few different combinations of the positions of the continents, dimensions of the continental shelves, and other characteristics that might affect the tidal friction process. Regardless of the configurations used, he could not get the lunar orbit much smaller than a circular orbit of about 30 earth radii over 4.6 Ga of time which relates to a rotation rate for earth of about 10 hr/d. Webb (1982) did a completely independent traceback calculation using somewhat different assumptions for the parameters of tidal friction and arrived at a similar conclusion. Is it ironic, or coincidental, that our (Malcuit et al. 1992; Malcuit 2015) most reasonable post-capture orbit, after about 600 Ma of post-capture orbit circularization to about 10% eccentricity has the angular momentum equivalent to a circular orbit of about 30 earth radii? Indeed, the angular momentum of a 30 earth radii lunar orbit plus the angular momentum of a primitive earth with a prograde rotation rate of 10 hr/day equals the angular momentum of the Earth-Moon system.

School Two proposes that any systematic (i.e., orderly) component in the planetary accretion process is "overwhelmed" by the random process of accumulation of bodies of various sizes. In this model it is the large impacting bodies that determine both the rotation rate and obliquity (tilt angle) of the resulting planetary unit. Some early calculations for this model are by Wetherill (1985), Tremaine (1991), and Dones and Tremaine (1993a, 1993b). Major support for this stochastic process model is from the promoters of the Giant Impact Model (GIM) for the origin of the Earth's Moon. A prograde rotation rate of about 4.5 hr/day is needed immediately following the tangential impact for the traditional model of the GIM that features a mars-mass body impacting obliquely onto the primitive earth. Thus, for nearly all of these stochastic impact models it is very important to have a large planetoid to impact obliquely onto an earth-like planet. But the question is: "WHERE DID SUCH A LARGE BODY COME FROM???" The place of origin of Theia (the hypothetical mars-mass impactor in the GIM model) has not been determined. Belbruno and Gott (2005) suggest an origin near either the L4 or L5 Lagrange point of the orbit of Earth as a source for the mars-mass body because of the similarity of oxygen isotopes between Earth and Moon. Herwartz et al. (2014)

suggest that Theia had the composition of enstatite chondrites which implies that the body formed in the vicinity of the earth's orbit. Meier et al. (2014) suggest that Theia had an earth-like composition and formed in the vicinity of the orbits of Earth and Venus.

Davies (2008) and Gillman et al. (2016) suggest that the present retrograde rotation of Venus may have been caused by way of impacts with planetoids associated with the Late Heavy Bombardment (LHB) of the inner part of the Solar System. [Note that the LHB is about to become abandoned because of lack of evidence in the rock records of the Earth and Moon (Mann 2018)]. The place of origin for a large impactor, or a covey of smaller impactors, for a GIM model for Venus to explain the retrograde rotation of Venus has not been suggested. The place of origin of the TWO EARTH-MASS BODY to impact on Uranus (Slattery et al. 1992) to cause the obliquity to increase instantaneously to 98 degrees has not been determined or proposed. Although there can be stochastic impactors postulated for nearly all planetary obliquity and planetary rotation problems, not one of these solutions explains the details of planet evolution or planet-satellite system evolution. Furthermore, other solutions have been proposed that do not involve a mysterious giant impactor. (Note: Various features of giant impact models for solving solar system problems will be discussed extensively in Chap. 6.)

In a short note in SCIENCE Dan Clery (2013) summarized the Royal Society of London Meeting on the Origin of the Earth's Moon. The title of the article is "THE IMPACT MODEL GOT WHACKED". In my view the promoters of the GIM have never recovered from that "whacking". According to Mann (2018), the concept of the Late Heavy Bombardment is heading toward the exit as well.

The stochastic impact model for planet formation looked like a very good solution when the Giant Impact Model was presented at the Kona Conference (see Hartmann et al. 1986, for the resulting Conference Volume) because the three traditional models (fission, co-formation, and capture) had "fatal flaws". The "fatal flaw" syndrome is now "infecting" the GIM as well as its replacement, the Syenestia model. As a historical note, it seems like the GIM is now having about as much success for solving Solar System Problems as the STABLE CONTINENT PARADIGM did in the earth science realm in the 1960's. The continental drift/plate tectonics model came up with much better solutions.

Returning to the problems of the evolution of the lunar orbit, Ross and Schubert (1989) presented a model for the "forward" evolution of the lunar orbit. This calculation was based on the evolution of the Earth-Moon system in the aftermath of the GIM. The initial conditions for the calculation were (1) a lunar orbit just beyond the Roche limit (about 3 earth radii) and (2) a rotation rate for earth of about 4.5 hr/d. Using their selected set of body deformation parameters they found that the lunar orbit evolved out to about 50 earth radii in about 2 billion years. But they found that it was very difficult to get the lunar orbit to expand beyond about 53 earth radii in 4.6 billion years. They speculated that the tidal friction associated with the ocean tidal regime must have increased significantly at about that time (i. e., when the lunar orbit was at about 50 to 53 earth radii and the rotation rate of earth was about 16.9 hr/d; *i. e.*, at about 50 earth radii). It is ironic that both Davies (1992) and Stern (2005) suggest the thermal conditions of the upper mantle of earth would be conducive to the initiation of "slab-pull" plate tectonics (i.e., modern style plate tectonics) at about 1.0 Ga ago. Perhaps deeper ocean basins and moderately wide continental shelves would be conducive to increased tidal friction in the 1.0 Ga to recent eras of earth history.

An interesting quote from Ross and Schubert (1989, p. 9540) is:

> With the solid tide dissipation we adapted here, the Moon would never move beyond about 53 R_E were it not for the ocean tide.

On page 9540–9541:

> In fact, numerical models of the ocean tide in backward evolution calculations suggest that when the semimajor axis becomes less than about 40–50 R_E, oceanic dissipation falls sharply and the Moon remains "stuck" at about 30 R_E [Hansen 1982; Webb 1982]. To circumvent the problem, Hansen [1982] proposed a capture origin for the Moon with capture at about 30 R_E! Our results show that such drastic proposals are not required to reconcile formation of the Moon near the Earth with inactive oceanic tides inward of about 30 R_E.

This last quote is a comment against the Lunar Capture Model for the origin of the Earth-Moon system and the authors refer to the Hansen (1982) and Webb (1982) calculations. Hansen (1982) did say that he could not get the lunar orbit beyond 30 earth radii in his trace-back calculation and suggested that the Moon was captured in the early history of planet earth. In my view, this speculation of lunar capture by Hansen (1982) was a good one and our calculations suggest that a highly elliptical post-capture lunar orbit has the orbital angular momentum of a 30 earth radii circular geocentric orbit. Such a prograde lunar orbit in association with an earth-like planet with a prograde rotation rate of 10 hr/d yields the angular momentum of the Earth-Moon system.

A note on the term "drastic proposal" for a capture origin deserves a brief comment. I point out that the standard Giant Impact Model is over 3000 times more energetic than any capture model and the Synestia model is much more energetic than the GIM. So which models are the "DRASTIC PROPOSALS"?

How do these contrasting models for initial planet rotation relate to Planet Mars? Well, the rotation rate of 24.6 hr/d for Mars is very consistent with the Specific Angular Momentum vs. Mass diagram of MacDonald (1963, 1964) in Fig. 1.2. This rotation rate is also very consistent with the concept of Mars being a very normal terrestrial planet (Nimmo and Tanaka 2005), essentially a prototype of a terrestrial planet. Where Mars fits in with the stochastic approach to the problem of planetary rotation is not at all clear.

A quote from Laskar is appropriate here (Laskar 1995, p. 104):

> Mars is far from the Sun, and its satellites Phobos and Diemos have masses far too small to slow its rotation, so that its present rotation period of 24 hours 37 minutes is likely to be close to its primordial rotation period.

1.3 The Scientific Method as It Applies to Models in Theoretical Planetology

The scientific method is a procedure for developing and testing new ideas in the natural sciences. It can also be used to make predictions of outcomes of simple and/or complex systems. There are five basic steps to the process:

(a) We start with a LIST OF FEATURES TO BE EXPLAINED (i.e., the facts to be explained by our hypothesis or model—a model simply being a somewhat more detailed explanation than a hypothesis)
(b) We FORMULATE A HYPOTHESIS (a hypothesis is simply an untested explanation)
(c) We TEST THE HYPOTHESIS (1) by making more observations of whatever is being investigated, (2) by doing relevant experiments, (3) by doing relevant calculations, and (4) by making predictions and then independently checking the predictions for accuracy
(d) The HYPOTHESIS is either VERIFIED or REJECTED based on the results of the tests. In many cases in "big picture" natural science issues, the verification or rejection can come decades after the hypothesis is proposed. And in some cases, a rejection is reversed to verification after a new test or new technology for testing the idea has been discovered or invented
(e) If VERIFIED, a HYPOTHESIS becomes a THEORY (a theory is simply a well-tested explanation)

Two recent articles on the SCIENTIFIC METHOD are by van Loon (2004) and Lipton (2005). These authors emphasize the progression from speculations to hypotheses to models as well as the concept of making predictions that can be independently tested (i. e., without prejudice).

Now let us discuss some of the special characteristics of the subject of this book: planet Mars, the Red Planet. Figure 1.3 shows a planar view of the geometry of inner part of the Solar System and Fig. 1.4 is a diagram illustrating the orbits of the outer planets. These diagrams will give the reader some orientation for the discussions in this introductory chapter. Figure 1.5 shows typical images of the terrestrial planets. We will begin our discussion by briefly discussing some special features of planet Mars and then compare and contrast some of these features with those of planet Earth. Then we can consider models in theoretical planetology and make predictions of what a mars-mass exoplanet would look like if it underwent a prograde gravitational capture scenario or a retrograde gravitational capture scenario. And, in my view, we already know the results of a "normal" evolutionary pathway for an essentially moonless mars-mass planet!

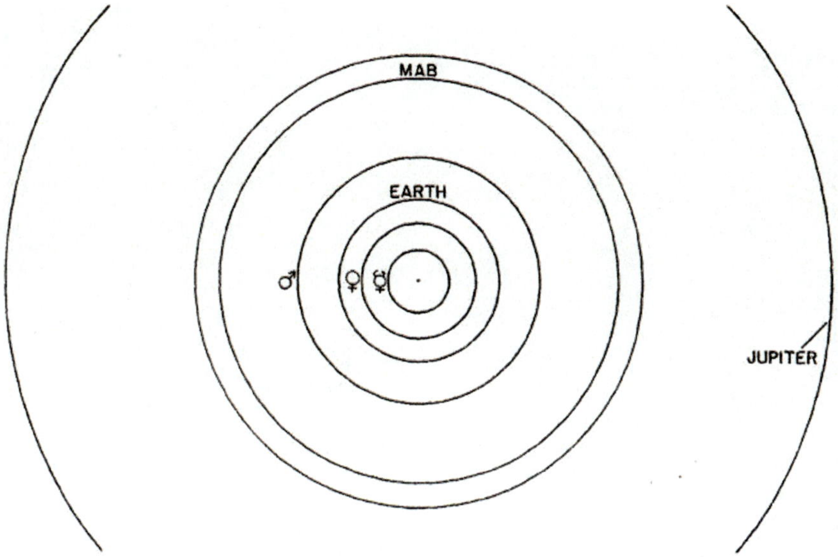

Fig. 1.3 Scale sketch of the orbits of the planets in the inner part of the Solar System. Planet Mars is between Earth and MAB (Main Asteroid Belt). Venus and Mercury, respectively, are interior to Earth. For scale, Earth is located at 1 astronomical unit (AU).

1.4 Some Special Features of Mars as a Planet

A discussion of these features is considered as a preface for all five scenarios listed in Sect. 1.1. I will select certain characteristics for discussion that have been emphasized by many investigators. I should point out that all of the proposed interpretations will not get equal consideration.

1.4.1 Does Mars Have a Metallic Core?

It is not clear from the data in Table 1.1 that planet Mars has a metallic core large enough to generate an internal magnetic field as a "stand alone" planet. The moment of inertia (MI) factor for Mars is lower than that for the Earth's Moon but much higher than the MI factor or Earth, Venus, or Mercury.

There is no general agreement on the heat source for planet differentiation including metallic core formation. It is some combination of (1) heat from accretion of bodies to form the planet, (2) heat from the decay of radioactive elements especially the short-lived isotopes of Al-26 and Fe-60, and (3) heat generated by gravitational emplacement of the metallic core. Roberts and Arkani-Hamed (2017)

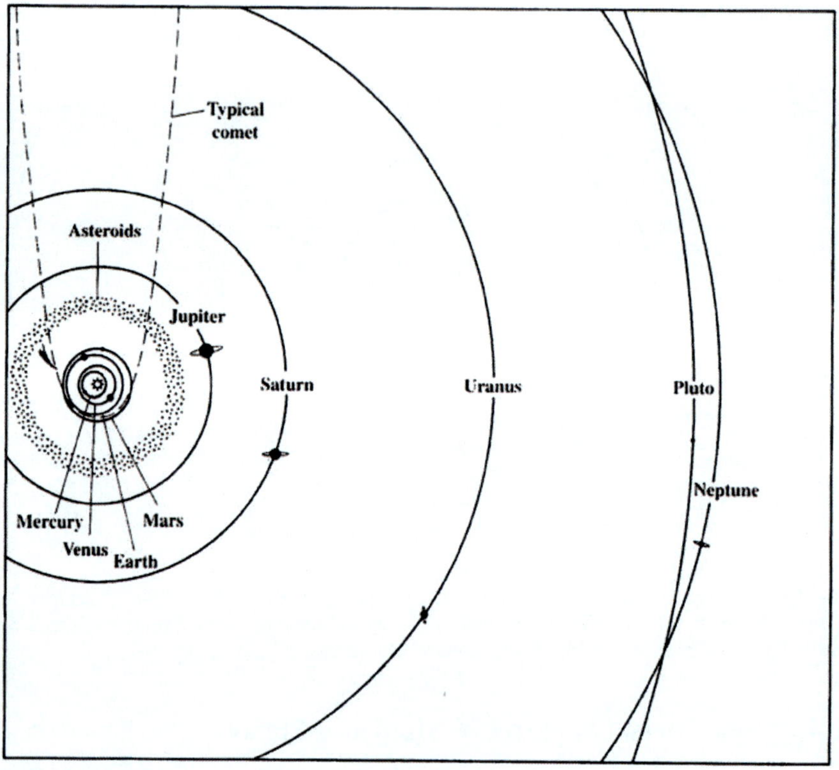

Fig. 1.4 Scale sketch of a broader view of the Solar System showing the relative positions of the orbits of the planets in the Solar System. The terrestrial planets are inside the belt of the Asteroids and the outer gaseous planet are outside of the belt of the Asteroids. The highly elliptical orbit of a typical comet is shown with a perihelion in the vicinity of the orbit of Mars. Note that Pluto is nearer to the Sun than is Neptune for a portion of its orbit. Note also that pictograms of the planets on the diagram are not to scale but the relative sizes of the pictograms of the planets on the diagram are nearly to scale. (Diagram is from Lunine, (1999, Fig. 1.2) with permission from Cambridge Univ. Press)

presented a model for core formation in planet Earth via accumulation of moon- and mars-sized planetoids. Certain parts of this model can be applied to Mars for the formation of a molten metallic ring or a molten inner core that would be useful for generating a short-lived global magnetic field. Zhang (1994) did an estimate of the radius of the martian core, based on the physical parameters that were known at that time, to be between 1520 and 1850 km. He assumed that the mean moment-of-inertia ratio was between 0.350 and 0.360. More recently Yoder et al. (2003) did an estimate of the radius of the core of Mars based on an analysis of the effects of the solar (rock) tide on the planet. Their analysis resulted in a core radius between 1520 and 1840 km and in their analysis at least the outer part of the core is liquid. Thus,

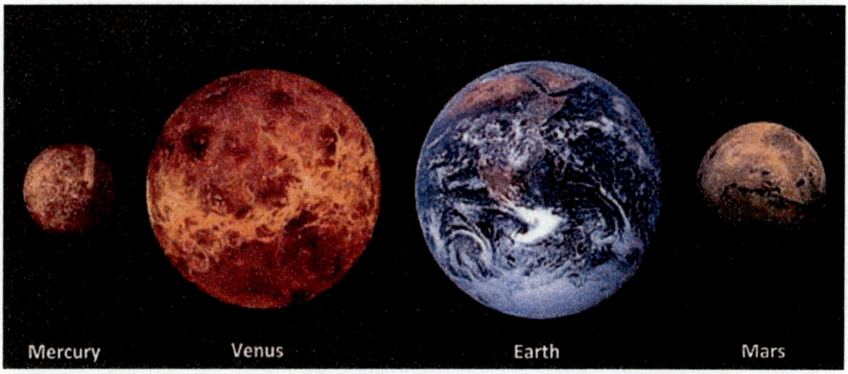

Fig. 1.5 Scale pictograms of the four terrestrial planets. Venus and Earth are very similar in diameter. Mars is about one-half the diameter of Earth and about one-tenth of the mass of Earth. The Earth's Moon, not shown in this diagram, is about one-half the diameter of Mars and about one-tenth the mass of Mars. (From Faure and Mensing (2007, Fig. 11.1) (NASA, LPI))

Table 1.1 Numerical values for the coefficient of the momentum of inertia (MI FACTOR) for the terrestrial planets and the Moon as well as the and uncompressed density for each of the bodies

Body	MI factor	Compressed density	Uncompressed density
Moon	0.393	3.34	3.34
Mars	0.366	3.93	3.74
Mercury	0.346	5.30	5.30
Earth	0.331	5.52	4.05
Venus	0.330	5.24	4.00

The values are arranged in decreasing numerical value of the MI FACTOR
Source: Lodders and Fegley (1998), p. 91–95

there seems to be general agreement on an estimate of the radius of the metallic core of Mars. Perhaps the INSIGHT mission to Mars (Voosen 2018) will give us more detailed information on the geometry and physical state of a martian core.

In contrast, there has never been general agreement on whether the Earth's Moon has a metallic core (see discussion in side-bar below). The coefficient of its moment of inertia is about 0.393 (Table 1.1) and that of a homogeneous sphere is 0.400. This number suggests that there is a fairly homogeneous distribution of mass within the Moon beneath the magma ocean zone. Perhaps the recent Chinese space mission which successfully landed a craft on the lunar backside highlands (Chang'e 4) will yield information for this lunar internal zonation issue.

Brief Discussion on the Issue of a Metallic Core in the Earth's Moon
The researchers promoting the Giant Impact Model (GIM) for the origin of the Moon favor a metallic core because, in their model, the body of the Moon was completely melted following the Giant Impact Event. Most investigators that attempt to explain the remanent magnetic patterns recorded in lunar rocks and/or lunar surface patterns favor a metallic core because they can conceive of no other way to generate an internal magnetic field (e.g., Weiss and Tikoo 2014). Scheinberg et al. suggest a basal magma ocean dynamo as an explanation for the generation of the lunar magnetic field. However, many years ago Levy (1972, 1974) concluded that, using the electrical conductivity properties of the earth's metallic core, a lunar like body with a 350 km radius metallic core would need to rotate so rapidly that it would reach the disintegration limit before it could generate a magnetic field of the strength that is recorded in lunar rocks.

Using a much different qualitative model, Malcuit (2015, Ch. 4) suggests that an internal magnetic field can be generated at two distinct eras in lunar history via a shallow shell magnetic amplifier operating in the lunar magma ocean zone because of differential rotation of the lunar crustal complex and the lunar mantle beneath the magma ocean zone. This mechanism was first proposed by Smoluchowski (1973a, 1973b). Malcuit (2015) suggests that these sets of magnetic amplifiers operated (1) during the magma ocean era (~4.5–4.4 Ga ago) and (2) during the lunar capture and subsequent orbit circularization era (~3.95 to 3.60 Ga ago). For this shallow-shell magnetic amplifier model the lunar interior beneath the magma ocean zone can be undifferentiated material and there is no need for a metallic core.

Planet Mars is about ten times more massive than the Earth's Moon and Mars is probably a marginal case for forming a central metallic core (see Fig. 1.6). Regardless of whether the electrically conducting metal is in a central core or in a metal ring, it would be useful for generating a magnetic field if there is a sufficient unidirectional force to cause differential rotation within a molten metallic zone. A scale model for the geometry of the two cases (Fig. 1.6a, b) and numerical simulations of the operation of these two cases may yield a definitive solution for the process of the generation of a martian magnetic field. A more definite solution for the size of the core of Mars could be gained by analysis of seismic waves that penetrate the martian core. This will be a task for the InSight lander mission launched in May 2018 and scheduled to be operational in 2020 (Voosen 2018).

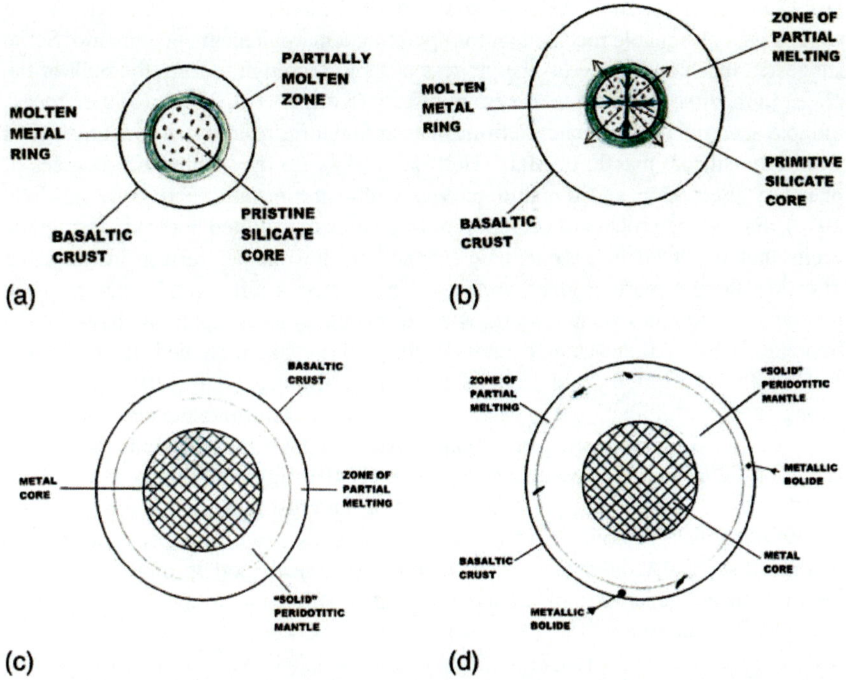

Fig. 1.6 Schematic diagrams, somewhat to scale, for how metallic cores may form within planetary bodies if various masses accreted from small bodies of chondritic meteoritic composition. (**a**) Thin molten metal ring forms around core of silicate material. (**b**) As mass is accreted to the planetary body and the confining pressure on the primitive silicate core increases, the silicate core becomes gravitationally unstable and is displaced outward as the more dense metallic material moves inward to form a metallic core. (**c**) As the accretion process continues the metallic core increases in mass as long as the metallic material can form inverse diapirs to penetrate through the mantle rock to get to the core complex. (**d**) Eventually, as confining pressure increases, the metal diapirs cannot penetrate the more "solid" mantle and any newly accreted metallic material is confined to circulate within the zone of partial melting of the upper part of the partially molten mantle. Note: (**a, b**) relate to a mars mass body; (**c, d**) relate to Venus and Earth mass bodies

1.4.2 Did Mars Have an Internally Generated Magnetic Field and When Did It Cease Operation?

There is general agreement that some martian surface rocks have a magnetic signature and that this signature resulted from an internally generated magnetic field in some short era within a period of perhaps 500 Ma (4.5 Ga to 4.0 Ga) (Acuna et al. 1999; Stevenson 2001; Nimmo and Tanaka 2005; Connerney et al. 2004; Soloman et al. 2005; Kobayashi and Dauphas 2013; Lillis et al. 2013; Zhang and O'Neil 2016; Vervelidou et al. 2017). There is also general agreement that the primitive andesitic/basaltic or andesitic crust formed over a period of about 50 Ma or less (Kruiger et al. 2017; Elkins-Tanton 2018; Bouvier et al. 2018). The debate centers on a reasonable mechanism for operating a martian magnetic dynamo. Some suggested mechanisms are (a) the process of rapid accretion to form the bulk of the planet in the first 10 Ma of solar system history (Kruiger et al. 2017), (b) the process of rapid accumulation of material from an accretion torus resulting in a magma ocean within the silicate mantle of Mars (Helffrich 2017), (c) the combined processes of planetary precession and dynamo activity within the metallic core of Mars (Wei 2016), and (4) the process of accretion of fragments of captured bodies in retrograde orbits that impacted in a systematic manner on the martian surface in the same direction along a martian great-circle pattern to cause a differential rotation of the martian core complex relative to the martian mantle-crust complex (Kobayashi and Sprenke 2010). Arkani-Hamed and Olson (2008), Arkani-Hamed et al. (2008), Arkani-Hamed (2009b), and Roberts and Arkani-Hamed (2014, 2017) present a strong case for unidirectional impacts in limited eras of time onto the martian surface as well as for polar wandering in the early history of Mars. In earlier papers Arkani-Hamad (2009a) and Kobayashi and Sprenke (2010) suggested that such unidirectional impacts caused both (a) westerly drift of the crust relative to the mantle and (b) the dynamics for powering a martian magnetic dynamo for a brief era of early martian history. Detailed numerical and/or mechanical models of the processes involved in the generation of a magnetic field of sufficient strength may lead to a more definite solution.

1.4.3 What Caused the Hemispherical Dichotomy on Mars?

Figure 1.7 shows a view of the hemispherical dichotomy on Mars. In general, the Northern Hemisphere is characterized by lower topography than the Southern Hemisphere. At present there are two categories of explanations for the great topography contrast of the two hemispheres. An early review of models is in McGill and Squyres (1991). One category of explanations for the northern lowlands is that it is due mainly to internal activity such as foundering of crust into the mantle due to a process of hot spot elevation of crust resulting in subduction along the edges of the slab and replacement by younger section of basaltic or andesitic-basaltic crust (Sleep 1994; Elkin-Tanton et al. 2003; Nimmo 2005; Nimmo et al. (2008); Scheinberg et al. 2014). The general concept is that as sections of a thickened cool crust became unstable and more prone to subduction, they subside through a magma ocean and eventually settle to near the core-mantle boundary. Such a subduction process can be triggered by a small impacting body, a local inhomogeneity in the crust, or a system of fractures (weaknesses) in the crust. Another suggestion for a cause of the hemispherical dichotomy is that it is due to the impact of a large bolide in what is now the northern polar region (Wilhelms and Squyres 1984; Kiefer 2008; Marinova et al. 2008, 2011; Nimmo et al. 2008; Andrews-Hanna et al. 2008; Leone et al. 2014; Rosenblatt et al. 2016). Still another suggestion involved a giant impact that caused convection activity in the antipodal region (Citron et al. 2018a). Regardless of the

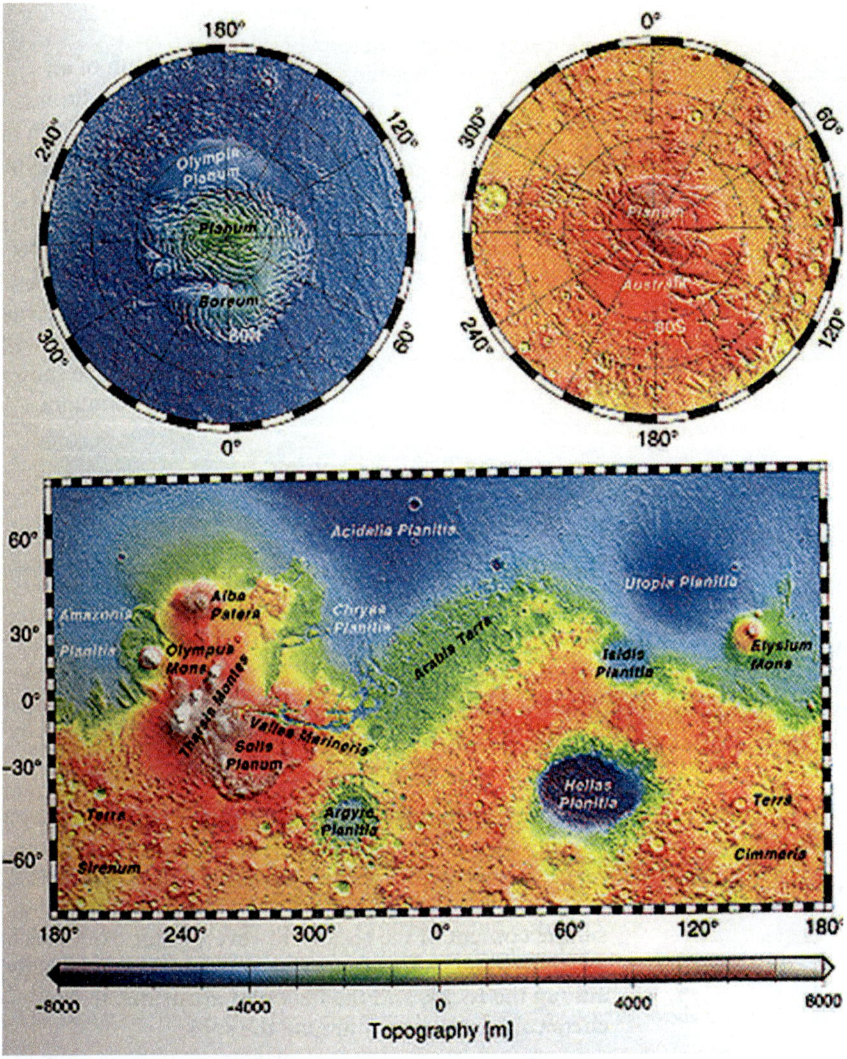

Fig. 1.7 Mars showing the hemispherical dichotomy of the martian surface. Top left: the northern polar region; top right; the southern polar region; bottom: Mercator projection of the entire surface of Mars. (Maps are from Bell 2008, colored photo section). (Courtesy of NASA/LPI)

cause of the missing primitive crust, the north polar area is resurfaced with rocks of basaltic-andesitic composition and the entire region is lower in elevation than other regions of the primitive crust. (Rosenblatt et al. 2016) suggests than the large body hit Mars when the north pole of the planet was located in the equatorial zone and that the shell of the planet reoriented itself via true polar wandering. Regardless of the mechanism that caused the martian crustal dichotomy, Cassata et al. (2018) suggest that it formed within the first 100 Ma of martian history because of the date and petrologic information obtained from martian breccia NWA 7034 (and associated stoney meteorites) which at the time of publication was the "oldest and most diverse of the martian meteorites".

At this point in time I favor endogenic processes for the Borealis basin resurfacing (e.g., Sleep 1994) as well as for the development of Valles Marineris and the Tharsis Rise and associated structural-volcanic complex. There are also several impact events mixed in with this scenario (e.g., Yin 2012).

1.4.4 Is There Evidence for Oceans of Water on Mars Early in Its History?

Figure 1.8 shows some features that may relate to the Noachian era of martian history. Baker et al. (1991) and Carr (1996) summarize abundant evidence for ancient oceans, ice sheets, and a hydrologic cycle on Mars for various times in the history of Mars. They claim that in contrast to the cold, dry climate of the recent geologic era of the planet, there is abundant evidence for the action of flowing ice and water from earlier eras. Baker (2005) also states that there is evidence that Mars had an active hydrological cycle in more recent times. Murray et al. (2005) present high resolution stereo camera evidence for their interpretation of a frozen sea close to the martian equator. Head et al. (2005), Brough et al. (2016), and Bramson et al. (2017) present convincing evidence for low to mid-latitude ice accumulation due to

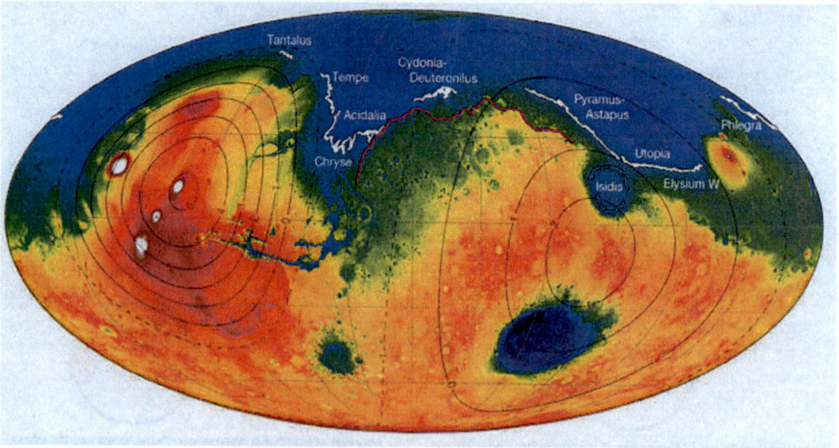

Fig. 1.8 Map showing some features that may relate to the Noachian era of martian history. The lowland areas may have been covered by ocean water at various time in the past. The irregular white lines on the northern portion of the map are thought to be paleo-shorelines. (Map is from Citron et al. 2018b, Extended version of Figure 2 in Methods section with permission from Springer-Nature)

glacial action on the planet in recent time and for a recent mass loss of glacial ice and significant retreat of regional glaciers. The large impact craters of Hellas and Argyre appear to have some paleoshorelines and other evidence of standing water in the early history of the craters. Hydrated minerals which were located by various rovers suggest that water was at or near the surface as well. An analysis of D/H ratios for several regions of the Martian surface by Villanueva et al. (2015) suggest that there were oceans of water on the surface of Mars in the early history of the planet. Wilson et al. (2016) summarize the evidence for cool-wet climate conditions in the mid-latitudes on Mars at various times in early martian history. A recent analysis of the Ar-38/Ar-36 ratio in the upper atmosphere of Mars by Jakosky et al. (2017) suggests that ~66% of the atmospheric argon has been lost to space. This study suggests that this escape of atmospheric gases may have caused the transition from an early warm, wet environment to the dry conditions of the martian climate today. Ramirez (2017) suggests that a warm and semi-arid climate appears to be the simpliest and most logical solution to martian paleoclimate in this early era.

More recently Zuber (2018) and Citron et al. (2018a, 2018b) present very convincing evidence for paleo-shorelines along the edges of the Borealis lowland areas on Mars (see Fig. 1.8). The shoreline features are not located at the same levels. This characteristic suggests to Citron et al. (2018a, 2018b) that crustal deformation was occurring during the development of the shoreline features and that the crustal uplift was caused by the development of the volcanic features of the Tharsis area. In a recent study of a section in the Aeolis Dorsa area of the paleo-shoreline, Hughes et al. (2019) identified features of probable exhumed deposits of fluvial sedimentary origin that he interpreted as a fluvial-deltaic sequence. These deposits are definitely associated with the northern ocean complex. In contrast similar deposits of sedimentary deposits of fluvial origin have been located in individual and isolated impact craters such as Bradbury Crater (Bramble et al. 2019).

A major unanswered question is the source of all that water for the Borealis Basin. Where did all that water come from and when did it appear on the surface of Mars? In an attempt to answer these questions perhaps we should consider the history of water on planet Earth.

There are two basic models that have been suggested for the accumulation of a large volume of water on Earth and these models may yield some insight into the source of water on early Mars.

Model 1: The water was outgassed from the original material involved in the accumulation of the planet Earth. The main argument against this model is presented by Taylor (2001). He suggested that the X-Wind model for the evolution of the Sun severely dehydrates the rocky material out to at least the vicinity of Earth and perhaps to the orbit of Mars. Thus this simple "outgassing model" (Drake and Righter 2002) is inconsistent with the X-Wind model for the early history of the Sun and the solar nebula.

Model 2: Albarede (2009) suggested that the materials that accreted to form planet Earth were severely dehydrated and devolatilized. His solution is to have water and associated volatiles "drift" back into the inner solar system in the form of

Themis-type asteroids. The model of asteroidal water and volatiles works well for Earth because the Deuterium to Hydrogen (D/H) ratio of asteroids is similar to that of the Earth's oceans. But the D/H ratio for Mars is more consistent with an origin from comets. An apparent solution to this dilemma is for Mars to get a significant portion of its water from Jupiter-captured comets. Jupiter-captured comets are comets that have close encounters with Jupiter and get inserted into asteroid-like orbits. Another possibility is that some of the water and volatiles could come via bodies from the Kuiper Belt, the region of the Solar System between the orbit of Neptune and the inner edge of the Oort Cloud (which is the source region of the comets). Albarede (2009) suggests a time-frame of 100 Ma (+ or − 50 Ma) for arrival of water for Earth and this time-frame may be appropriate for planet Mars as well.

Carr and Head (2019) discuss the origin and fate of a "frozen Hesperian-age ocean" in the Borealis Basin (northern lowland basin) of Mars. They suggest that a portion of the ocean basin was filled by catastrophic releases of groundwater from cryogenic deposits that were periodically melted during a progressive heating of the surface and near-surface layers of this northern area of Mars. Then after the deposition of this water in the Borealis Basin, most of that water evaporated to the atmosphere.

Carr and Head (2019) give an estimate of the total surface and near-surface water of Mars as a global equivalent layer (GEL) of 110 meters. Their estimate of the present inventory of surface and near surface water is about 30 GEL. Thus, about 80 GEL may be accounted for by a combination of atmospheric escape over marologic time and sequestration of hydrous material in the deeper layers of the crustal complex.

1.4.5 Was There a Fairly Dense Atmosphere on Mars Early in Its History?

First of all, most of the evidence concerning the density, composition, and mass of an earlier atmosphere is indirect (McKay et al. 1991). For example, if there was a significant quantity of surface water on Mars, then there would necessarily have been a massive quantity of atmospheric water. If there was massive volcanism in early Martian history, then there would have been significant quantities of CO_2 in the atmosphere. Both the H_2O and the CO_2 as well as any methane would produce some "greenhouse" effect on early Mars. Evidence of waxing and waning of ice caps in the present and past eras suggest a periodic change of greenhouse gases over time (Ramirez 2017). Thus, the evidence for a fairly massive atmosphere in the past is substantial (Villanueva et al. 2015; Chaffin et al. 2017; Slipski and Jakosky 2016) but most of these liquids and gases have either escaped from the planet or have been absorbed by chemical fixation in the sedimentary rocks and sediments on the surface of Mars. There seems to be an irregular periodicity to these "warm" and "cool" spells that probably relates to some combination of (1) volcanic activity (McKay et al. 1991), (2) orbital parameters of heliocentric orbit eccentricity and inclination as well as the planetary parameters of obliquity and precession over the history of the planet (Laskar et al. 2004), and (3) the relatively slow process of the escape of atmospheric gases over geologic time.

1.5 Summary

This chapter is an introduction to some of the features of planet Mars that many planetary scientists consider as the most "normal" of our set of terrestrial planets. Mars has a transparent atmosphere, ice deposits of a variety of compositions, a rotation rate similar to that of Earth, an obliquity (tilt) angle similar to that of Earth, conic volcanic forms, an extensive planetary rift system, and polar ice deposits that wax and wane with the seasonal cycle. Thus far we have no substantial evidence of life forms, either extant or fossil, in the rocks and sediments of the planet.

We briefly discussed the evidence for the presence of a metallic core in Mars. The coefficient of the moment of inertia for Mars is intermediate between that of Earth and the Moon. Thus, we can all agree that there is a concentration of a dense substance (mainly iron) in the central part of the body of Mars. Mars has a remanent magnetic signature that was detected via satellite magnetometer surveys. Most investigators agree that the martian magnetic field was generated in a molten layer within the core of the planet but the motive force is not clear. The time duration of the global magnetic field was brief but not yet precisely quantified. Most investigators agree that it ceased operation by 4.0 Ga before present.

The hemispheric dichotomy of the global topography is an outstanding characteristic of Mars. There are several hypotheses for the cause of the hemispherical differences but there is no consensus on the cause of these differences. There is substantial evidence for extensive bodies of water on the surface and most investigators agree that water covered much of the northern hemisphere lowlands. The ultimate source of the water as well as the timing of its disappearance from the martian surface are subjects of debate.

The final discussion of this introductory chapter was about the history of the atmosphere of Mars. There is general agreement that Mars had a fairly dense atmosphere early in the history of the planet especially in the era during which ocean water was abundant on the surface. However, the history of escape of the atmospheric gases is a subject of debate.

References

Acuna, M. H., et al. (1999). Global distribution of crustal magnetization discovered by the Mars Global Surveyor MAG/ER experiment. *Science, 284*, 790–793.

Albarede, F. (2009). Volatile accretion history of the terrestrial planets and dynamic implications. *Nature, 461*, 1227–1233.

Andrews-Hanna, J. C., Zuber, M. T., & Banerdt, B. (2008). The Borealis basin and the origin of the martian crustal dichotomy. *Nature, 453*, 1212–1215.

Arkani-Hamed, J. (2009a). Did tidal deformation power the core dynamo of Mars? *Icarus, 201*, 31–43.

Arkani-Hamed, J. (2009b). Polar wander of Mars: Evidence from giant impact basins. *Icarus, 204*, 489–498.

Arkani-Hamed, J., & Olson, P. (2008). Giant impacts, core stratification, and failure of the Martian dynamo. *Journal of Geophysical Research, 115*, 16. https://doi.org/10.1029/2010JE003579.

Arkani-Hamed, J., Seyed-Mahmoud, B., Aldredge, K. D., & Baker, R. E. (2008). Tidal excitation of elliptical instability on the Martian core: Possible mechanism for generating the core dynamo. *Journal of Geophysical Research, 113*, E06003 (14 p.) https://doi.org/10.1029/2007JE002982.

Baker, V. R. (2005). Picturing a recently active Mars. *Nature, 434*, 280–283.

Baker, V. R., Strom, R. G., Gulick, V. C., Kargel, J. S., Komatsu, G., & Kale, V. S. (1991). Ancient oceans, ice sheets and the hydrologic cycle on Mars. *Nature, 352*, 589–594.

Belbruno, E., & Gott, J. R., III. (2005). Where did the Moon come from? *The Astronomical Journal, 129*, 1724–1745.

Bell, J. (Ed.). (2008). *The martian surface: Composition, mineralogy, and physical properties*. Cambridge: Cambridge University Press, 636 p.

Bouvier, L. C., Costa, M. M., Connelly, J. N., et al. (2018). Evidence for extremely rapid magma ocean crystallization and crust formation on Mars. *Nature, 558*, 586–589.

Bramble, M. S., Goudge, T. A., Milliken, R. E., & Mustard, J. F. (2019). Testing the deltaic origin of fan deposits at Bradbury Crater, Mars. *Icarus, 319*, 363–366.

Bramson, A. M., Byrne, S., & Bapst, J. (2017). Preservation of midlatitude ice sheets on Mars. *Journal of Geophysical Research: Planets, 122*, 17. https://doi.org/10.1002/2017JE005357.

Brough, S., Hubbard, B., & Hubbard, A. (2016). Former extent of glacier-like forms on Mars. *Icarus, 274*, 37–49.

Carr, M., & Head, J. (2019). Mars: Formation and fate of a frozen Hesperian ocean. *Icarus, 319*, 433–343.

Carr, M. H. (1996). *Water on Mars*. New York: Oxford University Press, 229 p.

Cassata, W. S., Cohen, B. E., Mark, D. F., Trappitsch, R., Crow, C. A., Wimpenny, J., Lee, M. R., & Smith, C. L. (2018). Chronology of martian breccia NWA 7034 and the formation of the martian crustal dichotomy. *Science Advances, 4*, 11, p. (eaap8306).

Chaffin, M. S., Deighan, J., Schneider, N. M., & Stewart, A. I. F. (2017). Elevated atmospheric escape of atomic hydrogen from Mars induced by high-altitude water. *Nature Geoscience, 10*, 174–178.

Citron, R. I., Manga, M., & Hemingway, D. J. (2018b). Timing of oceans on Mars from shoreline deformation. *Nature, 555*, 643–646.

Citron, R. I., Manga, M., & Tan, E. (2018a). A hybrid origin of the martian crustal dichotomy: Degree-1 convection antipodal to a giant impact. *Earth and Planetary Science Letters, 491*, 58–66.

Clery, D. (2013). Impact theory gets whacked. *Science, 342*, 183–185.

Cloud, P. E., Jr. (1972). A working model of the primitive earth. *American Journal of Science, 272*, 537–548.

Cloud, P. E., Jr. (1974). Rubey conference on crustal evolution (meeting report). *Science, 183*, 878–881.

Connerney, J. E. P., Acuna, M. H., Ness, N. F., Spohn, T., & Schubert, G. (2004). Mars crustal magnetism. *Space Science Reviews, 111*, 1–32.

Davies, G. F. (1992). On the emergence of plate tectonics. *Geology, 20*, 963–966.

Davies, J. H. (2008). Did a mega-collision dry Venus' interior? *Earth and Planetary Science Letters, 268*, 376–383.

Dones, L., & Tremaine, S. (1993a). Why does the Earth spin forward? *Science, 259*, 350–354.

Dones, L., & Tremaine, S. (1993b). On the origin of planetary spins. *Icarus, 103*, 67–92.

Drake, M. J., & Righter, K. (2002). Determining the composition of the Earth. *Nature, 416*, 39–44.

Elkins-Tanton, L. T. (2018). Rapid formation of Mars. *Nature, 558*, 522–523.

Elkin-Tanton, L. T., Parmentier, E. M., & Hess, P. C. (2003). Magma ocean fractional crystallization and cumulate overturn in terrestrial planets: Implications for Mars. *Meteoritics and Planetary Science, 38*, 1753–1771.

Faure, G., & Mensing, T. M. (2007). *Introduction to planetary science: The geological perspective*. Dordrecht: Springer, 526 p.

Gillman, C., Golabek, G. J., & Tackley, P. J. (2016). Effect of a single large impact on the coupled atmosphere-interior evolution of Venus. *Icarus, 268*, 295–312.

Hansen, K. (1982). Secular effects of oceanic tidal dissipation on the Moon's orbit and the Earth's rotation. *Reviews of Geophysics, 20*, 457–480.

Hartmann, W. K., Phillips, R. J., & Taylor, G. J. (Eds.). (1986). *Origin of the Moon*. Houston: Lunar and Planetary Institute, 781 p.

Head, J. W., Neukum, G., Jaumann, R., Hiesinger, H., Hauber, E., Carr, M., Masson, P., Folng, B., Hoffmann, H., Kreslavsky, M., Werner, S., Milkovich, S., van Gasselt, S., & The HRSC Co-Investigation Team. (2005). Tropical to mid-latitude snow and ice accumulation, flow and glaciation on Mars. *Nature, 434*, 346–351.

Helffrich, G. (2017). Feasibility of a magma ocean dynamo on Mars. *Progress in Earth and Planetary Science, 4*, 21. , 6 p. https://doi.org/10.1186/s40645-017-0135-8.

Herrick, D. L., & Parmentier, E. M. (1994). Episodic large-scale overturn of two-layer mantles in terrestrial planets. *Journal of Geophysical Research, 99*, 2053–2062.

Herrick, R. R. (1994). Resurfacing history of Venus. *Geology, 22*, 703–706.

Herwartz, D., Pack, A., Freidrichs, B., & Bischoff, A. (2014). Identification of the giant impactor Theia in lunar rocks. *Science, 344*, 1146–1150.

Hughes, C. M., Cardenas, B. T., Goudge, T. A., & Mohrig, D. (2019). Deltaic deposits indicative of a paleo-coastline at Aeolis Dorsa, Mars. *Icarus, 317*, 442–453.

Jakosky, B. M., & Edwards, C. S. (2018). Inventory of CO_2 available for terraforming Mars. *Nature Astronomy, 2*, 634–639.

Jakosky, B. M., Slipski, M., Benna, M., Mahaffy, P., Elrod, M., Yelle, R., Stone, S., & Alsaeed, N. (2017). Mars' atmospheric history derived from upper-atmosphere measurements of $^{38}Ar/^{36}Ar$. *Science, 355*, 1408–1410.

Johansen, A., & Lambrechts, M. (2017). Forming planets via pebble accretion. *Annual Review of Earth and Planetary Sciences, 45*, 359–387.

Kiefer, W. S. (2008). Forming the martian great divide. *Nature, 453*, 1191–1192.

Kobayashi, D., & Dauphas, N. (2013). Small planetesimals in a massive disk formed Mars. *Icarus, 225*, 122–130.

Kobayashi, D., & Sprenke, K. F. (2010). Lithospheric drift on early Mars: Evidence in the magnetic field. *Icarus, 210*, 37–42.

Kruiger, T. S., Kleine, T., Borg, L. E., Brennecka, G. A., Irving, A. J., Bischoff, A., & Agee, C. B. (2017). The early differentiation of Mars inferred from Hf-W chronometry. *Earth and Planetary Science Letters, 474*, 345–354.

Laskar, J. (1995). *Large scale chaos and marginal stability in the solar system: XIth international congress of mathematical physics* (p. 120). Boston: International Press.

Laskar, J., Correia, A. C. M., Gastineau, M., Joutel, F., Lavard, B., & Robutel, P. (2004). Long term evolution and chaotic diffusion of the insolation quantities of Mars. *Icarus, 170*, 343–364.

Leone, G., Tackley, P. J., Gerya, T. V., May, D. A., & Zhu, G. (2014). Three-dimensional simulations of the southern polar giant impact hypothesis for the origin of the martian dichotomy. *Geophysical Research Letters, 41*, 8736–8743. https://doi.org/10.1002/2014GL062261.

Levy, E. H. (1972). Magnetic dynamo in the Moon: A comparison with the earth. *Science, 178*, 52–53.

Levy, E. H. (1974). Magnetic dynamo in the moon? *The Moon, 9*, 49–56.

Lillis, R. J., Robbins, S., Manga, M., Halekas, J. S., & Frey, H. V. (2013). Time history of the martian dynamo from crater magnetic field analysis. *Journal of Geophysical Research: Planets, 118*, 1488–1511. https://doi.org/10.1002/jgre.20105.

Lipton, P. (2005). Testing hypotheses: Prediction and prejudice. *Science, 307*, 219–221.

Lissauer, J. J., Berman, A. F., Greenzweig, Y., & Kary, D. M. (1997). Accretion of mass and spin angular momentum by a planet on an eccentric orbit. *Icarus, 127*, 65–92.

Lissauer, J. J., & Kary, D. M. (1991). The origin of the systematic component of planetary rotation. 1: Planet on a circular orbit. *Icarus, 94*, 126–159.

Lodders, K., & Fegley, B., Jr. (1998). *The planetary scientist's handbook*. New York: Oxford University Press, 371 p.

Lunine, J. I. (1999). *Earth: Evolution of a habitable world*. Cambridge: Cambridge University Press, 319 p.

MacDonald, J. G. F. (1963). The internal constitution of the inner planets and the Moon. *Space Science Review, 2*, 473–557.

MacDonald, J. G. F. (1964). Tidal friction. *Reviews of Geophysics, 2*, 467–541.

Malcuit, R. J., Mehringer, D. M., & Winters, R. R. (1992). A gravitational capture origin for the earth-moon system: Implications for the early history of earth and moon. In J. E. Glover & S. E. Ho (Eds.), *The Archaean: Terrains, processes and Metallogeny: Geology department (key center) and university extension* (Vol. 22, pp. 223–235). Perth: The University of Western Australia, Publication.

Malcuit, R. J. (2015). *The twin sister planets, Venus and Earth: Why are they so different?* (p. 401). Cham: Springer International Publishers.

Mann, A. (2018). Cataclysm's end. *Nature, 553*, 393–395.

Marinova, M. M., Aharonson, O., & Asphaug, E. (2008). Mega-impact formation of the Mars hemispheric dichotomy. *Nature, 453*, 1216–1219. https://doi.org/10.1038/nature07070.

Marinova, M. M., Aharonson, O., & Asphaug, E. (2011). Geophysical consequences of planetary-scale impacts into a Mars-like planet. *Icarus, 211*, 960–985.

McGill, G. E., & Squyres, S. W. (1991). Origin of the martian crustal dichotomy: Evaluating hypotheses. *Icarus, 93*, 386–393.

McKay, C. P., Toon, O. B., & Kasting, J. F. (1991). Making Mars habitable. *Nature, 352*, 489–496.

Meier, M. M. M., Reufer, A., & Wieler, R. (2014). On the origin and composition of Theia: Constraints from new models of the Giant Impact. *Icarus, 242*, 316–328.

Murray, J. B., Muller, J. P., Neukum, G., Werner, S. C., van Gasselt, S., Hauber, E., Marklewicz, W. J., Head, J. W., III, Foing, B. H., Page, D., Mitchell, K. L., Portyankina, G., & the HRSC Co-Investigator Team. (2005). Evidence from the Mars express high resolution stereo camera for a frozen sea close to Mars' equator. *Nature, 434*, 352–356.

Nimmo, F. (2005). Tectonic consequences of martian dichotomy modification by lower-crustal flow and erosion. *Geology, 33*, 533–536.

Nimmo, F., Hart, S. D., Korycansky, D. G., & Agnor, C. B. (2008). Implications of an impact origin for the martian hemispheric dichotomy. *Nature, 453*, 1220–1223. https://doi.org/10.1038/nature07025.

Nimmo, F., & Tanaka, K. (2005). Early crustal evolution of Mars. *Annual Reviews, Earth and Planetary Sciences, 33*, 133–161.

Ramirez, R. M. (2017). A warmer and wetter solution for early Mars and the challenges with transient warming. *Icarus, 297*, 71–82.

Roberts, J. H., & Arkani-Hamed, J. (2014). Impact heating and coupled core cooling and mantle dynamics on Mars. *Journal of Geophysical Research: Planets, 119*, 729–744. https://doi.org/10.1002/2013JE004603.

Roberts, J. H., & Arkani-Hamed, J. (2017). Effects of basin-forming impacts on the thermal evolution and magnetic field of Mars. *Earth and Planetary Science Letters, 478*, 192–202. https://doi.org/10.1016/j.epsl.2017.08.031.

Rosenblatt, P., Charnoz, S., Dunseath, K. M., Terao-Dunseath, M., Trinh, A., Hyodo, R., Glenda, H., & Toupin, S. (2016). Accretion of Phobos and Deimos in an extended debris disc stirred by transient moons. *Nature Geoscience, 9*, 581–583.

Ross, M. N., & Schubert, G. (1989). Evolution of the lunar orbit with temperature- and frequency-dependent dissipation. *Journal of Geophysical Research, 94*, 9533–9544.

Scheinberg, A., Elkins-Tanton, L. T., & Zhong, S. J. (2014). Timescale and morphology of martian mantle overturn immediately following magma ocean solidification. *Journal of Geophysical Research: Planets.*, v, *119*, 454–467.

Slattery, W. L., Benz, W., & Cameron, A. G. W. (1992). Giant impacts on a primitive Uranus. *Icarus, 99*, 167–174.

Sleep, N. H. (1994). Martian plate tectonics. *Journal of Geophysical Research, 99*, 5639–5655.

Slipski, M., & Jakosky, B. M. (2016). Argon isotopes as tracers for martian atmospheric loss. *Icarus, 272*, 212–227.

Smoluchowski, R. (1973a). Lunar tides and magnetism. *Nature, 242*, 516–517.

Smoluchowski, R. (1973b). Magnetism of the moon. *The Moon, 7*, 127–131.

Soloman, S. C., Aharonson, O., Aurnou, J. M., Banerdt, W. B., Carr, M. H., Dombard, A. J., Frey, H. V., Golombek, M. P., Hauck, S. A., II, Head, J. W., III, Jakosky, B. M., Johnson, C. L., McGovern, P. J., Neumann, G. A., Phillips, R. J., Smith, D. E., & Zuber, M. T. (2005). New perspectives on ancient Mars. *Science, 307*, 1214–1220.

Stern, R. (2005). Evidence from ophiolites, blueschists, and ultrahigh-pressure metamorphism terranes that the modern episode of subduction tectonics began in Neoproterozoic time. *Geology, 33*, 557–570.

Stevenson, D. J. (2001). Mars' core and magnetism. *Nature, 412*, 214–219.

Taylor, S. R. (2001). *Solar system evolution: A new perspective*. Cambridge: Cambridge University Press, 460 p.

Tremaine, S. (1991). On the origin of the obliquities of the outer planets. *Icarus, 89*, 85–92.

van Loon, A. J. (2004). From speculation to model: The challenge of launching new ideas in the earth sciences. *Earth-Science Reviews, 65*, 305–313.

Vervelidou, F., Lesur, V., Grott, M., Morschhauser, A., & Lillis, R. J. (2017). Constraining the date of the martian dynamo shutdown by means of crater magnetiztion signatures. *Journal of Geophysical Research: Planets, 122*, 2294–2311. https://doi.org/10.1002/2017JE005410.

Villanueva, G. L., Mumma, M. J., Novak, R. E., Kaufl, H. U., Hartogh, P., Encrenaz, T., Tokunaga, A., Khayat, A., & Smith, M. D. (2015). Strong water isotopic anomalies in the martian atmosphere: Probing current and ancient reservoirs. *Science, 348*, 218–221.

Voosen, P. (2018). NASA lander to probe interior of Mars. *Science, 360*, 247–248.

Webb, D. J. (1982). On the reduction in tidal dissipation produced by increases in the Earth's rotation rate and its effect on the long-term history of the Moon's orbit. In P. Brosche & J. Sundermann (Eds.), *Tidal friction and the earth's rotation II* (pp. 210–221). Berlin: Springer.

Wei, X. (2016). The combined effect of precession and convection on the dynamo action. *The Astrophysical Journal, 827*, 123. (6 p.).

Weiss, B. P., & Tikoo, S. M. (2014). The lunar dynamo. *Science, 346*, 1246753, 10 p.

Wetherill, G. W. (1985). Occurrence of giant impacts during the growth of the terrestrial planets. *Science, 228*, 877–879.

Wilhelms, D. E., & Squyres, S. W. (1984). The hemispheric dichotomy may be due to a giant impact. *Nature, 309*, 138–140.

Wilson, S. A., Howard, A. D., Moore, J. M., & Grant, J. A. (2016). A cold-wet middle-latitude environment on Mars during the Hesperian-Amazonian transition: Evidence from northern Arabia valleys and paleolakes. *Journal of Geophysical Research: Planets, 121*, 1667–1694.

Yin, A. (2012). Structural analysis of the Valles Marineris fault zone: Possible evidence for large-scale strike-slip faulting on Mars. *Lithosphere, 4*, 286–330. https://doi.org/10.1130/L192.1.

Yoder, C. F., Konopliv, A. S., Yuan, D. N., Standish, E. M., & Folkner, W. M. (2003). Fluid core size of Mars from detection of the solar tide. *Science, 300*, 2999–2303.

Zhang, C. Z. (1994). The martian mean moment-of-inertia and the size of the Mars' core. *Earth, Moon and Planets, 64*, 117–124.

Zhang, S., & O'Neill, C. (2016). The early geodynamic evolution of Mars-type planets. *Icarus, 265*, 187–208.

Zuber, M. T. (2018). Oceans on Mars formed early. *Nature, 555*, 590–591.

Chapter 2
The Origin of the Sun and the Early Evolution of the Solar System with Special Emphasis on Mars, Asteroids, and Meteorites

Fifty years ago in these pages, meteorite researcher John A. Wood observed, 'Only recently have we begun to study chrondrules as entities. They contain a wealth of information . . . .. about the processes that have acted on them. We may be able to learn about the nature and evolution of the solar nebula, the formation of the planets, some stages of the evolution of the sun and the time scales for all of these processes.' Half a century later scientists still have much to learn, but the picture provided by these primordial messengers of the solar system is at last coming into focus. — From Rubin (2013, p. 41).

Chrondrules would not be predicted to exist if they did not exist. This may seem like an odd statement, but it is true. The millimeter-sized igneous spheres known as chondrules are the dominant structural component of chondritic meteorites (Figure 1), the oldest rocks in our collections [Grossman 1988; Brearley and Jones 1998; Amelin et al. 2002; Connolly and Desch 2004; Lauretta et al. 2006; Jones 2012]. However, from the point of view of astrophysics, astronomy, planetary science, and even geology, no one would predict that the most abundant objects within the hypothesized building blocks of the Earth-like planets would have been processed into small molten silicate spheres before accumulating in the earliest planetesimals. But they were. For over 150 years, science has puzzled over what process (or processes) could have produced them?— From Connolly and Jones (2016, p. 1885–1899).

The focus of this book is on the origin and evolution of planet MARS. Mars is not the largest or next largest of the terrestrial planets but Mars appears to be in the most pristine condition of any of the terrestrial planets (Ehlmann et al. 2016; Wilson et al. 2016). It may have a rotation rate that has changed very little over geologic time (MacDonald 1963, 1964) and large portions of its surface may be inherited from the earliest era of the Solar System (Kobayashi and Sprenke 2010). To make much progress on the history of planet Mars, we must attempt to develop a history of at least the inner part of the Solar System. But then to get a more complete story of Mars we must reach out to the Asteroid Zone and planet Jupiter, the main controller of the dynamical history of bodies in the Asteroid Zone. And for certain episodes in the history of Mars, such as the source of water and other volatiles, we may need to reach out as far as the Oort Cloud for a source of cometary water because of the D/H ratio of atmospheric hydrogen is similar to that in comets (Robert 2001). So, in

R. Malcuit, *Geoforming Mars*, https://doi.org/10.1007/978-3-030-58876-2_2

reality, when we consider all of these factors in the story of the origin and evolution of planet MARS we are forced to develop a working model for the origin and evolution of the Solar System.

The origin of the Solar System has intrigued scientists for centuries. As recently as five decades ago the models were still very general (e. g., Cameron 1962) and were concerned mainly with the collapse of a cloud of stellar dust and gas of roughly solar composition and the transformation of that cloud into a rapidly rotating disk-shaped mass around a proto-sun. The next few decades were dominated by calculations of equilibrium chemical condensation models from a cooling nebula of solar composition (e.g., Lewis 1972, 1974; Grossman 1972) based mainly on the temperature and pressure conditions for the solar nebula from Cameron and Pine (1973). Identification of high-temperature condensates [calcium-aluminum inclusions (CAIs)] in the Allende meteorite was very important for the development of more sophisticated models for the evolution of the Solar System. After the discovery of CAIs it was important to develop models to explain (a) the origin of chondrules (the main constituent of chondritic meteorites), (b) the origin of CAIs, as well as (c) the origin of the very fine-grained matrix of the chondritic meteorites. Since there is a vast quantity of information on the subject, the author of this book, or of any book of this nature, must pick and choose portions of the various models and hypotheses that seem most logical for his or her concept of the origin and history of planet MARS.

2.1 List of Facts to Be Expalined by a Successful Model

In pursuit of a suitable working model let us start with a list of facts to be explained and use some relevant and generally accepted interpretations of these facts.

1. Chemical composition and body density (compressed and uncompressed) patterns of the Mars, Earth, Venus, Mercury, the Asteroids, Jupiter and the other outer planets and their satellites, and comets in addition to the various groups of chrondritic and achondritic meteorites. (See Table 2.1 and Fig. 2.1 for some body density patterns.)
2. Composition and dates of formation of calcium-aluminum inclusions (CAIs) that occur in chondritic meteorites as well as an explanation of the processing that results in the various types of CAIs.
3. Patterns of oxygen isotope ratios for bodies of the Solar System (see Fig. 2.2 for some trends of oxygen isotope ratio patterns).
4. Composition and dates of formation of chondrules and chondritic meteorites as well as the devolatization patterns associated with vulcanoid planetoids, planets, and chondritic meteorites. (See Fig. 2.3 for the potassium content relative to uranium for these bodies.)
5. Magma ocean development on the Moon, in particular, but this explanation may include the development of magma oceans on other bodies such as the Vulcanoid planetoids and Mercury.

Table 2.1 Values of compressed density and uncompressed density for various solar system entities

Body	Compressed density (g/cm^3)	Uncompressed density (g/cm^3)
Moon	3.34	3.34
Mercury	5.43	5.30
Venus	5.24	4.00
Earth	5.52	4.05
Mars	3.93	3.74
Jupiter	1.33	0.10
Saturn	0.69	0.10
Uranus	1.32	0.30
Neptune	1.64	0.30
Pluto	~2.0	~2.0
SUN	1.41	(0.10)
4 Vesta	3.3	3.3
Angra	~3.3	~3.3
Ceres	2.7	2.7
Pallas	2.6	2.6

Sources: Moon through Sun (Lodders and Fegley 1998, p. 91–95); Asteroids (Kowal 1996, p, 45); Angrite grain density (Britt et al. 2010, 1869.pdf)

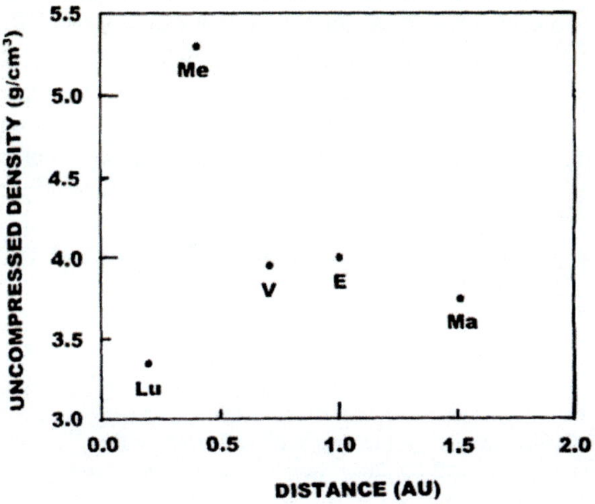

Fig. 2.1 Plot of uncompressed densities of the Moon (Lu), Mercury (Me), Venus V), Earth (E), and Mars (Ma). The exceptionally low uncompressed density of Luna as well as the exceptionally high uncompressed density of Mercury need to be explained by a successful model for the origin and evolution of the Solar System. (Numerical values are from Lodders and Fegley (1998, p. 91); diagram from Malcuit (2015, Fig. 2.2) with permission from Springer)

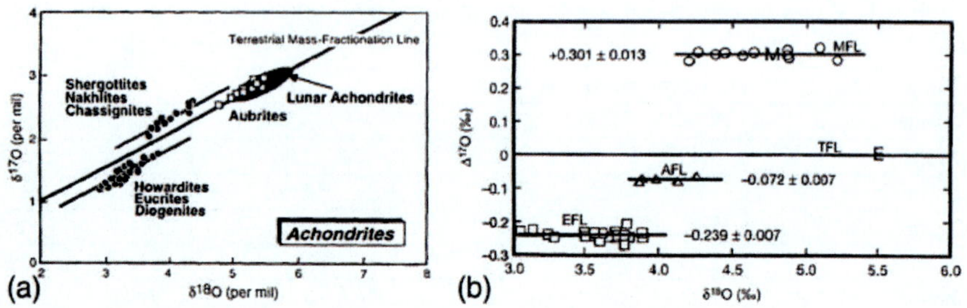

Fig. 2.2 Plots of some basic oxygen isotope information for some solar system bodies. (**a**) Three-oxygen isotope plot showing the position of the terrestrial fractionation line as well as the position of lunar achondrites (meteorites interpreted to be from the Moon), SNC meteorites (interpreted to be from Mars), and HED meteorites (interpreted to be from 4 Vesta). (Diagram from McSween, (1999, Fig. 4.4) with permission of Cambridge Univ. Press.) (**b**) A delta 17 oxygen vs. delta 18 oxygen plot showing the terrestrial fractionation line (TFL), the Mars fractionation line (MFL), the eucrite parent body fractionation line (EFL), and the angrite fractionation line (AFL). Note that both Earth (E) and Moon (not shown) are essentially on the terrestrial fractionation line. Again, all of this information needs to be explained by a successful model for Solar System origin and evolution. (Diagram from Greenwood et al. (2005, Fig. 2) with permission from Springer-Nature)

6. Patterns of magnetization of minerals, rocks, and planetary crusts on various solar system bodies as well as on meteorites and asteroids.
7. The redistribution of angular momentum from the Sun to the planets early in Solar System history. Over 90% of the mass of the Solar System is in the Sun and over 90% of the angular momentum is associated with the planets. For example, if the Sun had about 90% of the angular momentum, then the rotation rate of the Sun would be about 1 rotation per earth day. In contrast the rotation period of the Sun is about 28 earth days (i.e., about one rotation per earth month).

2.2 Possible Sequence of Events that Leads to the Solar System as We Know It

> Protoplanetary disks are the birthplace of planetary systems, including the solar system. The spatial distribution of both gas and dust in the disks provide the initial conditions for planet formation. The time evolution of the disks dictates when and where planet formation begins, and eventually determines the final architecture of planetary systems both in mass and in orbital distance (e.g., Ida and Lin 2004; Mordisini et al. 2009, 2016; Benz et al. 2014; Hasegawa 2016).
>
> Supportive evidence for the presence of magnetic fields in disks may exist in the form of magnetized chondrules found in chondrites (Fu et al. 2014). The results of the lab experiments imply that magnetic fields should have played an important role in both the distribution and the growth of planet-forming materials in the solar nebula (e.g., Shu et al. 1996; Desch and Cuzzi 2000; Ceisla 2007; Hasegawa et al. 2016; Wang et al. 2017). — From Hasegawa et al. (2017, p. 1).

There is no doubt that the origin and evolution of the Solar System is a major controversial topic in the planetary science community. Any successful model or combination of models has many facts to explain including all the complex chemical

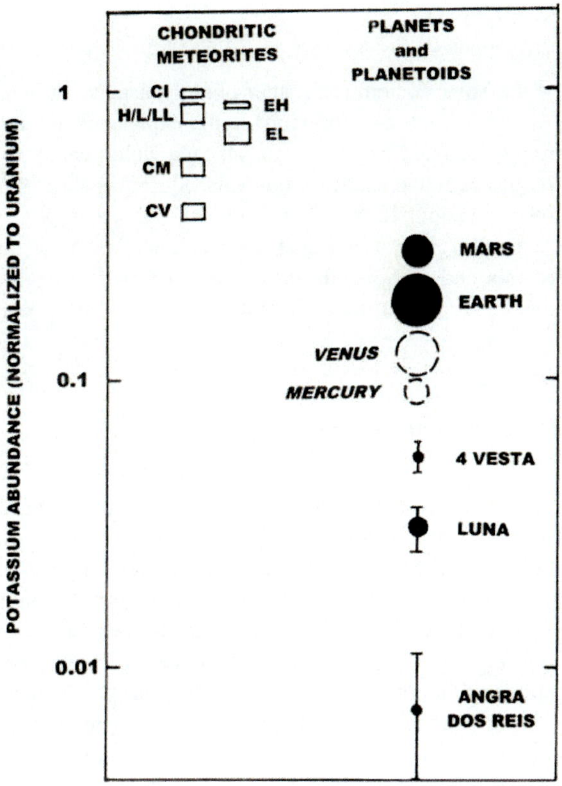

Fig. 2.3 Plot of potassium abundances (normalized to uranium) for various bodies of the solar system as well as for various classes of meteorites. Angra dos Reis represents the Angrite meteorites; Eucrites are basaltic meteorites that are thought to be from the Eucrite Parent Body which, in turn, is considered to be Asteroid 4 Vesta; *EH* and *EL* are enstatite chondritic meteorites; *H*, *L*, and *LL* are ordinary chondritic meteorites; *CV* and *CM* are types of carbonaceous chondrites, and the *CI* type is the most primitive carbonaceous chondritic meteorite (see McSween, 1999, for more details on classification). The two *open circle* patterns in the Planets and Planetoids column represent my prediction for the position of the potassium abundance (normalized to U) for planets Venus and Mercury. [Note: There is limited information from Venus via the Venera 8 mission. Vinogradov et al. (1973) report that the K/U ratio for magmatic rocks of Venus is very similar to magmatic rocks on Earth]. (Diagram adapted from Humayan and Clayton (1995, Fig. 1) with permission from Elsevier)

and physical features of meteorites. There is general agreement among students of the Solar System that we know much more about the relevant subjects now than we did at any time in the past. Indeed, a reasonable history of the Solar System is built as a succession of semi-successful models over time. In this treatment I am attempting to present a sequence of models that seem most logical to me at this time in the history of thought on the origin of the Solar System. I should also note that the chemistry for several of these stages of development is inherited from the preceding stage of development. This aspect of the models will become much more clear as we develop the story.

2.2.1 The Nebular Stage

The chemistry of the Solar System is a matter of happenstance. It is inherited in the form of a cloud of dust and gas from previous stellar explosions. It really is a mixture of small stellar explosions and large ones which each yield their special addition of elements and isotopes of elements. Since our Solar System is about 4.6 billion years old and the Universe is about 13.6 billion years old, many astrophysicists consider our Sun to be about a fifth generation star. Thus, our Solar System began as a mixture of chemical elements and compounds in the form of an extensive concentration (a cloud) of gas and dust in a particular location within the Milky Way Galaxy.

2.2.2 Gravitational Infall Stage

This extensive cloud of dust and gas would not be homogeneously distributed and eventually some of the denser portions would coagulate to form a more dense mass than surrounding regions. There would also be some, very slow rotational motion associated with this cloud. As more mass is concentrated near the center of the cloud of dust and gas, the rotation rate of the cloud would increase in the direction of the initial rotation. The denser central region would rotate much more rapidly than the outer parts but all would be rotating in the same direction. Since there is a physical law called "the law of the conservation of angular momentum", the quantity of angular momentum that was associated with the cloud initially would be associated with the cloud at later times unless and until some of the mass of the system is lost. Since angular momentum has only one form (i.e., it cannot change in form like energy can), it is much easier to handle mathematically. The cloud of dust and gas that eventually will form the Sun increases in rotation rate with the central portion of the cloud rotating much more rapidly than the outer part. This differential rotation of the concentric bands of the cloud leads to the development of shear zones and associated thermal activity.

2.2.3 X-Wind Stage

Eventually the temperature near the center of the cloud increases to the point where nuclear reactions begin to be triggered. This exponential increase in temperature along with the rapid rotation at the center of the cloud results in a very strong magnetic field that interacts dynamically with the surrounding cloud. This results in a transfer of angular momentum from the rapidly rotating proto-sun to the cloud by way of a mechanical coupling of the ionized cloud and the magnetic field flux lines. Thus, the central part of the cloud (the proto-sun) decreases in rotation rate as angular momentum is imparted to the surrounding cloud. Fig. 2.4 is a diagram

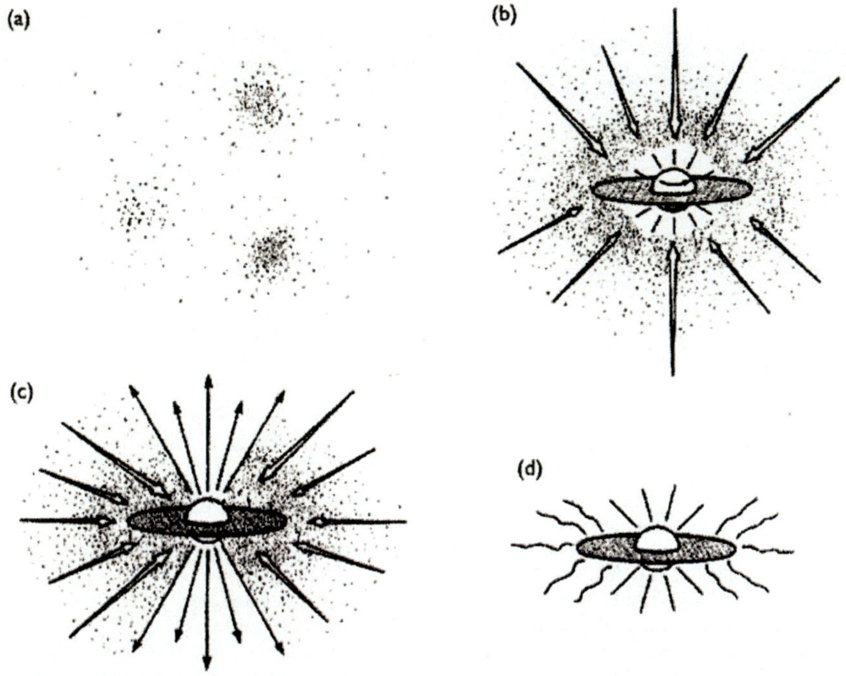

Fig. 2.4 Pictograms of four very general stages leading to the formation of the solar system: (**a**) Dust and gas from previous stellar explosions, (**b**) Infall of material and increase in rotation rate, (**c**) Infall continues but some material is ejected via bipolar outflow, (**d**) The star assumes a disk shape, becomes visible, and settles down to the main sequence of burning. (Diagram from Taylor (2001, Fig. 3.1) with permission from Cambridge University Press)

showing four generalized stages in the early formation of the Sun and the Solar System. Figure 2.5 shows a cross-sectional sketch showing some of the major features of the X-Wind model (Shu et al. 2001). The scale on the diagram is in astronomical units and the range is from the Sun to 5 AU (the orbital radius of planet Jupiter). Note the temperature increase from less than 500 degrees K (~222 degrees C) to over 1500 degrees K (~1222 degrees C) near the Sun. Calcium-Aluminum Inclusions (CAIs))are thought to form in the "hot disk" area. Then during very high energy pulsations, due to disruption of magnetic flux lines and subsequent reformation events, high-temperature mineral components are flash-melted within the "hot disk" area. Some of the newly formed CAIs on the outer edges of the inner disk are probably hurled above the midplane to the distance of the orbit of Jupiter and beyond. It is during this X-Wind Stage that much of the angular momentum of solar rotation is transferred to the proto-planetary disk.

The sequence of high-temperature pulses of the X-Wind leads to the formation of CAIs and also causes a progressive devolatilization of the inner part of the solar nebular disk. A quote form Taylor (1998, p. 176) is appropriate here:

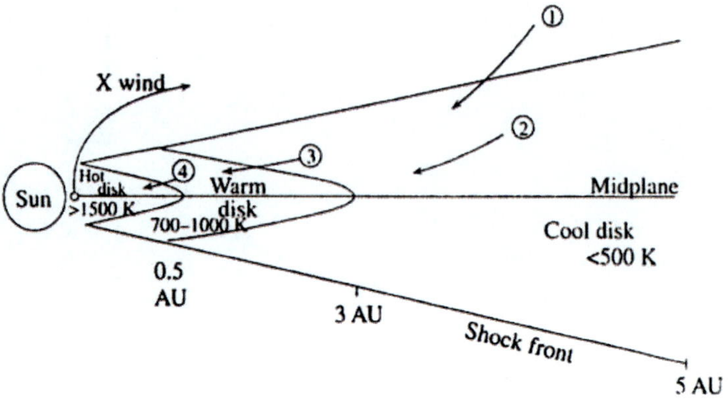

Fig. 2.5 Cross-sectional sketch illustrating some major features of the X-WIND model. Gravitational infall occurs symmetrically along the mid-plane of the nascent Solar System. The temperature increases as the proto-sun is approached. Inside ~0.5 AU the temperature is over 1500°K and there are high-temperature heat pulses caused by the breaking and reforming of magnetic field lines powered by the rapid rotation of the proto-sun. Bi-polar outflow is not shown on this diagram. In this model, CAIs can be formed only within a constantly shifting zone centered on 0.06 AU from the center of the Sun [the Reconnection Ring in this model (but not shown in this diagram)]. The X-Wind *arrow* indicates the trajectory by which CAIs can be propelled to outer regions of the solar system such as the zone of the asteroids and possibly beyond. The trajectory of the CAI debris is guided by a strong magnetic field that is rotating with the proto-sun and the proto-sun is rotating rapidly. All in all, this is a very dynamic scene but it may be a necessary condition for the formation of CAIs. (Figure is from Taylor (2001, Fig. 5.1) with permission from Cambridge University Press)

> All the evidence from the meteorites is that the primitive mineral components were dry. Water was driven out of the inner solar system and condensed as ice far away from the Sun, at a 'snow-line' in the vicinity of Jupiter. Thus, the terrestrial planets seem to have accumulated from dry rubble. Probably the only way for water to get to the inner planets is from comets coming from the outer reaches of the solar system.

Another quote from Taylor (1998, p. 48) completes the devolatilization scene for the early history of the Sun:

> The inner parts of the nebula were thus cleared of the gas and depleted in volatile elements very early, probably within about one million years of the formation of the Sun. The cause of this clearing of the inner regions of the nebula is twofold. Early on, gas was swept into the Sun. Then, after the nuclear furnace ignited, strong winds blew outward from the Sun and swept away any remaining gas. A few bodies, ranging from boulders to small mountains, survived even in the strong stellar winds. This was a fortunate circumstance, because we are standing on a planet that formed from this collection of rubble.

Figure 2.6 is a depiction of the volatile expulsion events proposed by Taylor (1998, 2001, 2012). The devolatization is characterized by very high temperatures near the Sun exponentially decreasing to the freezing point of water at the Frost Line. The large mass of Jupiter is thought to be a result of this overabundance of volatiles in the vicinity of the orbit of Jupiter. The transport of CAIs is thought to result from the X-Wind action. CAIs are found in various types of meteorites as well as in cometary material (Joswiak and Brownlee 2014; Joswiak et al. 2017).

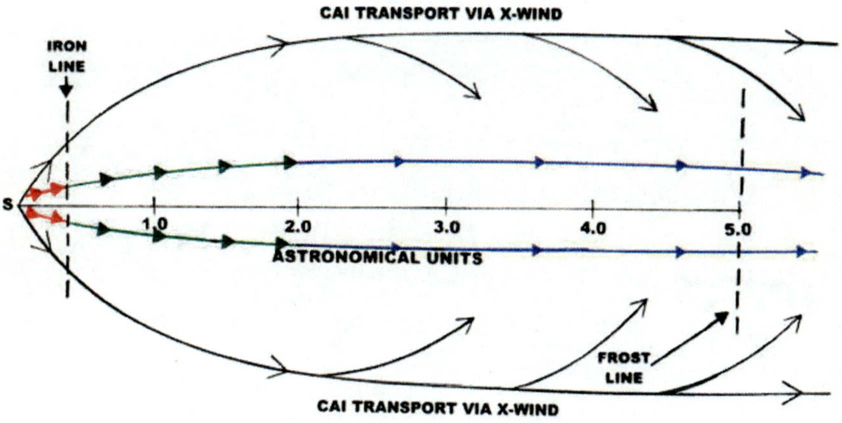

Fig. 2.6 A cartoon illustrating the possible effects of the X-Wind on the proto-planetary cloud. Progressive devolatilization is the key concept (red arrows extreme devolatilization; green arrows strong to moderate devolatilization; blue arrows weak devolatilization . CAIs are propelled outward above and below the disc to be assimilated within chondritic meteorites and some cometary materials (Joswiak and Brownlee 2014; Joswiak et al. 2017). Note: The position of the "Iron Line" and the "Frost Line" are stationary on the diagram but, in reality, these boundaries would fluctuate within limits as the X-Wind process evolves

The CAI materials in the vicinity of 0.1–0.2 AU accrete into planetesimals as the X-Wind action diminishes. These early formed planetesimals have the average chemistry of CAIs. It should be noted here that the chemical composition of the Moon (Wood 1974; Anderson 1975), Vesta (Ruzicka et al. 1999, 2001; Guterl 2008) and Angra (Taylor 2001) (the parent body of Angrite meteorites) have much in common and they are all very similar in chemistry to a composite CAI. I note here that this chemical relationship was noticed soon after lunar rocks were analyzed (e.g., Anderson 1973a, 1973b, 1975; Cameron 1972, 1973; Gast 1972; Wood 1974). A brief quote from Licandro et al. (2017, A126, p. 1) is appropriate here (Fig. 2.7):

> Recently, some V-type asteroids have been discovered far from the Vesta family supporting the hypothesis of the presence of multiple basaltic asteroids in the early solar system.

This information is consistent with several of these basaltic Vulcanoid planetoids surviving the journey from inside the orbit of Mercury to become "basaltic asteroids" which, in turn, become sources of "basaltic meteorites" similar to classical "vestoid" meteorites. DeMeo et al. (2019) did an extensive study of A-type asteroids (olivine-rich asteroids that are thought to be the mantle material of differentiated planetoids that have been fragmented by collision). Their conclusion was that most likely these A-type asteroids did not form locally (i.e., in the asteroid belt) but instead were implanted as collisional fragments of bodies formed elsewhere. My suggestion is that the parent bodies were formed as Vulcanoid planetoids along with Vesta, Angra, and Luna. Likewise, planetoids forming in the vicinity of the "Iron Line" (Figs. 2.7 and 2.8) are potential sources of "metallic meteorites" . The orbits of some of these

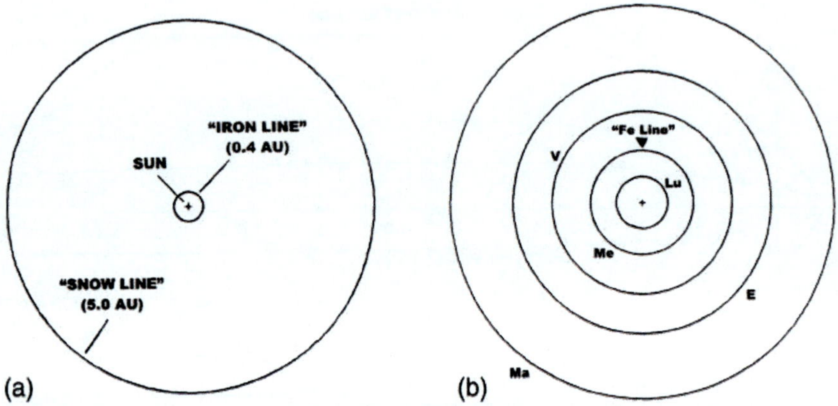

Fig. 2.7 (**a**) Scale diagrammatic sketch showing the approximate locations of the "snow *line*" (near the orbit of Jupiter) and the "iron line" (near the orbit of Mercury. (**b**) Position of the "*iron line*" relative to the proposed location of the orbit of Luna (Lu) and the current locations of planets Mercury (*Me*), Venus (*V*), Earth (*E*), and Mars (*Ma*). View for both diagram is from the north pole of the Solar System. (From Malcuit (2015, Fig. 2.7) with permission from Springer)

planetoids, whether basaltic or metallic, can be perturbed by a combination of planets Mercury, Venus, Earth, and Mars to become members of the asteroid population.

Concomitant with the diminishing X-Wind activity, the dust, ice, and gas at the "frost line" accretes into various shaped bodies in an accretion torus of Jupiter. Eventually these icy planetismals accumulate to form planet Jupiter. Saturn forms in a similar manner from a combination of materials coming in along the proto-solar mid-plane and volatile material that was propelled out from the inner zones of the Solar System. Wang et al. (2017) estimate that the X-Wind phase of solar evolution was over in ~1 Ma or less.

2.2.4 Disk-Wind Stage

> Chondrites are primitive meteorites derived from asteroids that have not undergone melting and chemical differentiation. As such, chondrites provide some of the most direct constraints on how solid material was formed, transported, and mixed within the solar protoplanetary disk.
>
> The origin of these once-molten silicate spherules remains debated, but the currently favored hypothesis states that chondrules formed by melting of dust aggregates in the solar nebula, induced by shock waves or current sheets (e.g., Desch et al. 2005; Morris et al. 2012; McNally et al. 2013).
>
> This complementarity implies that within a given chondrite, chondrules and matrix derived from a common reservoir of dust, and that after their formation neither appreciable chrondrules or matrix were lost. This suggests rapid accretion of chrondrite parent bodies after chrondrule formation and only limited exchange of chrondrules about different chrondrite formation regions (Budde et al. 2016a). — From Gerber et al. (2017, p. 1).

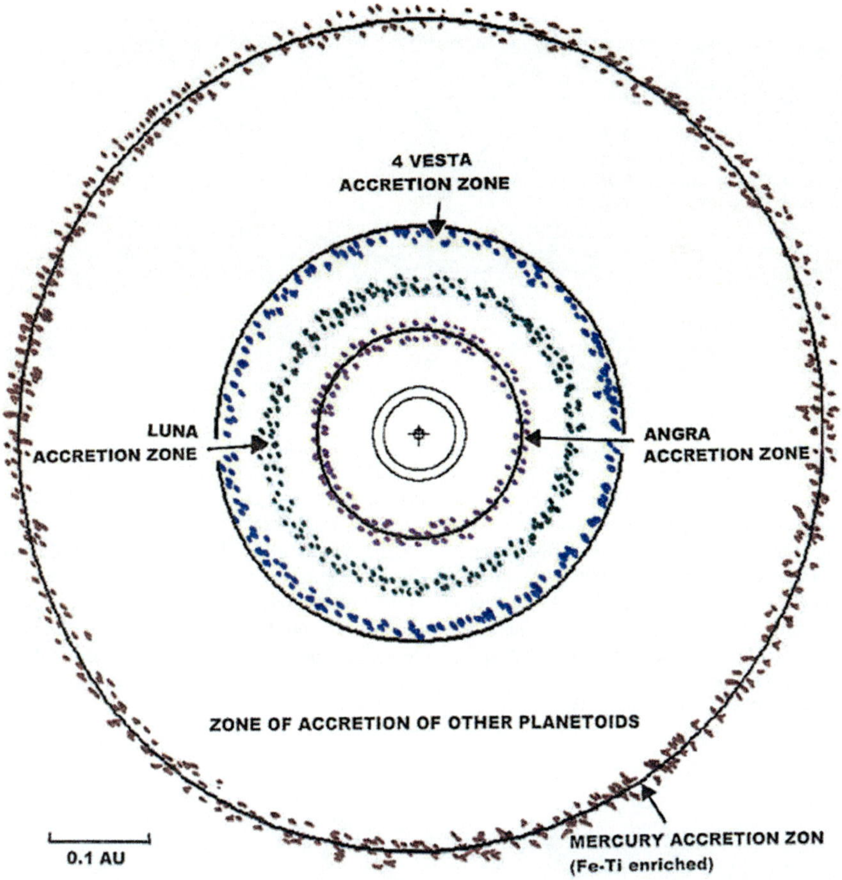

Fig. 2.8 Scale diagram viewed from the north pole of the Solar System showing my concept of the accretion zones that lead to formation of planet Mercury and the Vulcanoids. The debris zones are devolatilizated residue from the action of the X-Wind removing Fe, Ni, and other more volatile siderophile elements from the Vulcanoid Zone. This action causes a buildup of metal in the vicinity of the orbit of Mercury. The inner concentric circles mark the location of metal-poor CAI material resulting from the X-Wind action (Shu et al. 2001; Wood 2004; Liffman et al. 2012) which subsequently forms Vulcanoid planetoids. (Note that many more Vulcanoids would be formed than are shown in the diagram.) (From Malcuit (2015, Fig. 2.8) with permission from Springer)

The disk-wind model (Fig. 2.9) is characterized by lower energy pulsations of sheets of magnetic flux lines that periodically "flash-melt" concentrations of dust grains that have sufficient material density to make glass beads. These flash-melted beads of glass, with some unmelted inclusions, solidify very quickly, within minutes, to form chondrules. This "flash-melting and quench" process operates on the local chemistry of the disk material and may be responsible for the apparent chemical trends of the various groups of meteorites that were inherited from the X-Wind devolatilization process (Rubin 2013, 2015). Figures 2.10 and 2.11 are a representation of some products of the Disc-Wind model.

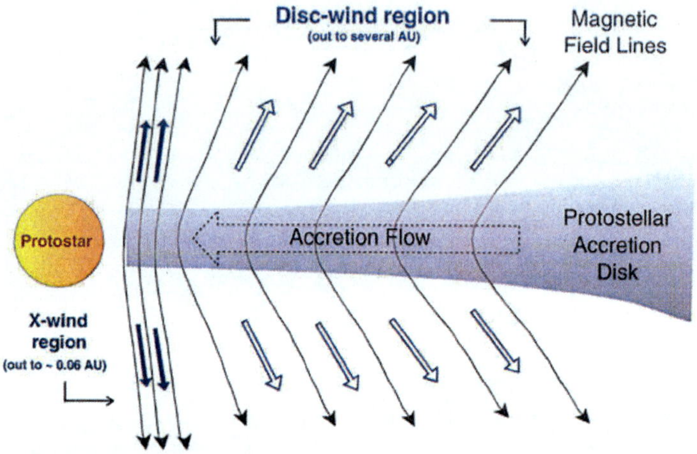

Fig. 2.9 Diagrammatic sketch showing some features of the X-wind model as well as the Disc-wind model. The general concept is that X-Winds operate very close to the protostar (i.e., near the disc truncation region) and that Disc Winds can operate to distances of a few astronomical units from the protostar. The X-Winds are thought to generate CAIs and related very high temperature minerals and mineral assemblages; the Disc Winds are thought to generate chondrules of a variety of compositions depending on the chemistry of the local region. Wang et al. (2017) estimate the duration of the disk-wind episode to be ~2 Ma or less. (Diagram from Salmeron and Ireland (2012, Fig. 1) with permission from Elsevier)

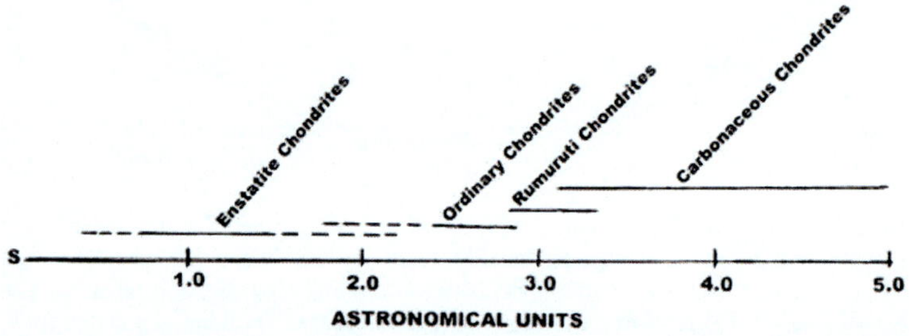

Fig. 2.10 A composite diagram showing the representation of the zonation of the different types of meteorites with distance from the Sun. (Simplified from Rubin 2013)

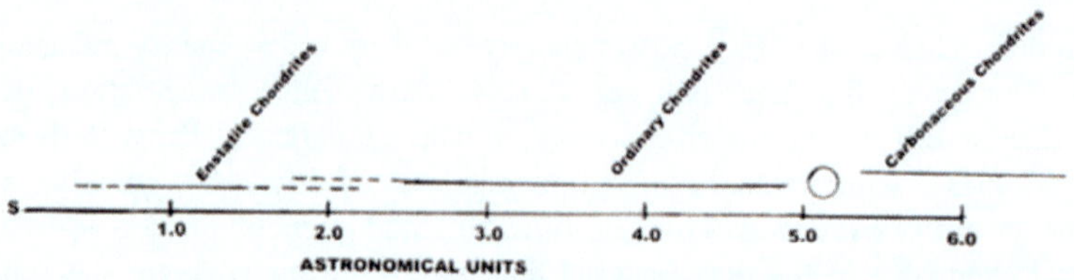

Fig. 2.11 A composite diagram showing the representation of the zonation of the different types of meteorites as suggested by Gerber et al. 2017) who think that there are significant differences between ordinary chondrites and carbonaceous chondrites. They suggest that the formation of Jupiter may have imposed a significant partition in meteorite distribution

Since the Disc-Wind sends its periodic pulses of magnetically-generated thermal energy (i.e., current sheets) into the mid-plane section of the proto-solar cloud, this model relates to the diversity in composition of chondrules (Krot et al. 2002; Lauretta et al. 2006). The dust in the cloud increases in volatile content systematically from the Venus-Earth region out to the Asteroid Zone because of the previous thermal pulses of the X-Wind. The volatile content of the chondrules should reflect this decrease in thermal energy with distance from the proto-sun.

2.2.5 Formation of Chondritic Meteorites Stage

There is general agreement in the meteoritic community that most chondrules were formed via disk-winds (Salmeron and Ireland 2012) but there is some disagreement concerning the resulting circumsolar distribution of the resulting chondritic meteorites. Currently there are two general models that explain many of the features of chondrules and chondrite meteorites. The main difference between the two is whether or not Jupiter is involved in causing a discontinuity in the succession of the different classes of chrondritic meteorites (Budde et al. 2016b). I will first discuss the model developed by Rubin (2013) in which there is no discontinuity and this will be followed by a model presented by Gerber et al. (2017) in which there is a discontinuity in the heliocentric distribution of chondrites caused by the presence of planet Jupiter.

Figure 2.10 is a composite diagram showing a representation of the zonation of the different types of meteorites (simplified from Rubin 2013).

Figure 2.11 is a composite diagram showing a representation of the zonation of the different type of meteorites according to Gerber et al. (2017).

Figure 2.12 is a scale sketch illustrating the places of origin of various materials in the solar system as well as the present location of the terrestrial planets.

2.2.6 Planetoid Accretion Stage

The terrestrial planets and asteroids begin to accrete by accumulation of material within their respective accretion tori. The composition of these bodies is essentially inherited from a combination of the composition of the chondrules and associated matrix material of the chrondritic meteorites in their zone of formation plus perhaps some late infalling material as well as traces of CAI material. Gravitational perturbations by Jupiter and Saturn modifies (or controls) the accretion process for Mars and the Asteroids. At this point in the model, there is very little water in the Venus-Earth space but more is in the Mars space because of a combination of increased

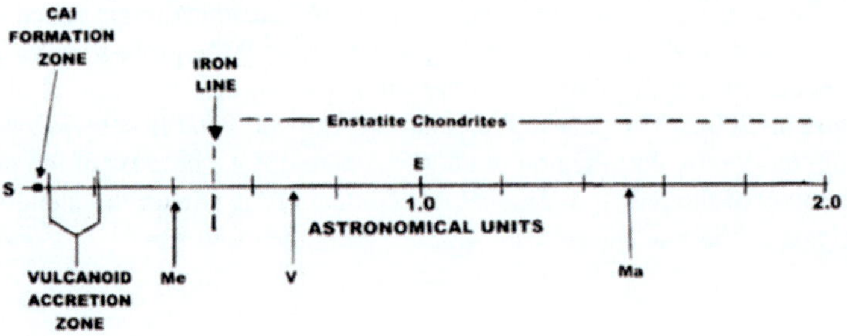

Fig. 2.12 Probable places of formation of various planets, planetoids, and other entities within 2.0 AU. The X-Wind model suggests that CAI can only form in a restricted zone near the Sun. The Vulcanoid planetoids form mainly from the various types of CAIs. The Iron Line fluctuates over a significant range in the vicinity of the orbit of Mercury. Most meteoriticists agree that planets Venus, Earth, and Mars have a large component of Enstatite Chondrites

volatile element content (because X-Wind action decreases exponentially with radial distance) and the addition of some volatiles from infalling material along the accretion mid-plane.

Liebske and Khan (2019) did an extensive study of several important isotopic measurements for both primitive chondrites and differentiated achondrites and mixed them stochastically in an attempt of reproduce the isotopic signatures of Earth and Mars. The best result for Earth was to use a large proportion of enstatite chondrite-like material. For Mars the result suggested that oxidized material in addition to that provided by chondritic mixtures is required. And for Mercury they suggest that because of its large core the planet has a significantly higher metal to silicate ratio than can be explained by accretion of enstatite chondrite material. My suggestion is that the IRON LINE model shown in Fig. 2.7 and Malcuit (2015, p. 56, Fig. 4.2) may be a reasonable solution for the extremely high iron content of planet Mercury. Two quotes from Liebske and Khan (2019, p. 131) sum up the apparent complexity of the terrestrial planet accretion problem:

> ... it is our contention that considering geophysical properties and redox ranges of isochemical mixtures of planetary building blocks provides important insights on planetary accretion in the Solar System.
>
> This contribution has shown that consideration of isotope data alone is insufficient as a means of distinguishing between terrestrial planet building blocks.

Figure 2.13 is the author's concept of the accretion tori of planets Venus and Earth.

Figure 2.14 is a pictorial summary of the orbital calculations of Evans and Tabachnik (1999, 2002). A major "surprise" for most planetary scientists is that the classical zone of Vulcanoids (Weidenschilling 1978; Leake et al. 1987) is characterized by very stable planetoid orbits.

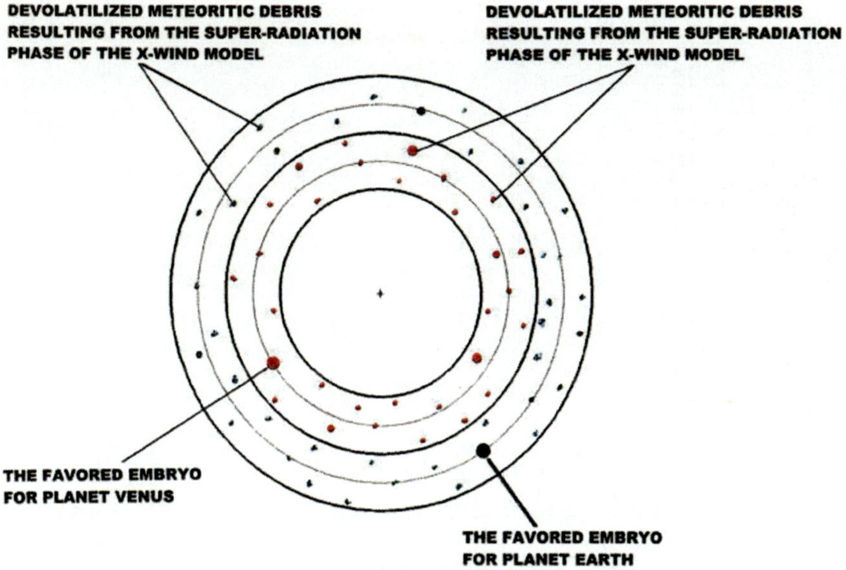

Fig. 2.13 General geometry of the accretion tori for planets Venus and Earth. The favored planetary embryos accumulate whatever particles are in their sphere of influence

2.2.7 FU Orionis Stage

This stage involves pulsations of microwave heating/melting events that affect the refractory planetoids that formed near the sun in the Vulcanoid Zone as well as those as far out as the orbit of planet Mercury. The outer portion of the Vulcanoids would be melted by FU Orionis heating to form a magma ocean zone. As the magma ocean cools an anorthositic crust forms via fractional crystallization of calcium-rich plagioclase in the magma ocean and subsequent floatation to the surface. Then a series of magnetic amplifiers form within the magma ocean zone of the planetoid because of solar gravitational tides causing differential rotation of the newly formed refractory crustal complex and the Vulcanoid planetoid interior. Such a convective mechanism in a lunar magma ocean zone was proposed by Smoluchowski (1973a, b) for the cause of the magnetic patterns registered in the primitive crust of the Moon. More details on the process of tidal vorticity induction is in Bostrom (2000).

Information from DeMeo et al. (2019) and Steenstra et al. (2019) can also be explained by the FU Ori heating model. They suggest that asteroid Vesta has a metallic core and that several volatile elements are sequestered in that metallic core. However, there is no physical/mineralogical evidence for metallic core formation in the basaltic asteroid Vesta.

The Iron Line concept, illustrated in Figs. 2.6 and 2.7, can be used to explain most, if not all, of the features of metallic meteorites. Planetoids formed near the Iron Line should have a metal-rich interior and an anorthositic basalt exterior formed

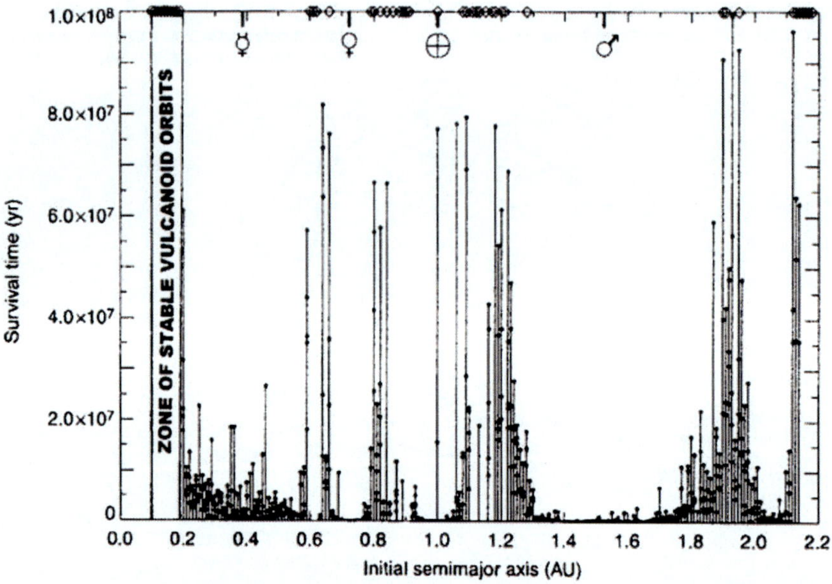

Fig. 2.14 Results of extensive calculations by Evans and Tabachnik (1999, 2002) illustrating the stability of orbits of planetoids in the inner Solar System. The positions of the terrestrial planets are shown by symbols at the top of the diagram. Open diamond symbols at the top of the diagram denote orbits of particles that are stable for over 100 Ma. The stable orbits between 0.1 and 0.2 were too numerous to be plotted as lines. (Diagram modified from Evans and Tabachnik (1999, Fig. 1) with permission from Springer-Nature)

above the magma ocean zone. If some of the crustal material is stripped away via minor collisions with other planetoids, then a metal-rich body with the composition of metallic asteroids could result.

Steenstra et al. (2019) state that “Vesta is believed to be a prime example of a rocky protoplanet”. They also state that eucrites (basaltic fragments from Vesta) and diogenites (coarser-grained fragments from Vesta) are depleted in many volatile elements. In my view, this complement of volatile elements was depleted by X-Wind activity even before the body of Vesta was formed. In the view of Steenstra et al. (2019) these volatiles are sequestered in a metallic core of Vesta. Steenstra et al. (2019) also suggest that the lunar VSEs went to a lunar core. With a moment of inertia factor of 0.393, there may be no lunar metallic core. Perhaps future lunar missions sponsored by China will do the relevant seismic experiments to determine whether or not there is a lunar metallic core. If there is no metallic lunar core, the question of the “missing” VSEs is settled. If there is evidence of a lunar metallic core, then the question is centered on the volatile content of the metallic core.

There is some evidence that planet Mercury may have been involved in the FU Ori phase of solar evolution. The investigators of the data from the Messenger Mission to Mercury report a widespread surface sulfur content of ~4%. There is also large quantities of potassium on the surface in certain locations. Sulfur could be a temporary atmospheric gas. Then as the surface cooled following the FU Ori sequence of events, the atmospheric sulfur precipitated as a surface layer of sulfur. The lava flows that are potassium rich could be related to a volcanic eruption of "KREEP"-like material from a magma ocean of this iron-rich planet.

There could be some effects from the FU Ori heating events in the Venus-Earth zone but evidence of these heating events may not be preserved because the final stages of accretion of Venus and Earth may have not been completed at this time. Future work on the timing of these events will furnish more details on probable effects. Figure 2.15 is an illustration of the possible thermal pulses associated with FU Orionis as well as T-Tauri thermal outbursts and Fig. 2.16 shows some of the possible effects for bodies within the orbit of planet Mercury.

2.2.8 T-Tauri Stage

The T-Tauri stage (Fig. 2.15) is characterized by longer wavelength radiation from the sun resulting in thermal and hydrothermal metamorphism (alteration) of bodies in the terrestrial planet realm and the asteroid zone. As in the case of the FU Orionis radiation, the radiant-thermal energy decreases exponentially in intensity away from the solar source. This radiation stage may be registered in the extensive hydrothermal metamorphism of asteroids on the inner edge of the asteroid zone. Herbert et al. (1991) suggest that there is a progressive decrease in the intensity of hydrothermal metamorphism in asteroids with distance from the Sun.

2.3 A Question: What Was Happening in the Outer Solar System While All of This Intense Action Was Occuring in the Inner Solar System?

> In all the integrations which are reported here, the large planet motions are always very regular, while the diffusion of the inner planet' orbits in the chaotic zone is much larger than what was already seen over 200 Myr (Laskar 1990; Laskar et al. 1992). —From Laskar (1994, p. L12).

The main event occurring in the outer solar system is the continuing infall of dust and gas from previous stellar explosions. The infalling materials experience some "push-back" from the various pulsations of energy emanating from the proto-sun, but the region beyond Jupiter and Saturn is not greatly affected by the push-back action. Planet Jupiter accumulates (accretes) material being propelled from the inner part of the solar system as well as material coming in toward the sun from the outer regions. Saturn probably gets some from both the inner part of the Solar System as well as from the infalling regime. Uranus and Neptune accumulate (accrete) ice and associated gas-rich bodies that formed mainly from infalling material. These locally formed planetoids along with some planetoids from the Kuiper Belt eventually accrete to form Uranus and Neptune. In my model, two of these planetoids (which

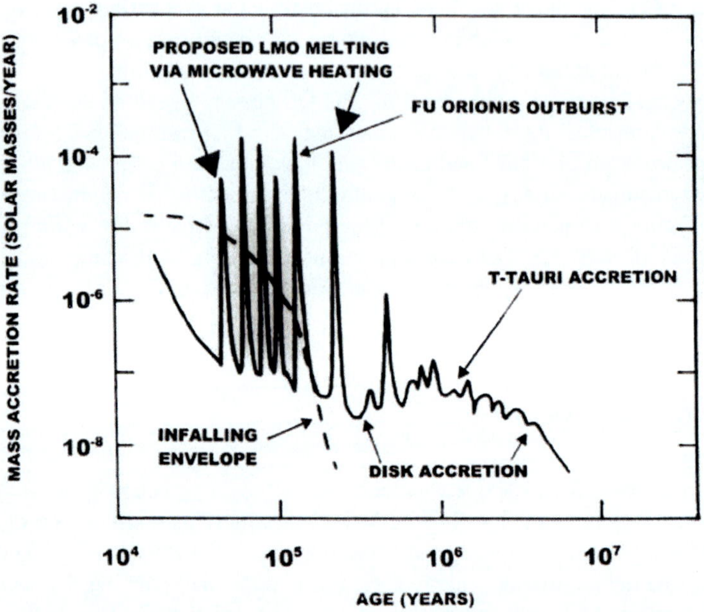

Fig. 2.15 Diagram illustrating the spasmodic nature of the FU Orionis stage of solar evolution. Note that there are intense energy "spikes" associated with the FU Ori stage and that these "spikes" are absent during the T-Tauri stage. The author suggests that the melting of the lunar magma ocean as well as the generation of a lunar magnetic amplifier in the LMO were powered, in part, by the thermal energy of the FU Ori stages of solar development. (Diagram adapted from Calvet et al. (2000, Fig. 7), with permission from University of Arizona Press)

either formed locally or in the Kuiper Belt) get captured, one by Uranus and the other by Neptune into retrograde orbits. These stories will be told in Chap. 6.

There is general agreement that the source of comets is the Oort Cloud (de Pater and Lissauer 2001, 2015). It is also well known that comets come into the inner part of the solar system from any direction so they are thought to reside in a spherical belt surrounding the solar system which is located beyond the Kuiper Belt. Figure 2.17 is a diagram showing the orbits of the outer planets and the Kuiper Belt.

Note that the "orbital stability" model discussed above contrasts greatly with the "orbital chaos" models that other groups of solar system scientists have been developing (*e.g.*, Batygin et al. 2016). These "orbital chaos" models, which now appear to be in mild "disfavor" (Mann 2018) will be discussed in Chap. 6.

2.4 Summary

The reader at this early point in this book may get the feeling that these events associated with the origin and evolution of the Solar System represent a "long chain of complications" as suggested by Alfven (1969), Alfven and Alfven (1972) Cloud

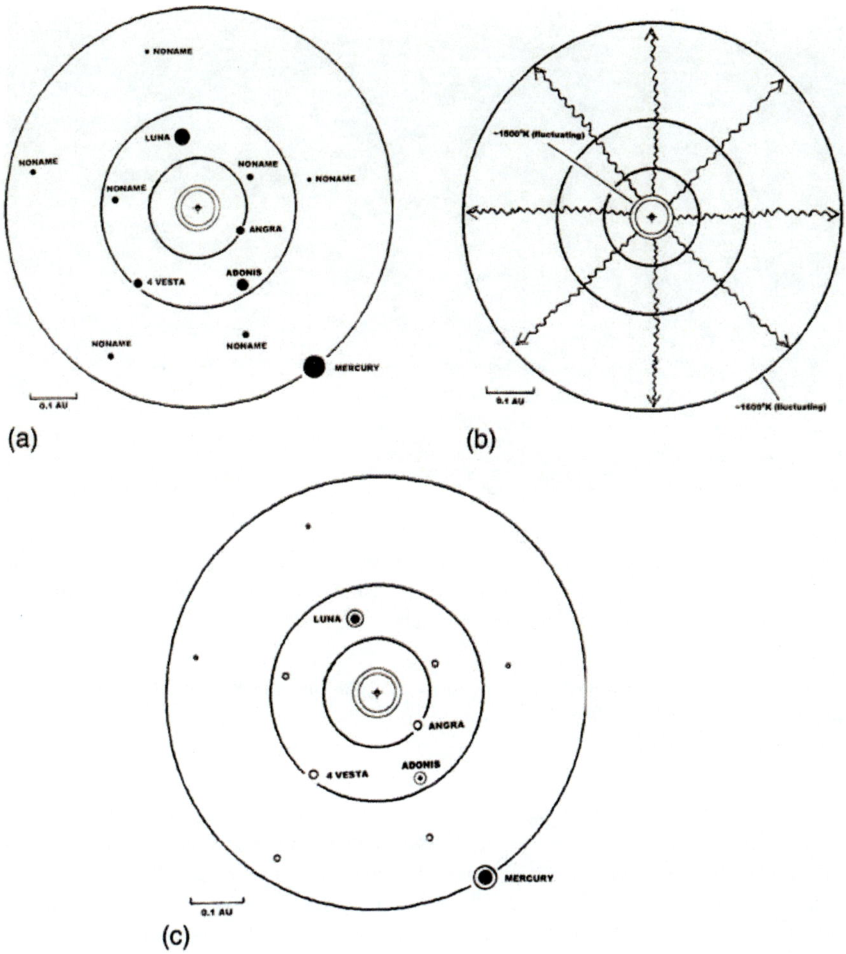

Fig. 2.16 Stages in the "microwaving" process for planet Mercury and the Vulcanoid Planetoids. (**a**) Before the "microwaving" process, (**b**) During the "microwaving" process, (**c**) After the "microwaving" process. View is from the north pole of the Solar System. (From Malcuit (2015, Fig. 2.11) with permission of Springer)

(1978, 1988), Taylor (1998, 2012), Laskar (1995, 1996), and others. Indeed, I think there is good reason to think this way. Although there may be a number of unusual events affecting the processing of a proto-solar cloud of roughly solar composition into the planetary bodies we have today, I cannot apologize for these complications. One event leads to another to another in a somewhat logical chain of related events. There are very few, if any *ad hoc* assumptions in this model. Once the planets and associated planetoids are formed by planetoid accretion (*i.e.*, collision) processes, there are still many sub-plots to unravel.

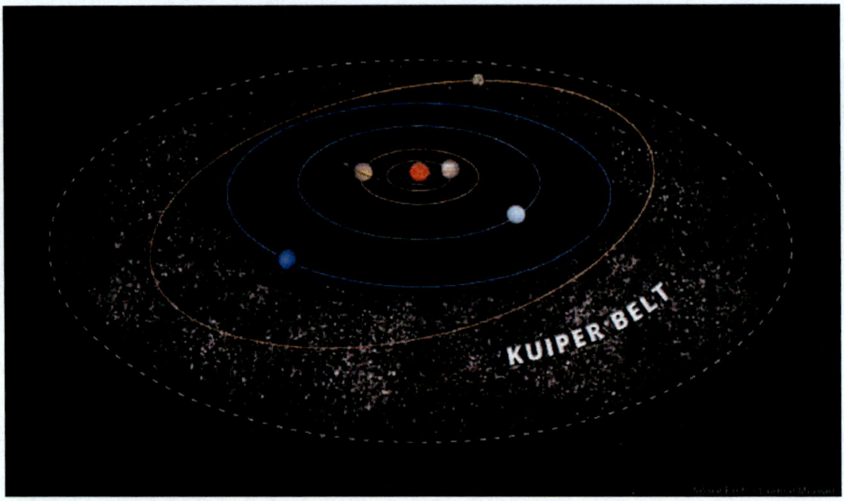

Fig. 2.17 Illustration of the Sun (center), the orbits of the four outer gaseous/icy planets (Jupiter, Saturn, Uranus, and Neptune), as well as the orbit of Pluto (yellow). The orbit of Pluto is depicted as a planetary body in the Kuiper Belt, a reservoir of icy bodies which can be gravitationally perturbed so that they can interact with the four gaseous/icy planets. (Artwork is courtesy of Laurine Moreau)

For example, in the story that is unfolding in this book I am suggesting four important gravitational planetoid capture episodes in Solar System history. Two of these captured satellites are still in existence – one in prograde orbit (i.e., Earth's moon, Luna) and one in retrograde orbit (i.e., Neptune's moon, Triton). The other two are planetoids that were captured in retrograde orbit to become planet-satellite systems but these satellites no longer exist because they eventually coalesced with the planets that captured them (i.e., phantom satellites: meaning that there is physical evidence that can be explained by their former existence). These are (1) a former retrograde satellite of Uranus of about two mars masses (20 moon-masses) (Kubo-Oka and Nakazawa 1995) and (2) a former retrograde satellite of Venus of about one-half moon mass (Malcuit 2015, Chap. 6). There will be a further development of the Uranus story in Chap. 7.

References

Alfven, H. (1969). *Atom, man, and the universe: The long chain of complications*. San Francisco: WH. Freeman and, 110 p.

Alfven, H., & Alfven, K. (1972). *Living on the third planet*. San Francisco: WH. Freeman and Company, 187 p.

Amelin, Y., Krot, A. N., Hutcheon, I. D., & Ulyanov, A. A. (2002). Lead isotopic ages of chondrules and calcium-aluminum—Rich inclusions. *Science, 297*, 1678–1683.

Anderson, D. L. (1973a). The moon as a high temperature condensate. *The Moon, 8*, 33–57.

Anderson, D. L. (1973b). The composition and origin of the moon. *Earth and Planetary Science Letters, 18*, 301–316.

Anderson, D. L. (1975). On the composition of the lunar interior. *Journal of Geophysical Research, 80*, 1555–1557.

Batygin, K., Laughlin, G., & Morbidelli, A. (2016). Born of chaos. *Scientific American, 314*(5), 30–37.

Benz, W., Ida, S., Alibert, Y., Lin, D., & Mordasini, C. (2014). Planet population synthesis. In H. Beuther et al. (Eds.), *Protostars and planets VI* (pp. 691–713). Tucson: University of Arizona Press.

Bostrom, R. C. (2000). *Tectonic consequences of Earth's rotation*. Oxford: Oxford University Press, 266 p.

Brearley, A. J., & Jones, R. H. (1998). Chapter 3: Chrondritic meteorites. In J. J. Papike (Ed.), *Planetary materials: Reviews of mineralogy* (Vol. 36, pp. 1–398). Washington, D.C.: Mineralogical Society of America.

Britt, D. T., Macke, R. J., Kiefer, W., & Consolmagno. (2010). An overview of achondritic density, porosity and magnetic susceptibility: Abstracts, 41st Lunar and Planetary Science Conference, 1869. pdf.

Budde, G., Kleine, T., Kruijer, T. S., Burkhardt, C., & Metzler, K. (2016a). Tungsten isotopic constraints on the age and origin of chondrules. *Proceedings of the National Academy of Sciences, 113*, 2886–2891.

Budde, G., Burkhardt, C., Brennecka, G. A., Fischer-Goode, M., Kruijer, T. S., & Kleine, T. (2016b). Molybdenum isotopic evidence for the origin of chondrules and a distinct genetic heritage of carbonaceous and non-carbonaceous meteorites. *Earth and Planetary Science Letters, 454*, 293–303.

Calvet, N., Hartmann, L., & Strom, S. E. (2000). Evolution of disk accretion. In V. Manning, A. P. Boss, & S. S. Russell (Eds.), *Protostars and planets, IV* (pp. 377–399). Tuscon: University of Arizona Press.

Cameron, A. G. W. (1962). The formation of the sun and the planets. *Icarus, 1*, 13–69.

Cameron, A. G. W. (1972). Orbital eccentricity of mercury and the origin of the moon. *Nature, 240*, 299–300.

Cameron, A. G. W. (1973). Properties of the solar nebula and the origin of the moon. *The Moon, 7*, 377–383.

Cameron, A. G. W., & Pine, M. R. (1973). Numerical models of the primitive solar nebula. *Icarus, 18*, 377–406.

Ceisla, F. J. (2007). Outward transport of high temperature materials around the midplane of the solar nebula. *Science, 318*, 613–615.

Cloud, P. E. (1978). *Cosmos, earth and man*. New Haven: Yale University Press, 371 p.

Cloud, P. E. (1988). *Oasis in space: Earth history from the beginning*. New York: W. W. Norton and Company, 508 p.

Connolly, H. C., & Desch, S. J. (2004). On the origin of the "klein Kugelchen" called chondrules. *Chemie der Erde/Geochemistry, 64*, 95–125.

Connolly, H. C., Jr., & Jones, R. H. (2016). Chondrules: The canonical and noncanonical views. *Journal of Geophysical Research: Planets, 121*, 1885–1899. https://doi.org/10.1002/2016JE005113.

DeMeo, F. E., Polishook, D., Carry, B., Burt, B. J., Hsieh, H. H., Benzel, R. P., Moskovitz, & Burbine, T. H. (2019). Olivine-dominated A-type asteroids in the main belt: Distribution, abundance and relation to families. *Icarus, 322*, 13–30.

de Pater, I., & Lissauer, J. J. (2001). *Planetary sciences*. Cambridge: Cambridge University Press, 528 p.

de Pater, L., & Lissauer, J. J. (2015). *Planetary sciences* (2nd ed.). Cambridge: Cambridge University Press, 688 p.

Desch, S. J., & Cuzzi, J. N. (2000). The generation of lightning in the solar nebula. *Icarus, 143*, 87–105.

Desch, S. J., Ciesla, F. J., Hood, L. L., & Nakamoto, T. (2005). Heating of chondritic materials in solar nebula shocks. In A. N. Krot, E. R. D. Scott, & B. Reipurth (Eds.), *Chrondrites and the protoplanetary disk* (Vol. 341, pp. 849–872). San Francisco: Astronomical Society of the Pacific.

Ehlmann, B. L., Anderson, F. S., et al. (2016). The sustainability of habitability on terrestrial planets: Insights, questions, and needed measurements from mars for understanding the evolution of earth-like worlds. *Journal of Geophysical Research: Planets, 121*, 1927–1961.

Evans, N. W., & Tabachnik, S. (1999). Possible long-lived asteroid belts in the inner solar system. *Nature, 399*, 41–43.

Evans, N. W., & Tabachnik, S. (2002). Structure of possible long-lived asteroid belts. *Monthly Notices of the Royal Astronomical Society, 333*, L1–L5.

Fu, R. R., Weiss, B. P., Lima, E. A., et al. (2014). Solar nebula magnetic fields recorded in the Semarkona meteorite. *Science, 346*, 1089–1092.

Gast, P. W. (1972). The chemical composition and structure of the moon. *The Moon, 5*, 121–148.

Gerber, S., Burkhardt, C., Bidde, G., Metzler, K., & Kleine, T. (2017). Mixing and transport of dust in the early solar nebula as inferred from titanium isotope variation among chrondrules. *The Astrophysical Journal Letters, 841*, L17. (7 p.). https://doi.org/10.3847/2041-8213/aa72a2.

Greenwood, R. C., Franchi, I. A., Jambon, A., & Buchanan, P. C. (2005). Widespread magma oceans on asteroidal bodies in the early Solar System. *Nature, 435*, 916–918.

Grossman, L. (1972). Condensation in the primitive solar nebula. *Geochimica et Cosmochimica Acta, 36*, 597–619.

Grossman, L. (1988). Formation of chondrules. In J. F. Kerridge & M. S. Mathews (Eds.), *Meteorites and the early solar system* (pp. 680–696). Tucson: University of Arizona Press.

Guterl, F. (2008). Mission to the forgotten planets: Discover (Feb. issue), p. 48–52.

Hasegawa, Y. (2016). Super-earths as failed cores in orbital migration traps. *The Astrophysical Journal, 832*, 83. (18 p.). https://doi.org/10.3847/0004-637X/832/1/83.

Hasegawa, Y., Turner, N. J., Masiero, J., Wakita, S., Matsumoto, Y., & Oshino, S. (2016). Forming chrondrites in a solar nebula with magnetically induced turbulence. *The Astrophysical Journal Letters, 820*, L12–L18. https://doi.org/10.3847/2041-8205/820/L12.

Hasegawa, Y., Okuzumi, S., Flock, M., & Turner, N. J. (2017). Magnetically induced disk winds and transport in the HL Tau disk. *The Astrophysical Journal, 845*, 31. , (13 p.). https://doi.org/10.3847/1538-4357/aa7d55.

Herbert, F., Sonett, C. P., & Gaffey, M. J. (1991). Protoplanetary thermal metamorphism: The hypothesis of electromagnetic induction in the protosolar wind. In C. P. Sonett, M. S. Giampapa, & M. S. Mathews (Eds.), *The sun in time* (pp. 710–739). Tuscon: University of Arizona Press.

Humayun, M., & Clayton, R. N. (1995). Potassium isotope cosmochemistry: genetic implications of volatile element depletion. *Geochimica et Cosmochimica Acta, 59*, 2131–2148.

Ida, S., & Lin, D. N. C. (2004). Toward a deterministic model of planetary formation. I. A desert in the mass and semimajor axis distributions of extrasolar planets. *The Astrophysical Journal, 604*, 388–413.

Jones, R. H. (2012). Petrographic constraints on the diversity of chondrule reservoirs in the protoplanetary disk. *Meteoritics and Planetary Science, 47*, 1176–1190. https://doi.org/10.1111/j.1945-5100.2011.01327.x.

Joswiak, D. J., & Brownlee, D. E. (2014). Refractory-rich materials in comets: CAIs, Al-rich chondrules and AOAs from comet wild 2 and a giant cluster interplanetary dust particle (IPD) of probable cometary origin and comparison to refractory-rich objects in chondrites: 45th lunar and planetary science conference, Lunar and Planetary Institute, Houston, 2282 pdf.

Joswiak, D. J., Brownlee, D. E., Nguyen, A. N., & Messenger, S. (2017). Refractory materials in comet samples. *Meteoritics and Planetary Science, 52*, 1612–1648.

Kobayashi, D., & Sprenke, K. F. (2010). Lithospheric drift on early mars: Evidence in the magnetic field. *Icarus, 210*, 37–42.

Kowal, C. T. (1996). *Asteroids – Their nature and utilization* (2nd ed.). Chichester: Praxis Publishing Company (Wiley), 153 p.

Krot, A. N., McKeegan, K. D., Leshin, L. A., MacPherson, G. J., & Scott, E. R. D. (2002). Existence of an ^{16}O-rich gaseous reservoir in the solar nebula. *Science, 295*, 1051–1054.

Kubo-Oka, T., & Nakazawa, K. (1995). Gradual increase in the obliquity of Uranus due tidal interaction with a hypothetical retrograde satellite. *Icarus, 114*, 21–32.

Laskar, J. (1990). The chaotic motion of the solar system: A Numerical estimate of the size of the chaotic zones. *Icarus, 88*, 266–291.

Laskar, J., Froeschle, C., & Cellitti, A. (1992). The measure of chaos by the numerical analysis of the fundamental frequencies. *Application to the Standard Mapping: Physica D, 56*, 253–269.

Laskar, J. (1994). Large-scale chaos in the solar system. *Astronomy and Astrophysics, 287*, L9–L12.

Laskar, J. (1995). *Large scale chaos and marginal stability in the solar system: XIth international congress of mathematical physics*. Boston: International Press, 120 p.

Laskar, J. (1996). Large scale chaos and marginal stability in the solar system. *Celestial Mechanics and Dynamical Astronomy, 64*, 115–162.

Lauretta, D. S., Nagahara, H., & Alexander, C. M. O.'. D. (2006). Petrology and origin of ferromagnesian silicate chondrules. In D. S. Lauretta & H. Y. McSween (Eds.), *Meteoritics and the early solar system, II* (pp. 431–459). Tucson: University of Arizona Press.

Leake, M. A., Chapman, C. R., Weidenschilling, S. J., Davis, D. R., & Greenberg, R. (1987). The chronology of Mercury's geological and geophysical evolution: The Vulcanoid hypothesis. *Icarus, 71*, 350–375.

Lewis, J. S. (1972). Metal/silicate fractionation in the solar system. *Earth and Planetary Science Letters, 15*, 286–290.

Lewis, J. S. (1974). The chemistry of the solar system. *Scientific American, 230*(3), 51–65.

Licandro, J., Popescu, M., Morate, D., & de Leon, J. (2017). V-type candidates and Vesta family asteroids in the Moving Objects Vista (Movis) catalogue. *Astronomy and Astrophysics, 600*, A126. , (9 p.). https://doi.org/10.1051/0004-6361/201629465.

Liebske, C., & Khan, A. (2019). On the principal building blocks of mars and earth. *Icarus, 322*, 121–134.

Liffman, K., Pignatale, F. C., Maddison, S., & Brooks, G. (2012). Refractory metal nuggets—Formation of the first condensates in the solar nebula. *Icarus, 221*, 89–105.

Lodders, K., & Fegley, B., Jr. (1998). *The planetary scientist's companion* (371 p). New York: Oxford University Press.

MacDonald, G. J. F. (1963). The internal constitutions of the inner planets and the moon. *Space Science Reviews, 2*, 473–557.

MacDonald, G. J. F. (1964). Tidal friction. *Reviews of Geophysics, 2*, 467–541.

Malcuit, R. J. (2015). *The twin sister planet, Venus and earth: Why are they so different?* (401 p). Cham: Springer International Publishers.

Mann, A. (2018). Cataclysm's end. *Nature, 553*, 393–395.

McNally, C. P., Hubbard, A., MacLow, M. M., Ebel, D. S., & D'Alessio. (2013). Mineral processing by short circuits in protoplanetary disks. *The Astrophysical Journal Letters, 767*, L2. (6 p.).

McSween, H. Y., Jr. (1999). *Meteorites and their parent bodies*. Cambridge: Cambridge University Press, 310 p.

Mordisini, C., Alibert, Y., & Benz, W. (2009). Extrasolar planet population synthesis. I. Method, formation tracks, and mass-distance distribution. *Astronomy and Astrophysics, 501*, 1139–1160. https://doi.org/10.1051/0004-6361/200810301.

Mordisini, C., van Boekel, R., Molliere, P., Henning, T., & Benneke, B. (2016). The imprint of exoplanet formation history on observable present-day spectra of hot jupiters. *The Astrophysical Journal, 832*, 41. (32 p.).

Moreau, L. (@laurinemoreau.com).

Morris, M. A., Boley, A. C., Desch, S. J., & Athanassladou, T. (2012). Chrondrule formation in bow shocks around eccentric planetary embryos. *The Astrophysical Journal, 752*, 27–44. https://doi.org/10.1088/0004-637X/752/1/27.

Robert, F. (2001). The origin of water on earth. *Science, 293*, 1056–1058.

Rubin, A. E. (2013). Secrets of primitive meteorites. *Scientific American, 308*(2), 36–41.

Rubin, A. E. (2015). Maskelynite in asteroidal, lunar and planetary meteorites: An indicator of shock pressure during impact ejection from their parent bodies. *Icarus, 257*, 221–229.

Ruzicka, A., Snyder, G. A., & Taylor, L. A. (1999). Giant impact and fission hypotheses for the origin of the moon: A critical review of some geochemical evidence. In G. A. Snyder, C. R. Neal, & W. G. Ernst (Eds.), *Planetary petrology and geochemistry* (Vol. 2, pp. 121–134). Boulder: Geological Society of America, International Book Series.

Ruzicka, A., Snyder, G. A., & Taylor, L. A. (2001). Comparative geochemistry of basalts for the moon, earth, HED asteroid, and mars: Implications for the origin of the Moon. *Geochimica et Cosmochimica Acta, 65*, 979–997.

Salmeron, R., & Ireland, T. R. (2012). Formation of chondrules in magnetic winds blowing through the proto-asteroid belt. *Earth and Planetary Science Letters, 327-328*, 61–67.

Shu, F. H., Shang, H., & Lee, T. (1996). Toward an astrophysical theory of chondrites. *Science, 271*, 1545–1552.

Shu, F. H., Shang, H., Gounelle, M., Glassgold, A. E., & Lee, T. (2001). The origin of chondrules and refractory inclusions in chondritic meteorites. *Astrophysical Journal, 548*, 1029–1050.

Smoluchowski, R. (1973a). Lunar tides and magnetism. *Nature, 242*, 516–517.

Smoluchowski, R. (1973b). Magnetism of the Moon. *The Moon, 7*, 127–131.

Steenstra, E. S., Dankers, D., Berndt, J., Klemme, S., Matveev, S., & van Westrenen, W. (2019). Significant depletion of volatile elements in the mantle of asteroid Vesta due to core formation. *Icarus, 317*, 669–681.

Taylor, S. R. (1998). *Destiny or chance: Our solar system and its place in the cosmos*. Cambridge: Cambridge University Press, 229 p.

Taylor, S. R. (2001). *Solar system evolution: A new perspective*. Cambridge: Cambridge University Press, 460 p.

Taylor, S. R. (2012). *Destiny or chance revisited: Planets and their place in the Cosmos*. Cambridge: Cambridge University Press, 291 p.

Vinogradov, A. P., Surkov, Y. A., & Kirnozov, F. F. (1973). The content of uranium, thorium, and potassium in the rocks of Venus as measured by Venera 8. *Icarus, 20*, 253–259.

Wang, H., Weiss, B. P., Bai, X.-N., et al. (2017). Lifetime of the solar nebula constrained by meteorite paleomagnetism. *Science, 355*, 623–627.

Weidenschilling, S. J. (1978). Iron/silicate fractionation and the origin of Mercury. *Icarus, 35*, 99–111.

Wilson, S. A., Howard, A. D., Moore, J. M., & Grant, J. A. (2016). A cold-wet middle latitude environment on mars during the Hesperian-Amazonian transition: Evidence from northern Arabia valleys and paleolakes. *Journal of Geophysical Research: Planets, 121*, 1667–1694.

Wood, J. A. (1974). Summary of the 5th lunar science conference: Constraints on structure and composition of the lunar interior: Geotimes, June Issue, p. 16–17.

Wood, J. A. (2004). Formation of chondritic refractory inclusions: The astrophysical setting. *Geochimica et Cosmochimica Acta, 68*, 4007–4021.

Chapter 3
Models for the Origin of the Current Martian Satellites

The recent evidence that Phobos is made of carbonaceous chondritic material suggests that the Martian moons are captured objects and not locally formed satellites. However, as pointed out by Burns (1972, 1977, 1978) and emphasized by Pollack (1977), the low inclinations of these satellites' orbits are very difficult to account for if they were tidally captured. And thus we propose instead that they were acquired as a result of gas drag effects in a primordial Martian nebula.—From Pollack et al. (1979, p. 608).

The meager available information that is pertinent to the origin and evolution of the Martian satellites is contradictory. The known physical properties of the Martian moons (density, albedo, color and spectral reflectivity) are similar to those of many C-type asteroids, the dark 'carbonaceous' objects abundant in the outer belt but scarce near Mars; thus this line of physical evidence suggests that Phobos and Diemos are captured bodies. In contrast, calculated histories of orbital evolution due to tides in the planet and in the satellites indicate that these small craggy moons originated on nearly circular, uninclined orbits not far from their current positions; hence dynamicists prefer an origin in circum-Martian orbit.—From Burns (1992, p. 283).

Here we use numerical simulations to suggest that Phobos and Diemos accreted from the outer portion of a debris disc formed after a giant impact on Mars. In our simulations, larger moons form from material in the denser inner disc and migrate outwards due to gravitational interactions with the disc. The resulting orbital resonances spread outwards and gather dispersed outer disc debris, facilitating accretion into two satellites of sizes similar to Phobos and Deimos. The larger inner moons fall back to Mars after about 5 million years due to the tidal pull of the planet, after which the two outer satellites evolve into Phobos- and Deimos-like orbits. The proposed scenario can explain why Mars has two small satellites instead of one large moon. Our model predicts that Phobos and Deimos are composed of a mixture of material from Mars and the impactor.—From Rosenblatt et al. (2016, p. 481).

It has been proposed that Mars' moons formed from a disk produced by a large impact with the planet. However, whether such an event could produce tiny Phobos and Diemos remains unclear. Using a hybrid N-body model of moon accumulation that includes a full treatment of moon-moon dynamical interactions, we first identify new constraints on the disk properties needed to produce Phobos and Deimos. We then simulate the impact formation of disks using smoothed particle hydrodynamics, including a novel approach that resolves the impact ejecta with order-of-magnitude finer mass resolution than existing methods. We find that

R. Malcuit, *Geoforming Mars*, https://doi.org/10.1007/978-3-030-58876-2_3

forming Phobos-Deimos requires an oblique impact by a Vesta-to-Ceres sized object with ~10^{-3} times Mar's mass, a much less massive impactor than previously considered.—From Canup and Salmon (2018, p. 1).

The presence of the martian satellites is probably not very important to the evolution of Mars as a planet but the development of explanations for their origin may lead to explanations of features on the martian surface that are important. For example, some of these features may relate to impact craters on the surface of Mars that tell us something about the presence and masses of former satellites. First of all we will consider the classical explanations presented by Pollack et al. (1979) and Burns (1992) and then we will discuss the newest ideas on the formation of martian satellites via a Giant Impact on Mars (Rosenblatt et al. 2016) as well as models featuring a minor body impact proposed by Canup and Salmon (2018). Then we will consider models of encounters with Mars by larger bodies such as asteroids, which drift inward to the martian orbit, as well as other bodies of substantial mass formed in a zone in or near the boundary of the Earth-Mars accretion tori. Such bodies could survive for several hundred million years in quasi-stable orbits as calculated by Evans and Tabachnik (1999, 2002) and eventually impact on the surface of Mars. Many of these bodies could impact directly onto the martian surface but others could be tidally disrupted during close encounters with Mars. Such tidal disruption processes could result in martian satellites in quasi-stable orbits. These tidally disrupted bodies can have long orbital histories or short ones but most, if not all, will impact onto the surface of Mars to form either single impact craters, doublet impact craters (if disrupted again just before impact), or patterns of impact craters. Some of these patterns may form linear arrays on the surface of Mars (i.e., great-circle patterns) as described by Arkani-Hamed (2005, 2009a, b) and Kobayashi and Sprenke (2010). Figure 3.1 is a diagram of the geometry of the satellite orbits relative to Mars and Fig. 3.2 shows photos of Phobos and Deimos.

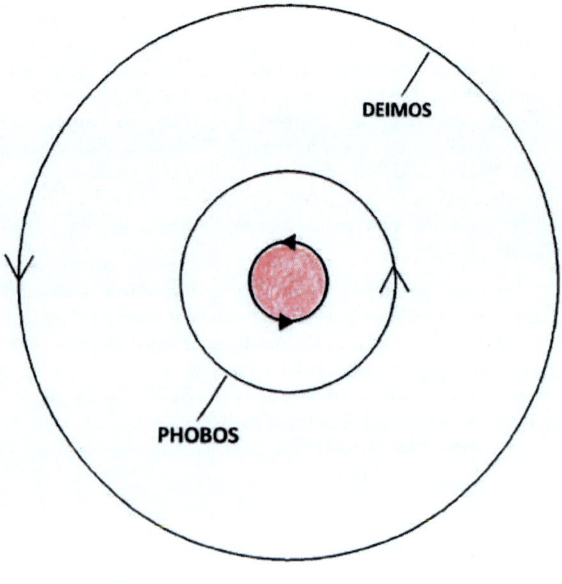

Fig. 3.1 A scale diagram of the current satellite orbits relative to a cross-section of Mars. View is from the north pole of the solar system

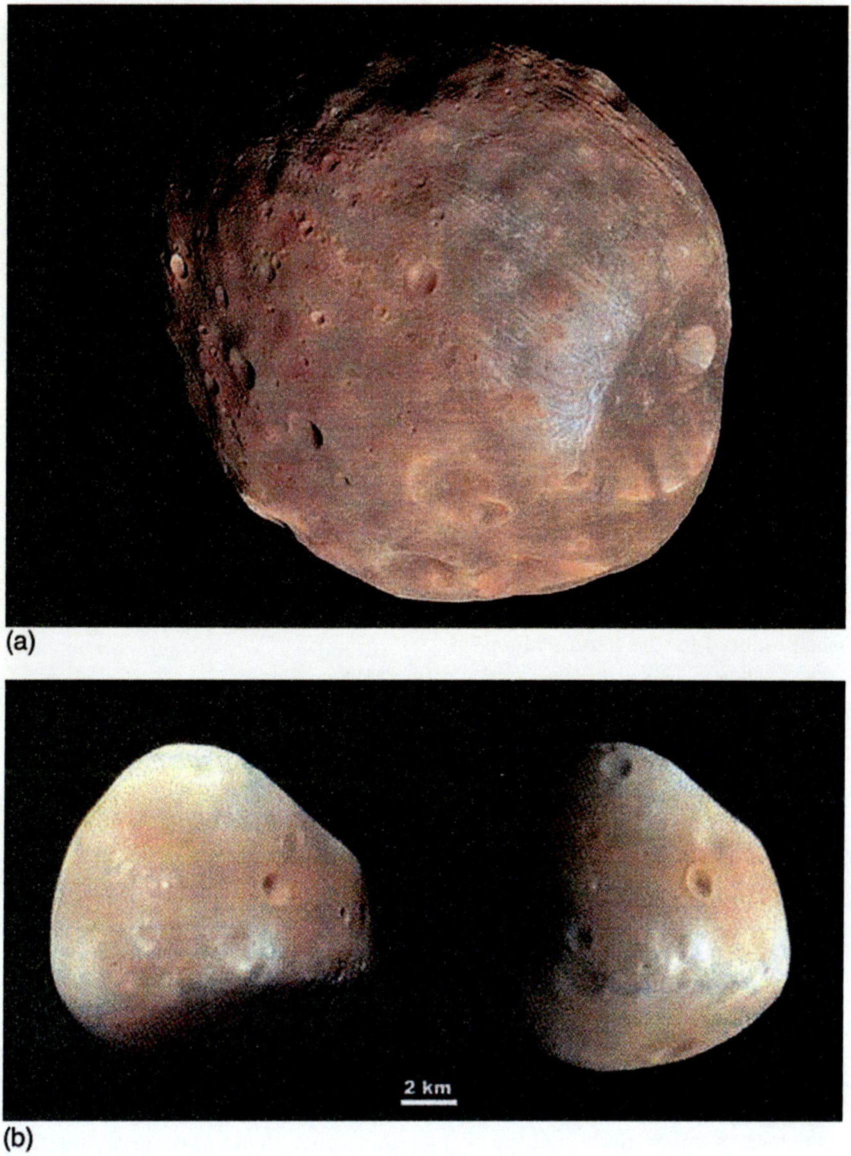

Fig. 3.2 (**a**) Photo of Phobos. (**b**) Photos of Deimos. (Both photos are courtesy of NASA; Reconnaissance Orbiter Photos)

3.1 List of Facts to Be Explained by a Successful Model

Table 3.1 Some physical and orbital properties of Phobos and Diemos

Feature	Phobos	Diemos
Orbital radius	2.76 R_{ma}	6.92 R_{ma}
Period	7 hr, 39 min	30 hr, 18 min
Orbit inclination	~1.068 degrees	~1.789 degrees
Orbit eccentricity	~0.01515	~0.000196
Mass	~1.08 E16 kg	~1.80 E15 kg
Density	~1.90 g/cm^3	~1.7 g/cm^3

From Burns (1992, p. 1284)

3.2 Some Models that Have Been Suggested for the Origin of the Martian Satellites

Since Mars is a terrestrial planet and Mars is the only other terrestrial planet to have extant satellites, we should frame our discussion with the general models that have been proposed for the Earth's Moon.

1. A fission model for martian satellites would feature fission of material from the planet due to rapid rotation of the planet early in its history.
2. A co-formation model for martian satellites would feature formation in martian orbit from space debris associated with the accretion of the planet.
3. Capture models for martian satellites would feature the insertion of the body of the satellite into a stable marocentric orbit. Capture models can be subdivided into at least three categories:

   (a) Gravitational capture features dissipation of the energy for orbit insertion via tidal deformation processes within one, or both, of the interacting bodies during a close encounter to the planet.
   (b) Collisional capture features dissipation of the energy for orbit insertion by collision of the candidate planetoid with a smaller body during a close encounter to the planet.
   (c) Gas-drag capture features dissipation of the energy for orbit insertion during one or more close encounters through a fairly dense "primitive" martian atmosphere.

4. A giant impact model features formation of a satellite, or satellites, by condensation from an extensive cloud of dust and gas resulting from a collision between proto-mars and a fairly massive planetoid in a mars-crossing heliocentric orbit.

The major facts to be explained by a successful model for the origin of the martian satellites are:

- The regularity of the orbits of Phobos and Deimos; both have low eccentricity as well as low inclination orbits relative to the equatorial plane of mars
- The small orbital radii of the satellites: Phobos = 2.76 mars radii; Deimos = 6.92 mars radii
- The small masses of the bodies (see Table 3.1) relative to the mass of Mars
- The low density of the bodies relative to the uncompressed density of Mars (~3.74 g/cm^3); both of the satellites have a density in the range of 1.7–2.2 g/cm^2

Some generalized comments on satellite origin models for planet Mars are as follows. The low density suggests to some scientists that they are captured asteroids but the capture mechanisms for inserting a small asteroid into a marocentric orbit from an original heliocentric orbit are not easily justified (Peale 1999). Furthermore, the expected result for most capture scenarios is a post-capture orbit of significant eccentricity. After a typical capture, a highly elliptical orbit must be circularized into an orbit of low eccentricity and small major axis.

Because of the recent space missions to the Red Planet, interest in the origin of the satellites has increased and a variety of models have been proposed. I will give a general description of some of these models and the interested reader can go to the original sources for more details.

3.2.1 *Fission Models*

I know of no "pure" fission model for the origin of the martial satellites. However, some giant-impact suggestions for the origin of the martian satellites have elements of a "fission" origin.

3.2.2 *Co-Formation Models*

I know of no "pure" co-formation model for the origin of the martial satellites, but several investigators have suggested that the satellites may be simply "left-over" from the accretion process.

3.2.3 *Capture Models*

3.2.3.1 Gravitational Capture via Tidal Dissipation Processes

Malcuit (2011) suggested in an abstract that Mars could have been made more habitable for organisms if it had gravitationally captured a sizeable satellite (0.01–0.02 mars-mass body). In Chaps. 4 and 5 of this book I present detailed models for prograde gravitational capture and retrograde gravitational capture, respectively, for planet Mars that would have changed the course of martian history. [Note: Gravitational capture of small bodies (i.e., Phobos and Diemos mass) is impossible by tidal dissipation processes.]

3.2.3.2 Gravitational Capture via Collisional Processes

Although such models have been mentioned in the literature, I know of no detailed treatment of such a collisional capture model.

3.2.3.3 Gravitational Capture via Gas Drag in an Extensive Primitive Atmosphere

A quote by Pollack et al. (1979, p. 608):

> A single parent body was captured at a point well beyond Diemos' current orbital distance and fractured into at least two large pieces, Phobos and Deimos, which subsequently separated from one another. Continued gas drag caused their orbital eccentricities and inclinations to approach zero, while their orbital semimajor axes decreased to a value close to Deimos' current value, at which point their Keplerian velocities approximately matched the nebula's velocity and little further orbital evolution occurred due to gas drag. Then the Martian nebula dissipated. Subsequently the more massive moon, Phobos, underwent much tidal evolution that changed its semimajor axis from one close to that of Deimos to its present, much smaller value.

This model by Pollack et al. (1979) looks favorable for the origin of Phobos and Diemos The details depend on the atmospheric density soon after accretion when Themis-type asteroids are interacting with Mars. Hunten (1979) also supports the concept of capture by atmospheric drag in a primitive martian atmosphere.

3.2.4 *Giant Impact Models*

A quote from Asphaug (2016, p. 568):

> The potato-shaped moons of Mars, Phobos and Diemos, were once believed to be captured asteroids, but that idea is called into question by their circular equatorial orbits and their relationship to the planet's rotation. The alternative explanation – that the two small moons, 22 km and 12 km in diameter, formed out of debris from a giant impact faces its own serious challenges.

Several authors have investigated the possibility that the two small martian moons were formed via large body impact processes (Marinova et al. 2011; Rosenblatt 2011; Rosenblatt and Charnoz 2012; Citron et al. 2015; Canup and Salmon 2018). In the Rosenblatt et al. (2016) model the present orbital locations are the result of interactions with much larger transient (or phantom) satellites that no longer exist. These larger bodies were postulated to be formed within a debris disk of martian surface and near-surface material by a large impacting body or bodies. More details of this giant impact scheme are in Hyodo et al. (2017a, b), and Hesselbrock and Minton (2017). In general, this large body impact model is fairly complex. The modelers want an impact location (now the north pole Borealis Basin) to be in the equatorial zone and the impact angle to be ~45 degrees, the mass of the body to be 0.03 mars-mass, and the impact velocity to be ~6 km/sec. Hyodo et al. (2017a) claim that there is enough angular momentum to increase the rotation rate of Mars from 0.0 hr/day to the present rotation rate of ~24.6 hr/day. Hyodo et al. (2017b) predict (1) that the building blocks of Phobos and Deimos should contain both impactor and martian materials and (2) that most of the material (about 50%) would come from a depth of about 50–150 km in the martian mantle.

The model of Canup and Salmon (2018) features an oblique impact of a Vesta-to-Ceres sized impactor (a body much smaller than 0.03 mars-mass) in the prograde direction. The impact debris, then, goes into orbit and some of the debris accumulates into Phobos and Diemos type bodies in prograde orbit. The main predictions from this model are that the impact debris would be a mixture of impactor and near surface martian material, both of which would be partially dehydrated by the heat generated by the impact.

In contrast, the predictions for the case of Mars being the most normal of the terrestrial planets are (1) that the rotation rate is essentially primordial at ~24.6 hr/day and (2) that the crust in the north polar area is simply foundered or subducted crust and was replaced by a younger mantle derived basaltic-gabbroic complex. In this model the tiny martian satellites have an origin that may be independent of any large or mid-sized body impact process or processes.

The composition of the martian moons will help us focus in on an origin model for these martian moons. JAXA (Japan Aerospace eXploration Agency) is planning a Martian Moons eXplorer (MMX) mission which will attempt to land on the moons and return samples to Earth for detailed analysis.

Like all other models for the two small martian satellites, these large impact schemes must be processed through the scientific method of problem solving. In the author's opinion these large impact models appear to be much more "energetic" than is needed to explain two small satellites in prograde orbit around what is probably the most normal terrestrial planet (MacDonald 1963, 1964; Nimmo and Tanaka 2005) in our system of rocky planets.

3.2.5 *Combinations of Models*

A broader appreciation of the problem of the origin of the martian satellites can be gained by discussions of the models. The following is an example from Mignard (1981, p. 378):

At first glance the solutions pictured in Figs. 3 and 4 lead us to think that both Phobos and Diemos were probably captured bodies because of the possibility of having been in a significantly eccentric orbit. But if such a prediction may be valid for Phobos it is ruled out for Deimos which orbits too far from the planet and is too slight a satellite to be influenced by the tides. Its capture in its present state remains unlikely because its orbit lies nearly in the equatorial plane of Mars. So either Deimos was formed in situ or mechanisms other than tides have acted to modify its orbit.

As regards Phobos, no definitive conclusions can be drawn from the conclusions achieved through this paper, various orbits being possible in the past according to the importance of the dissipation within the satellite.

A large dissipation allows us to consider that an origin by capture is possible for two reasons. (i) The time needed to evolve inward from a highly eccentric orbit can be less than the age of the Solar System, (ii) the gravitational perturbations of the Sun have caused the inclination of the planetary orbit to have been small, the latter explanation being more restrictive than the former, very large A being required.

Nevertheless, with this solution other difficulties arise, like the low probability of capture which requires a small approach velocity to Mars because of the low efficiency of tides to remove energy. Moreover, in such a case the subsequent evolution of the satellite would lead Phobos in the vicinity of the Deimos's orbit in a nearly equatorial plane (Figs 4 and 6) with an eccentricity of 0.6, which makes that probability of an encounter large. Smith and Tolson (1977) estimate the collision time-scale as 10^3–10^4 yr.

The following are a few quotes from the Discussion Section of an article by Andert et al. (2010, p. 3):

The porosity, the ratio of the bulk density and the grain density of an object, represents the percentage of the volume occupied by voids. The porosity of Phobos is computed from its mass, its bulk density and known grain densities of the hydrous chondrites of the CM group and the Tagish-Lake meteorite samples. The result of 30 $\pm$ 5 percent suggests that the interior of Phobos contains large voids. Similar large porosities and low bulk densities have been found in C-type asteroid such as the asteroid Mathide (Yeomans et al. 1997) and the Jupiter's small inner moon Amalthea (Anderson et al. 2005). A similar formation process of these porous bodies, however, is totally unclear.

The interior structure of Phobos could well be the result of its complete shattering and subsequent reassembly, as is thought to have occurred in the history of many asteroids subjected to violent collisions (Richardson et al. 2002). The existence of the Stickney crater on Phobos would support the conclusion that Phobos contains large voids throughout its interior.

The origin of Phobos can be discussed in terms of its orbital history. Several scenarios ranging from possible to speculative have been proposed. The surface of Phobos shows some spectral similarities to those of various asteroids types. Based in those similarities it was suggested that Phobos is a former asteroid, formed in the outer asteroid belt and later captured by Mars (Burns 1992). This scenario, however, does not explain how the energy loss required to change the incoming hyperbolic orbit into an elliptical orbit bound to Mars is accounted for (Burns 1992; Peale 2007). Models of orbit evolution based on tidal interactions between Mars and Phobos cannot account for the current near-circular and near-equatorial orbit (Mignard 1981). Scenarios of evolution to the current circular orbit require an additional drag by, e.g., the primitive planetary nebulae or the Martian atmosphere (Sasaki 1990). An alternative is a Phobos formed in an orbit around Mars. Phobos and Deimos could be remnants of an early, larger body that was broken into parts by gravitational gradient forces during Mars capture (Singer 2007). Or Phobos could have formed by the re-accretion of impact debris lifted into Mars' orbit (Craddock 1994). If Phobos were a remnant of a larger moon, it is not expected to be as porous as is reported here. If Phobos were formed from the re-accretion of impact debris lifted into Mars' orbit, the disc would be composed of a mixture of Martian crust and impactor material. The spectral properties, however, of the Phobos surface and the Martian crust do not match very well.

This inconsistency is resolved by a collision between a body already orbiting Mars but formed from the debris disc remaining after formation of Mars and a second body originating from the asteroid belt (Peale 2007). This scenario is consistent with the high porosity of Pbobos and its spectral properties.

COMMENT: This appears to be a collisional capture scenario.

3.2.6 *The Magnetic Rock Patterns on Mars and Implications for the Origin of Phobos and Deimos*

This quote is the Abstract of the article by Kobayashi and Sprenke (2010, p. 37). The title of the paper is "Lithospheric drift on early Mars: Evidence in the magnetic field".

The crustal magnetic anomalies on Mars may represent hot spot tracks resulting from lithospheric drift on ancient Mars. As evidence, an analysis of lineation patterns derived from the deltaBr magnetic map is presented. The deltaBr map, largely free of external magnetic field effects allows excellent detail of the magnetic anomaly pattern, particularly in areas of Mars where the field is relatively weak. Using cluster analysis, we show that the elongated anomalies in the martian magnetic field form concentric small circles (parallels of latitude) about two distinct north pole locations. If these pole locations represent ancient spin axes, then tidal force on the early lithosphere by former satellites in retrograde orbits may have pulled the lithosphere in an east-west direction over hot mantle plumes. With an active martian core dynamo, this may have resulted in the observed magnetic anomaly pattern of concentric small circles. As further evidence, we observed that, of the 15 martian giant impact basins that were possibly formed while the dynamo was active, seven lie along the equators of our two proposed paleopoles. We also find that four other re-magnetized giant impact basins lie along a great circle about the mean magnetic paleopole of Mars. These 11 impact basins, likely the result of fallen retrograde satellite fragments, indicate that Mars once had moons large enough to cause tidal drag on the early martian lithosphere. The results of this study suggest that the magnetic signatures of this tidal interaction have been preserved to the present day.

A quote of the first three sentences from the Abstract of a paper by Arkani-Hamed (2009a, p. 489). The title of the paper is "Polar wander of Mars: Evidence from giant impact basins".

We investigate the polar wonder of Mars in the last ~4.2 Ga. We identify two sets of basins from the 20 giant impact basins reported by Frey [Frey, H., 2008, Geophys. Res. Lett. 35, L13203] which trace great circles on Mars, and propose that the great circles were the prevailing equators of Mars at the impact times. Monte Carlo tests are conducted to demonstrate that the two sets of basins are most likely not created by random impacts.

Another quote from p. 491:

If several giant impact basins on a planet trace a great circle, that great circle must be the equatorial plane of the planet at the impact time. It is highly unlikely that several asteroids on different orbital planes, or on a single plane that is appreciably inclined relative to the equator of a planet, just happen to impact at the equator of the planet (Arkani-Hamed 2009a).

> Likewise, impact basins on a planet created by several individual satellites may not trace a great circle despite the fact that the orbital plane of an individual satellite coincides with the prevailing equatorial plane of the planet. The main reason is that a satellite may orbit a planet for several thousand to several hundred million years before impacting on it (e.g., Arkani-Hamed 2009b). Therefore, several satellites may impact over several thousand to several million years, during which the equatorial plane of a planet may vary considerably due to the precession of its rotation axis and wobbling of its orbital plane. For example, the martian rotation axis presently precesses with an obliquity of ~24° and a period of ~170,000 years (Folkner et al. 1997; Yoder and Standish 1997). Also, the obliquity has changed by several tens of degrees and the orbit normal of Mars has changed by more than 10° during its history (e.g., Ward 1979; Spada and Alfonsi 1998; Bills and James 1999; Laskar et al. 2004).

A quote from another article by Jafar Arkani-Hamed (2009a, p. 31). The title of the paper is: "Did tidal deformation power the core dynamo of Mars?"

> A strong core dynamo of Mars that existed in the early history of the planet ceased at around 4 Ga (e.g., Acuna et al. 1999; Arkani-Hamed 2004; Johnson and Phillips 2005; Lillis et al. 2008). One possible scenario is that the core dynamo was maintained by vigorous thermal convection in a liquid core driven by appreciable convection in the mantle (e.g., Nimmo and Stevenson 2000; Stevenson 2001; Breuer and Spohn 2003). Once a stagnant lithosphere developed on the surface and Mars became a one-plate planet, the mantle convection became sluggish and reduced heat loss form the core, thus decreasing the vigor of core convection and killed the dynamo. An alternative scenario recently put forward by Arkani-Hamed et al. (2008) suggests that the core dynamo was likely powered by elliptical instability in the martian core that was excited by tidal forces of a large asteroid, ~2.9×10^{20} kg, which was orbiting Mars on the equatorial plane. The spin-orbit coupling of Mars and the asteroid brought them closer, and once the asteroid approached the Roche limit of Mars, it was ruptured into many fragments. The large fragments created the giant basins Argyre, Hellas, Isidis, and Utopia upon impacting Mars at around 4 Ga. As the asteroid disintegrated and the fragments impacted on Mars, the tidal force vanished and the core dynamo ceases. The close proximity of the formation time of the basins and the cessation time of the core dynamo (e.g., Arkani-Hamed 2004; Johnson and Phillips 2005) may not be coincidental, rather it may imply a causal relationship between the two events.

Side-Bar Discussion on "The Power of Earth Tides at the Present Time and Perhaps Throughout the History of the Earth" as a Prelude to a Discussion of Low Angle, Grazing Impacts on Mars as a Mechanism for Powering the Short-Lived Magnetic Field of mars

There has been a general neglect of the tidal influence of the Moon in the operation of planet Earth, especially in the realm of Plate Tectonics. I have a feeling that the influence of the solid earth tides (rock tides raised by a combined gravitational action of the Moon and Sun) was not even mentioned by the participants of the Penrose Conference on plate tectonics with the title: "WHEN DID PLATE TECTONICS BEGIN?" which was held in Lander, Wyoming, June 13–18, 2006). (Note: The earth tide mechanism was not mentioned in either the Summary in Nature (2006, v. 442, p. 128–131) or in the Penrose Conference Report to the Geological Society of America, 2006) The concept that earth tides are involved in the Plate Tectonics operation

(continued)

appears to be given an even lower ranking than the concept of Stern (2005) that the modern style of plate tectonics started about 1 Ga ago.

Nonetheless, there is a growing body of literature on the subject of EARTH TIDES AND PLATE TECTONICS. Perhaps the next Penrose Conference on the subject of Plate Tectonics will have a title: "EARTH TIDES AND PLATE TECTONICS". This topic will be pursued in more detail in Chap. 10.

3.3 How Does this Digression on Earth Tide Phenomena Fit in with Our Discussion of the Origin of Martian Satellites?

The concept of a very-low viscosity hydrous iron-rich mantle rocks can be applied to a low-viscosity zone in the martian mantle. The driving force for differential rotation of the martian crust relative to the mantle is some combination of (1) a mars-satellite tidal coupling in the retrograde direction, (2) tidal torque of any tidal bulge raised by the orbiting bodies, and (3) impacts of bodies of various masses onto the surface of Mars in the retrograde direction. The suggestion by Doglioni and Panza (2015, p. 146) that a layer or layers of very-low-viscosity mantle material in the upper mantle of Earth is very applicable to the model proposed by Kobayashi and Sprenke (2010) for causing differential rotation of the martian mantle relative to the crust of planet Mars. The same mechanism can cause differential rotation of a martian core relative to the martian mantle. In the case of Mars, the tidal force is caused by interaction with planetoids orbiting and eventually impacting in the retrograde direction.

3.4 A Model Featuring Tidal Disruption of Ceres-Type Asteroids During Close Non-Capture Encounters With Mars Early in its History: Implications for Magnetic Field Generation and Origin of the Martian Satellites

This complex model features encounters of Ceres-type asteroids encountering Mars. This model may relate to (1) the driving mechanism for the martian magnetic field (Kobayashi and Sprenke 2010), (2) the great-circle patterns of impact features identified by Arkani-Hamed (2009a, b), and (3) perhaps the outstanding problem of the origin of Phobos and Deimos.

In Chaps. 4 and 5, I present gravitational capture models for both and retrograde orientations for capturing a satellite that is about 1/100 the mass of Mars (i.e., about 8 times the mass of the asteroid, Ceres). This type of body is sufficiently large to be gravitationally captured by tidal deformation processes if the tidal deformation properties are within certain limits. Using the same numerical simulation code, I present the results of non-capture encounters for Ceres-mass bodies in heliocentric orbit in the vicinity of the martian orbit for both prograde and retrograde orientations. The major difference of the simulations for this section and that of Chaps. 4 and 5 is that these Ceres-mass planetoids have essentially no chance to be gravitationally captured by tidal deformation processes. However, these bodies can be partially tidally disrupted as they pass through the Weightlessness Limit (Malcuit et al. 1975; Malcuit 2015) of the Mars—Ceres-type planetoid system. The main body that yields the tidally disrupted masses then escapes back into a mars-like heliocentric orbit. The tidally disrupted bodies, of various shapes and sizes, have lost energy during the disruption process and are captured into marocentric orbits with a variety of eccentricities. The orbits of these disrupted bodies would eventually be effected by (1) heliocentric eccentricity variations of the martian orbit over time, (2) obliquity changes of planet Mars over time, (3) precession motion of planet Mars over time, (4) orbital energy dissipation via tidal action with mars, and (5) orbital energy dissipation via atmospheric drag when encounters are sufficiently close to Mars. Bodies in retrograde orbits about Mars tend to be less stable than bodies in prograde orbits and thus would be expected to impact on the surface over a much shorter period of geologic time.

Such a tidal disruption model for generating bodies of impactors that cause impact features may relate to the martian crustal surface record of grazing impact features (Schultz and Lutz-Garihan 1982) as well as the generation of a short-lived martian magnetic field via a model presented by Kobayashi and Sprenke (2010) which features a preponderance of low angle impacts in a retrograde direction.

3.4.1 Orientation Information for Close Encounters that may Result in Partial Tidal Disruption of a Ceres-type Asteroid

Figure 3.3 shows the geometry of stable capture orientations for both prograde and retrograde encounters between Ceres-type asteroids and planet Mars. Since the displacement Love number threshold for stable capture by tidal processes is generally beyond the realm of possibility, stable capture of the Ceres type bodies is physically improbable and essentially impossible. But the Ceres-type asteroid bodies can be partially tidally disrupted during a very close, possibly slightly grazing, encounters with Mars.

A stable capture zone can be defined as a region of space in which the orientation of the orbit of the encountering planetoid is favorable for a stable post-capture orbit if sufficient energy is dissipated for its capture into a planetocentric orbit (Malcuit and Winters 1996; Malcuit 2015). Although the geometry of stable capture zones is very useful for assessing the probability of stable capture, it can also be used to assess the results of non-capture encounters during which insufficient energy is dissipated

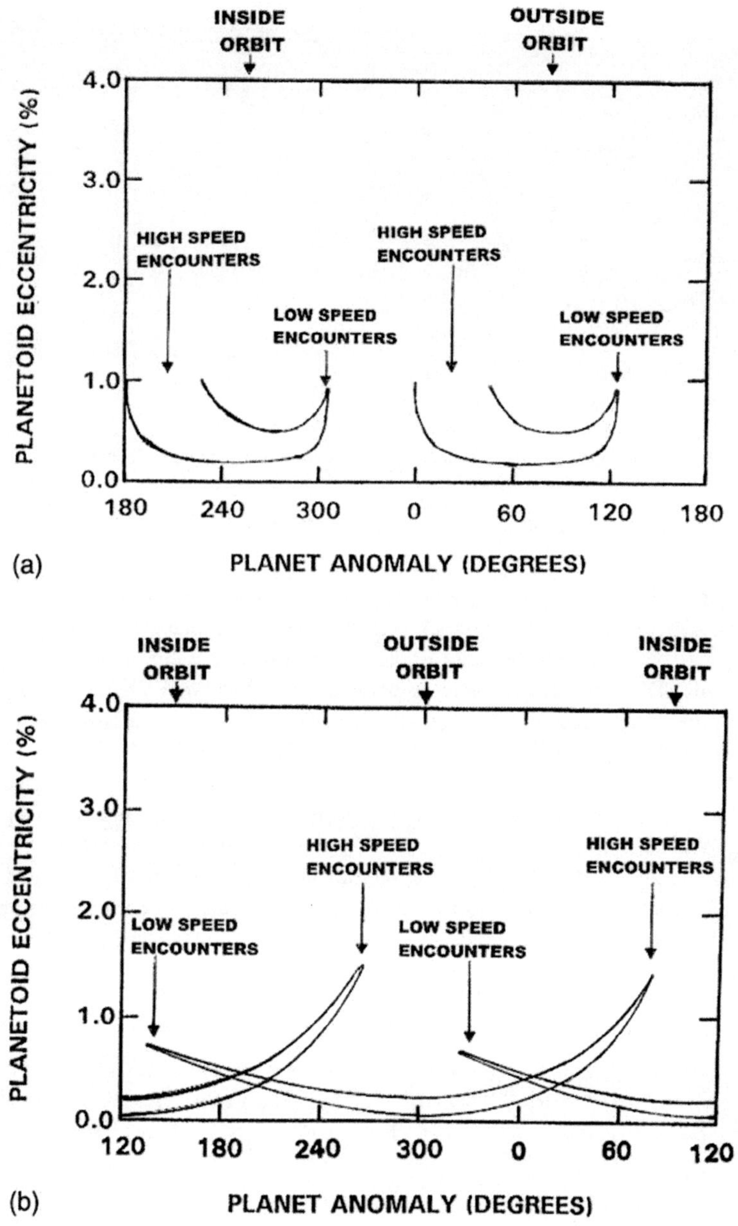

Fig. 3.3 Prograde (**a**) and retrograde (**b**) stable capture zones for a Ceres-type asteroid encountering Mars. Since the planetoids are not amendable to capture by normal tidal energy dissipation processes, the outlines of the SCZs are left blank. In general, the right ends of the prograde stable capture zone are the low energy positions and the left ends of the retrograde stable capture zones are the low energy positions. (Note: The procedure for two-dimensional mapping of SCZs is deferred to Chap. 4)

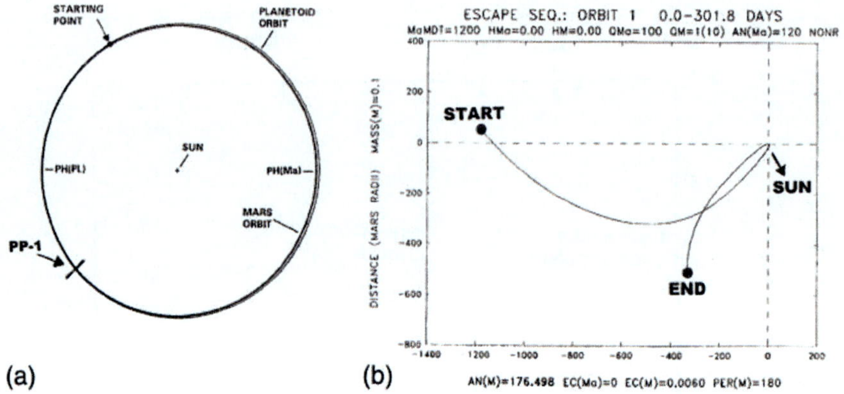

Fig. 3.4 Heliocentric and marocentric orientations that are necessary for stable prograde capture of a planetoid from a heliocentric orbit by Mars. (**a**) The orientation of the first encounter at PP-1 must be nearly perpendicular to the tangent of the heliocentric orbit. (**b**) The marocentric diagram shows such an orientation for the marocentric encounter. Other details for stable prograde capture conditions are stated in the text

within the interacting bodies to facilitate capture. For planetoid encounters in the PROGRADE direction (counterclockwise relative to the planet) there is a fairly sharp boundary between planetoid escape scenarios and stable captures. But for RETROGRADE encounters there is an extensive zone of collision encounters between the escape scenario (during which very little energy is dissipated within the interacting bodies) and the STABLE CAPTURE CONDITION. This extensive zone of collision encounters is caused by excessive solar gravitational perturbations that are associated with planetoids in retrograde motion around a planet in a prograde heliocentric orbit.

Figure 3.4 shows an example of a prograde planetary encounter that would result in a stable capture orientation if sufficient energy is dissipated for capture. Figure 3.5 shows an example of a retrograde encounter that would result in a stable capture orientation if sufficient energy were dissipated for capture.

3.4.2 Definition and Discussion of the Weightlessness Limit for Ceres-type Asteroids Encountering a Mars-Mass Planet

The Weightlessness Limit (W-limit) for this particular problem is the center-to-center distance between Mars and a Ceres-type asteroid at which weightlessness occurs at the sub-mars point on the asteroid surface (see Fig. 3.6 for geometric details). The W-limit is at about 1.60 mars radii for a particle at the sub-mars point on a spherical body. The W-limit is at a greater distance from mars for a tidally

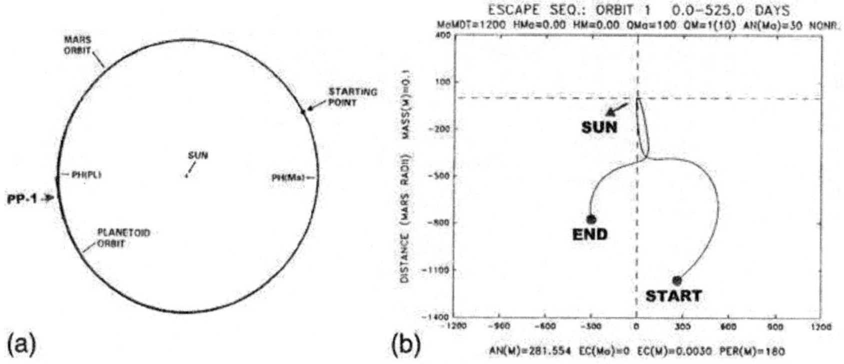

Fig. 3.5 Heliocentric and marocentric orientations that are necessary for stable retrograde capture of a planetoid from a heliocentric orbit. (**a**) The orientation of the first encounter (PP-1) much be nearly parallel to a tangent of the heliocentric orbit. (**b**) The marocentric diagram shows such an orientation for the marocentric encounter. Other details for stable retrograde capture conditions are stated in the text

deformed asteroid. A Ceres-type asteroid body can also have a weightless condition for a particle at the anti-sub-mars point during an encounter with mars but this condition occurs at a smaller mars-planetoid distance of separation and will not be discussed further here because tidal disruption in the sub-mars zone is the important process for this model.

An equation of the W-limit for the sub-mars point is:

$$\frac{GM_{ma}}{r^2} = \frac{GM_{ma}}{(r - R_{cta})^2} - \frac{GM_{cta}}{{R_{cta}}^2} \tag{3.1}$$

where

G	gravitational constant
M_{ma}	mass of Mars
M_{cta}	mass of asteroid
r	distance of separation of centers of Mars and Asteroid
R_{cta}	radius of ceres-type asteroid

The first term in the equation is the force on the body of the asteroid toward mars; the second term is the force on a particle at the sub-mars point on the asteroid toward mars; the third term is the force on the particle by the gravity of the asteroid. If a particle on the surface of the asteroid along the mars-asteroid center line is inside the W-limit, it will be lifted off the surface of the asteroid and move toward mars. If a particle on the surface of the asteroid along the mars-asteroid center line is beyond the W-limit, then it will remain at its normal position on the asteroid surface.

The position of the W-limit, however, is displaced to a greater distance from mars as the body of the asteroid is gravitationally deformed during a close encounter. For example, with an h value of 0.2, a value that could be associated with a Ceres-type

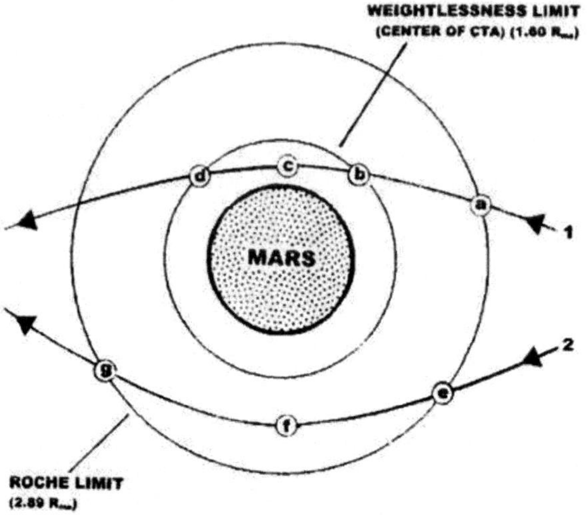

Fig. 3.6 Scale sketch of the weightlessness limit for the case of a Ceres-type asteroid encountering a mars-mass planet with neither body being deformed. Note that the classical Roche limit is located at a much greater distance from the planet than the weightlessness limit. Scenario 1: The body of the asteroid enters the classical Roche limit at position a. At position b the body enters the W-limit. Position c is the closest approach distance and the body of the asteroid goes out through the W-limit at position d. From positions b to d the body of the asteroid is within the W-limit and material can be lifted from the surface of the asteroid or extracted from the near-surface sub-mars portion of the asteroid. This space born material can take one of three pathways (1) it can be gravitationally propelled to mars, or (2) it can return to the body of the asteroid as the asteroid moves away from mars, or (3) it can be inserted into a marocentric orbit that may or may not be stable relative to solar gravitational perturbations. The details of the resulting particle orbits depend on the relative positions of the body of the asteroid, mars, and the sun at the time of gravitational liftoff. Scenario 2: The body of the asteroid enters the classical Roche limit at position e. The closest approach is at position f and the asteroid passes out through the Roche limit at position g. No particles would be gravitationally lifted off the surface during this scenario

asteroid, the center of the asteroid is at 1.60 mars radii for weightlessness at the sub-mars point. This shift in W-limit with increasing deformation of the asteroid body during an encounter is very important for analyzing the effects of tidal disruption of an a Ceres-type asteroidal body during a very close encounter. With even a small amount of tidal deformation, crack propagation can occur within the outer part of the asteroid body in the sub-mars volume of the asteroid. The crack-propagation leads to separation of portions of the asteroidal body. These detached particles tend to separate into smaller particles of irregular shape because of the differential gravitational forces acting upon them. These detached particles of asteroid composition are gravitationally attracted by mars and tend to lose orbital energy and fall into somewhat different marocentric orbits as the parent asteroidal body escapes back into heliocentric orbit. Bodies in retrograde marocentric orbit tend to be more unstable (i.e., have a greater variation in planetocentric eccentricity) than those in prograde marocentric orbits because of solar gravitational perturbation and, in some cases, because of tidal energy dissipation and/or atmospheric drag with the planet (in this case mars).

The asteroid encounters with mars, which are mainly in the plane of the planets, are about equally distributed between prograde and retrograde orientations. Also, the tidal disruption mechanism within the W-limit will not discriminate between prograde and retrograde encounters. Thus, because of this characteristic instability of retrograde planetocentric orbits relative to prograde orbits, there should be a bias for retrograde directions of impacting asteroidal debris on the martian surface.

3.4.3 Numerical Simulations of Ceres-type Asteroids Encountering Planet Mars

Figure 3.7 shows two types of plots that are useful for illustrating the results of numerical simulations of planetoid-planet encounters. The sun is located at the center of each of the diagrams. Figure 3.7a illustrates a non-rotating coordinate system in which the computer plot theoretically revolves around the sun but the plot itself does not rotate. Figure 3.7b illustrates a rotating coordinate system in which the plot theoretically revolves around the sun and the plot rotates as it revolves around the sun keeping the left side of the plot forever pointing toward the sun.

Figure 3.8 shows the results of prograde gravitational encounters between a Ceres-type asteroid and a mars-like planet. Figure 3.8a is the heliocentric orientation for the encounters. Figure 3.8b is the marocentric orientation for the encounters. Figure 3.8c, d show close-up views of close encounters with close approach distances of about 1.43 mars radii and about 1.15 mars radii, respectively. The position of the W-limit for the case of no tidal deformation is marked on the close encounter plots. Such encounters within the W-limit could not be expected to result in capture by tidal deformation processes. However, debris from the tidal disruption of Ceres-type bodies in prograde orbit may yield insight into the problem of the origin of Diemos and Phobos.

3.4.4 A Scheme for Getting Disrupted Material from a Ceres-Type Asteroid into Martian Orbit

Figure 3.9 shows the results of retrograde gravitational encounters between a Ceres-type asteroid and a mars-like planet. Figure 3.9a is the heliocentric orientation for the encounter. Figure 3.9b is the marocentric orientation for the encounters. Figure 3.9c shows a close-up of an encounter with a closest approach distance of ~1.43 mars radii which is within the W-limit. Figure 3.9d shows a close-up of an encounter with a closest approach distance of ~1.15 Mars radii which is well-within the W-limit and is a near grazing encounter. Note again that the position of the W-limit that is marked on the diagram is for the case of no tidal deformation.

Note that the encounter simulations in Figs. 3.8 and 3.9 are for encountering bodies characterized by no tidal deformation (i.e., displacement Love numbers = 0).

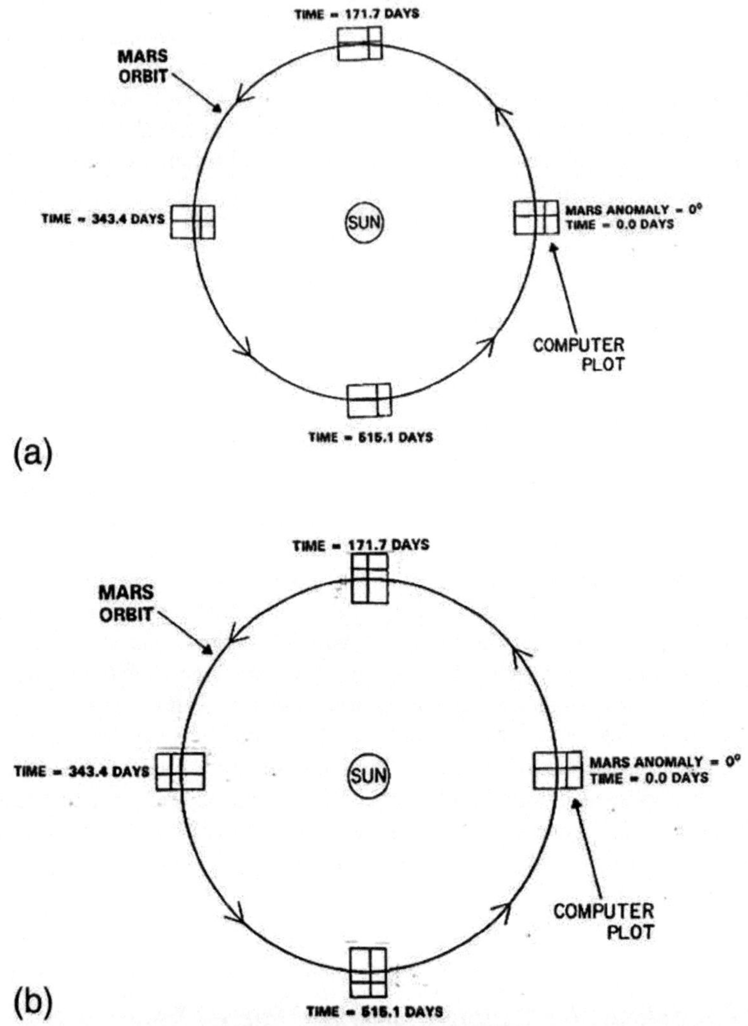

Fig. 3.7 Some orientation information for understanding the plots. (**a**) Diagram illustrating a non-rotating coordinate system for plotting the results of the simulations. The computer plot does not rotate as it theoretically revolves around the Sun. Planet Mars is at the origin of the non-rotating plot. (**b**) Diagram illustrating a rotating coordinate system for plotting the results of the numerical simulations. In this case the computer plot rotates as it theoretically revolves around the Sun. Special symmetry information can be gained from this type of plot but the plotted information is more difficult to understand. Thus all plots in this section of the chapter use the non-rotating coordinate system and all orbital plots are viewed form the north pole of the solar system

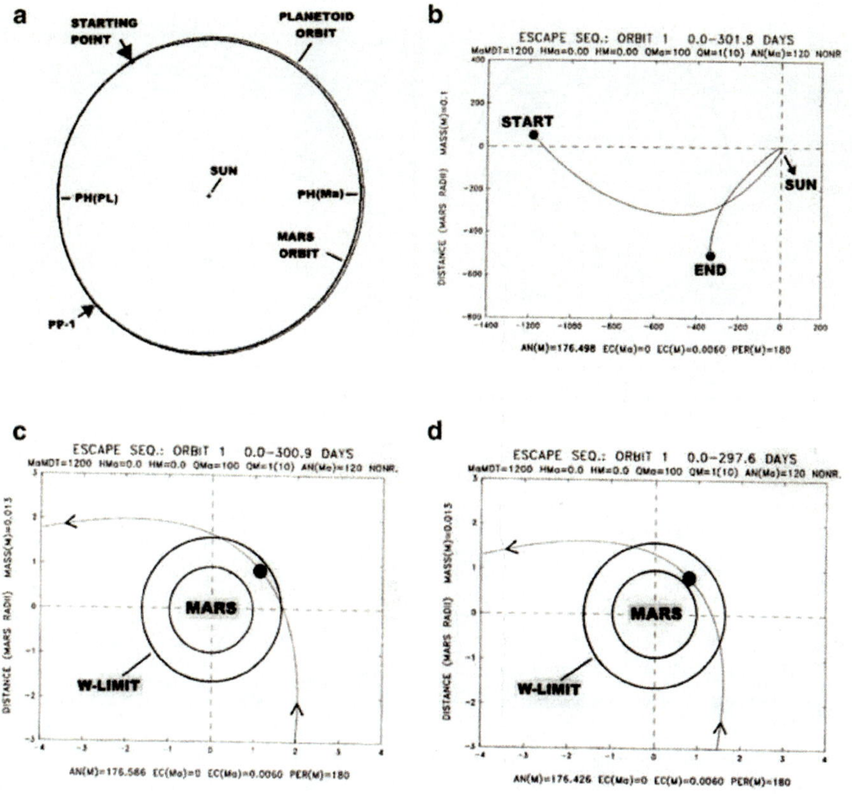

Fig. 3.8 (**a**) Plot of the heliocentric orientation for this set of simulations. The orbit of the asteroid is outside the orbit of the mars. The simulation begins at the STARTING POINT; the first and only encounter occurs at PP-1 (~0.544 earth year; ~0.290 mars year). The asteroid escapes back into heliocentric orbit. (**b**) Marocentric orientation for this set of simulations. The START point marks the beginning of the simulation, END marks to ending point, and the position of the SUN at the beginning of the simulation is indicated by the arrow. (**c**) Geometry of an encounter at ~1.43 mars radii. (**d**) Geometry of an encounter at ~ 1.15 mars radii. The position of the W-limit is shown on each of the close encounter diagrams. Only a very close encounter (a near grazing encounter) would be expected to result in any significant quantity of tidally disrupted bodies

Such simulations could not result in capture by tidal processes alone for Ceres-type asteroids. The only orbital situations that can come close to a stable capture result are near grazing encounters in retrograde orientations. Such encounters are also strong candidates for partial tidal disruption of a Ceres-type asteroid.

Such a series of tidal disruption scenarios may lead to an explanation of (1) great-circle patterns of impacts on the surface of mars as described by Arkani-Hamed (2009b) as well as (2) the mechanism for the generation of a martian magnetic field (Arkani-Hamed 2009a).

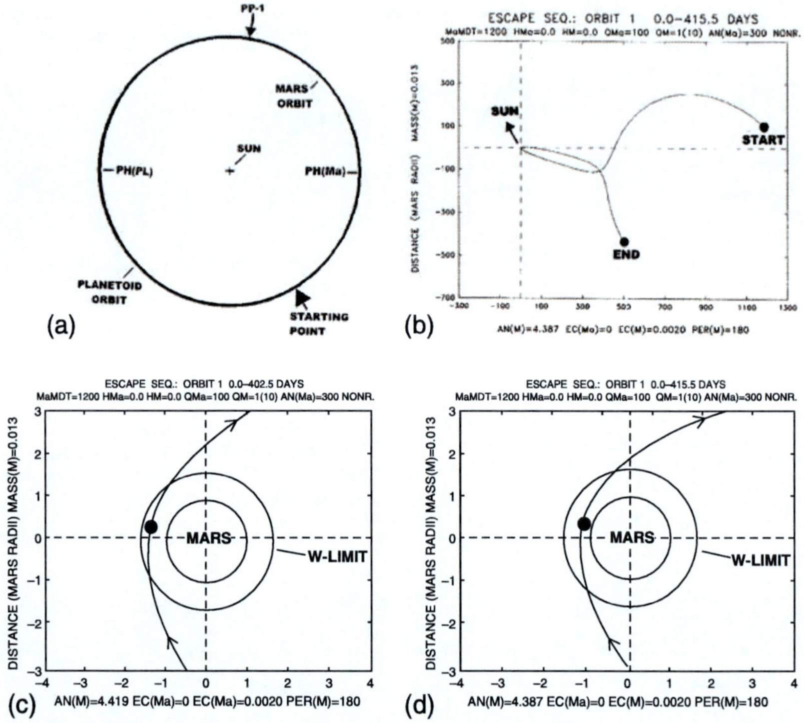

Fig. 3.9 (a) Plot of the heliocentric orientation for this set of simulations. The orbit of the asteroid is outside the orbit of mars. The simulation begins at the STARTING POINT; the first and only encounter occurs at PP-1 (~0.725 earth year; ~386 mars year). The asteroid escapes back into heliocentric orbit. (b) Marocentric orientation for this set of simulations. The START point marks the beginning of the simulation, END marks the ending point, and the position of the SUN at the beginning of the simulation is indicated by the arrow. (c) Geometry of an encounter at ~1.43 mars radii. (d) Geometry of an encounter at ~1.15 mars radii. The position of the W-limit is shown on each of the close encounter diagrams. Only a very close encounter (a near grazing encounter) would be a candidate to result in any significant quantity of tidally disrupted bodies

Figures 3.10 and 3.11 illustrate how asteroids undergoing partial disruption can yield fragments being inserted into marocentric orbits following a close tidally disruptive encounter. Such disrupted material, large and small, can result from either close prograde or retrograde encounters. During such a partial disruption scenario some parts of the Ceres-mass asteroid is inserted into a marocentric orbit but the main body of the encountering asteroid escapes back into a heliocentric orbit. Some of the disrupted fragments could impact over a brief period of geologic time to form a great-circle pattern of impactors and other bodies may impact along this some great-circle pattern over an extended period of geologic time. These tidal disruption scenarios, in both prograde and retrograde orientations, could happen several times with the retrograde fragments being the favored ones for impacts because of the less stable characteristics of their marocentric orbits.

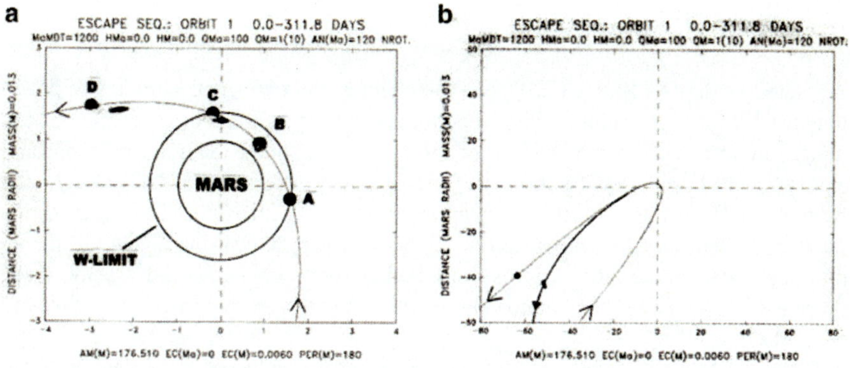

Fig. 3.10 Possible scheme for partial tidal disruption during a close encounter in the prograde direction. (**a**) Close-up view on an encounter well within the weightlessness limit of the system resulting in gravitational extraction of a portion of the near-side of the encountering body. (**b**) More distant view showing a major portion of the encountering body on an escape orbit relative to mars as well as the orbit of a small portion of the encountering body remaining in an elliptical orbit around Mars

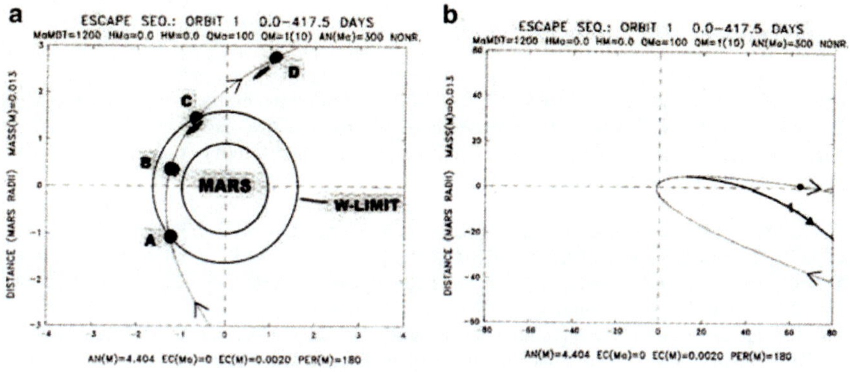

Fig. 3.11 Possible scheme for partial tidal disruption of a Ceres-type asteroid during a close encounter in the retrograde direction. (**a**) Close-up view of an encounter well within the weightlessness limit of the system resulting in gravitational extraction of a portion of the near-side of the encountering body. (**b**) More distant view showing a major portion of the encountering body on an escape orbit relative to Mars as well as the orbit of a small portion of the encountering body remaining in an elliptical orbit around Mars

In Figs. 3.10 and 3.11 I present some pictograms for how disrupted material could be inserted into marocentric orbits during close encounters with mars.

3.5 Summary

As stated in the INTRODUCTION to the chapter, the two small satellites of Mars may not be very important to the planetological history of Mars but speculations on their origin may lead to explanations of some (or many) of the features of the martian environment. In my view Phobos and Diemos could have been spalled off a Ceres-type asteroid during a close non-capture encounter with Mars. This spalled-off material is then inserted into an orbit of low eccentricity near the equatorial plane of Mars. The orbits of these spalled-off bodies then circularize and assume their present orbits using the atmospheric drag processes proposed by Pollack et al. (1979).

Thus we can entertain two points of view on the value of the extant martian satellites. As Burns (1992) suggests, the origin of these satellites may be as suggested in a Shakespearian play, "Much Ado About Nothing". On the other hand it is possible that the satellites may yield valuable insights into other processes affecting origin of surface and interior features of planet Mars.

References Cited

Acuna, M. H., & 12 colleagues. (1999). Global distribution of crustal magnetization discovered by the Mars Global Surveyor MAG/ER experiment. *Science, 284*, 790–793.

Anderson, J. D., Johnson, T. V., Schubert, G., & 9 more co-authors. (2005). Amalthea's density is less than that of water. *Science, 308*, 1291–1293.

Andert, T. P., Rosenblatt, P., Patzold, M., Hausler, B., Dehaunt, V., Tyler, G. L., & Marty, J. C. (2010). Precise mass determination and the nature of Phobos. *Geophysical Research Letters, 37*, L09202. https://doi.org/10.1029/2009Gl041829.

Arkani-Hamed, J. (2004). Timing of the martian core dynamo. *Journal of Geophysical Research, 109*, E03006, 1–12. https://doi.org/10.1029/2003JE002195109.

Arkani-Hamed, J. (2005). Giant impact basins trace the ancient equator of Mars. *Journal of Geophysical Research, 110*, E04012. https://doi.org/10.1029/2004JE002342.

Arkani-Hamed, J. (2009a). Did tidal deformation power the core dynamo of Mars? *Icarus, 201*, 31–43.

Arkani-Hamed, J. (2009b). Polar wander of Mars: Evidence from giant impact basins. *Icarus, 204*, 489–498.

Arkani-Hamed, J., Seyed-Mahmoud, B., Aldridge, K. D., & Baker, R. E. (2008). Tidal excitation of elliptical instability in the martian core: Possible mechanism for generating the core dynamo. *Journal of Geophysical Research, 113*, E06003.

Asphaug, E. (2016). Rise and fall of the martian moons. *Nature Geoscience, 9*, 568–569.

Bills, B. G., & James, T. S. (1999). Moments of inertia and rotational stability of Mars: Lithospheric support of subhydrostatic rotational flattening. *Journal of Geophysical Research, 104*, 9081–9096.

Breuer, D., & Spohn, T. (2003). Early plate tectonics versus single-plate tectonics on Mars: Evidence from magnetic field history and crust evolution. *Journal of Geophysical Research, 108*, 5072. https://doi.org/10.1029/2002JE001999.

Burns, J. A. (1972). The dynamical characteristics of Phobos and Deimos. *Reviews of Geophysics and Space Physics, 10*, 463–483.

Burns, J. A. (1977). Orbital evolution. In J. A. Burns (Ed.), *Planetary satellites* (pp. 113–156). Tucson: University of Arizona Press.

Burns, J. A. (1978). The dynamical evolution and origin of the martian moons. *Vistas in Astronomy, 22*, 193–210.

Burns, J. A. (1992). Contradictory clues as to the origin of the martian moons. In H. H. Kieffer, B. M. Jakosky, C. W. Snyder, & M. S. Matthews (Eds.), *Mars* (pp. 1283–1301). Tucson: University of Arizona Press.

Canup, R., & Salmon, J. (2018). Origin of Phobos and Deimos by the impact of a Vesta-to-Ceres sized body with Mars. *Science Advances, 4*, eaar6887, 6 p.

Citron, F. I., Genda, H., & Ida, S. (2015). Formation of Phobos and Diemos via a giant impact. *Icarus, 252*, 334–338.

Craddock, R. A. (1994). Are Phobos and Diemos the result of a giant impact? *Icarus, 211*, 1150–1161.

Doglioni, C., & Panza, G. (2015). Polarized plate tectonics. *Advances in Geophysics, 56*, 1–167.

Evans, N. W., & Tabachnik, S. (1999). Possible long-lived asteroid belts in the inner solar system. *Nature, 399*, 41–43.

Evans, N. W., & Tabachnik, S. (2002). Structure of possible long-lived asteroid belts. *Monthly Notices of the Royal Astronomical Society, 333*, L1–L5.

Folkner, W. M., Kahn, R. D., Preston, R. A., Yoder, C. F., Standish, E. M., Williams, J. G., Edwards, C. D., & Hellings, R. W. (1997). Mars dynamics from Earth-based tracking of the Mars Pathfinder lander. *Journal of Geophysical Research, 102*, 4057–4064.

Frey, H. (2008). Ages of very large impact basins on Mars: Implications for the late heavy bombardment in the inner Solar System. *Geophysical Research Letters, 35*, L13203.

Hesselbrock, A. J., & Minton, D. A. (2017). An ongoing satellite-ring cycle of Mars and the origins of Phobos and Deimos. *Nature Geoscience, 10*, 266–269.

Hunten, D. M. (1979). Capture of Phobos and Deimos by protoatmospheric drag. *Icarus, 37*, 113–123.

Hyodo, R., Genda, H., Charnoz, S., & Rosenblatt, P. (2017a). On the impact origin of Phobos and Diemos. I. Thermodynamic and physical aspects. *The Astrophysical Journal, 845*, 125, 8 p. https://doi.org/10.3847/1538-4357/aa81c4.

Hyodo, R., Rosenblatt, P., Genda, H., & Charnoz, S. (2017b). On the impact origin of Phobos and Deimos: II. True polar wander and disk evolution. *The Astrophysical Journal, 851*, 122, 9 p. https://doi.org/10.3847/1538-4357/aa9984.

Johnson, C. L., & Phillips, R. J. (2005). Evolution of the Tharsis region of Mars: Insights from magnetic field observations. *Earth and Planetary Science Letters, 230*, 241–154.

Kobayashi, D., & Sprenke, K. F. (2010). Lithospheric drift on early Mars: Evidence in the magnetic field. *Icarus, 210*, 37–42.

Laskar, J., Correia, A. C. M., Gastineau, M., Joutel, F., Levrard, B., & Robutel, P. (2004). Long term evolution and chaotic diffusion of the insolation quantities of Mars. *Icarus, 170*, 343–364.

Lillis, R. J., Frey, H. V., & Manga, M. (2008). Rapid decrease in martian crustal magnetization in the Noachian era: Implications for the dynamo and climate of early Mars. *Geophysical Research Letters, 35*, L14203. https://doi.org/10.1029/2008/GL034338.

MacDonald, G. J. F. (1963). The internal constitutions of the inner planets and the Moon. *Space Science Reviews, 2*, 473–557.

MacDonald, G. J. F. (1964). Tidal friction. *Reviews of Geophysics, 2*, 467–541.

Malcuit, R. J. (2011). Terraforming Mars by tidal processes. *Geological Society of America, Abstracts with Programs, 45*(4), 3.

Malcuit, R. J. (2015). *The twin sister planet, Venus and Earth: Why are they so different?* (p. 401). Cham: Springer International Publishers.

Malcuit, R. J., & Winters, R. R. (1996). *Geometry of stable capture zones for planet Earth and implications for estimating the probability of stable gravitational capture of planetoids from heliocentric orbit: Abstracts Volume, 27th Lunar and Planetary Science Conference* (pp. 799–800). Houston: Lunar and Planetary Institute.

Malcuit, R. J., Byerly, G. R., Vogel, T. A., & Stoeckley, T. R. (1975). The great-circle pattern of large circular maria: Product of an Earth-Moon encounter. *The Moon, 12*, 55–62.

Marinova, M. M., Aharonson, O., & Asphaug, E. (2011). Geophysical consequences of planetary-scale impacts into a mars-like planet. *Icarus, 211*, 960–985.
Mignard, F. (1981). Evolution of the martian satellites. *Monthly Notices of the Royal Astronomical Society, 194*, 365–379.
Nimmo, F., & Stevenson, D. J. (2000). Influence of early plate tectonics on the thermal evolution and magnetic field of Mars. *Journal of Geophysical Research, 105*, 11,969–11,979.
Nimmo, F., & Tanaka, K. (2005). Early crustal evolution of Mars. *Annual Reviews of the Earth and Planetary Sciences, 33*, 133–161.
Peale, S. J. (1999). Origin and evolution of the natural satellites. *Annual Reviews of Astronomy and Astrophysics, 37*, 533–602.
Peale, S. J. (2007). The origin of the natural satellites. In T. Spohn (Ed.), *Planets and moon treatise on geophysics* (Vol. 10, pp. 465–508). Amsterdam: Elsevier.
Pollack, J. B. (1977). Phobos and deimos. In J. A. Burns (Ed.), *Planetary satellites* (pp. 319–345). Tucson: University of Arizona Press.
Pollack, J. B., Burns, J. A., & Tauber, M. E. (1979). Gas drag in primordial circumplanetary envelopes: A mechanism for satellite capture. *Icarus, 37*, 587–611.
Richardson, D. C., Leinhardt, Z. M., Melosh, H. J., Bottke, W. F., Jr., & Asphaug, E. (2002). *Asteroids III* (pp. 501–515). Tucson: University of Arizona Press.
Rosenblatt, P. (2011). The origin of the martian moons revisited. *Astronomy and Astrophysics Review, 19*, 1. https://doi.org/10.1007/s00159-011-0044-6.
Rosenblatt, P., & Charnoz, S. (2012). On the formation of the martian moons from a circum-martian accretion disk. *Icarus, 221*, 806–815.
Rosenblatt, P., Charnoz, S., Dunseath, K. M., Terao-Dunseath, M., Trinh, A., Hyodo, R., Genda, H., & Toupin, S. (2016). Accretion of Phobos and Deimos in an extended debris disc stirred by transient moons. *Nature Geoscience, 9*, 581–583.
Sasaki, S. (1990). Origin of phobos – Aerodynamic drag capture by the primary atmosphere of Mars. *LPI Contribution, 1543*, 1069.
Schultz, P. H., & Lutz-Garihan, A. B. (1982). Grazing impacts on Mars: A record of lost satellites. *Journal of Geophysical Research, 87*(supplement), A84–A96.
Singer, S. F. (2007). Origin of the martian satellites Phobos and Deimos. *LPI Contribution, 1377*, 36.
Smith, H. R., & Tolson, R. H. (1977). The Q of Mars and the early orbit of Phobos (abstract). *Transactions of the American Geophysical Union, 58*, 1181.
Spada, G., & Alfonsi, L. (1998). Obliquity variations due to climate friction on Mars' Darwin vs layered models. *Journal of Geophysical Research, 103*(E12), 29599N. –28605N.
Stern, R. J. (2005). Evidence from ophiolites, blueschists, and ultrahigh-pressure metamorphic terranes that the modern episode of subduction tectonics began in Neoproterozoic time. *Geology, 33*, 557–560.
Stevenson, D. J. (2001). Mars core and magnetism. *Nature, 412*, 214–219.
Ward, W. R. (1979). Present obliquity oscillations of Mars' fourth-order accuracy in orbital *e* and *i*. *Journal of Geophysical Research, 84*, 237–241.
Yeomans, D. K., Barriot, J. P., Dunham, D. W., Farquhar, R. W., Giorgini, J. D., Helfrich, C. E., Konopliv, A. S., McAdams, J. V., Miller, J. K., Owen, W. M., Jr., Scheeres, D. J., Synnott, S. P., & Williams, B. G. (1997). Estimating the mass of Asteroid 253 Mathilde from tracking data during the NEAR flyby. *Science, 278*, 2106–2109.
Yoder, C. F., & Standish, E. M. (1997). Martian precession and rotation from Viking lander. *Journal of Geophysical Research, 102*(E2), 4056–4080.

Chapter 4
A Prograde Gravitational Capture Model for a Sizeable Volcanoid Planetoid (or Asteroid) for Mars

> We find that there are two possible long-lived belts of asteroids. The first region lies between the Sun and Mercury, in the range 0.09–0.21 astronomical units, where remnant planetesimals may survive for the age of the Solar System provided that their radii are greater than ~0.1 kilometers. The second region of stability is between Earth and Mars (range 1.08–1.28 astronomical units), where a population of bodies that are on circular orbits may survive.—From Evans and Tabachnik (1999).

In this book we are exploring the advantages and disadvantages of acquiring a large (0.1–0.2 moon-mass) satellite for Mars. When considering the information in the quote above, there are two major source regions for candidate planetoids for capture: (1) the Vulcanoid Zone (Weidenschilling 1978; Leake et al. 1987; Evans and Tabachnik 1999) between 0.09 and 0.21 AU and (2) the zone between the Earth and Mars (Evans and Tabachnik 1999) located between 1.08 and 1.28 AU. In the first few 100 million years of solar system evolution there may have been many other planetoids on planet crossing orbits that would also be candidates for capture. Malcuit (2015) suggested that planets Venus and Earth sampled the Vulcanoid Zone. He suggested that the vulcanoid planetoid Adonis was captured into a stable retrograde orbit by Venus and that the gravitational capture episode eventually led to the transformation of Venus into the "Hades" planet that it is today. In contrast, Earth captured the vulcanoid planetoid Luna into a stable prograde orbit and evolution of that system resulted in the "paradise" planet we enjoy today.

4.1 Place of Origin of Candidate Planetoids for Capture

There are many possibilities for a place of origin for candidate planetoids for planet Mars. The range of options is anywhere from the Vulcanoid Zone (Evans and Tabachnik 1999, 2002) to the Outer Asteroid Belt (Fig. 4.1). A critical issue is that a candidate planetoid (1) must be massive enough to store the energy for capture and

R. Malcuit, *Geoforming Mars*, https://doi.org/10.1007/978-3-030-58876-2_4

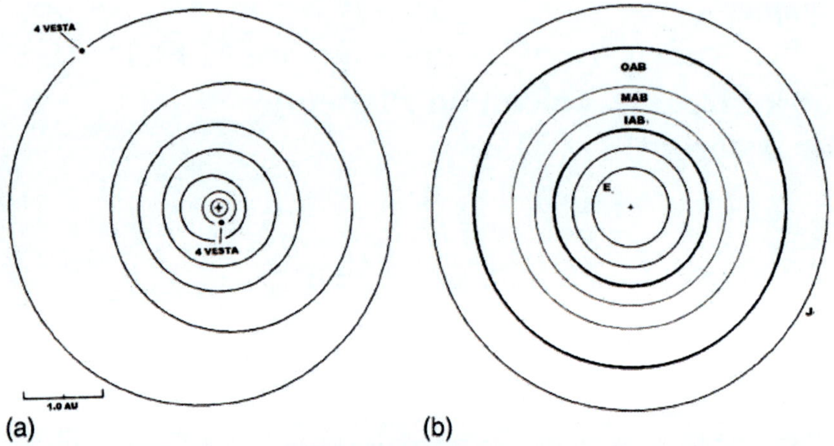

Fig. 4.1 A few possible sources for candidate planetoids for capture by planet Mars. (**a**) The Vulcanoid Zone (Evans and Tabachnik 1999) as a source of Vulcanoid Planetoids. In my view asteroid 4 Vesta is a Vulcanoid planetoid from that zone and any planetoid coming from that zone would need to work its way outward to get into an orbital configuration very similar to planet Mars (on this diagram the martian orbit is the next orbit inside that of asteroid 4 Vesta). (**b**) The Asteroid Belt is a source of asteroids with a "chondritic" composition. The most likely source would be the IAB (on this diagram the orbit of Mars is between the orbit of Earth and the IAB) because of its proximity of the martian orbit and its greater distance from planet Jupiter (J). Other locations such as the middle asteroid belt (MAB) and outer asteroid belt (OAB) are possible as well the belt of quasi-stable planetoid orbits between the orbits of Earth and Mars

(2) must be thermally and structurally suitable for capture via tidal deformation processes at the time of capture.

First let us consider the features of Vulcanoid planetoids as presented by Malcuit (2015, Chap. 2). These planetoids are formed from mainly calcium-aluminum inclusion material (CAI chemistry) and thus are very refractory bodies. To explain the body of the Earth's Moon as a Vulcanoid planetoid with an anorthositic crust and a magma ocean but with no metallic core, Malcuit (2015) suggested that the outer 600 km of the body of the Moon was remelted by a series of FU Orionis heating events. If the body of the Moon started out as a Vulcanoid planetoid and had this remelting and subsequent partial differentiation event, then all sibling Vulcanoid Planetoids would go through at least the same heating events, and depending on size of the planetoid, would undergo some sort of differentiation as well.

After formation, remelting, and cooling, any Vulcanoid Planetoid that becomes a candidate for capture by planet Mars would necessarily undergo a long-term (100s of million years) trek featuring many planetary gravitational perturbations to work its way past Mercury, Venus, and Earth. Then its orbit would need to be gravitationally "trimmed" to be in an orbit of "low" inclination and moderate eccentricity relative to the martian orbit. The smaller the candidate planetoid, the more difficult it is to capture. Thus only planetoids above a certain threshold of mass can be captured by tidal dissipation processes.

Now let us go to the opposite end of the spectrum for the source of candidate planetoids – the Zone of the Asteroids. Most asteroid scholars agree that asteroids are composed mainly of chondritic meteorite material (McSween 1999). Impact processes during accretion of an asteroidal body would melt and fuse some of the material in the interior of the body but many asteroids appear to have the composition of chondritic meteorites on their surfaces (McSween 1999). The Japanese mission to asteroid Ryugu, thought to be a ruble-pile asteroid of carbonaceous chondrite composition, sampled the surface of this asteroid and the sample return to Earth is scheduled for next year (2020) (Castelvecchi 2019; Mitchikami et al. 2019). Asteroid Ryugu has a semi-major axis of 1.19 AU which places it between the orbits of Earth and Mars (but it was probably not formed there). The extreme limit of an asteroidal source for candidate planetoids for a large martian satellite is the Outer Asteroid Belt (OAB) (Fig. 4.1a). Such an asteroidal body would then be gravitationally perturbed by planet Jupiter to get it out of its orbit of origin. Then via a series of distant encounters, close encounters, and minor collisions with smaller asteroids, the body may end up in an orbit of low inclination and low eccentricity relative to planet Mars (within about 3% for both parameters), from which it could be gravitationally captured by Mars.

Asteroids are generally low-density bodies and it may be a challenge for such a body to dissipate the energy to facilitate its own capture. The planet is mainly a passive bystander so that most of the energy for capture must be dissipated within the encountering body in essentially one pass.

Now let us consider a third source for candidate planetoids, the region between 1.08 and 1.28 AU, which was identified by Evans and Tabachnik (1999) as a somewhat stable zone for planetoids. The uncompressed density of bodies forming here should be from about 3.7 to 4.0 g/cm^3 (see Table 1.1). Thus any planetoid for collision or tidal disruption encounters should have an isotopic signature similar or Earth or Mars.

Since there are very few chemical restrictions on the bodies of candidate planetoids, the planetoid can originate anywhere from the VULCANOID ZONE to the OUTER ASTEROID BELT. Thus the candidates can come from such places as the Middle Asteroid Belt (MAB), Inner Asteroid Belt (IAB), as well as from the zone between the orbits of Mars and planet Earth, as well as from the zone between planets Earth and Venus as well as from a zone between planets Venus and Mercury. Evans and Tabachnik (1999, 2002) show the results of calculations to estimate the stability of orbits in the Solar System (see Fig. 2.2). The plot shows the location and relative stability of planetoid orbits from 0.1 to 2.2 AU (see Figs. 2.6, 2.7, 2.15, 2.16, and 2.17 also). These calculations and plots shed light on possible sources for candidate planetoids in the region of interest for capture by planet Mars.

4.2 Prograde Gravitational Capture of a 0.1 Moon-Mass Planetoid and the Subsequent Orbit Evolution: An Attempt to Geoform Mars

If the Earth's Moon is a captured body, it was captured in the prograde direction. Thus, when considering a capture origin of a satellite for Mars, let us consider in this chapter the merits and demerits of a prograde gravitational capture scenario for a mars-planetoid system. A quote from Counselman (1973, p. 307) helps to set the stage for the outcome of a prograde capture scenario:

> ... the orbit may decay outward toward escape, although at an ever decreasing rate; or the satellite's orbital and planet's spin periods may approach stable synchronism.

Let us also give this hypothetical satellite a name. A reasonable name is CONCORDIA , a Roman goddess of agreement, understanding, and marital harmony.

4.2.1 A Two-Body Analysis and a Discussion of the Paradoxes Associated with the Capture Process

Figure 4.2 shows a scale sketch of a two-body rendition of prograde gravitational capture of a planetoid for planet Mars. The essentials of such a capture are that sufficient energy must be dissipated to insert the planetoid into an orbit that is within the stability limit [the Hill sphere (Lissauer et al. 1997; dePater and Lissauer 2001)] for planet Mars and is stable against future gravitational perturbations. There are four

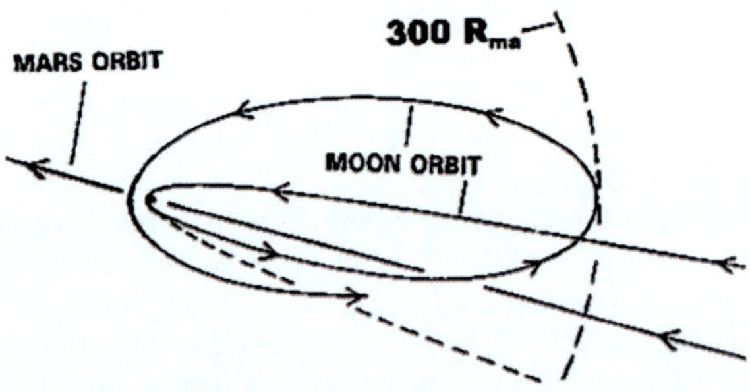

Fig. 4.2 A two-body representation of gravitational capture of a 0.1 moon-mass body by Mars. Both the planet and planetoid are in very similar, but not identical, prograde heliocentric orbits. The planetoid has a close encounter with the planet. If no energy is dissipated within the two bodies, then the planetoid departs from the planet along the *dashed line* and attains an orbit very similar to its pre-encounter orbit. If sufficient energy is dissipated in a combination of the two bodies, then the planetoid is inserted into a highly elliptical orbit that stays within the stability limit for the Sun-Mars-planetoid system [i.e., the Hill sphere for the system (Roy 1965; dePater and Lissauer 2001)]. View is from the north pole of the solar system

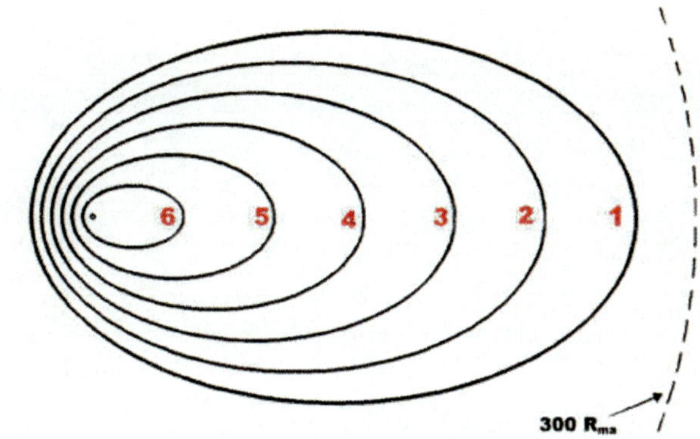

Fig. 4.3 Graphical representation of the Marocentric Orbits in Table 4.1. Orbit 1 = 300 Rma; orbit 2 = 250 Rma; orbit 3 = 200 Rma; orbit 4 = 150 Rma; orbit 5 = 100 Rma; orbit 6 = 50 Rma. All orbits have eccentricity of 0.8

paradoxes associated with any attempt at gravitational capture of a planetoid from a heliocentric orbit into a planetocentric orbit (in this case a marocentric orbit) and these are discussed below.

The dictionary definition of a PARADOX, for our purposes, is "a statement that seems contradictory, unbelievable, or absurd but that may be actually true in fact". I will list the first three paradoxes here and then explain them in the following sections of this chapter.

PARADOX 1: The encountering planetoid (the smaller body) must absorb most of the energy for its own capture.
PARADOX 2: Larger planetoids are more capturable than smaller planetoids.
PARADOX 3: "Cool" planetoids are more capturable than "hot" planetoids. (One must distinguish between "cool" and "cold" for this paradox.)

All of these statements seem counterintuitive but, again, they are all true. Such is the nature of a paradox.

Figure 4.3 is a scale sketch of a close gravitational encounter that is a candidate for a successful capture. The essential features for successful gravitational capture by tidal deformation processes are:

1. *A MARS-LIKE HELIOCENTRIC ORBIT*: For successful capture a 0.1 moon-mass planetoid has to be in a heliocentric orbit very similar in geometry to that of the mars-like planet (i.e., within 3% in heliocentric orbit eccentricity).
2. *A VERY CLOSE ENCOUNTER* : For successful capture a 0.1 moon-mass planetoid must have a very close encounter with the mars-like planet (i.e., within 1.5 mars radii, center-to-center distance).
3. *ENERGY DISSIPATION VIA PLANETOID ROCK TIDES*: For successful capture a 0.1 moon-mass planetoid must be sufficiently deformable in order to store the energy for capture (about 1.4 E26 J); the planetoid body must also be able to dissipate a large fraction of the tidally stored energy during the timescale of the close encounter (about 30 min).
4. *INSERTION INTO A STABLE MAROCENTRIC ORBIT*: For successful capture the semi-major axis of the post-capture orbit of the planetoid must be as large as

Table 4.1 Energy dissipation needed to insert a 0.1 moon-mass planetoid into orbits with specific dimensions

Major axis (Mars radii)	Energy necessary for capture (Joules)
300	Ecapture = 0 – (−3.09 E 26 J) = 3.09 E26 J
250	Ecapture = 0 – (−3.71 E 26 J) = 3.71 E26 J
200	Ecapture = 0 – (−4.64 E 26 J) = 4.64 E26 J
150	Ecapture = 0 – (−6.18 E 26 J) = 6.18 E26 J
100	Ecapture = 0 – (−9.27 E 26 J) = 9.27 E26 J
50	Ecapture = 0 – (−18.54 E26 J) = 18.54 E26 J

possible but must be within the stability zone of the Sun-Mars-planetoid dynamical system [i.e., within the Hill sphere (within ~300 mars radii)].

Equation 4.1 shows the energy considerations for gravitational capture of a 0.1 moon-mass planetoid from a mars-like heliocentric orbit into a large post-capture marocentric orbit. As a first case let us assume, for simplicity, that the orbit of the candidate planetoid has zero eccentricity and inclination and is located in the common plane of the planets; it is also in an orbit that is coincident with the orbit of Mars. In this case the first term of Eq. 4.1 is zero.

Energy to be dissipated for capture by Mars is:

$$\Delta E_{capture} = \frac{1}{2} M_{pl} V_{\infty}^2 - \left(\frac{-G\, M_{ma} M_{pl}}{2a} \right) \tag{4.1}$$

Where M_{pl} is the mass of the planetoid, v_{∞} is the velocity of the planetoid at infinity, G is the gravitational constant, M_{ma} is the mass of Mars, and 2a is the major axis of the planetocentric orbit.

The values in Table 4.1 relate to the second term of this equation. The 300 M_{ma} (mars radii) value is for the largest capture orbit possible and this represents the lowest energy dissipation possible for capture (Fig. 4.3, orbit state 1). If we want to capture the planetoid into a smaller orbit, then significantly more energy must be dissipated during the initial encounter for successful capture (Fig. 4.3, orbit states 2, 3, 4, 5, and 6). In this two-body case, capture into this largest orbit is just marginally possible if 3.09 E26 Joules of energy are dissipated by some process or combination of processes.

Now let us consider the first term of Eq. 4.1 for the case where the heliocentric orbit of the planetoid is not coincident with the orbit or the planet. In this case there is some value to the velocity at infinity (i.e., the velocity of the planetoid relative to that of the planet when the bodies are well beyond the boundaries of the Hill sphere). Table 4.2 gives some values for the velocity at infinity (v_{∞}), for the encountering body as well as values for the extra energy dissipation needed for successful capture into the largest marocentric orbit possible. Figure 4.4 shows two heliocentric orbits similar to those listed in Table 4.2. One can notice that the energy dissipation requirements increase rapidly as the encounter velocity at infinity increases because

Table 4.2 Encounters with velocity at infinity values added to the basic energy of capture for planetoid orbits that are slightly larger than the martian orbit

ECC (%)	Vel (infinity) (km/sec)	Energy necessary for capture (Joules)
~0.01	0.10	Ecapture = 0.37 E26 J – (–3.09 E26 J) = 3.46 E26 J
~0.02	0.20	Ecapture = 1.47 E26 J – (–3.09 E26 J) = 4.56 E26 J
~0.04	0.30	Ecapture = 3.31 E26 J – (–3.09 E26 J) = 6.40 E26 J
~0.05	0.40	Ecapture = 5.88 E26 J – (–3.09 E26 J) = 8.97 E26 J
~0.06	0.50	Ecapture = 9.19 E26 J – (–3.09 E26 J) = 12.28 E26 J
~0.08	0.75	Ecapture = 20.67 E26 J – (3.09 E26 J) = 23.76 E26 J
~0.11	1.00	Ecapture = 36.75 E26 J – (3.09 E26 J) = 39.84 E26 J

The approximate value of the eccentricity of the planetoid orbit is shown in the first column

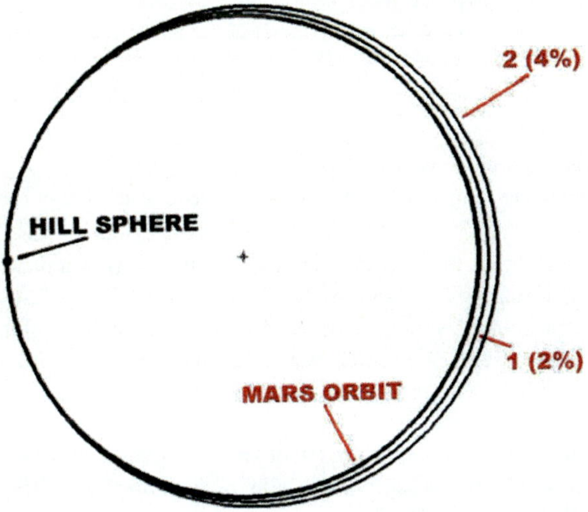

Fig. 4.4 Heliocentric representation of velocity at infinity

of the "velocity squared" dependence. Even with a value of 0.2 km/s, nearly 50% more energy dissipation is necessary for capture than for the zero v_∞ case.

The conclusion from this discussion is that we need a 0.1 moon-mass planetoid in essentially in a co-planar mars-like heliocentric orbit to be inserted into the largest possible stable marocentric orbit (i.e., the planetoid is just barely captured). *These are the necessary conditions for a successful gravitational capture scenario in a two-body context!*

As noted earlier in this section, there are four paradoxes that seem to have clouded a reasonable assessment of the GRAVITATIONAL CAPTURE OF SATELLITES BY PLANETS. The first three will be discussed in this segment and the fourth will be discussed a bit later. *The first paradox is that the encountering body must absorb most of the energy for its own capture.* To understand this paradox we need some background information on the concept of the displacement Love number for tidally deformed bodies. A reasonable displacement Love number for early Mars is 0.7.

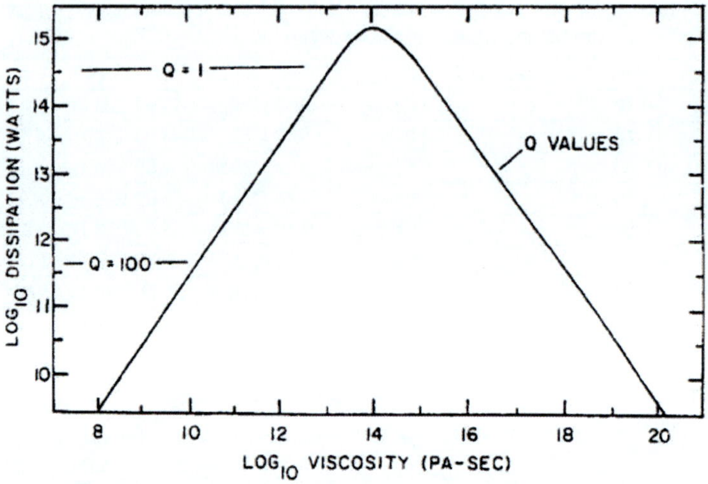

Fig. 4.5 Plot of energy dissipation vs. body viscosity for a lunar-like body. (From Ross and Schubert 1986, Fig. 3, p. D450, with permission from the American Geophysical Union). Note that a Q value near one is associated with an intermediate value of body viscosity

In brief, the displacement Love number (h) of a planet or planetoid relates to the deformability of the body. The Love numbers are dimensionless parameters representing various aspects of elastic tidal deformation (Love 1911, 1927). Peale and Cassen (1978) show that the displacement Love number (h) is the important one for analyzing energy storage by radial tidal deformation. The displacement Love number is defined by Stacey (1977) as the ratio of the height of the body tide to the height of the equilibrium (static) marine tide (i.e., the height of the ocean tide if it had time to come to equilibrium with the tidal potential). The displacement Love number of the Moon at present (a cold rigid body) is estimated to be ~0.033 (Kaula and Harris 1973; Ross and Schubert 1986). The displacement Love number for Earth is estimated to be about 0.54 (Munk and MacDonald 1960; MacDonald 1964). The displacement Love number for the water on a planet covered with a significant thickness of ocean water is 1.

The second paradox is that larger planetoids are more capturable than smaller planetoids. To understand this paradox we need information on the concept of the specific dissipation factor (Q) in addition to the h factor. 1/Q is the fraction of stored energy that is subsequently dissipated during an encounter cycle. For example, if an effective planetary Q equals 10, then about one-tenth of the energy stored during an encounter cycle will be dissipated (mainly in the form of heat within the body). The present effective planetary Q of Earth is estimated to be about 13 (mainly because of energy dissipation by ocean tides) (Munk and MacDonald 1960). The present planetary Q for the solid Earth is estimated to be 100 or higher (Melchior 1978). The present Q value of Io (the innermost Galilean satellite of planet Jupiter) is estimated to be about 1 (Ross and Schubert 1986). Figure 4.5 shows a plot of Energy Dissipation vs. Body Viscosity with the associated Q values. In the analysis of Ross

Table 4.3 Values for energy stored in Mars during a single encounter ($h_{ma} = 1$)

Perimar radius	Type of encounter	Energy stored (Joules) (Joules)
For 1.30 mars radii	(near grazing)	1.32 E26
For 1.35 mars radii	(non-grazing)	1.05 E26
For 1.40 mars radii	(non-grazing)	0.846 E26
For 1.43 mars radii	(non-grazing)	0.745 E26

Note that with a Q factor of 100, only 1/100 of the energy stored is dissipated within the body

and Schubert (1986) very low Q values are associated with intermediate viscosity values (~10 E14 Pa-sec) (pascal-seconds). Q values of 1, 2, or 3 are necessary for capture of a lunar-like planetoid by Earth. Similar values are necessary for capture of a 0.1 moon-mass planetoid by planet Mars.

The displacement Love number (h) is a very important factor for gravitational capture and the h value is normally fairly closely related to the thermal content of the body. Small bodies tend to cool much more rapidly than the large bodies. Thus, small bodies become rigid much more rapidly than large bodies and it would be difficult to keep a small planetoid at the viscosity value that would be associated with a low Q value (i.e., it would be difficult to keep them in a physical state to be capturable). Thus, successful gravitational capture of a 0.1 moon-mass planetoid must occur early in solar system history (i. e., within a few hundred million years after planetoid origin).

A third paradox is that "cool" planetoids are more capturable than "hot" planetoids Most of the energy for capture (~90% or more) must be tidally stored and subsequently dissipated during the first close encounter of an encounter sequence if the planetoid is to be captured into a stable planetocentric orbit. The reason for this is that the viscosity will be considerably lower (and thus the Q factor higher) after significant melting due to energy dissipation associated with the earlier encounter. Distant encounters are not associated with much energy dissipation, relative to that needed for capture, because of the $1/r^6$ dependence. For example, an encounter beyond about 2 planet radii does not lead to much energy dissipation. But if a candidate planetoid has too high a viscosity value to have a low Q value, then a close encounter (or several more distant encounters) can prepare a planetoid for capture by bringing it into the viscosity "zone" for dissipation of the energy for capture.

Equation 4.2 is for energy storage in Mars during one close encounter. Again, the closeness of the encounter is important here because of the $1/r^6$ dependence. With h (mars) = 1.0, the energy stored in mars = the numerator of the equation divided by the denominator. Table 4.3 shows the results for various distances of close approach.

Table 4.4 Values for energy stored in a 0.1 moon-mass planetoid during a single encounter ($h_{pl} = 1$)

Perimar radius	Type of encounter	Energy stored (Joules) (Joules)
For 1.30 mars radii	(near grazing)	11.13 E26
For 1.35 mars radii	(non-grazing)	8.88 E26
For 1.40 mars radii	(non-grazing)	7.14 E26
For 1.43 mars radii	(non-grazing)	6.29 E26

Note that the values of energy stored in the planetoid is based on an h value of 1.0. A realistic value for a planetoid is 0.3. Thus only about 0.3 of that value shown in column 3 can be stored and subsequently dissipated in a planetoid with a Q value of 1

$$\Delta E_{\text{stored-mars}} = \frac{3\, h_{ms}\, G\, M_{pl}^{2}\, R_{ma}^{5}}{5\, r_p^{6}} \tag{4.2}$$

Where G is the gravitational constant, M_{pl} is the mass of the planetoid (a 0.1 moon-mass planetoid in this case), R_{ma} is the radius of Mars, h_{ma} is the displacement Love number of Mars, and r_p is the distance of closest approach during the encounter.

From the above numbers, even using a Q value of 100 (low for Mars) and h = 0.7, these numbers should be discouraging for tidal capture enthusiasts (i.e., no more than 1 E24 J of energy is dissipated per encounter). It would take over 100 encounters at a very close perigee passage distance to dissipate the energy for stable capture in a two-body context. Singer (1968, 1970) and nearly everyone else that looked at the problem of gravitational capture, including Kaula and Harris (1973), considered the planet to be the primary energy sink for capture. *Such a capture scenario would be very implausible by orders of magnitude.*

Energy storage in a 0.1 moon-mass planetoid during a single encounter is:

$$\Delta E_{\text{stored-planetoid}} = \frac{3\, h_{pl}\, G\, M_{ma}^{2}\, R_{pl}^{5}}{5\, r_p^{6}} \tag{4.3}$$

Where M_{ma} is the mass of Mars, R_{pl} is the radius of the planetoid (the 0.1 moon-mass body in this case), h_{pl} is the displacement Love number of the planetoid, and r_p is the distance of closest approach during the encounter.

Table 4.4 shows values of energy stored for specific values of close encounters.

The energy needed for capture for this two-body analysis is ~3.09 E26 J. The closeness of the encounter is very important here because of the $1/r^6$ dependence. The displacement Love number (h_{pl}) of the planetoid = 1.0. Now let us look at the numbers (Table 4.4) for the planetoid as we did before for Mars.

From the above numbers, even with a Q value of 1 for the planetoid and an h for the planetoid of about 0.3 (a reasonable value for the encountering planetoid), the energy dissipated in the planetoid during one encounter at 1.43 mars radii would be ~1.89 E26 J (i.e., about 60% of the energy dissipation necessary for gravitational capture). This calculation simply makes gravitational capture via energy dissipation in the planetoid marginally attractive in a two-body context.

Table 4.5 Values of energy dissipated in both bodies during an encounter between a 0.1 moon-mass body and a mars-mass body

Perimar radius	Ediss – Mars (h = 0.7; Q = 100)	Ediss – planetoid (h = 0.3, Q = 1)
1.43 Rma	5.22 E23 Joules	1.89 E26 Joules
1.53 Rma	3.48 E23 Joules	1.26 E26 Joules

Note that less than 0.3% of the energy is dissipated in the planet using these very reasonable parameters for the initial encounter of an encounter sequence. Note that the h and Q values for both the planet and planetoid are shown in the table

Energy dissipated in both bodies during a close gravitational encounter is:

$$\Delta\, E_{diss} = \frac{3G}{5r_p^6}\left[\frac{h_{pl}M_{ma}^2R_{pl}^5}{Q_{pl}} + \frac{h_{ma}M_{pl}^2R_{ma}^5}{Q_{ma}}\right] \tag{4.4}$$

where Q_{pl} is the specific dissipation factor for the planetoid and Q_{ma} is the specific dissipation factor for Mars.

Table 4.5 gives values of total energy dissipation during typical close encounters.

The energy needed for capture is ~3.09 E26 J. For a close encounter at 1.43 mars radii and with a Q for the planetoid = 1 and a Q for Mars = 100, capture from a mars-like orbit (velocity at infinity = 0) would require an h for the planetoid of 0.5 but a more reasonable value for h of the planetoid is 0.3). The displacement Love number for Mars is not very important for dissipating the energy for capture. Even using a h for Mars of 0.7 and a Q value of 100, only about 0.3% of the energy for capture is dissipated in the planet. These results appear to make gravitational capture of the candidate planetoid just marginally possible in a two-body context! This fact alone should be enough to discourage most capture enthusiasts!!!

In summary, even after understanding the paradoxes associated with gravitational capture, there is an unusual history surrounding the development of Eqs. 4.2, 4.3, and 4.4, the energy storage and energy dissipation equations. When Kaula and Harris (1973) derived the equations, they had the numerical coefficient value at 3/10 because they considered only the energy stored, and subsequently dissipated, to be associated with the tidal deformation of the body (and not with the subsequent relaxation of the tidal distortion). Winters and Malcuit (1977) showed that a second quantity, which is equal to the first, is associated with the tidal relaxation process. This changed the numerical coefficient to 3/5, a change which doubled the energy stored, and subsequently dissipated, during a close gravitational encounter. Peale and Cassen (1978) independently arrived at the same conclusion that the numerical coefficient should be 3/5. The bottom line is that these changes to this fundamental equation brought gravitational capture into the realm of the physically possible!!!

4.2.2 *Post-Capture Orbit Circularization Process*

Figure 4.6 shows a general view of the orbit circularization process.

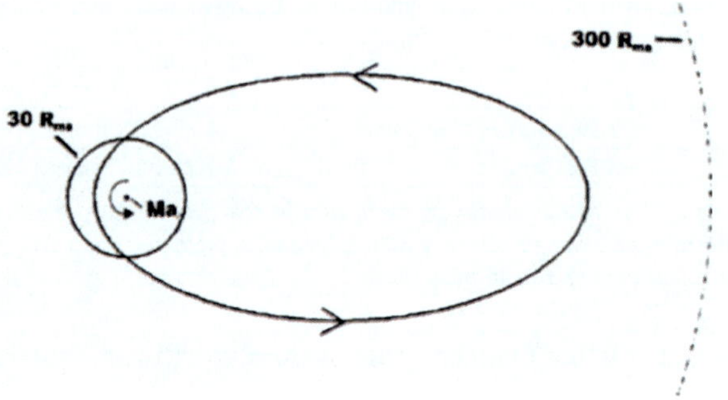

Fig. 4.6 Diagram illustrating the general scheme of the orbit circularization process. We start with the largest prograde post-capture orbit possible Semi-major axis = 126.9 mars radii and eccentricity ~0.8. This large major axis orbit then undergoes progressive circularization via tidal energy dissipation in a combination of the planetoid and Mars. If no angular momentum is imparted to the orbit from Mars' rotation, then the large orbit evolves into a 30 mars radii circular orbit. In the more realistic case, angular momentum is transferred to the planetoid orbit as Mars loses rotational angular momentum and the final circular orbit has a semi-major axis somewhat larger than the 30 mars radii shown on the diagram, the details depending on the h and Q values of the planetoid and Mars. A numerical simulation of the orbit circularization process is presented in the next section of this chapter

4.3 Numerical Simulation of Prograde Gravitational Capture of a 0.1 Moon-Mass Planetoid by Planet Mars

We had our first successful numerical simulations of the capture process in September of 1987 and our first report on the simulations for the Earth-Moon system was Malcuit et al. (1988) at the LPSC. More capture simulations are in Malcuit et al. (1989, 1992). There were two very favorable surprises associated with the capture simulations.

Surprise # 1: Solar gravitational perturbations during a capture encounter can decrease the required energy dissipation for capture by up to 50% relative to what was expected from a two-body calculation.

Surprise # 2: There are such things as STABLE CAPTURE ZONES (Malcuit and Winters 1996) and the probability of capture for any one encounter can be measured directly from the geometry of the appropriate stable capture zone.

4.3.1 Computer Code Information

The computer code for this general co-planar, three-body calculation uses a fourth-order Runge-Kutta numerical integration method. The accuracy is a few parts in

10 E5 for energy and a few parts in 10 E8 for angular momentum. The energy dissipation subroutine is activated for all encounters within 20 mars radii. The density of time-steps (calculation points for orbital elements and energy extraction from the orbit) is inversely proportional to the distance of separation of the planet and planetoid. If the perimar distance is outside 20 mars radii, then the energy is taken from the orbit impulsively at the perimar point.

There are two general orientation schemes for this co-planar calculation. One is a non-rotating coordinate system (Fig. 4.7) in which the marocentric computer plot theoretically revolves around the Sun in 1 martian year but the plot does not rotate. There is also a rotating coordinate system (not shown here but explained in a sidebar in Chap. 8) which is useful for analysis of certain orbital characteristics but it is not as easily understood as the non-rotating plot. The hallmark of a rotating coordinate system is that the computer plot rotates as it revolves around the Sun. Some orientation information for the numerical simulations is shown in Fig. 4.8.

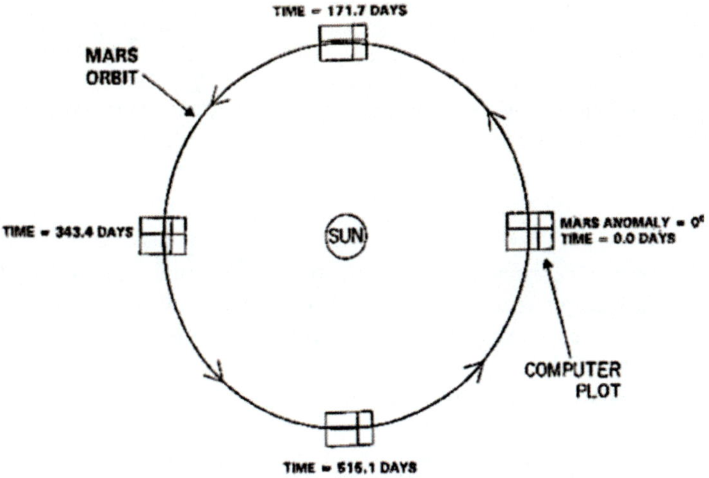

Fig. 4.7 Some orientation information for understanding the computer plots. Diagram illustrating a not-rotating coordinate system for plotting the results of numerical simulations. The computer plot does not rotate as it theoretically revolves around the Sun. Mars is at the origin of the rotating plot. All plots in this chapter use the non-rotating coordinate system and all orbital plots are viewed from the north pole of the solar system

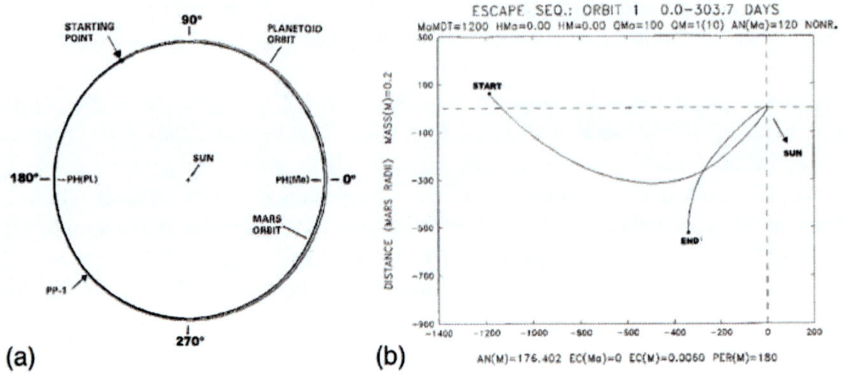

Fig. 4.8 Some terminology for the heliocentric and marocentric plots used in this chapter and in the book. (**a**) Planet anomaly is the position of the planet on the heliocentric orbit. The zero point for the planet anomaly the left of the diagram and is measured in degrees in a counter-clockwise direction. PH(Ma) is the perihelion position of the martian orbit; PH(PL) is the perihelion position for the orbit of the planetoid. Starting position is the position of the planet at the beginning of the simulation. PP-1 is the position of the first close encounter. (**b**) Organization of the information on the marocentric computer plot is the following: *First title line*: results of simulation (escape, collision, or capture); number of orbits; duration of simulation in earth days. *Second title line*: *MaMDT* initial mars-planetoid distance of separation; *HMa* h for mars; *HM* h for planetoid; *QE* Q of mars; *QM* Q of planetoid; *AN(Ma)* mars anomaly at beginning of calculation. Bottom line: *AN(M)* planetoid anomaly; *EC(M)* eccentricity of the orbit of the planetoid; *PER(M)* pericenter radius position for orbit of planetoid. Left side line: Scale for both axes in planet radii; fractional mass of planetoid relative to mass of earth's moon

4.3.2 Development of the Computer Code

The N-body numerical simulation code was borrowed from Tom Stoeckley (Michigan State University, Astronomy Dept.) by Bob Malcuit in 1976. This code was modified by computer science/physics students David Mehringer (DU, '88) Wentao Chen (DU, '92), and Alber Liau (DU, '93) under the supervision of Ronald Winters (Physics-Astronomy Dept.) and myself from 1987–1993. A two-body evolution code was developed by David Mehringer, Ronald Winters, and myself in the 1985–87 era. Plotting packages for planetocentric orbits, tidal amplitudes and tidal ranges were developed by David Mehringer, Wentao Chen, and Albert Liau form 1987 to 1993 under the supervision of Ronald Winters and myself.

4.3.3 A Sequence of Typical Orbital Encounters Leading to a Stable Capture Scenario

Figure 4.9 shows the results of a simulation in which there is no energy dissipation within either body. The starting positions as well as the ending positions are shown

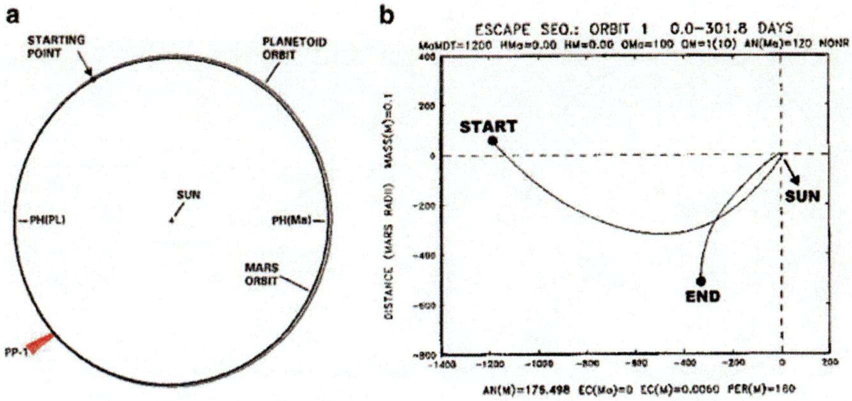

Fig. 4.9 Results of a close encounter in which no energy is dissipated. (**a**) Heliocentric orientation for the encounter. The simulation begins at the starting point (START). The close encounter occurs about 191.0 earth days (~0.277 mars year) later at PP-1 (perimar passage number 1. No energy is dissipated by tidal processes and the planetoid escapes to a heliocentric orbit about 110.8 earth days (~0.162 mars years) after PP-1. (**b**) Marocentric orientation for the encounter. Mars is at the origin of the plot. The simulation begins at 1200 mars radii, which is well beyond the boundary of the Hill Sphere (which is about 300 mars radii)

on the marocentric framework. The heliocentric orientation diagram shows the starting position relative to the martian orbit and PP-1 is the location of the first (and only) perimar passage in this orbital encounter sequence. The closest encounter distance is 1.43 mars radii; there is no energy dissipated during the encounter. The planetoid returns to the mars-like heliocentric orbit essentially unchanged. The starting position, as well as the ending position, are shown on marocentric plot. During the first encounter, the close approach distance is 1.43 mars radii. This bring the surface of the planetoid within about 600 km of the surface of Mars. I should note that the first encounter is focused on the value of 1.43 mars radii in order to eliminate a variable in the calculation.

Figure 4.10 shows the results of a simulation in which there is significant energy dissipation within the interacting bodies and the planetoid returns for two additional passages after the initial encounter. The encountering body then returns to a slightly changed mars-like heliocentric orbit. On the heliocentric orientation diagram the starting position is the same as in Fig. 4.10a as well as the position of the first perimar passage (encounter). There is not enough energy dissipated for capture and after three perimar passages the planetoid returns to a somewhat changed heliocentric orbit from which it can eventually have another encounter with Mars.

Figure 4.11 shows the results of a simulation in which there is sufficient energy dissipation for stable capture with displacement Love number of the planetoid elevated to 0.33. The first 40 orbits of the stable capture scenario are shown but the calculation was extended to over 100 orbits and it showed no signs of instability. On the heliocentric orientation diagram the starting position is the same and the position of the first six perimar passages are shown. Note that only 1.42 E26 J of

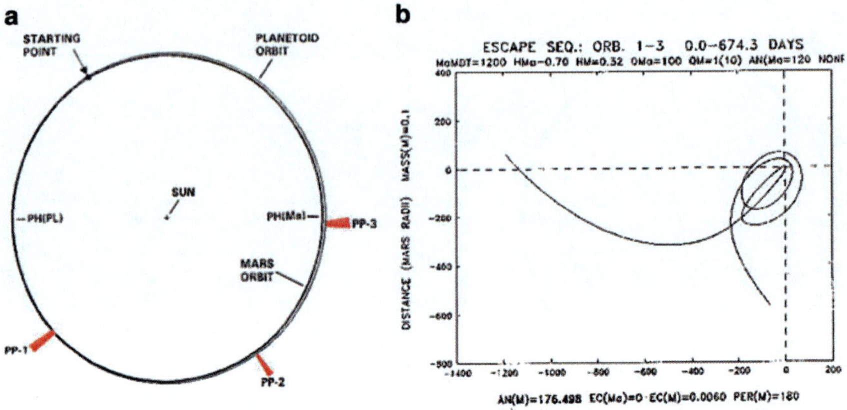

Fig. 4.10 Results of a three-orbit escape scenario. (**a**) The heliocentric orientation shows the starting point and the positions of the three close encounters of this encounter sequence. (**b**) The marocentric orientation shows the pattern of close encounters. Solar gravitational perturbations cause the variation in the orbital path. A significant quantity of energy is dissipated (~98% of the energy necessary for capture) but not quite enough for stable capture

energy dissipation was sufficient for stable capture for this particular set of co-planar orbital parameters. Some orbital resonance activity is obvious in the pattern of the orbital geometry. The bottom line is that a favorable orientation of the Sun during the timeframe of the initial encounter can reduce the energy dissipation requirements to about one-half.

The pattern of equilibrium tidal amplitudes for Mars are shown in Fig. 4.11d for the first 8 earth years (about 4 martian years). The tidal amplitudes for each body are calculated directly from the orbital files. Equilibrium rock tides of up to 10 km occur on the planet during the initial encounter and in this simulation the initial encounter is the closest. But this is not a general case. Many simulations will give similar results but each one is somewhat different. In some cases, just a slight change in input parameters can yield a much different result. In this particular simulation while Mars is experiencing 9 km rock tidal amplitudes, the rock tides for the planetoid for the first encounter are about 144 km (~16% radial tidal deformation). Within the first 8 years there are 3 perimar passages that raise over 5 km tidal amplitudes on Mars and over 72 km tidal amplitudes on the lunar body. Thus it is clear that in addition to the planetoid dissipating nearly all of the energy for its own capture, it is undergoing severe tidal distortions during the early stages of a stable capture scenario. Table 4.6 is a summary of the results of encounters in which the value of h (displacement Love number) is systematically increased.

An equation for tidal amplitudes on Mars is:

$$\text{Mars Tidal Amplitude} = \frac{M_{pl}}{M_{ma}} \left(\frac{R_{ma}}{r_p}\right)^3 R_{ma}\, h_{ma} \tag{4.5}$$

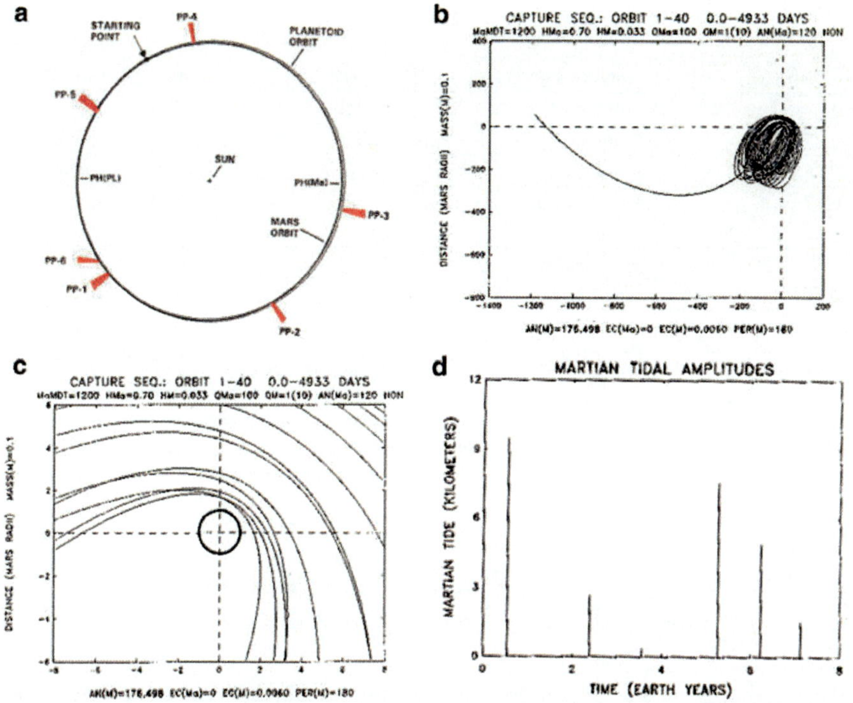

Fig. 4.11 Results of a stable capture scenario. (**a**) Positions of the first six perimar passages are shown on the heliocentric orientation. (**b**) Orbital pattern of the first 40 orbits (~13.5 earth years; ~7.2 mars years) are shown on the marocentric plot. After 100 orbits the major axis of the orbit is beginning to decrease. Note that about 1.42 E26 J of energy dissipation is sufficient for capture for this set of orbital parameters. (**c**) Expanded view of the pattern of close encounters for this stable capture scenario. In this scenario, the initial encounter was the closest encounter at 1.43 mars-radii. (**d**) Pattern of martial rock tidal amplitudes during the first 8 earth years (~4.25 martian years) for this orbital scenario. (Note: $h_{ma} = 0.33$ and not 0.033 as indicated on the second line on the plot)

Where M_{pl} is the mass of the planetoid (a 0.1 moon-mass body in this case), M_{ma} is the mass of Mars, R_{ma} is the radius of Mars, r_p is the distance of separation between the planet and planetoid, and h_{ma} is the displacement Love number of Mars.

An equation for tidal amplitudes on the 0.1 moon-mass planetoid is:

$$\text{Planetoid Tidal Amplitude} = \frac{M_{ma}}{M_{pl}} \left(\frac{R_{pl}}{r_p}\right)^3 R_{pl}\, h_{pl} \tag{4.6}$$

Where R_{pl} is the radius of the planetoid and h_{pl} is the displacement Love number of the planetoid.

Table 4.6 Summary of encounter results for a 0.1 moon-mass planetoid encountering a mars-mass planet in a prograde direction

h(planetoid)	Result of simulation	h(planetoid)	Result of simulation
0.0	1-orbit escape	0.34	40-orbit capture
0.25	1-orbit escape	0.35	40-orbit capture
0.26	2-orbit escape	0.36	40-orbit capture
0.27	2-orbit escape	0.37	40-orbit capture
0.28	3-orbit escape	0.38	40-orbit capture
0.29	3-orbit escape	0.39	40-orbit capture
0.30	3-orbit escape	0.40	40-orbit capture
0.31	3-orbit escape	0.41	40-orbit capture
0.32	3-orbit escape	0.42	40-orbit capture
0.33	40-orbit capture	0.43	40-orbit capture

The only variable for this series of encounters is the h value of the encountering planetoid. Some other constant values for this series of encounters are h for Mars = 0.7; Q for Mars = 100; Q values for the 0.1 moon-mass planetoid are 1 for the first encounter and 10 for all subsequent encounters. Note that the initial encounter distance for the encounter sequence is 1.43 mars radii

4.3.4 Geometry of Stable Prograde Capture Zones for Planetoids Being Captured by Mars

Each planet-planetoid combination has certain zones in space from which the planetoid can be captured into a STABLE PLANETOCENTRIC ORBIT (Malcuit and Winters 1996). A STABLE CAPTURE ZONE (SCZ) is defined as a region of parameter space (planetoid anomaly vs. planet eccentricity) in which the orientation of the orbit of the encountering planet is favorable for a stable post-capture orbit if sufficient energy is dissipated for its capture into a planetocentric orbit. In this book the SCZ concept is for co-planar encounters: i.e., all encounters are in the common plane of the planets of the Solar System and all of the encounter geometry is viewed from the north pole of the Solar System. Any encounter that is out of this plane would result in a less stable orbital configuration. Eventually, some adventurous character can define these SCZs in three dimensions.

Figure 4.12 shows the geometry of the Stable Capture Zones for encounters in the prograde direction. There is an SCZ for planetoids encountering a planet from an orbit that is slightly smaller than the orbit of the planet (inside orbits) and there is an SCZ for close encounters by planetoids that are in an orbit that is slightly larger than the orbit of the planet (outside orbit). The zones shaded in red require the lowest h_{pl} for stable capture. The zones shaded in blue require more energy dissipation for stable capture. The enclosed white zones yield a stable capture orientation but some process in addition to tidal energy dissipation (i.e., a non-tidal energy sink) is required for stable capture. Note that encounters outside the SCZs are possible but result is an orbital collision or an escape into a somewhat changed heliocentric orbit. Note also that there are no stable capture possibilities from orbits of very low eccentricity that are very similar to the orbit of the planet which is zero eccentricity.

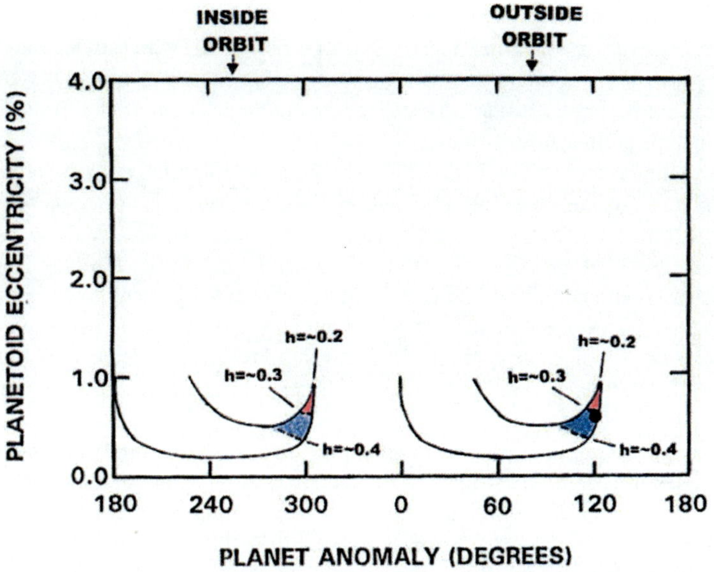

Fig. 4.12 Geometry of stable prograde capture zones for planet Mars and a 0.1 moon-mass planetoid encountering planet Mars form an orbit that is inside the orbit of Mars (*left*) and for the planetoid encountering Mars from an orbit that is outside the orbit or Mars (*right*). Note that the morphology of the SCZs for a 0.1 moon-mass planetoid are essentially the same but the h for capture is somewhat higher for the outside orbit case. In any case, the probability for prograde capture for a 0.1 moon-mass planetoid in the eccentricity zone of 0.6–0.9% range is 1/180, which is very low chance for successful capture

There are three important items to consider for determining the geometry of a STABLE CAPTURE ZONE: (1) the ORIENTATION of the orbit of the planetoid relative to the orbit of the planet; (2) ENERGY STORAGE during the timeframe of a close encounter; and (3) ENERGY DISSIPATION during the timeframe of a close encounter. If any of these items is outside acceptable boundaries, then STABLE CAPTURE will not occur. These concepts will become more lucid when reading subsequent sections of this chapter.

An additional advantage of SCZs is that the PROBABILITY OF CAPTURE can be determined directly from the geometry of an SCZ. For any given combination of planetoid orbital eccentricity and planet anomaly, the probability of successful capture can be determined *by the length of the line intercept within the SCZ divided by the total length of the line intercept on the plot* (i.e., the 360° of planet anomaly). Cursory examination of the prograde SCZs suggests that prograde capture has a very low probability of capture even when the displacement Love numbers are favorable of capture. The geometry of retrograde SCZs will be discussed in Chap. 5 and the geometry of SCZs for several other planets is discussed in Chap. 8.

A Procedure for Mapping Stable Capture Zones in Two Dimensions
There are three important items for mapping stable capture zones: (1) orientation of the encounter, (2) energy storage during the encounter, and (3) energy dissipation during the encounter.

The importance of ORIENTATION of the planetoid relative to the orbit of the planet can not be overly stressed. Without an encounter oriented within narrow limits, there is no successful capture. For successful prograde capture, the major axis of the initial planetoid orbit must be within about + or – 3° to a line that is perpendicular to the tangent of the orbit of the planet. For successful retrograde capture, the major axis of the initial planetoid orbit must be within about + or – 3° of a line parallel to the tangent of the orbit of the planet. These rules are valid for encounters of planetoids from orbits that are either interior or exterior to the orbit of the planet.

For gravitational capture, energy much be stored by tidal deformation processes in a combination of the interacting bodies. Thus the displacement Love numbers of the combination of the encountering bodies must be large enough to temporarily store the energy for capture. If the body of the planetoid is too rigid, then there will be no capture via gravitational (tidal) processes.

And lastly, a very large percentage of stored energy must be dissipated during the time-frame of a gravitational encounter from a heliocentric orbit. Ross and Schubert (1986) demonstrated that bodies of roughly lunar-size can have extremely low Q values when they have values of body viscosity within a certain range of values. In this special case of low Q values, a very high percentage of the tidally stored energy can be dissipated during the time-frame of the encounter.

Now let us consider a procedure for mapping the geometry of an SCZ in two dimensions (i.e., planetoid anomaly along the horizontal axis and planetoid orbit eccentricity along the vertical axis). The process, as I am describing it, is laborious and time-consuming. Defining the boundaries of one SCZ can entail thousands of individual runs. First, one must find a tentative stable capture position. Then to be sure that a capture is stable, a simulation of at least 40 orbits (and preferably 100 orbits) is necessary. Once all (or most) of the simulations are stable captures for that position of planetoid anomaly and planetoid eccentricity, the computer operator can move a certain increment of planetoid anomaly or planetoid eccentricity and repeat the process. Eventually all useful locations are to be explored and evaluated and the boundaries of a stable capture zone are defined. Each point entails between 10 and 20 simulations in order to determine if the satellite orbit is stable. One hundred calculation points can give one a sense of the field of stable capture, but it can take about three times more calculation points to define the boundaries in sufficient detail. Thus, 20 simulations per point times 300 points gives over 6000

(continued)

individual calculations. Six thousand times 4 SCZs yields 24,000 computer runs.

Much of this process can be automated by an industrious physics and/or computer science student. In my view, an automated system would be very important for defining Stable Capture Zones in three dimensions.

4.3.5 Post-Capture Orbit Evolution

A computer program has been developed to calculate a time scale for the post-capture orbit circularization in a two-body context. The computer program averages, then makes orbit corrections, then calculates, then averages, makes orbit corrections adiabatically as the orbit progressively circularizes. Depending on the h values and Q values that one uses as input, the resulting time scale is different. Figure 4.13 and Table 4.7 give the results of one such calculation.

4.3.6 Summary and Statement of the Fourth Paradox

We have established that stable prograde capture of a 0.1 moon-mass planetoid is physically possible in a three-body interaction context. But a related question is: Where do the candidate planetoids for capture come from?

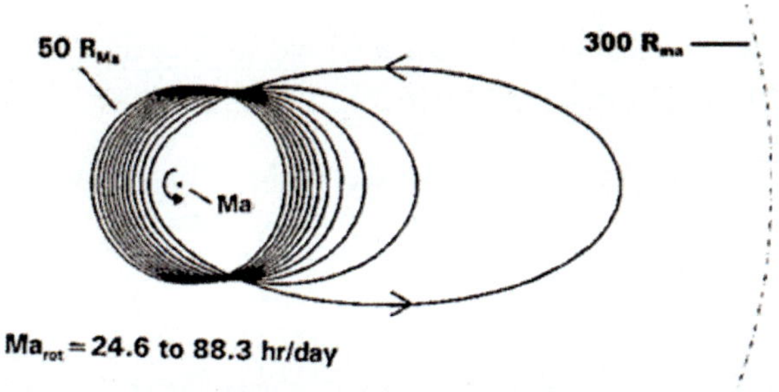

Fig. 4.13 Diagram showing the results of a post-capture orbit evolution calculation for a prograde capture scenario of a 0.1 moon-mass planetoid and a mars-mass planet. Initial rotation rate for Mars is 24.6 hr/day. Planetoid orbit eccentricity = 0.8699; Semi-major axis (SMA) = 126.9 mars radii. Deformation parameters: h (Mars) = 0.42; Q (Mars) = 100; h (planetoid) = 0.5; Q (planetoid) = 1000. The orbit circularizes to 10% eccentricity in about 4 billion years. The resulting circular orbit has a majoı axis of about 50 mars radii. In this calculation the angular momentum of planetary spin is transferred via tangential tidal action to the satellite orbit as the calculation proceeds. The orbit of the satellite becomes circularized mainly because of radial tidal action within the satellite during perigee passages. The satellite orbit expands from a semi-major axis of 30–50 R_{ma} because of tangential tidal action caused by the rock tides raised on the planet by the satellite. Note that all orbits are viewed from the north pole of the solar system

Table 4.7 Some numerical values for a number of time steps for the calculated post-capture prograde marocentric orbit evolution shown in Fig. 4.13

Orbit eccentricity	Perimar distance (Rma)	Major axis (Rma)	Time (Ga)	Rotation rate (hr/day)
0.870	16.5	126.9	0.0	24.6
0.685	22.1	70.1	0.5	32.4
0.554	26.7	59.9	1.0	40.4
0.457	30.5	56.1	1.5	48.3
0.381	33.6	54.4	2.0	55.9
0.313	36.7	53.4	2.5	63.4
0.248	39.6	52.7	3.0	71.3
0.182	42.9	52.3	3.5	79.5
0.098	47.0	52.1	4.0	88.3

PARADOX # 4: A planetoid can be captured from a mars-like orbit but it probably was not formed in that near mars-like heliocentric orbit.

If it is from the Asteroid Zone, would such a loosely structured asteroidal body be capable of dissipating the energy for capture in essentially one encounter without being tidally disrupted? And would it be capable of dissipating the energy for capture in one pass?

If the candidate planetoid is from the Vulcanoid Zone, then the body would have an igneous history. If the Vulcanoid planetoid had an igneous history via melting during FU Orionis events then after a few non-capture encounters with Mars, it might have a body viscosity compatible with a low Q value that would permit tidal capture if the eccentricity and inclination of the planetoid orbit are within certain limits.

The other major question of this chapter is: Would prograde capture of a sizable satellite have made conditions more favorable for the origin and development of a life system on Mars?

We all know what happened to planet Earth. It has a recorded history of continuous life with increasing complexity going forward in time (Ward and Kirschvink 2015). The history of life forms on Earth can be traced back to at least 3.5 billion years and possibly to before 4.0 billion years. Mars, in contrast, may contain in its rocks a record of microbes that lived in the earliest history of the planet, perhaps in the first 500 million years. Some planetologists are becoming more confident that planet Mars, in those first 500 Ma or so, had all (or most) of the ingredients necessary for the development of a biological system: liquid water on the surface, carbon dioxide and water vapor (both greenhouse gasses) in the atmosphere, as well as significant quantities of the ingredients for methane and ammonia in some combination of the atmosphere and hydrosphere (Villanueva et al. 2015).

A reasonable response to the question of whether there would be favorable conditions for life is "YES", but for only a limited amount of time. The rock tidal activity associated with prograde capture of a 0.1 moon-mass planetoid and the subsequent orbit evolution would have pumped energy into the lithosphere and would have caused significant outgassing of the martian mantle. Martian ocean tidal activity would have caused significant agitation along any shorelines, especially in the lower latitudes of the planet. Outgassed carbon dioxide and water vapor would have added to any pre-capture greenhouse effect on the planet. In general, a prograde gravitational capture scenario for planet Mars could have added a few 10 E8 years of habitable conditions for planet Mars (i. e., "GEOFORMED" the planet for a while).

Habitable conditions on Mars, however, would be limited by the initial rotation rate of the planet. Since the prograde capture scenario starts with a Mars rotation rate of 24.6 hr/day (see Table 4.6), at 250 Ma the rotation rate is at 30 hr/day, and at 480 Ma the rotation rate is at 32.4 hr/day. Eventually the nights would be too long and too frigid for organisms to survive in the shallow water and near surface environments of the planet.

Figures 4.14 and 4.15 illustrate a sequence of probable surface conditions during a gravitational capture episode and the subsequent marocentric orbit evolution.

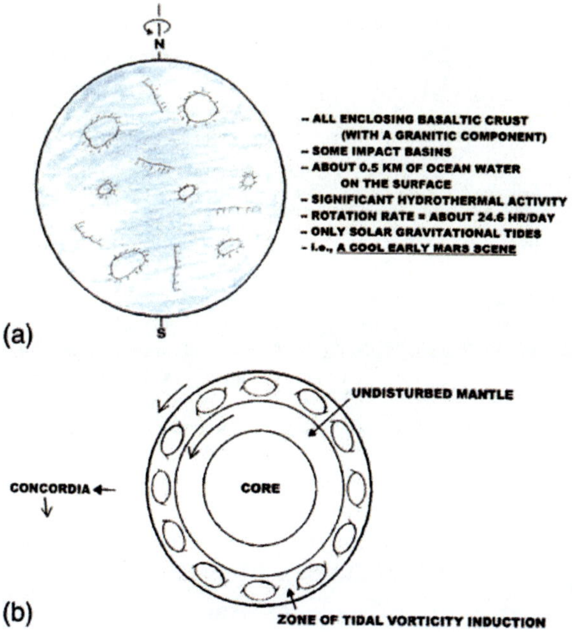

Fig. 4.14 (**a**) Probable condition of a mars-like planet before the capture of a 0.1 moon-mass satellite (~4.0–4.3 Ga ago). After a hot accretion and core formation, a chill crust would form over the planet. Some of the early crust would founder but eventually a basaltic-andesitic crust with pockets of granite would form. At this time a mars-like planet would be very similar to planet Mars today (a one-plate planet with a "stagnant lid"). (**b**) Equatorial cross-section of a mars-like planet about 1 Ka (~5 Ma on the two-body evolution scale) after capture of a prograde satellite that I have named Concordia. Note that the mantle circulation cells rotate in the opposite direction to the rotation direction of the planet. Concept of tidal vorticity induction is from Bostrom (2000). View is from the north pole of the solar system

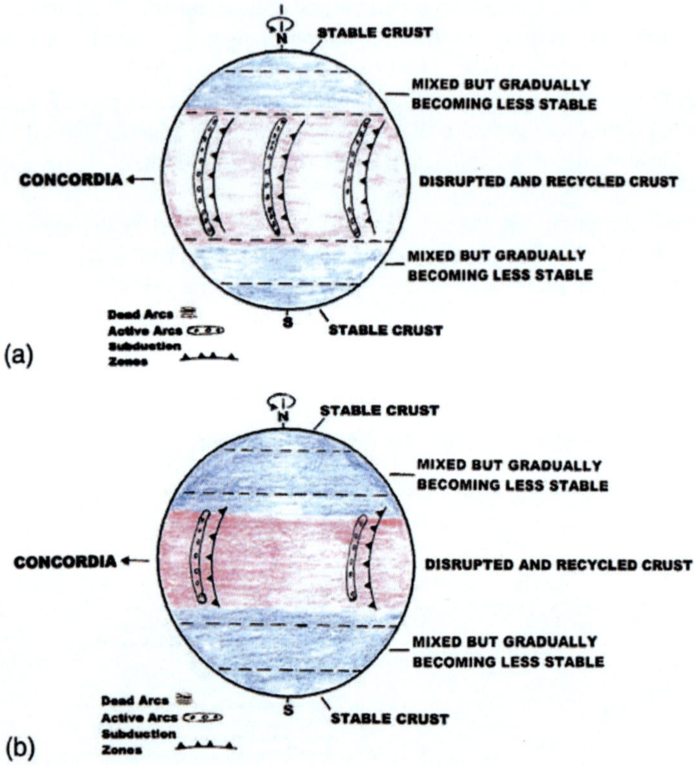

Fig. 4.15 Schematic diagrams showing the zonation of probable surface features (sans ocean water) on a mars-like planet soon after capture. The equatorial zone of the planet would experience 10–5 km rock tides for a geologically short (a few tens of thousands of years) and elevated tidal amplitudes for many millions of years. Since the very high tides effect mainly the equatorial zone, most of the recycling of the primitive crust would occur there and the polar zones would be only weakly effected. (**a**) The author's concept of surface conditions a few thousand years after capture. (**b**) The author's concept of surface conditions a few million years into the circularization sequence

4.4 Numerical Simulation of Prograde Gravitational Capture of a 0.2 Moon-Mass Planetoid by Planet Mars: Another Attempt to Geoform Mars

In this section of the chapter we explore the possibility of capturing a larger planetoid to see if our chances of developing a more habitable planet are increased. After all, the rock tides on Mars would be twice as high and any mantle reworking would be more vigorous and this would lead to more volcanic outgassing from the mantle of the planet.

For our treatment of the prograde capture of larger satellite we can skip a few of the explanations that are common to both prograde capture scenarios such as: computer code information and development of the computer code.

4.4.1 A Sequence of Typical Orbital Encounters Leading to a Stable Capture Scenario of a 0.2 Moon-Mass Planetoid

Figure 4.16 shows the results of a simulation in which there is no energy dissipation within either body. The heliocentric orientation diagram shows the starting position relative to the martian orbit and PP-1 is the location of the first (and only) perimar

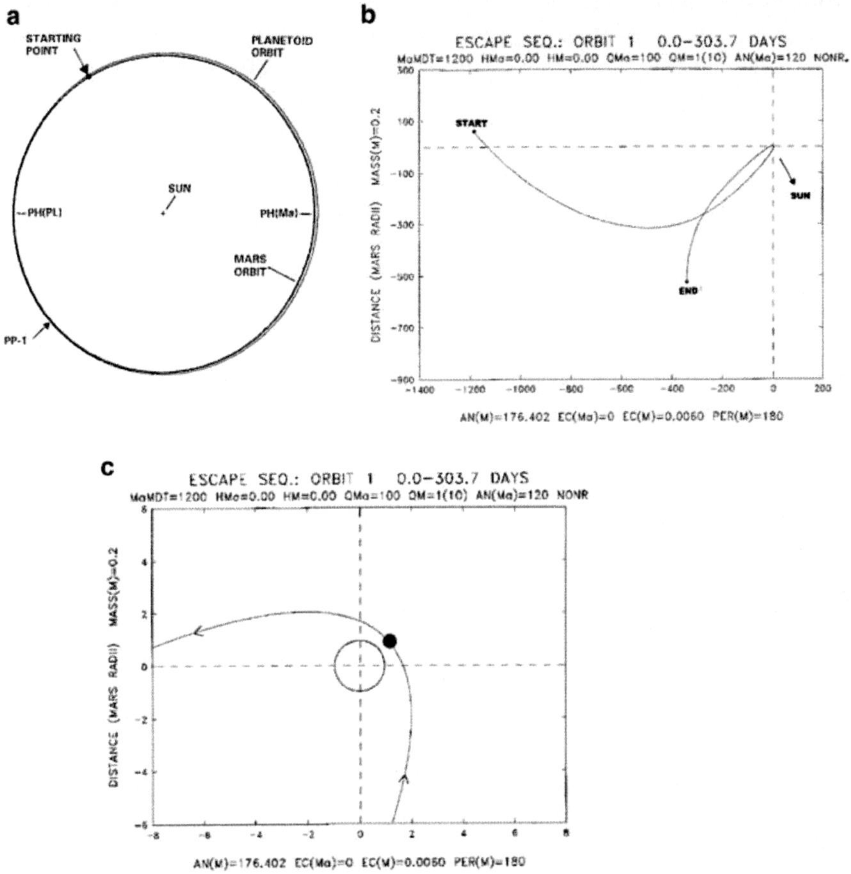

Fig. 4.16 Results of a close encounter in which no energy was dissipated. (**a**) Heliocentric orientation for the encounter. The simulation begins at the starting point and the planetoid orbit is outside that of the planet. The close encounter occurs at PP-1 (perimar passage number 1). No energy is dissipated by tidal processes and the planetoid escapes to a heliocentric orbit. (**b**) Marocentric orientation for the encounter. Mars is at the origin of the plot. The arrow on the plot indicates the position of the sun at the beginning of the simulation. The simulation begins at 1200 mars radii, which is well beyond the boundary of the Hill sphere. (**c**) Close-up of the close encounter showing the diameters of the bodies at closest approach. The distance of separation at the time of the closest encounter is about 1000 km

passage in this orbital encounter sequence. The starting positions as well as the ending positions are shown in the marocentric framework. The closest encounter distance is 1.43 mars radii (the same as the case with a 0.1 moon-mass planetoid); there is no energy dissipated during the encounter. During this encounter the surface of the planetoid approaches to within about 600 km of the surface of Mars. The planetoid returns to a mars-like heliocentric orbit essentially unchanged. (Note that all of these diagrams are viewed from the north pole of the Solar System unless noted otherwise.)

Figure 4.17 shows the results of a simulation in which there is significant energy dissipation within the interacting bodies and the planetoid returns for two additional passages following the initial encounter. On the heliocentric orientation diagram the starting position is the same as in Fig. 4.17a as well as the position of the first perimar passage. There is not enough energy dissipated for capture and after three perimar passages the planetoid returns to a somewhat changed heliocentric orbit from which it can eventually have another encounter with the planet.

Figure 4.18 shows the results of a stable capture scenario. Note that the tidal amplitude for the initial encounter is about 2 times that of the scenario featuring a 0.1 moon-mass body.

4.4.2 *Summary for the Prograde Capture Scenario for a 0.2 Moon-Mass Planetoid*

In summary, there are advantages and disadvantages associated with prograde gravitational capture of a 0.2 moon-mass planetoid by a mars-like planet. Although

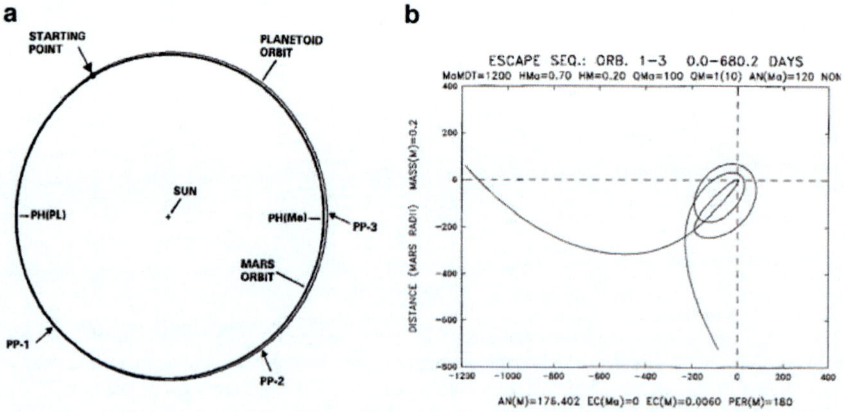

Fig. 4.17 Results of a three orbit escape scenario. (**a**) Heliocentric orientation diagram showing the starting point and the positions of the three close encounters of this encounter sequence. (**b**) The marocentric orientation showing the pattern of close encounters. Solar gravitational perturbations cause the variation in the orbital path. A significant quantity of energy (2.72 E26 J) is dissipated but not quite enough for stable capture

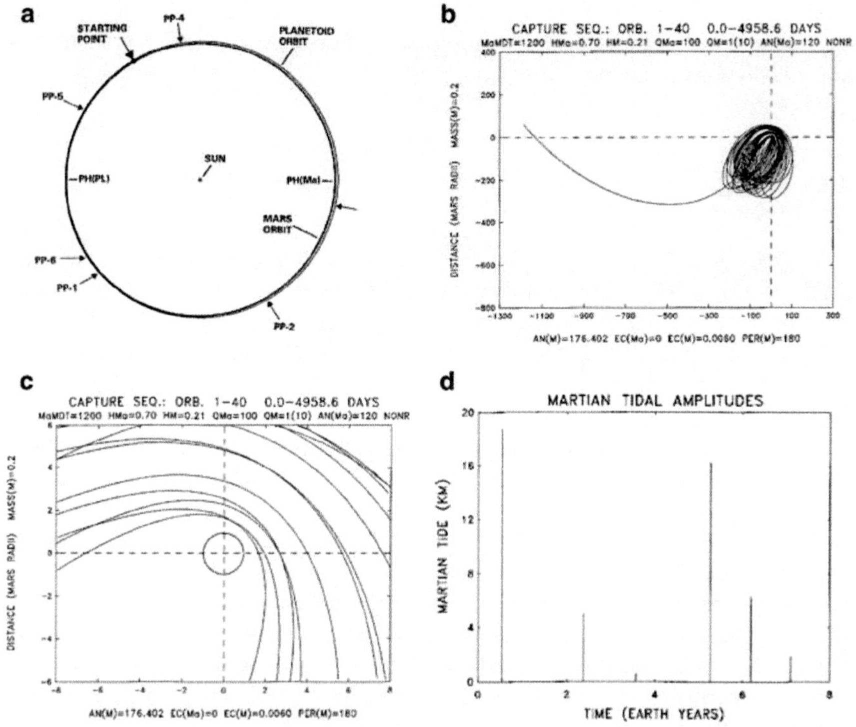

Fig. 4.18 Results of a stable capture scenario. (**a**) Positions of the first six perimar passages are shown on the heliocentric orientation diagram. (**b**) Orbital pattern of the first 40 orbits are shown on this marocentric plot. The time for the 40 orbit scenario is ~13.6 earth years (~7.2 mars years). After 100 orbits the major axis of the orbit is beginning to decrease. Note that about 2.85 E26 J of energy is sufficient for capture for this set of orbital parameters. The total energy dissipated within both bodies during the 40 orbits is 3.50 E26 J. (**c**) Close-up view of the pattern of close encounters. In this scenario the initial encounter of the sequence is the closest encounter. (**d**) Tidal amplitude plot for the first 8 earth years (~4.25 martian years). Note that the tidal amplitude for PP-1 is ~18 km (~2× that of a 0.1 moon-mass planetoid)

the energy dissipation requirements are higher, the displacement Love number of a planetoid can be numerically lower for stable capture to occur (0.21 vs. 0.33). In general, larger planetoids are more easily captured than smaller planetoids (see discussion of Paradox # 2). The rock tides for a 0.2 moon-mass planetoid are about twice as high as for the 0.1 moon-mass case and the energy storage and dissipation in both bodies is about twice as high after adjusting for the lower value of the planetoid Love number necessary for capture.

The disadvantages of capturing the larger planetoid are that the circularization sequence occurs much more rapidly (*i.e.*, in about one-half the time) and, more importantly, the planet loses its rotational angular momentum over a much shorter period of time. The orbit circularization sequence would be similar to that in Fig. 4.13 but the result is a satellite orbit that is nearly spin-orbit synchronized near the end of the circularization sequence. Thus the martian month and sidereal day would be about equal. As briefly discussed before, long days and long nights are not friendly for life forms!

4.5 Summary

A prograde gravitational capture model looks favorable for the Earth-Moon system (Malcuit 2015). We are comfortable with a 24 hr day. The rock and ocean tidal amplitudes and ranges are moderate on planet Earth mainly because our satellite is in a near circular orbit at a comfortable distance of about 60 earth radii. One of the key features for our system is that the original (pre-capture) rotation rate was about 10 hr/day (MacDonald 1963). Furthermore, we inhabitants of Earth are lucky (or fortunate) to have our Moon in a prograde orbit because the geometry of stable capture zones for both earth-like and mars-like planets strongly favors capture into a retrograde orbit.

The case for prograde planetoid capture for planet Mars is a good bit different than for Earth. Mars was at a great disadvantage from the beginning because of its primordial prograde rotation rate of about 24.6 hr/day. As we have seen in this chapter, prograde capture scenarios for Mars slow the rotation rate of the planet to unfavorable values quickly. The tidal activity for the early phases of a prograde capture scenario for either the 0.1 or 0.2 moon-mass planetoid for Mars are favorable for the development of life forms but these favorable conditions are short-lived relative to the history of planet Mars.

In the next chapter we will examine the consequence of retrograde gravitational capture of both 0.1 and 0.2 moon-mass planetoids for planet Mars. Perhaps the reader will be surprised to know that RETROGRADE capture scenarios are much more favorable for developing habitable conditions for biota than are prograde capture scenarios. This conclusion is valid for most terrestrial exoplanets as well!

References Cited

Bostrom, R. C. (2000). *Tectonic consequences of Earth's rotation* (p. 266). Oxford: Cambridge University Press.

Castelvecchi, D. (2019). Japanese spacecraft probes asteroid's guts for first time. *Nature, 571*, 306–307.

Counselman, C. C. I. I. I. (1973). Outcomes of tidal evolution. *The Astrophysical Journal, 180*, 307–314.

dePater, I., & Lissauer, J. J. (2001). *Planetary sciences* (p. 528). Cambridge: Cambridge University Press.

Evans, N. W., & Tabachnik, S. (1999). Possible long-lived asteroid belts in the inner Solar System. *Nature, 399*, 41–43.

Evans, W. N., & Tabachnik, S. (2002). Structure of possible long-lived asteroids belts. *Monthly Notices of the Royal Astronomical Society, 333*, L1–L5.

Kaula, W. M., & Harris, A. W. (1973). Dynamically plausible hypotheses of lunar origin. *Nature, 245*, 367–369.

Leake, M. A., Chapman, C. R., Weidenschilling, S. J., Davis, D. R., & Greenberg, R. (1987). The chronology of Mercury's geological and geophysical evolution: The Vulcanoid hypothesis. *Icarus, 71*, 350–375.

Lissauer, J. J., Berman, A. F., Greenzweig, Y., & Kary, D. M. (1997). Accretion of mass and spin angular momentum by a planet on an eccentric orbit. *Icarus, 127*, 65–92.

Love, A. E. H. (1911). *Some problems in geodynamics* (p. 180). Cambridge: Cambridge University Press, (reprinted by Dover, 1967).

Love, A. E. H. (1927). *A treatise on the mathematical theory of elasticity* (4th ed., p. 643). Cambridge: Cambridge University Press.

MacDonald, G. J. F. (1963). The internal constitutions of the inner planets and the Moon. *Space Science Reviews, 2*, 473–557.

MacDonald, G. J. F. (1964). Tidal friction. *Reviews of Geophysics, 2*, 467–541.

Malcuit, R. J. (2015). *The twin sister planets, Venus and Earth: Why are they so different?* (p. 401). Cham: Springer International Publishing.

Malcuit, R. J., & Winters, R. R. (1996). Geometry of stable capture zones for planet Earth and implications for estimating the probability of stable gravitational capture of planetoids from heliocentric orbit. In *Abstracts Volume, 27th Lunar and Planetary Science Conference* (pp. 799–800). Houston: Lunar and Planetary Institute.

Malcuit, R. J., Mehringer, D. M., & Winters, R. R. (1988). Computer simulation of "intact" gravitational capture of a lunar-like body by an earth-like body. In *Abstracts Volume, 19th Lunar and Planetary Science Conference* (pp. 718–719). Houston: Lunar and Planetary Institute.

Malcuit, R. J., Mehringer, D. M., & Winters, R. R. (1989). Numerical simulation of gravitational capture of a lunar-like body by Earth. In *Proceedings of the 19th Lunar and Planetary Science Conference* (pp. 581–591). Houston: Lunar and Planetary Institute.

Malcuit, R. J., Mehringer, D. M., & Winters, R. R. (1992). A gravitational capture origin for the Earth-Moon system: Implications for the early history of the Earth and Moon. In J. E. Glover & S. E. Ho (Eds.), *Proceedings Volume, 3rd International Archaean Symposium* (Vol. 22, pp. 223–0235). The University of Western Australia Publication.

McSween, H. Y., Jr. (1999). *Meteorites and their parent planets* (2nd ed., p. 310). Cambridge: Cambridge University Press.

Melchior, P. J. (1978). *The tides of planet Earth* (p. 609). New York: Pergamon Press.

Mitchikami, T., Honda, C., Miyamoto, H., & (and 33 more co-authors). (2019). Boulder size and shape distributions on asteroid Ryugu. *Icarus, 331*, 179–191.

Munk, W. H., & MacDonald, G. J. F. (1960). *The rotation of the Earth* (p. 323). Cambridge: Cambridge University Press.

Peale, S. J., & Cassen, P. (1978). Contributions of tidal dissipation to lunar thermal history. *Icarus, 36*, 245–269.

Ross, M., & Schubert, G. (1986). Tidal dissipation in a viscoelastic planet: Proceedings of the 16th Lunar and Planetary Science Conference. *Journal of Geophysical Research, 91*, D447–D452.

Roy, A. E. (1965). *The foundations of astrodynamics* (p. 385). New York: The Macmillan Company.

Singer, S. F. (1968). The origin of the Moon and geophysical consequences. *Geophysical Journal, Royal Astronomical Society, 15*, 205–226.

Singer, S. F. (1970). The origin of the Moon and its consequences. *Transactions, American Geophysical Union, 51*, 637–641.

Stacey, F. D. (1977). *Physics of the earth* (2nd ed., p. 414). Hoboken: Wiley.

Villanueva, G. L., Mumma, M. J., Novak, R. E., Kaufl, H. U., Hartogh, P., Encrenaz, T., Tokunaga, A., Khayat, A., & Smith, M. D. (2015). Strong water isotopic anomalies in the martian atmosphere: Probing current and ancient reservoirs. *Science, 348*, 218–221.

Ward, P., & Kirschvink, J. (2015). *A new history of life* (p. 391). New York: Bloomsbury Press.

Weidenschilling, S. J. (1978). Iron/silicate fractionation and the origin of Mercury. *Icarus, 35*, 99–111.

Winters, R. R., & Malcuit, R. J. (1977). The lunar capture hypothesis revisited. *The Moon, 17*, 353–358.

Chapter 5
A Retrograde Gravitational Capture Model for a Sizeable Satellite for Mars

> Among the more perplexing features of the Solar System's morphology is the scarcity of satellites in the inner Solar System: neither Mercury nor Venus seems to have satellites, the Earth has only one, and Mars is orbited merely by the tiny Phobos and Deimos.—From Burns (1973).

> When tides raised by a satellite dissipate energy in a spinning planet, the planet-satellite system evolves dynamically toward one of three possible final conditions. The satellite orbit may decay inward until the satellite is destroyed; the orbit may decay outward toward escape although at an ever decreasing rate; or the satellite's orbital and planet's spin periods may approach stable synchronism. Which of these outcomes of tidal evolution is approached depends on the satellite-planet mass ratio and on the initial spin and orbital angular velocities.—From Counselman III (1973).

The subject matter of this chapter relates to both of the quotes above. From the quote from Burns (1973) we get a feeling that the Earth-Moon system may be special. In my opinion, the Earth without the Moon would not be a very pleasant place about 4.6 billion years after its formation. In the second quote Counselman III (1973) gives us three possible final conditions. The first one (i.e., "the satellite orbit may decay inward until the satellite is destroyed") is "spot on" for this chapter.

5.1 Retrograde Gravitational Capture of a 0.1 Moon-Mass Satellite and Subsequent Orbit Circularization: An Attempt to Venoform Mars

We begin this chapter with a two-body analysis of retrograde capture and then we will proceed to three-body numerical simulations of capture of a 0.1 moon-mass planetoid into a retrograde orbit around a mars-mass planet in a mars-like heliocentric orbit. We also need to come up with a reasonable name for this theoretical retrograde satellite. I am proposing the name of Ubertas (a Roman goddess of

R. Malcuit, *Geoforming Mars*, https://doi.org/10.1007/978-3-030-58876-2_5

agriculture who personified fruitfulness of soil and plants and abundance in general) because a successful retrograde capture of a large satellite would result in long-term habitable conditions on a mars-like planet.

5.1.1 A Two-Body Analysis of Retrograde Capture

Gravitational capture of a moderate-sized planetoid can only occur when the orbit of the candidate planetoid is in a near mars-like heliocentric orbit. Figure 5.1 is a somewhat to scale sketch of what is involved in a retrograde capture process.

5.1.2 Post-Capture Orbit Circularization Process

Figure 5.2 shows a general view of the orbit circularization process. Under ideal conditions of no angular momentum transfer from the rotating planet and the planetoid in an elliptical orbit, most of the orbital energy of the post-capture elliptical orbit is deposited as thermal energy via tidal deformation processes in the bodies of the planet and planetoid. In this case the orbital energy decreases as the semi-major axis of the planetoid decreases but the angular momentum of the system does not change. Thus the angular momentum of the large elliptical orbit is the same as the 30 mars-

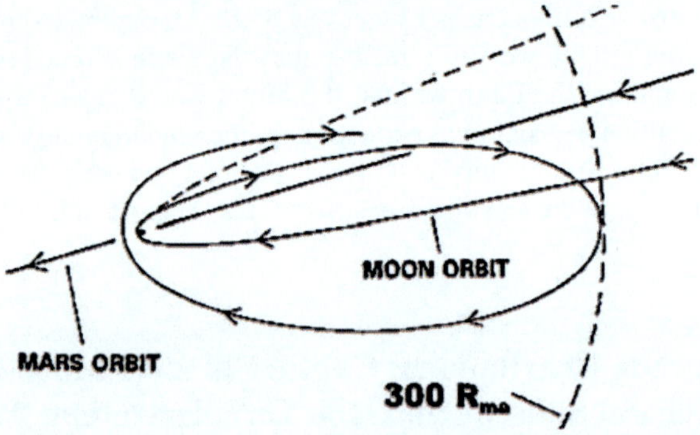

Fig. 5.1 A two-body representation of retrograde gravitational capture of a 0.1 moon-mass body by Mars. Both the planet and planetoid are in very similar, but not identical, prograde heliocentric orbits. The planetoid has a close encounter with the planet. If no energy is dissipated within the two bodies, then the planetoid departs from the planet along the *dashed line* and attains an orbit very similar to its pre-encounter orbit. If sufficient energy is dissipated in a combination of the two bodies, then the planetoid is inserted into a highly elliptical orbit that stays within the stability limit for the Sun-Mars-planetoid system [i.e., the Hill sphere for the system (Roy 1965)]. View is from the north pole of the Solar System

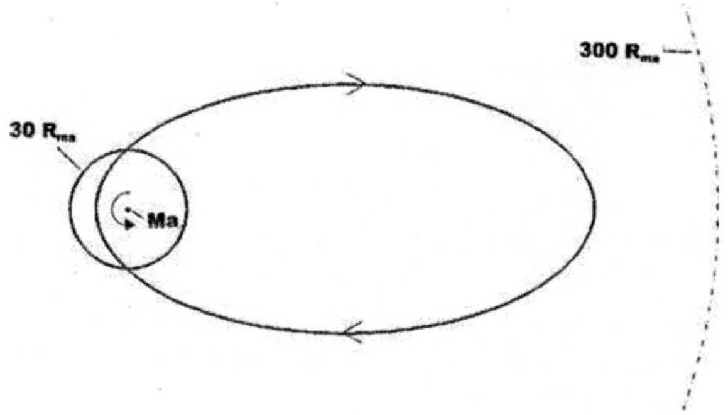

Fig. 5.2 Diagram illustrating the general scheme of the orbit circularization process. We start with the largest retrograde post-capture orbit possible, semi-major axis = 126.9 mars radii. This large major axis orbit then undergoes progressive circularization via tidal energy dissipation in a combination of the planetoid and Mars. If no angular momentum is imparted to the orbit from Mars' rotation, then the large orbit evolves into a 30 mars radii circular orbit (which is the case shown in this figure). In the more realistic case, angular momentum is transferred to the planetoid orbit as Mars loses rotational angular momentum (the angular momentum vector of the orbit is negative relative to the positive angular momentum vector of the spin of the planet) and the final circular orbit has a semi-major axis somewhat smaller than the 30 mars radii shown on the diagram, the details depending in the body deformation properties of the planetoid and Mars. The results of a numerical simulation of the orbit circularization process is presented later in this chapter

radii circular orbit. In a more realistic model the prograde rotation rate of the mars-like planet decreases as the elliptical orbit of the newly captured satellite decreases.

5.1.3 Circular Orbit Evolution

Following the orbit circularization process, the circular retrograde orbit gradually decreases in major axis as the prograde rotation rate of the planet decreases. The major controls on the timescale of circular orbit evolution are the body deformation properties of the two interacting bodies: i. e. the displacement Love numbers and the specific dissipation factors of the tidally interacting bodies. Most of the energy dissipation for the circular orbit evolution is deposited in the crust and upper mantle portion of the planet.

5.2 Numerical Simulation of Retrograde Planetoid Capture for Mars and a 0.1 Moon-Mass Planetoid

In this section, some results of numerical simulations of retrograde encounters are presented in order to give the reader a better understanding of the retrograde capture process.

5.2.1 A Sequence of Orbital Encounters Leading to Stable Retrograde Capture

Figure 5.3 shows the results of a simulation in which there is no energy dissipation within either body. The starting position as well as the ending position, relative to the martian orbit, are shown on the marocentric framework. The trajectory on the right of the marocentric diagram is the approach limb and that on the left is the departing limb of the encounter. Closest approach distance is 1.43 mars radii and there is no energy dissipated during the encounter. The planetoid returns to a mars-like heliocentric orbit essentially unchanged. During the first encounter, the planetoid comes within about 600 km of the surface of Mars. I should note that the first encounter is focused on the value of 1.43 mars radii in order to eliminate a variable in the calculation.

Figure 5.4 shows the results of a simulation in which there is significant energy dissipation within the interacting bodies and the planetoid returns for eight perimar passages before escaping back into a somewhat changed mars-like heliocentric orbit. The positions of the perimar passages are shown on the heliocentric orientation plot. Note that the eighth encounter of this sequence of encounters results in an escape in the prograde direction. Only about one-half of the energy for capture is dissipated during the initial encounter of the encounter sequence. After escape back into a mars-like heliocentric orbit, the planetoid could have future interactions with the mars-like planet.

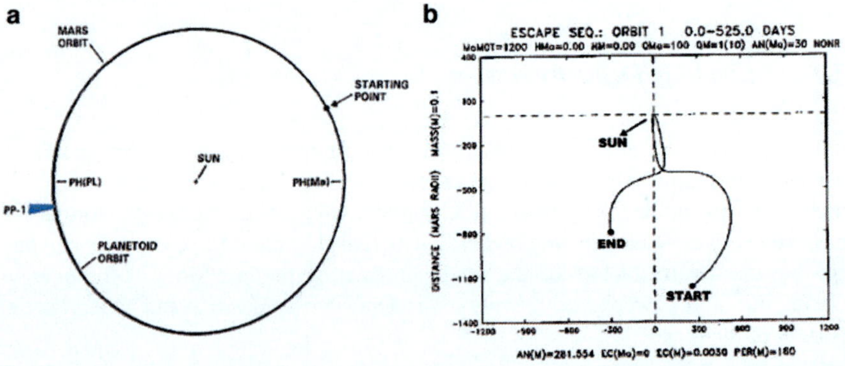

Fig. 5.3 Results of a close encounter in which no energy is dissipated. (**a**) Heliocentric orientation for the encounter. The simulation begins at the STARTING POINT. The close encounter occurs nearly one-half (0.43) of a martian year (295 earth days) later at PP-1. No energy is dissipated by tidal processes and the planetoid escaped back into a heliocentric orbit about 230 earth days after PP-1. (**b**) Marocentric orientation for the encounter. Mars is at the origin of the plot. The simulation begins at 1200 mars radii, which is well beyond the boundary of the Hill sphere (which is at about 300 mars radii)

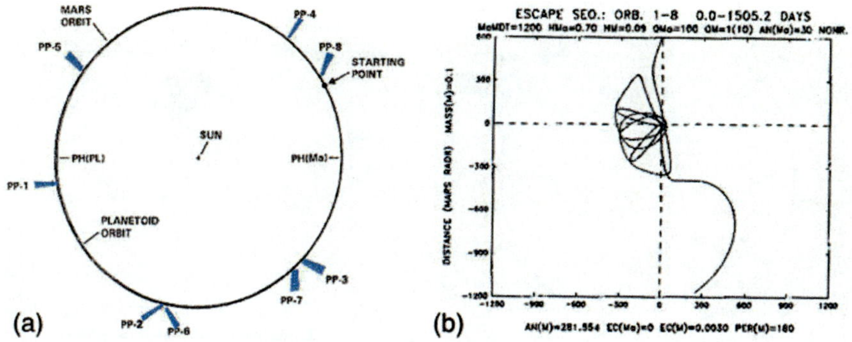

Fig. 5.4 Results of a 8-orbit escape scenario. (**a**) The heliocentric orientation shows the STARTING POINT and the positions of the eight close encounters of this encounter sequence. (**b**) The marocentric orientation shows the pattern of close encounters. Solar gravitational perturbations cause the variation in the orbital path. About one-half of the energy needed for retrograde capture is dissipated during the first encounter. Note that the first seven encounters are in the retrograde direction but that the eighth encounter (the encounter that results in escape into heliocentric orbit) is in the prograde direction

Figure 5.5 shows the results of a simulation in which there is sufficient energy dissipation for stable capture. The first 40 orbits of this stable retrograde capture scenario are shown but the calculation was extended to over 100 orbits and it showed no signs of instability. On the heliocentric orientation diagram the starting position for the scenario is the same as in the previous figures and the positions of the first five perimar passages are shown. Note than only 1.09 E26 J of energy dissipation was sufficient for stable capture for this particular set of co-planar orbital parameters. This is less than 40% of what was expected from the two-body analysis. The bottom line is that favorable orientation of the Sun during the timeframe of the initial encounter can reduce the energy dissipation requirements significantly.

Figure 5.5a shows the pattern of close encounters for this successful retrograde capture scenario (~13 earth years of orbits; ~7.4 mars years). Figure 5.5b shows the associated martian tidal amplitudes for the first 8 earth years (~4.25 mars years) of orbits. Equilibrium rock tides of up to 10 km occur on the planet during the initial encounter and during the seventh earth year there is an encounter that raises an equilibrium rock tide over 11 km. The rock tidal patterns for retrograde sequences generally are less regular than for prograde encounter sequences. Many simulations will give roughly similar results but each one is somewhat different. In some cases, just a slight change in input parameters can yield a much different result. The rock tides on the encountering planetoid are much higher than on the mars-like planet. Thus it is clear that in addition to the planetoid dissipating nearly all of the energy for its own capture, it is undergoing severe tidal distortions during the early stages of a stable capture scenario.

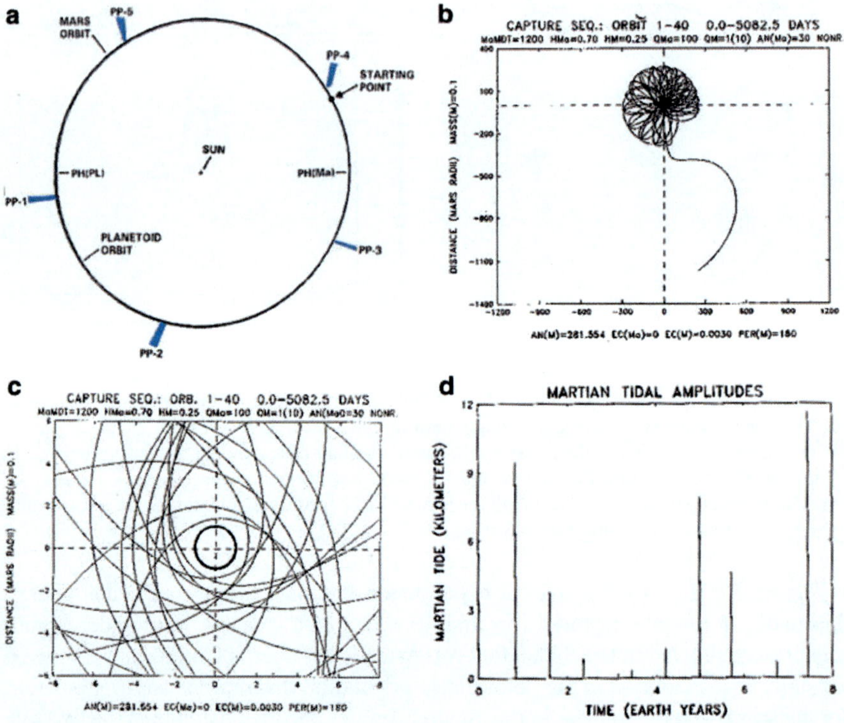

Fig. 5.5 Results of a stable retrograde capture scenario. (**a**) Position of the first five perimar encounters that occur in the first martian years are shown on the heliocentric orientation diagram. (**b**) Orbital pattern of the first 40 orbits (~13.9 earth years of orbits; ~7.4 mars years). (**c**) Expanded view of close encounters associated with orbits 1–40. (**d**) Pattern of tidal amplitudes for 8 earth years (~4.25 mars years) of orbits

5.2.2 *Geometry of Retrograde SCZs for Planetoids Being Captured by Mars*

The geometry of the two stable retrograde capture zones (SCZ) for planet Mars and a 0.1 moon-mass planetoid is shown in Fig. 5.6. The SCZ for a planetoid encountering Mars from an orbit that is slight larger than that of the planet (outside orbit) is centered in the diagram. The SCZ for a planetoid encountering Mars from an orbit that is slightly smaller than that of the satellite is shown as a "split" form on the diagram. The location of the parameters for the sequence of diagrams in Figs. 5.3, 5.4, and 5.5 is planet anomaly 30° and the eccentricity of the planetoid orbit is 0.003. This is near the left end of the inside SCZ.

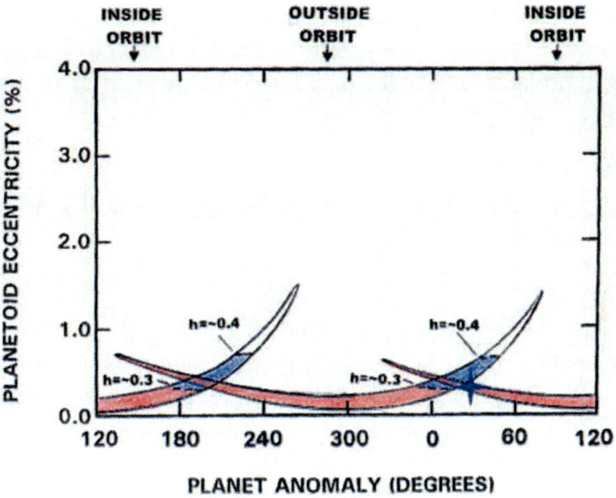

Fig. 5.6 Geometry of the retrograde stable capture zones for planet Mars for the case where the orbit of planet Mars has zero eccentricity. The four-pointed star marks the approximate location of the initial conditions for this sequence of encounters leading to stable retrograde capture

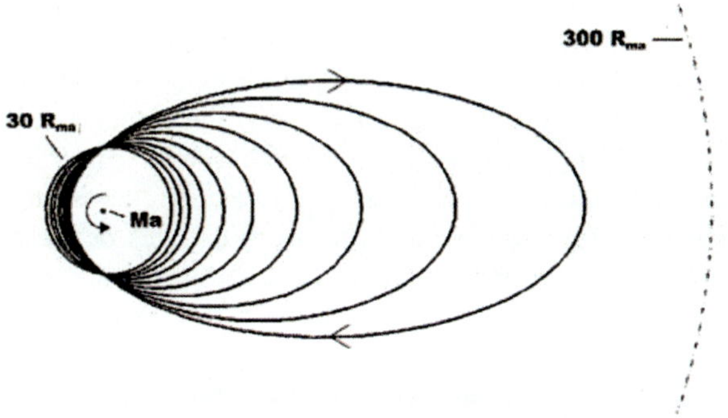

Fig. 5.7 Diagram showing the calculated time scale for retrograde orbit circularization. The input parameters are: h of Mars = 0.60, Q for Mars = 100, h for planetoid = 0.50, Q for planetoid = 1000. Solar gravitational perturbations would cause the energy dissipation to be increased and thus the time-scale to be shortened (probably to about 600 million years)

5.2.3 Post-Capture Orbit Circularization Era

Once the planetoid is captured into a stable retrograde orbit, the post-capture orbit begins a circularization sequence of events. Figure 5.7 shows the results of a numerical simulation of progressive circularization of the orbit. The program does a series of orbital calculations, then averages the results, resets the parameters, then calculates another series of orbital calculations and "adiabatically" calculates the orbital evolution. This calculation has none of the solar-induced orbital perturbations, so the time-scale of the calculation is longer than it should be. The calculated time scale is about 900 million years. The orbits in this diagram are spaced at 100 Ma intervals. My guesstimate of the time-scale after corrections for three-body interactions is about 600 Ma. During this time the orbital energy decreases from ~3.638 E26 joules to ~0.146 E26 joules. Thus about 3.490 E26 joules are dissipated in a combination of the two interacting bodies. Although over 90% of the energy is dissipated in the captured planetoid, the mars-like planet undergoes significant tidal deformation. This dynamic activity would probably result in crustal fracturing in the equatorial zone of the planet and possible subduction activity akin to plate tectonic activity on Earth.

5.2.4 *Circular Orbit Evolution for a 0.1 Moon-Mass Satellite in Retrograde Orbit*

Once the orbit is circularized then it undergoes a decrease in semi-major axis as the planetary rotation slows because of the cancellation of rotational angular momentum of the planet. The satellite orbital radius at the time of zero sidereal rotation rate of Mars is ~12 mars radii and the satellite is revolving in the retrograde direction. The rotation rate for Mars when all of the angular momentum is transferred to Mars rotation is ~21 hr/day retrograde. This spin-orbit scenario is very similar to that of planet Venus and its satellite Adonis as described in Malcuit (2015). The difference is that Mars is torqued up by rock tidal action to a reasonably high retrograde rotation rate, somewhat faster than its primordial prograde rotation rate. The similarity with the Venus-Adonis scenario is that the satellite breaks up in orbit near the Roche Limit for a solid body (about 1.5 mars radii) and the particles, large and small, coalesce with planet Mars. Although Mars is much smaller than Venus, this model predicts that such a scenario would result in a smoldering mess on the surface of Mars and a dense carbon dioxide-rich atmosphere like that of planet Venus. The difference is that a Mars-like body would lose the carbon dioxide atmosphere more rapidly because of the lower mass of planet Mars. Most, if not all, of the surface of the planet, in this scenario, would be covered by basaltic lava flows.

Although the circular orbit evolution for a 0.1 moon-mass planetoid is interesting, let us consider the case of capturing a 0.2 moon-mass body in a retrograde orbit by Mars. Although the results are similar, the timescale for orbital evolution is shorter and the final rotation rate for Mars is much more rapid in the retrograde direction.

5.3 Numerical Simulation of Retrograde Planetoid Capture for Mars and a 0.2 Moon-Mass Planetoid

This set of simulations of retrograde capture are very similar to the previous set of simulations. They have the same planet anomaly and planetoid eccentricity values and they plot in the same position on the SCZ plot in Fig. 5.6. There are, however, four distinct differences:

1. The displacement Love number of the planetoid is much lower for successful capture.
2. The rock and ocean tides on the planet are about twice as high and the cumulative effects on the surface environment of the mars-like planet are significantly greater.

3. The timescale for the circular orbit evolution is decreased by about one-half.
4. The final retrograde rotation rate for the planet is increased significantly over the 0.1 moon-mass case.

5.3.1 *A Sequence of Encounters Leading to Stable Retrograde Capture*

Figures 5.8, 5.9, 5.10, and 5.11 show a sequence of encounter plots resulting in a stable retrograde capture scenario for a 0.2 moon-mass planetoid by a mars-mass planet. The encounter simulation shown in Fig. 5.8 results in no energy dissipation

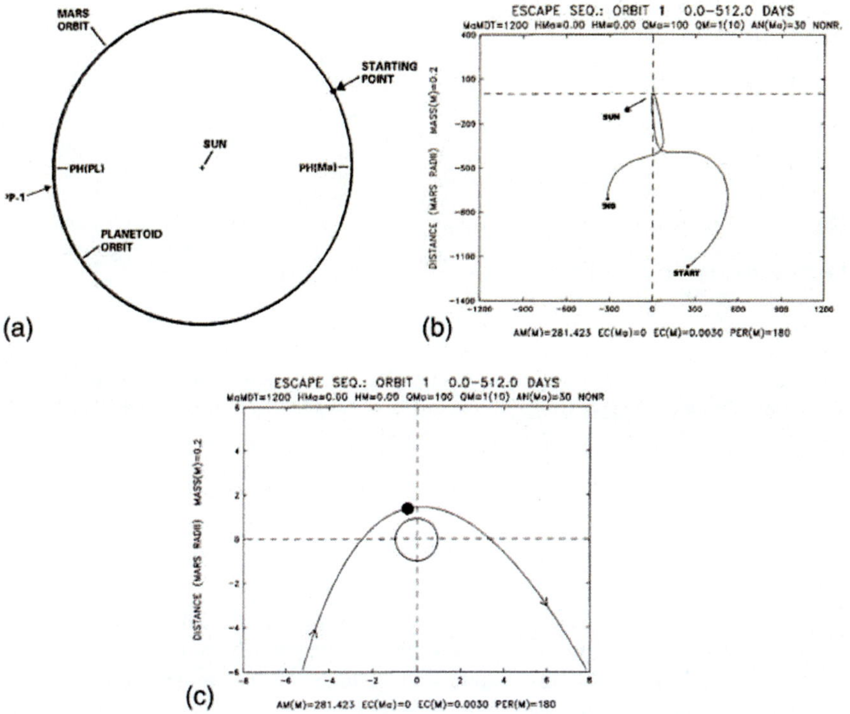

Fig. 5.8 Results of a close encounter in which no energy is dissipated. (**a**) Heliocentric orientation for the encounter. The simulation begins at the STARTING POINT. The close encounter occurs nearly one-half of a martian year (305 earth days) later at PP-1. No energy is dissipated by tidal processes and the planetoid escaped back into a heliocentric orbit about 207 earth days after PP-1. (**b**) Marocentric orientation for the encounter. Mars is at the origin of the plot. The simulation begins at 1200 mars radii, which is well beyond the boundary of the Hill sphere (which is at about 300 mars radii). (**c**) Close-up view of this non-capture, no energy dissipation encounter showing the relative diameters of the interacting bodies

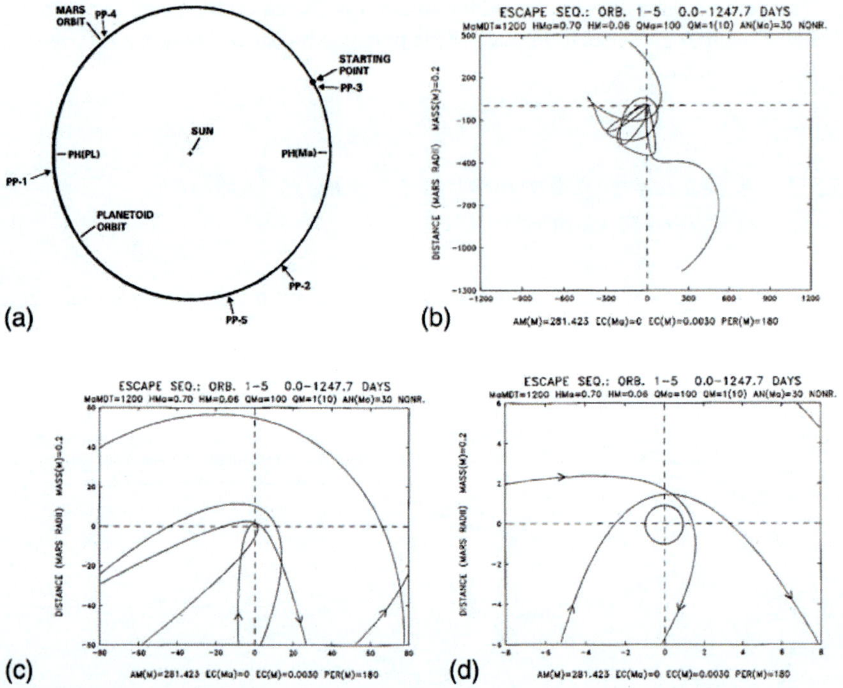

Fig. 5.9 Results of a 5-orbit escape scenario. (**a**) The heliocentric orientation shows the STARTING POINT and the positions of the five close encounters of this encounter sequence. (**b**) The marocentric orientation shows the pattern of close encounters. Solar gravitational perturbations cause the variation in the orbital path. About 40% of the energy needed for retrograde capture is dissipated during the first encounter. (**c**) Close-up of the close encounters illustrating that the fifth encounter is a distant prograde encounter that results in escape to a heliocentric orbit. (**d**) A very close-up view of two closest encounters of the encounter sequence

within the interacting bodies. Figure 5.9 shows an example of an encounter sequence in which enough energy is dissipated to bring the planetoid back for 5 encounters before it escapes back into heliocentric orbit. The encounter sequence in Fig. 5.10 results in a collision on the 19th orbit. Figure 5.11 shows a stable retrograde capture scenario as well as a plot of the tidal regime of Mars for 24 earth years (~12.8 martian years). Table 5.1 is a summary of encounter results with a systematic increase in h (displacement Love number) values for the 0.2 moon-mass planetoid.

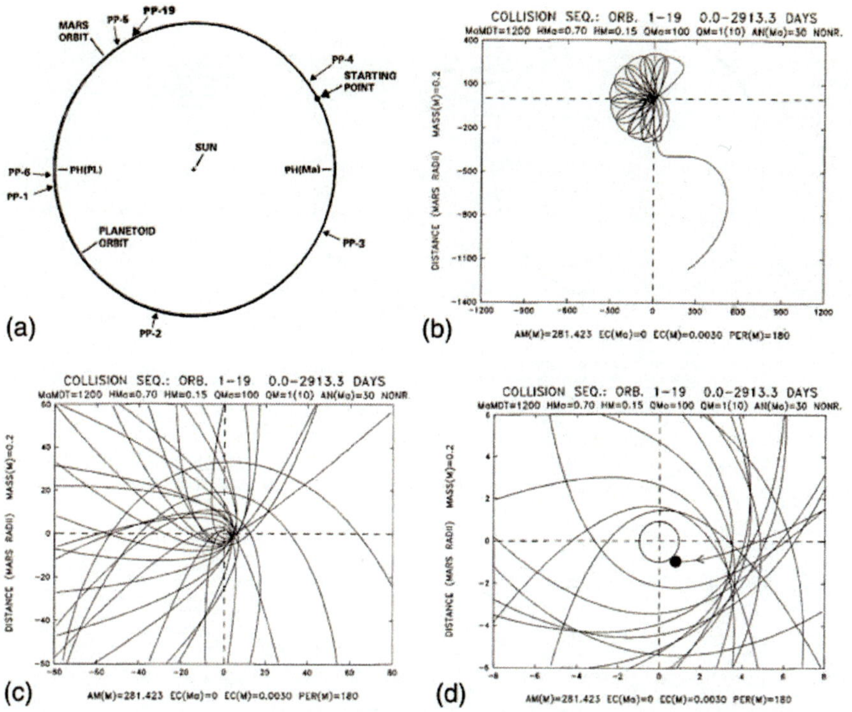

Fig. 5.10 Results of a 19-orbit encounter scenario that results in a collision. (**a**) The heliocentric orientation showing the positions of the first six encounters as well as the position of encounter 19, the collision encounter. (**b**) A marocentric orientation showing the general pattern of close encounters. About 94% of the energy necessary for capture is dissipated during the first encounter. (**c**) A closer view of the morphology of the 19 encounters. (**d**) An even closer view of the encounters and the collision of the planetoid with the mars-like planet in the retrograde direction. The perimar of the collision encounter is calculated to be at 0.67 mars radii (collision with no tidal deformation is at 1.30 mars radii). Thus the planetoid would probably be consumed by the mars-mass planet

5.3.2 *Post-Capture Orbit Circularization Era*

After a stable retrograde capture episode the highly elliptical orbit of Ubertas evolves, over a period of time, into a circular orbit as illustrated in Fig. 5.12.

5.3.2.1 The Orbit Circularization Process

Figure 5.13 shows the probable condition of the mars-like planet before the capture episode. The rock and ocean tidal regime on Mars would be very unusual soon after capture. There would be very high tidal amplitudes and ranges for a few days during perimar passages and then essentially no tidal effects for many days as the newly acquired satellite moves to the apomar position and makes its way back toward Mars. This tidal regime would persist for many millions of years until the orbit becomes circularized.

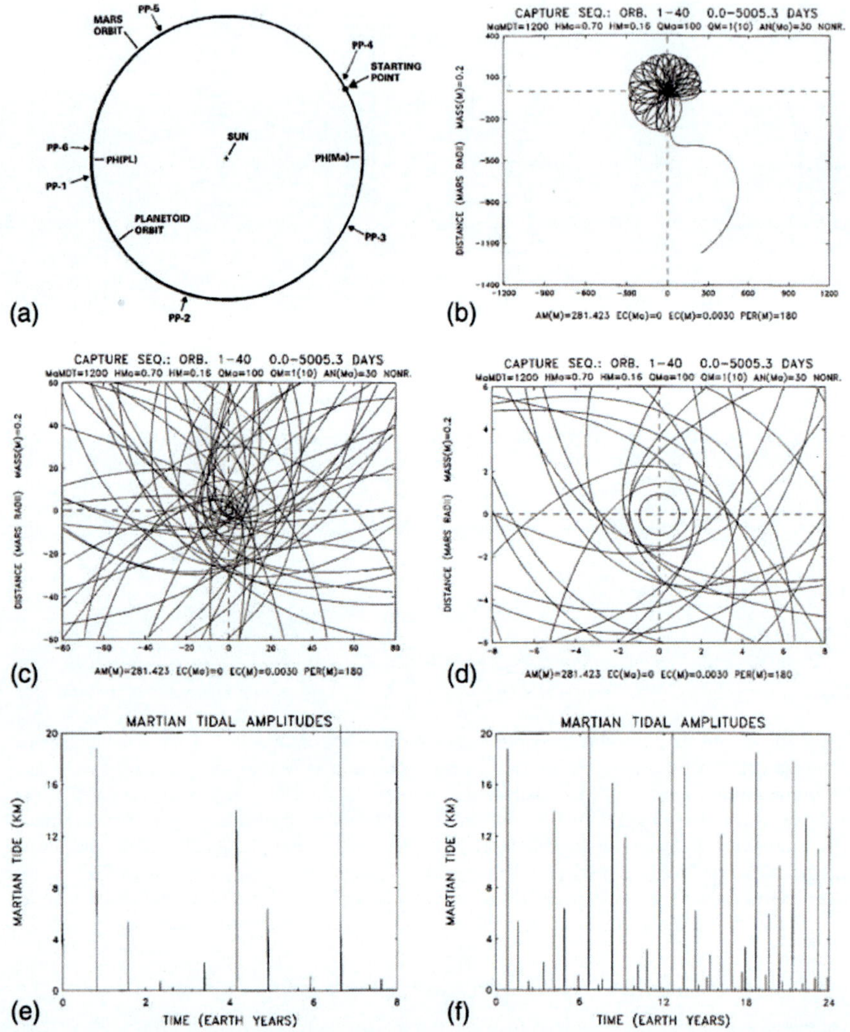

Fig. 5.11 Results of a stable capture scenario. (**a**) Position of the first five perimar encounters that occur in the first martian year are shown on this heliocentric orientation diagram. (**b**) Orbital pattern of the first 40 orbits (~13.7 earth years; ~7.3 martian years). (**c**) Close-up of the complex pattern of close encounters. (**d**) An even closer view of the close encounters relative to the surface of a mars-like planet. (**e**) Pattern of tidal amplitudes for the first 8 earth years of orbits (~4.3 martian years. (**f**) Pattern of the tidal amplitudes for the first 24 earth years of orbits (~12.8 martian years). Note that there are a few encounters that are closer than the initial encounter of the sequence

Since stable capture scenarios are broadly confined to the plane of the planets, most of the high-amplitude rock and ocean tidal activity is confined to the broadly defined equatorial zone. Table 5.2 gives values for equilibrium tidal amplitudes raised on a mars-like planet by a 0.2 moon-mass planet. (Note: The values would be about one-half as much for a 0.1 moon-mass planetoid.)

Table 5.1 Displacement Love number summary for simulations associated with the encounter parameters for the stable capture scenario in Fig. 5.11 (a somewhat typical scenario within a Retrograde Stable Capture Zone)

h of planetoid	Result	h of planetoid	Result
0.00	1-orbit escape	0.14	3-orbit collision
0.02	1-orbit escape	0.15	19-orbit collision
0.03	1-orbit escape	0.16	40-orbit capture
0.04	2-orbit escape	0.17	40-orbit capture
0.05	3-orbit escape	0.18	40-orbit capture
0.06	5-orbit escape	0.19	40-orbit capture
0.07	85-orbit collision	0.20	40-orbit capture
0.08	86-orbit collision	0.21	40-orbit capture
0.09	56-orbit collision	0.22	40-orbit capture
0.10	41-orbit collision	0.23	40-orbit capture
0.11	19-orbit collision	0.24	40-orbit capture
0.12	5-orbit collision	0.25	40-orbit capture
0.13	3-orbit collision	0.26	40-orbit capture

Orbital parameters are: Planet anomaly = 30 degrees; planetoid anomaly = 281.423 degrees; planetary orbit eccentricity = 0.0; planetoid orbit eccentricity = 0.003; pericenter radius for martian orbit = 0.0 degrees; pericenter radius for planetoid orbit = 180 degrees; perimar of the initial encounter = 1.43 mars radii retrograde; distance of separation at beginning of simulation = 1200 mars radii; orbital energy; displacement Love number for mars = 0.70; Q for mars =100, Q for planetoid = 1 for first encounter and 10 for all subsequent encounters. Note that all displacement Love numbers for the planetoid lower than 0.06 result in an escape and all of those between 0.07 and 0.15 result in a collision scenarios

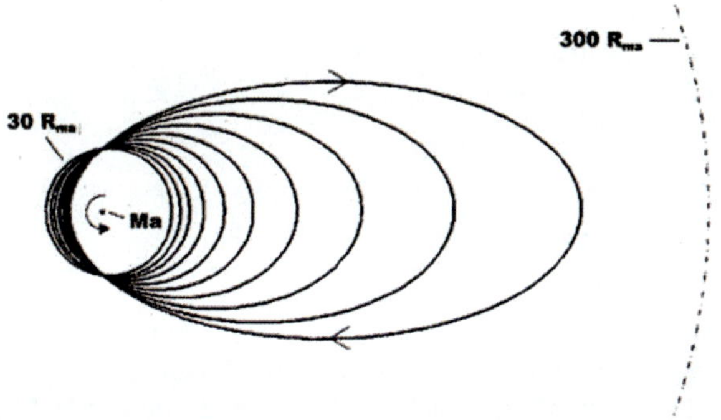

Fig. 5.12 Diagram showing general development of the post-capture orbit for a 0.2 moon-mass planetoid in a retrograde orbit. The diagram is the same as in Fig. 5.7 in that (1) the direction of revolution of the satellite is in the opposite direction to that of the planet but (2) the time scale for orbit circularization is much shorter. My guesstimate for the time scale to 10% eccentricity for the 0.1 moon-mass case was about 600 million years. For the 0.2 moon-mass planetoid, my estimate for the time scale to circularization is about 400 million years. Thus, there would be a significant increase in energy dissipation over this interval of time for the case of this larger planetoid

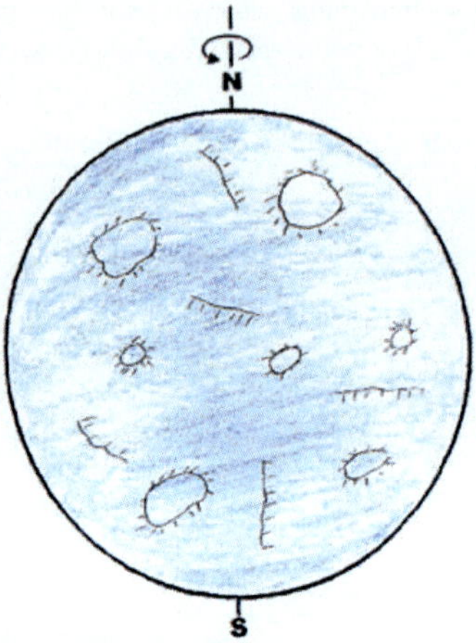

Fig. 5.13 Probable condition of a mars-like planet before the capture of a 0.2 moon-mass satellite (~4.0 Ga ago). After a hot accretion and core formation, a chill crust would form over the planet. Some of the early crust would founder but eventually a basaltic--andesitic crust with pockets of granitoids would form. At this time the mars-like planet would be very similar to planet Mars today (a one-plate planet with a "stagnant lid")

5.3.2.2 Sequence of Diagrams Illustrating Some of the Surface and Interior Effects Due to Post-Capture Orbital Circularization

Figure 5.14 depicts the probable interior and surface conditions of the mars-like planet soon after the stable retrograde capture episode. Table 5.2 gives values for tidal amplitudes for a mars-mass planet at various distances of separation of the planet and satellite.

5.3.3 Circular Orbit Evolution

Once the satellite orbit is circularized then the interacting bodies undergo a lengthy circular orbit evolution. Early on in the circular orbit evolution the tidal amplitudes are very low but as the orbit gets smaller and smaller, the tidal amplitudes get higher

Fig. 5.14 (continued) Bostrom (2000). View is from the north pole of the Solar System. (**b**) Schematic diagram showing the authors concept of the surface conditions of a mars-like planet soon after capture as well as the zonation of probable surface features (sans ocean water) on a mars-like planet. The equatorial zone of the planet would experience 20–10 km rock tides for a geologically short time (a few tens of thousands of years) and elevated tidal amplitudes for many millions of years. Since the very high tides effect mainly the equatorial zone, most of the recycling of the primitive crust would occur there and the polar zones would be only weakly effected. (**c**) The author's concept of surface conditions a few million years into the circularization sequence when the rock tides have decreased in amplitude

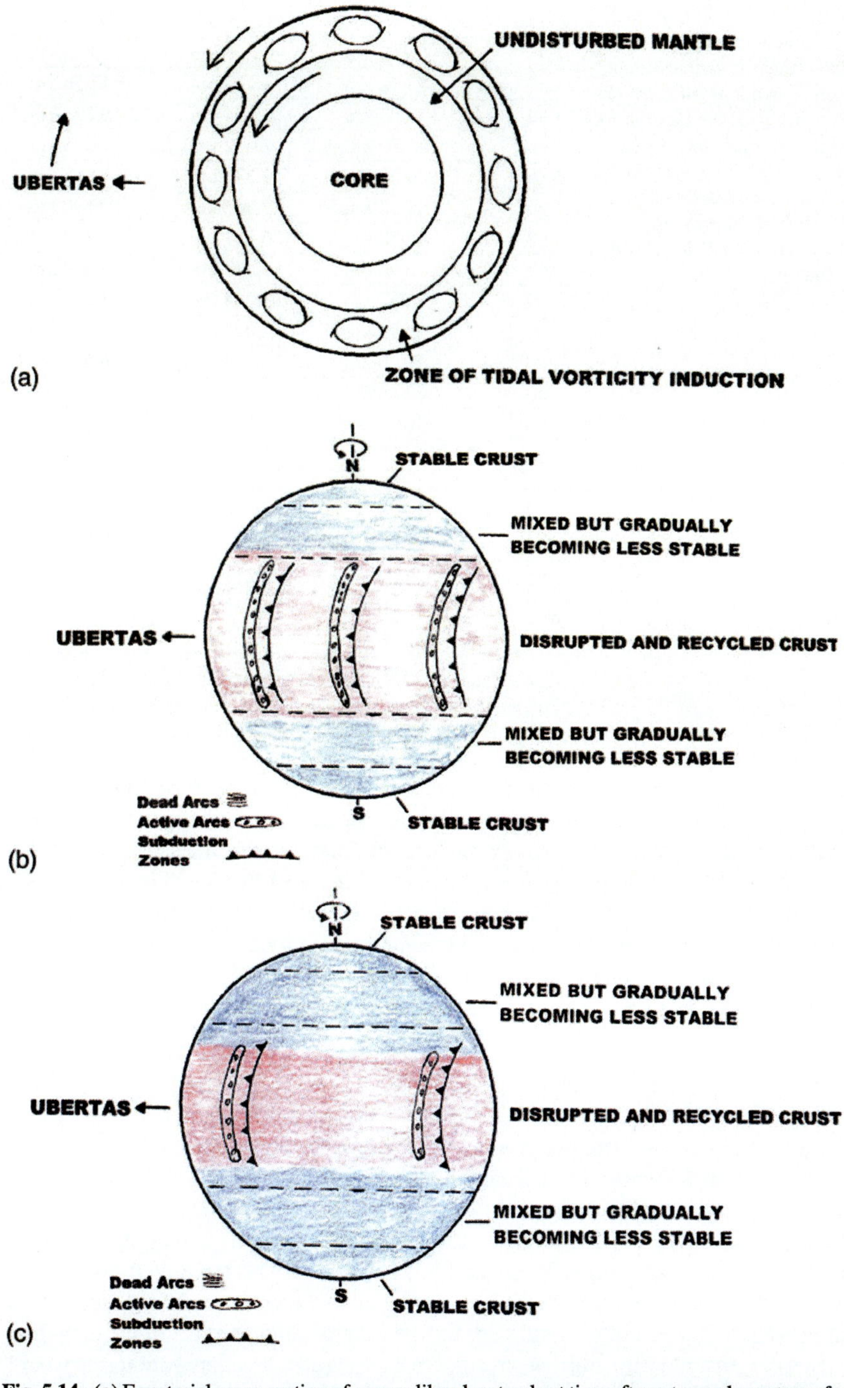

Fig. 5.14 (**a**) Equatorial cross-section of a mars-like planet a short time after retrograde capture of a 0.2 moon-mass satellite named UbertasUbertas. Note that the mantle circularization cells rotate in

Table 5.2 Equilibrium rock tidal amplitudes on a mars-like planet for various orbital radii for a 0.2 moon-mass satellite in a marocentric orbit for various displacement Love numbers. These values can be used for the analysis of both the orbit circularization sequence and the circular orbit evolution

r_p (R_{ma})	$h_{ma} = 1.0$	$h_{ma} = 0.7$	$h_{ma} = 0.5$
1.43	26.70 km	18.69 km	13.35 km
1.50	23.12 km	16.15 km	11.56 km
1.60	19.05 km	13.34 km	9.525 km
1.70	15.89 km	11.12 km	7.945 km
1.80	13.39 km	9.373 km	6.695 km
1.90	11.38 km	7.966 km	5.690 km
2.00	9.758 km	6.831 km	4.879 km
2.50	4.996 km	3.497 km	2.498 km
2.89	3.234 km	2.264 km	1.616 km
3.00	2.892 km	2.024 km	1.446 km
4.00	1.220 km	0.8540 km	0.6100 km
5.00	624.4 m	437.4 m	312.2 m
6.00	361.4 m	253.0 m	180.7 m
7.00	228.0 m	159.6 m	114.0 m
8.00	152.2 m	106.6 m	76.11 m
9.00	106.9 m	74.86 m	53.47 m
10.0	78.06 m	54.64 m	39.03 m
15.0	23.02 m	16.11 m	11.51 m
20.0	9.760 m	6.832 m	4.880 m
25.0	5.000 m	3.500 m	2.500 m
30.0	2.900 m	2.030 m	1.450 m

Note that polar zones are sheltered from severe tidal action

and higher and eventually, given enough time the tangential component of the rock tides as well as the energy dissipation within the planet reach super high values. Table 5.2 shows equilibrium rock tidal amplitudes on the mars-like planet for various orbital radii in the circularization sequence. Note that at 30 mars radii, the equilibrium rock tides with an h value of 0.7 is a mere 2.03 m, at 10 mars radii the value is over 50 m, and at 4 mars radii the equilibrium rock tide is near 1 km. Thus the circular orbit evolution is a self-accelerating process in time because the frequency of the tidal activity as well as the tidal amplitudes are monotonically increasing.

5.3.3.1 Sequence of Diagrams Showing the Possible Surface and Interior Effects During the Circular Orbit Era

Figures 5.15, 5.16, and 5.17 show a sequence of diagrams depicting placid conditions on the mars-like planet for about 3 billion years. This is the time following the orbit circularization era and before the severe tidal consequences of rapid contraction of the orbit of the satellite. During this era the prograde rotation rate of the mars-like planet decreases to zero and then changes to a retrograde direction. Life forms would

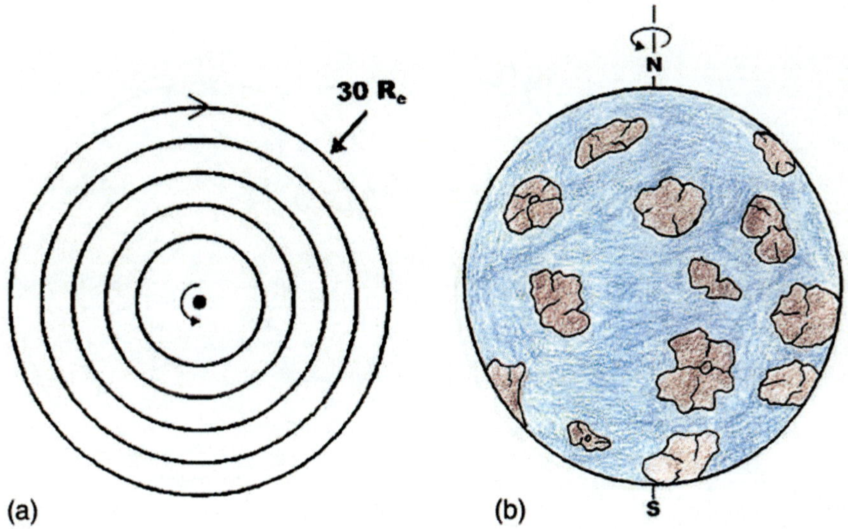

Fig. 5.15 (**a**) Scale sketch of the geometry of orbits for the early phase of the circular orbit era. View is from the north pole of the Solar System. (**b**) Hemispherical sketch of the probable surface conditions of the mars-like planet during this era with land masses and shallow seas. Equilibrium rock tidal amplitude on the mars-like planet ~2.0 m (ocean tidal amplitudes are about 1.4 times higher than the rock tidal amplitudes). Mars-like planet rotation rate ~ 24.6 hr/day prograde. Sidereal days/martian year ~ 670.2. Sidereal months/martian year ~60.7 with the satellite moving in the retrograde direction

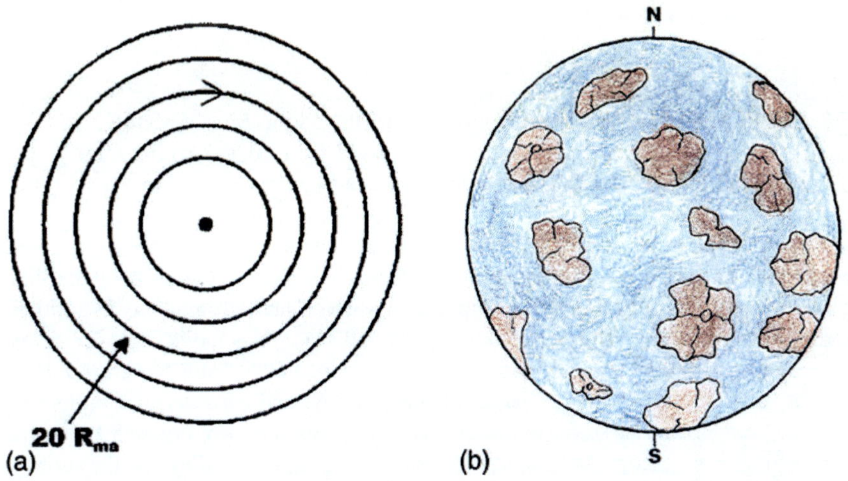

Fig. 5.16 (**a**) Scale sketch of the geometry of orbits for the intermediate phase of the circular orbit era when the satellite is at about 20 R_{ma}. View is from the north pole of the Solar System. (**b**) Hemispherical sketch of the probable surface conditions of the mars-like planet during this era (similar to those in Fig. 5.15). Equilibrium rock tidal amplitude on the mars-like planet ~11.7 m. Mars-like planet rotation rate is ~800 hr/day prograde. Sidereal days/martian year ~20.6. Sidereal months/martian year ~112.6 retrograde. Note that the martian planetary rotation rate changes from prograde to retrograde when the satellite radius is at about 20 mars radii. Also note that there are no planet rotation arrows in (**b**)

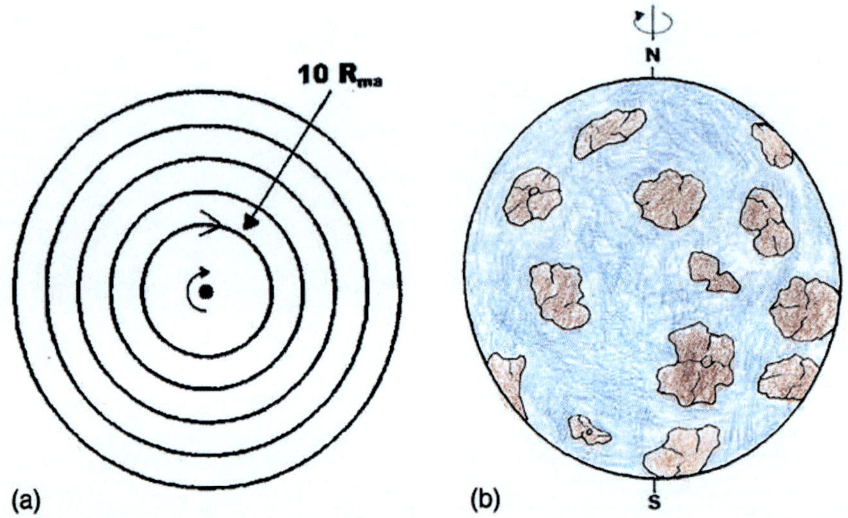

Fig. 5.17 (**a**) Scale sketch of the geometry of orbits for the late phase of the circular orbit era. View is from the north pole of the Solar System. (**b**) Hemispherical sketch of the probable surface conditions of the mars-like planet during this era. Equilibrium rock tidal amplitude on the mars-like planet ~54.6 m. Mars-like planet rotation rate ~ 20 hr/day retrograde. Sidereal days/martian year ~824.4. Sidereal months/martian year ~312.3 retrograde. There could be some limited subduction occurring in the equatorial zone at this phase of the orbital evolution

have a difficult time in the equatorial zone during this era but perhaps they could survive in the polar regions.

5.3.3.2 A Model for the Final Demise of the 0.2 Moon-Mass Satellite, Ubertas, from the Roche Limit to Solid-Body Breakup in Orbit and Eventual Coalescence with Planet Mars

Figure 5.18 shows a scale orbital diagram for the late stages of orbital contraction phase for this retrograde orbit evolution scenario. The rotation rate of the mars-like planet changes from ~20 hr/day retrograde to ~8.4 hr/day retrograde as the orbital radius of the planetoid decreases from 10 mars radii to 2 mars radii. Table 5.3 gives values of martian tidal amplitudes for a martian h value of 0.7. Table 5.4 gives values for (a) satellite orbit radii, (b) length of martian day, and (c) sidereal months per martian year for various stages in this orbital contraction scenario. Figure 5.19 shows the major zones of action for this orbital scenario. An attempt to illustrate the surface changes for this orbital contraction scenario is in Figs. 5.20, 5.21, 5.22, 5.23, and 5.24. Note that as Ubertas gets closer the planet, the surface rocks are progressively subducted into the martian mantle and the planet is progressively resurfaced with younger igneous rocks from the martian mantle.

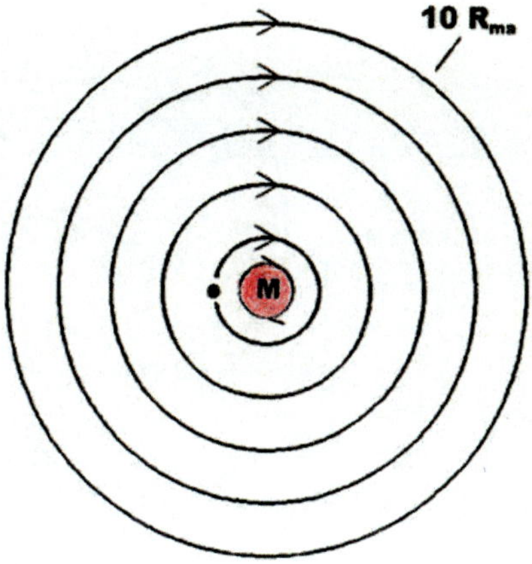

Fig. 5.18 Scale diagram illustrating the geometry of the "rapid" contraction phase of this retrograde orbital evolution scenario. Note that the planet is rotating in the retrograde direction and the planetoid is revolving about the planet in the retrograde direction. The angular momentum is being gradually transferred to the planet by way of the rock tidal action

Table 5.3 Equilibrium rock tidal amplitudes on Mars for various orbital radii for a 0.2 moon-mass satellite in an evolving retrograde orbit scenario ($h_{ma} = 0.7$)

Orbital radius (R_{ma})	Tidal amplitude (Mars)
30	2.03 m
25	3.50 m
20	6.83 m
15	16.1 m
10	54.6 m
8	107 m
6	253 m
4	854 m
2	6831 m

Table 5.4 Table showing values for satellite orbital radius, length of martian day, and sidereal months per martian year for planet mars and a 0.2 moon-mass satellite

Orbital radius (R_{ma})	Hours/Mars Day	Months/Mars Year
30	24.6	−60.7
25	46.0	−79.8
20	700	−112.6
15	−46.0	−171.2
10	−20.0	−312.3
8	−16.2	−440.3
6	−13.4	−680.2
4	−10.8	−1247.3
2	−9.1	−3528.5

To give the reader a more lucid concept of what happens during this late stage of orbital contraction, Fig. 5.19 shows two orientation diagrams: (1) an equatorial cross-sectional diagram of a mars-like planet and (2) a map view of one hemisphere of a mars-like planet. In Figs. 5.20, 5.21, 5.22, 5.23, and 5.24 we will be examining conditions at satellite orbital radii of 10, 8, 6, 4, and 2 mars radii, respectively.

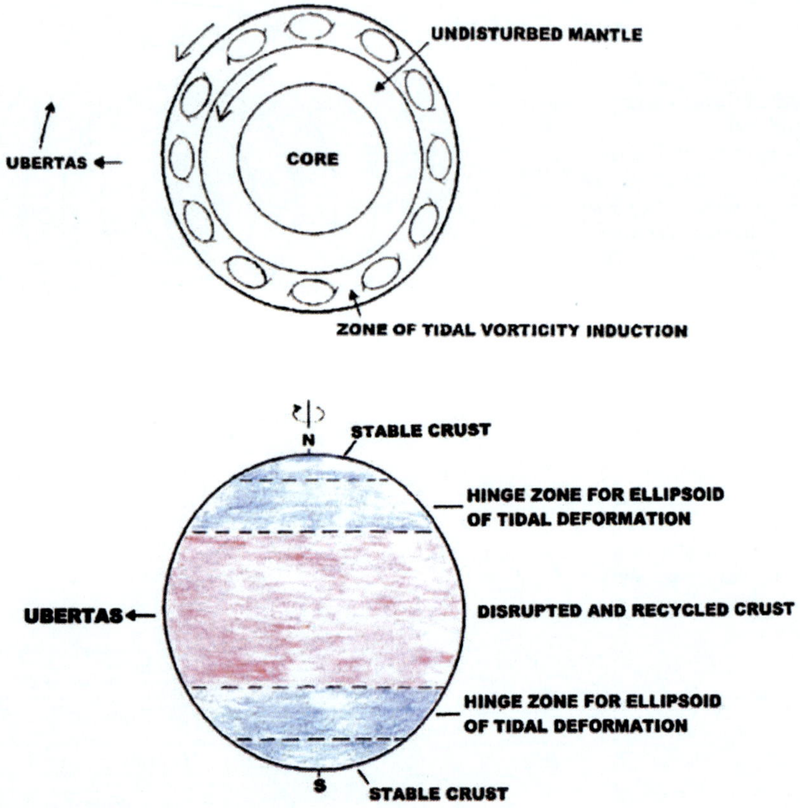

Fig. 5.19 (**a**) Equatorial cross-section of a mars-like planet showing the various zones of action during this orbital contraction scenario. Ubertas is in a retrograde orbit and is revolving at a faster rate than the planet is rotating. Thus, angular momentum is being transferred from the satellite orbit to the rotating planet via the rock tides. The crustal complex is rotating faster than the undisturbed mantle. If the core complex has an outer molten or partially molten zone, then a dipolar magnetic dynamo could be operating within the planet. The circulation cells in the "zone of tidal vorticity induction" are operating in the retrograde direction. (**b**) Hemispherical view of the surface of a mars-like planet showing the relative zones of crustal stability assuming that the obliquity (tilt angle) of the planet is fairly low. The polar zones experience the least rock tidal action, the hinge zone experiences more rock tidal action, and the equatorial zone is the zone of maximum rock tidal activity

Perhaps the reader has had enough global volcanism by this time in this orbital contraction scenario. BUT THE SHOW IS NOT OVER! Now we come to what I call "THE FINAL DEMISE SEQUENCE OF EVENTS OF THE SATELLITE". But it is not the Final Demise for the mars-like planet, it survives but the mantle of the planet is heavily degassed by all the rock tidal activity.

The next set of diagrams (Figs. 5.25, 5.26, 5.27, 5.28, 5.29, 5.30, 5.31, 5.32, 5.33, and 5.34) will give the reader some idea of some details of what happens to the

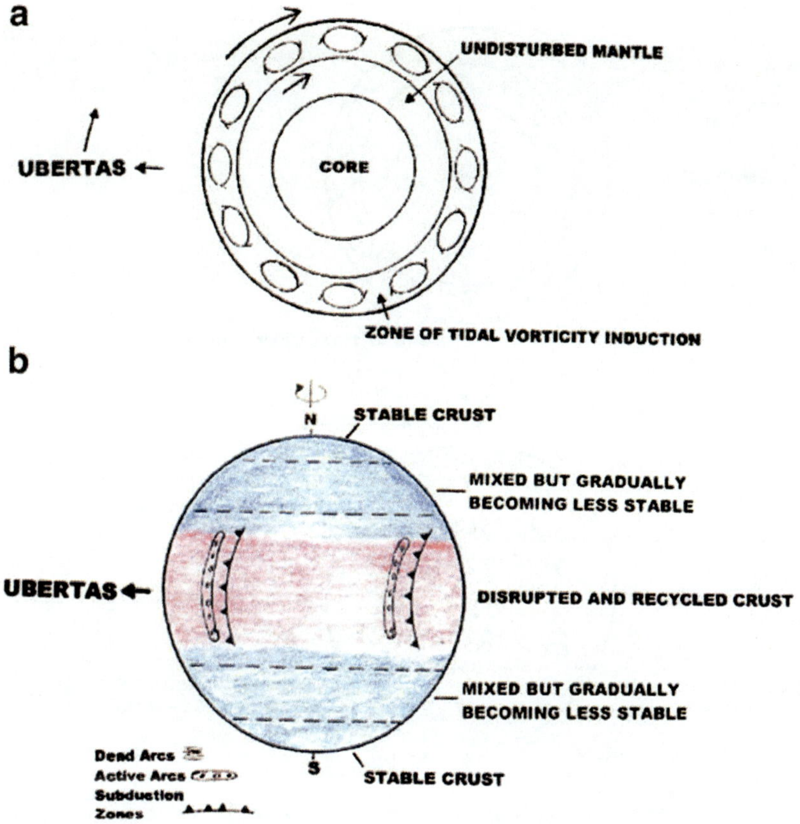

Fig. 5.20 Diagrams relating to the activities when Ubertas is in a circular orbit at 10 mars radii. Equilibrium tidal amplitude at this stage is 54.6 m. **(a)** View of the equatorial cross section of the planet. **(b)** Hemispherical surface view of the planet. Note that tidally induced volcanic arc activity is being initiated in the equatorial zone

satellite and the planet. But before we get into the details, let us discuss the concept of the Roche Limits for Satellites of Terrestrial Planets.

The CLASSICAL ROCHE LIMIT (Jeffrey 1947) is the radial distance at which a liquid (molten rock) satellite would become unstable because of stretching along the line of centers connecting the planet and the satellite. Inside the Roche Limit the satellite would separate into two or more masses. The Classical Roche Limit is located near 2.89 planet radii (~2.89 mars radii for our particular scenario).

The ROCHE LIMIT FOR A SOLID SATELLITE is the radial distance at which a solid body becomes unstable due to tidal deformation and begins to break up due to ductile faulting (cracking and crack propagation) into two or more parts (Jeffrey 1947; Aggarwal and Oberbeck 1974; Holsapple and Michel 2006, 2008). Any magma in a subcrustal magma zone within the satellite would facilitate the

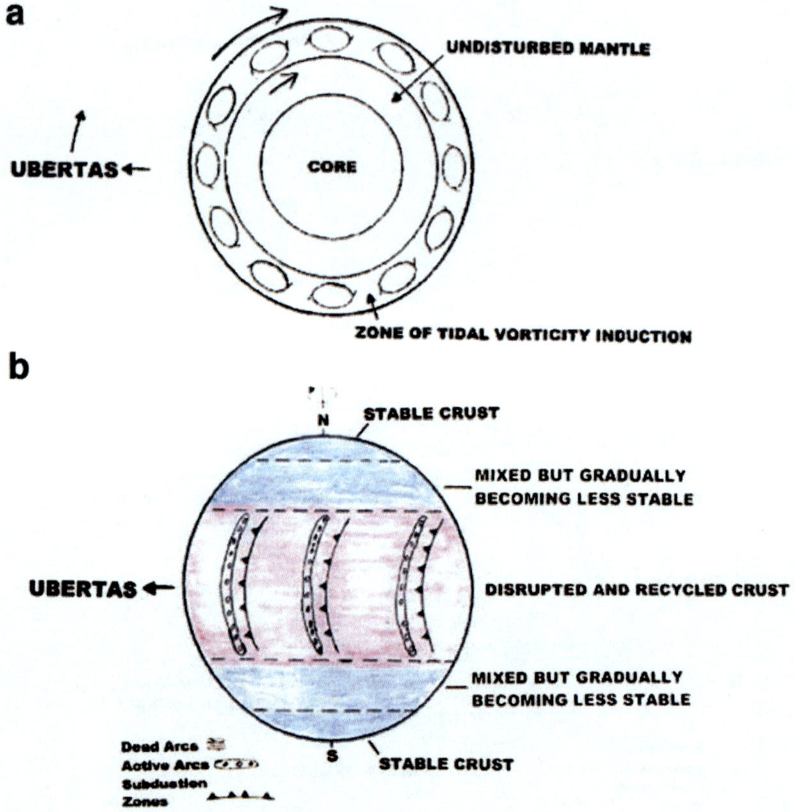

Fig. 5.21 Diagrams relating to the activities when Ubertas is in a circular orbit at 8 mars radii. Equilibrium tidal amplitude at this stage is 107 m. (**a**) View of the equatorial cross section of the planet. (**b**) Hemispherical surface view of the planet. Note that the zone of tidally induced volcanic arc activity is expanding

disintegration process. The Roche Limit for a Solid Body is located at about 1.6 planet radii (~1.6 mars radii for our particular scenario).

Next there is an era of chaotic orbital action around the mars-like planet. After the tripartite breakup of Ubertas, the parts would undergo mutual gravitational perturbations. These interactions consist of close encounters, grazing collisions, and occasional solid collisions. Because of the very low encounter velocities during collisions there would not be much vaporization but the collisions would produce many smaller (10's of meters to multi-kilometer-sized objects). Mutual gravitational interactions would lead to much scattering of the orbits of the particles so that they would probably occupy much of the retrograde orbit space. Some could be perturbed into prograde orbit space as well. Figure 5.31 may give the reader some idea of what this chaotic orbital motion would look like in two dimensions. Your imagination must put it in three dimensions.

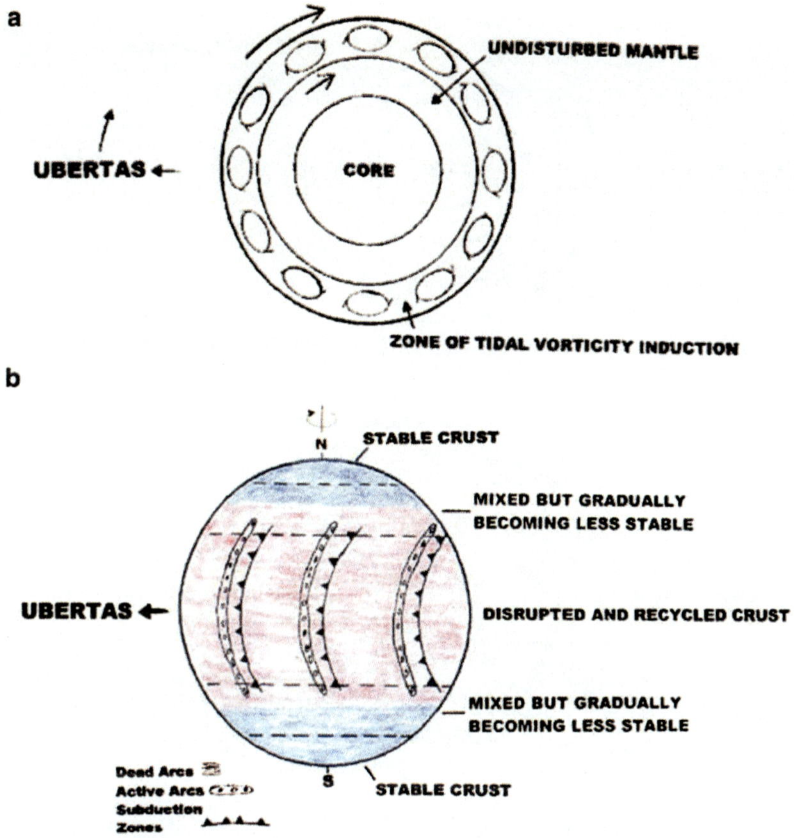

Fig. 5.22 Diagrams relating to the activities when Ubertas is in a circular orbit at 6 mars radii. Equilibrium tidal amplitude at this stage is 148 m. (**a**) View of the equatorial cross section of the planet. (**b**) Hemispherical surface view of the planet. Note that the zone of tidally induced volcanic arc activity is expanding. Most of the ocean water on the planet is transferred to atmosphere at this point in time

5.3.4 Summary of the Post-Capture Orbital Evolution of a 0.2 Moon-Mass Satellite (Ubertas) in Retrograde Orbit About a Mars-Like Planet in a Mars-Like Heliocentric Orbit

At 2.0 mars radii the equilibrium rock tidal amplitudes on the planet are over 6 km but the equilibrium tidal amplitudes on the satellite are about 12% of the radius of the satellite. The basaltic upper mantle of the planet is convecting but the tidal bulges on the satellite are essentially stationary and the rotational period of the satellite is synchronized with the orbital period of the satellite. Most of the energy dissipation for this system is due to the rock tidal action in the upper mantle of the planet.

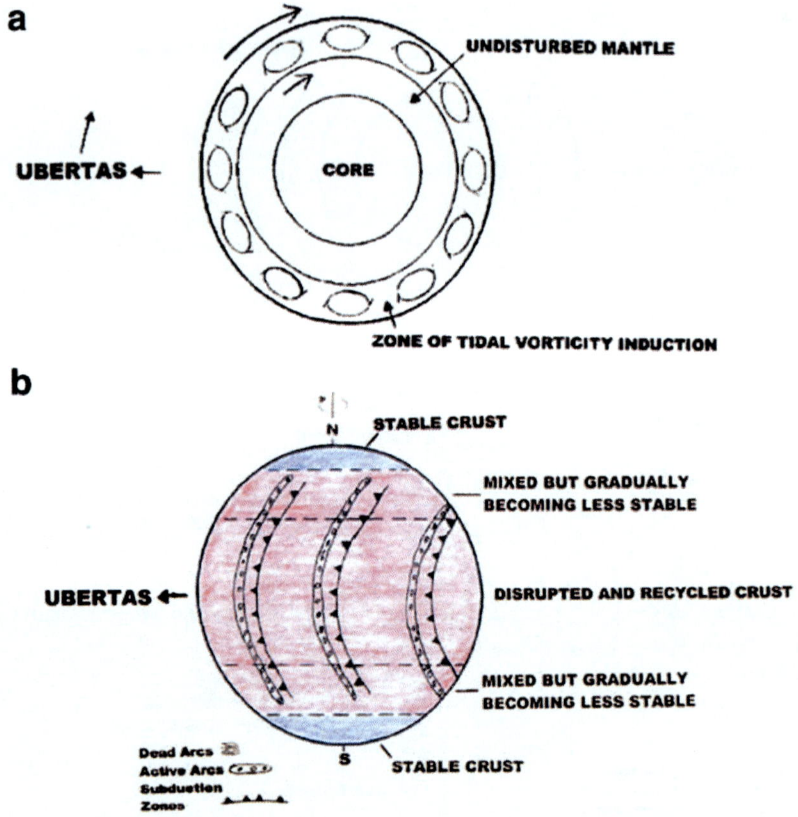

Fig. 5.23 Diagrams relating to the activities when Ubertas is in a circular orbit at 4 mars radii. Equilibrium tidal amplitude at this stage is 854 m. (**a**) View of the equatorial cross section of the planet. (**b**) Hemispherical surface view of the planet. Note that the zone of tidally induced volcanic arc activity is continually expanding

When Ubertas gets to 1.6 mars radii (see Fig. 5.27), then the satellite begins to break up in orbit. Since the satellite is in a semi-elastic state, the front-side crust separates from the backside crust and the core of the satellite is left behind. Now we have three separate orbiting satellites, each with a somewhat different orbital period. Eventually the orbiting bodies undergo some combination of grazing collisions and solid collisions and these collisions would generate many smaller bodies in orbits fairly close to the planet.

The orbits of the three larger pieces of Ubertas evolve via a combination of tidal interaction with the planet and gravitational perturbations due to interactions with the remaining large bodies. These larger bodies undergo oblique impacts onto the surface of the planet over a "brief" period of time in a marologic time framework (see Fig. 5.34). The smaller bodies, however, can remain in orbit for some time and

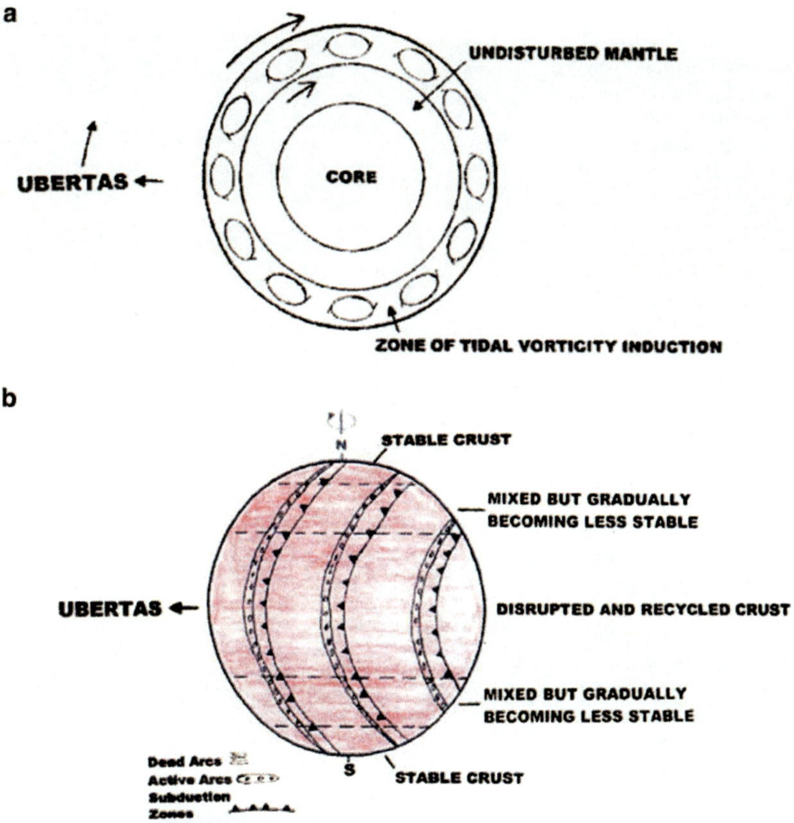

Fig. 5.24 Diagrams relating to the activities when Ubertas is in a circular orbit at 2 mars radii. Equilibrium tidal amplitude at this stage is 6831 m. (**a**) View of the equatorial cross section of the planet. (**b**) Hemispherical surface view of the planet. Note that the zone of tidally induced volcanic activity has expanded to engulf the entire planetary surface

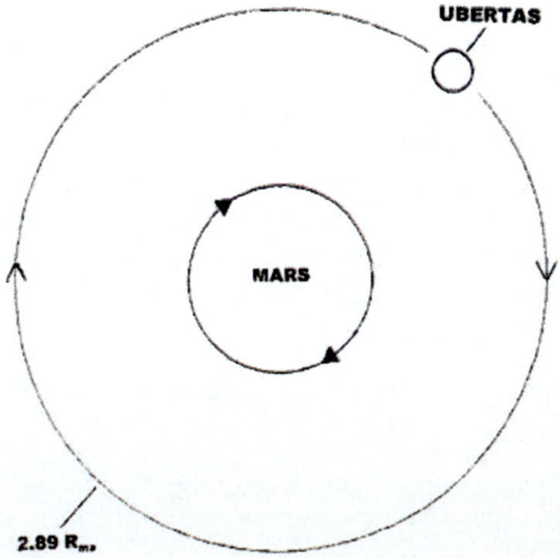

Fig. 5.25 Scale sketch of some features of the classical Roche limit for a Mars-Ubertas system. Mars rotation rate = 9.4 hr/day retrograde. There are 1753.2 days per martian year. Ubertas is orbiting the planet at about 2031.3 orbits per martian year in the retrograde direction in the equatorial plane of the planet. Equilibrium tidal deformation of Ubertas is =14.85 km [~1.5% of the radius ($h_{Ubertas} = 0.3$)]

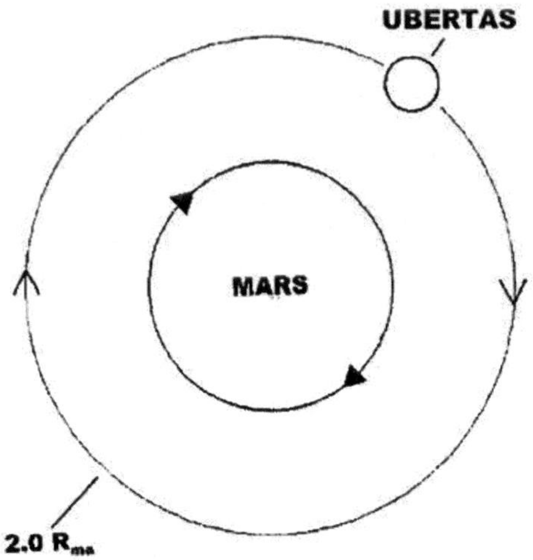

Fig. 5.26 Scale sketch of the Mars-Ubertas system between the Classical Roche Limit and the Roche Limit for a Solid Body. Mars rotation rate = 8.4 hr/day retrograde. There are 1961.9 days per martian year. Ubertas is orbiting Mars at about 3528.4 orbits per martian year in the retrograde direction in the equatorial plane of the planet. Equilibrium tidal deformation of Ubertas = 44.8 km [~4.4% of the radius ($h_{Ubertas}$ = 0.3)]

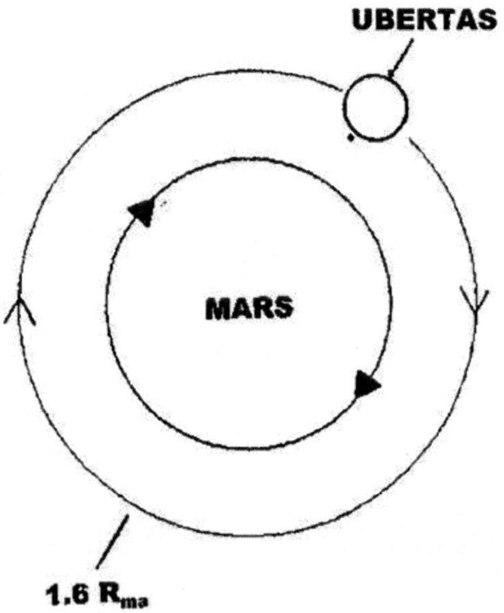

Fig. 5.27 Scale sketch of some features of Ubertas as it approaches the Roche Limit for a Solid Body for the Mars-Ubertas system. Mars rotation rate = 8 hr/day retrograde. There are 2060.0 days per martian year. Ubertas is orbiting Mars at about 4930.6 orbits per martian year in the retrograde direction in the equatorial plane of the planet. Equilibrium tidal deformation of Ubertas = 87.5 km [~8.59% of the radius (h = 0.3)]

would tend to impact at somewhat higher angles because of the gravitational perturbations and forced eccentricities caused by gravitational interactions with the larger bodies. The final scene is a convecting cauldron of basalt which then solidifies and cools and begins to register impacts of the remaining particles in orbit about the mars-like planet.

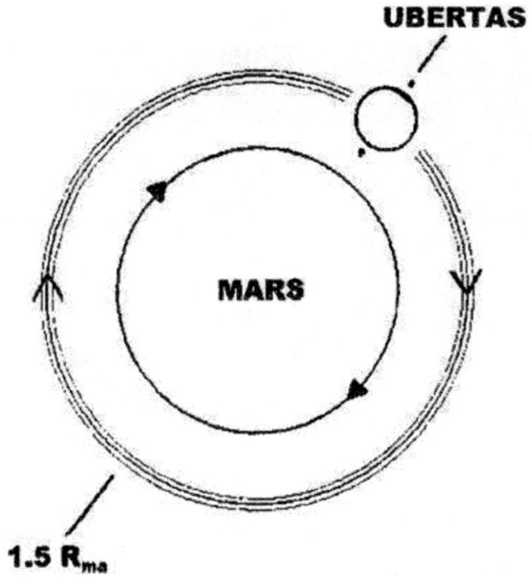

Fig. 5.28 Scale sketch of Ubertas slightly within the Roche Limit for a Solid Body in the Mars-Ubertas system. Mars rotation rate = 7.9 hr/day retrograde. There are 2086.1 days per martian year. Ubertas is orbiting Mars at about 5432.6 orbits per martian year in the retrograde direction in the equatorial plane of the planet. Equilibrium tidal deformation of Ubertas = 106.2 km [~10.4% of the radius ($h_{Ubertas} = 0.3$)]. Note that the dots along the line of centers show the approximate maximum tidal deformation of Ubertas

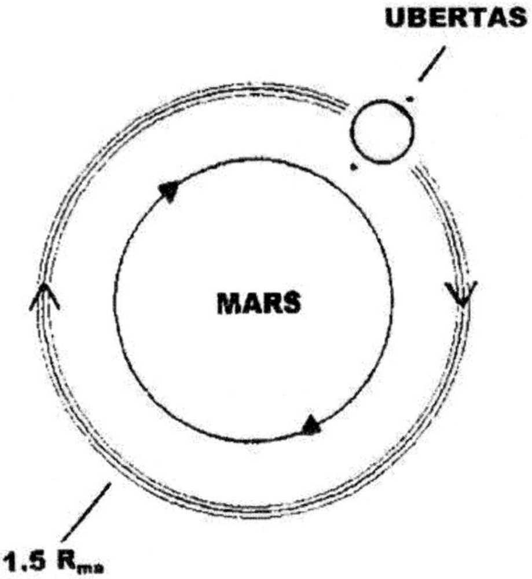

Fig. 5.29 Scale sketch of Ubertas slightly within the Roche Limit for a Solid Body in the Mars-Ubertas system. Mars rotation rate = 7.9 hr/day (retrograde). There are 2086.1 martian days per martian year. Ubertas is orbiting Mars at about 5432.6 orbits per martian year in the retrograde direction in the equatorial plane of the planet. Equilibrium tidal deformation of Ubertas = 176.0 km [~17.4% of the radius ($h_{Ubertas} = 0.5$)]. Note that the dots along the line of centers show the approximate maximum tidal deformation

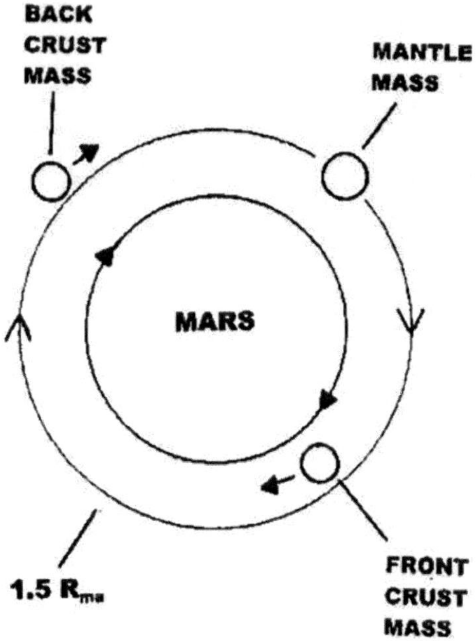

Fig. 5.30 Scale sketch of the initial breakup of Ubertas in the equatorial plane of Mars. The Front-Side Crust and the Back-Side Crust separate from the Mantle Complex and all three go into their individual retrograde orbits. Mars rotation rate = 7.9 hr/day (retrograde). There are 2086.1 martian days per martian year. The mantle mass is orbiting Mars at about 5432.6 orbits per martian year in the retrograde direction in the equatorial zone of the planet. The Front-Side Crustal Complex orbits the planet at a greater speed than the Mantle Mass; the Back-Side Crustal Complex orbits the planet more slowly than the Mantle Mass

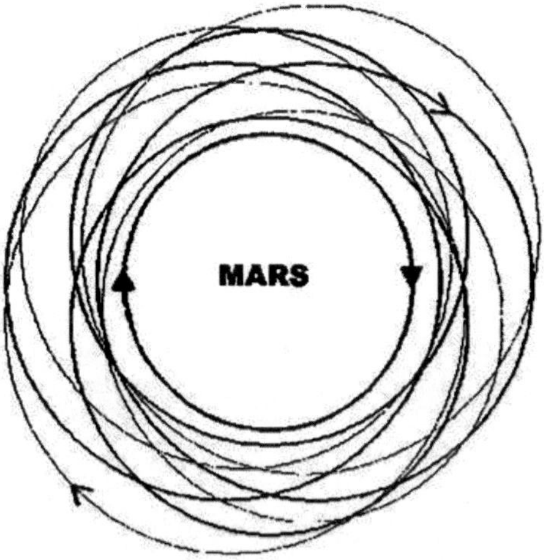

Fig. 5.31 Scale sketch of a two-dimensional (simplified) version of a swarm of orbiting objects in the equatorial plane about planet Mars. Mars rotation rate = 7.9 hr/day retrograde. Mantle and crustal "chips" from Ubertas are in an array of orbits about planet Mars. In this schematic diagram the darker lines represent mantle parts and the lighter lines represent crustal parts or fragments. Most, if not all, of these fragments are eventually perturbed so that they enter the dense atmosphere of Mars and impact onto the basaltic surface of Mars

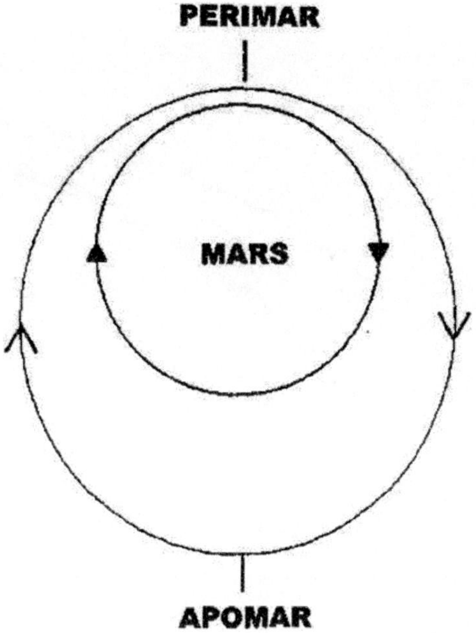

Fig. 5.32 Scale sketch of a particle (mantle chip or crustal chip) in an elliptical orbit in the equatorial plane about Mars. Mars rotation rate = 7.9 hr/day retrograde. Speed at perimar ~3.5 km/sec; speed at apomar ~2.0 km/sec. Note that the impact speed for a particle falling in from infinity to the surface of Mars is ~5.0 km/sec

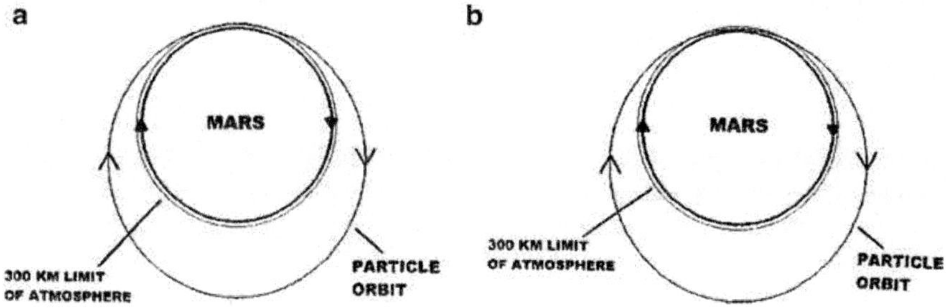

Fig. 5.33 Two scale sketches of particles in an elliptical orbit in the equatorial plane of Mars. Mars rotation rate for both diagrams = 7.9 hr/day retrograde. Speed of particle at perimar is about the same for both particles at ~3.0 km/sec. Major axis for both particles is the same at 1.4 mars radii. Eccentricity of orbit in diagram on left = 0.25; eccentricity of orbit in diagram on right = 0.26. Thus, the particle on the trajectory on the diagram on the right experiences much more atmospheric drag than the one in the left diagram

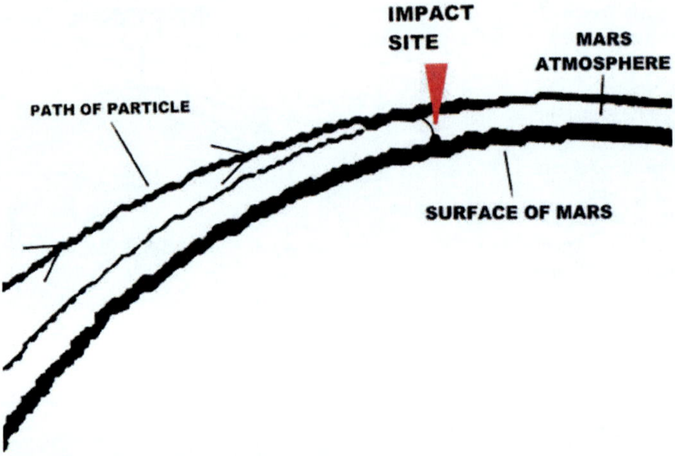

Fig. 5.34 Scale sketch of the impact side of a particle after entering the dense atmosphere of Mars. Since this Martian atmosphere would be about 80 times as dense as Earth's atmosphere at present, the particle would begin to decelerate and burn at about 200 km above the surface and would probably be falling vertically at the time of impact

The final result is a mars-like planet the surface of which looks very similar to that of planet Venus today!

5.4 Summary

In this chapter we explored the merits of a retrograde gravitational capture model as a possibility for the development of habitable conditions for biota for some length of marologic time on a mars-like planet in a mars-like heliocentric orbit. A two-body model suggests that gravitational capture into a highly elliptical retrograde orbit is physically possible if the satellite is in a physical state to store the energy for capture and to subsequently dissipate a high percentage of this stored energy in essentially one encounter (i.e., the capture encounter). Numerical simulations of 0.1 and 0.2 moon-mass satellites suggest that the larger satellite (the 0.2 moon-mass satellite) has a somewhat higher probability of capture because the displacement Love number (h) can be significantly lower for capture. (h for capture for 0.1 moon-mass planetoid is 0.25 and h for capture of a 0.2 moon-mass satellite is 0.16 for stable capture into a marocentric orbit). The remainder of this evolutionary scenario is for a 0.2 moon-mass satellite.

Following stable capture, the large major axis, high eccentricity marocentric orbit undergoes an orbit circularization sequence of events. Early on the orbits are strongly perturbed by solar influence causing several encounters to be close, energy dissipating encounters. The closer encounters cause the orbit to circularize fairly rapidly but then, as the orbit circularization progresses, the energy dissipation settles into a pattern more akin to a two-body orbit circularization sequence. After ~100 Ma or so the orbit becomes circularized to about 30 mars radii. Some crustal recycling via tidal vorticity induction in the mantle and resulting plate tectonic activity can occur during the circularization process. Much hydrothermal activity can occur also.

There is very little energy dissipation in the earlier phases of the circular orbit era. This era is characterized by stabilized crustal complexes, shallow seas, and low amplitude ocean tidal activity. Whatever life forms were present in the pre-capture era can survive the retrograde capture scenario and thrive during the circular orbit era. There can be up to 3 billion years of habitable conditions before the later phases of the circular orbit era make life on the mars-like planet uncomfortable because of the major tidal activity in the equatorial zone.

The other adverse factor affecting the biota is that the day length increases monotonically as the circular satellite orbit decreases in radius. The rotation rate of the planet changes from prograde to retrograde when the satellite orbit is at about 15 mars radii. The rock and ocean tidal activity would be present but the sun-lit days as well as the sun-less nights would get exceeding long. If there are a few 10s of degrees of obliquity (tilt angle), life forms could survive in the polar zones.

When the circular orbit decreases to about 10 mars radii, then tidally induced crustal recycling commences causing significant volcanism to occur. Degassing of the mantle via volcanism adds volatiles to the atmosphere. The retrograde rotation rate of the planet at 10 mars radii is ~20 hr/day and reasonable habitable conditions for biota have returned to the planet. Then, over the next few 100s of thousands of years the rock and ocean tidal amplitudes get higher and more frequent. As environmental conditions in the equatorial zone get unpleasant for biota, the organisms can migrate poleward into more favorable conditions.

As the satellite orbit moves ever more closely to the planet in a circular orbit, rock tidal activity gradually becomes very intense, the accumulation of greenhouse gases becomes greater, and the global atmospheric temperature reaches a point that leads to significant evaporation of ocean water to the atmosphere. As the biota move poleward seeking more habitable conditions, the satellite moves ever closer to the planet causing even higher rock and water tides. After the biota become extinct, the satellite eventually enters the Roche Limit for a Solid Body and breaks up in orbit by necking off two crustal complexes from the mantle complex. After a period of collisional diminution of the particles of the crustal and core complexes, the particles rain down through the dense atmosphere to impact on the surface of the planet. As the volcanic lava flows solidify they become crater counters. The final predictable result is a mars-mass planet with a rapid retrograde rotation rate and a dense greenhouse gas atmosphere with no satellite, no ocean water on the surface, and no life. In other words, this is a model for VENOFORMING MARS!

References

Aggarwal, H. R., & Oberbeck, V. R. (1974). Roche limit for a solid body. *Astrophysical Journal, 191*, 577–588.

Bostrom, R. C. (2000). *Tectonic consequences of the Earth's rotation* (p. 266). Oxford: Oxford University Press.

Burns, J. A. (1973). Where are the satellites of the inner planets? *Nature Physical Science, 242*, 23–25.

Counselman, C. C., III. (1973). Outcomes of tidal evolution. *The Astrophysical Journal, 180*, 307–314.

Holsapple, K. A., & Michel, P. (2006). Tidal disruptions: A continuum theory for solid bodies. *Icarus, 183*, 331–348.

Holsapple, K. A., & Michel, P. (2008). Tidal disruptions: II. A continuum theory for solid bodies with strength, with application to the Solar System. *Icarus, 193*, 283–301.

Jeffrey, H. (1947). The relation of cohesion to Roche's limit. *Monthly Notices of the Royal Astronomical Society, 3*, 260–262.

Malcuit, R. J. (2015). *The twin sister planets, Venus and Earth: Why are they so different?* (p. 401). Cham: Springer International Publishers.

Roy, A. E. (1965). *The foundations of astrodynamics* (p. 385). New York: The Macmillan Company.

Chapter 6
A History of a Ruling Paradigm in the Earth and Planetary Sciences That Guided Research for Three Decades: The Giant Impact Model for the Origin of the Moon and the Earth-Moon System

> To be successful any attempt to tell how and when the moon was born must be based on a true idea as to the origin of the solar system – an unsolved problem which is not only outside the scope of laboratory experiments but also, at critical points, beyond mathematical control, the second major aid to scientific progress. . . . . . There is only one other method left for examining the worth of the catastrophe hypothesis: to develop all promising guesses about a possible cause, with the hope of lighting on one that best explains the facts known about the moon itself, its relation to the earth, and the relation of both to the rest of the solar system.

Reginald A Daly 1946, Origin of the Moon and its Topography: Proceedings of the American Philosophical Society, v. 90, p. 118.

> We now turn our attention to major collisions between bodies of comparable size, and use as an example the major collision that may have been responsible for the formation of the Earth's Moon. The more dynamical aspects of this problem have previously been discussed by Cameron and Ward (1976) and Ward and Cameron (1978).
>
> Our original motivation for considering this problem came from asking how large a projectile would be needed to give the protoearth an angular momentum equal to that now possessed by the Earth-Moon system.
>
> This is nearly the mass of Mars.

Alistair G. W. Cameron 1983, Origin of the Atmospheres of the Terrestrial Planets: Icarus, v. 56, p. 198.

In 1975–1976, two independent groups proposed an alternative model. William Hartmann and Donald Davis (both at the Planetary Science Institute) suggested that the impact of a lunar-sized object with the early Earth had ejected into Earth orbit material from which the Moon then formed. If such material were derived primarily from the outer mantles of the colliding objects, then an iron-depleted moon might result. Alastair Cameron (now at the University of Arizona) and William Ward (now at the Southwest Research Institute) further recognized that if the collision had been a grazing one by a much larger, planet-sized impactor – one roughly the size of Mars, containing $\geq$10% of Earth's mass – the angular momentum delivered by the impact could account for Earth's rapid initial rotation. The concepts described in those researchers' works contain the basic elements of the now favored giant impact theory of lunar origin.

R. Malcuit, *Geoforming Mars*, https://doi.org/10.1007/978-3-030-58876-2_6

Robin M. Canup (2004a), Origin of the Terrestrial Planets and the Earth-Moon System: Physics Today, April 2004, p. 60.

The Giant-Impact Model got a firm foothold in the earth and planetary science community during the Kona Conference on THE ORIGIN OF THE MOON (13–16 October 1984). Since that time the promoters of the concept have obtained copious quantities of research funds at the federal level to pursue their work. These researchers have created, or supervised the creation of, some spectacular numerical simulations and from 1984 to the present many of these investigators have claimed to be working on a model that looks very promising for explaining the ORIGIN OF THE MOON AND THE EARTH-MOON SYSTEM as well as for the solution to other problems in the Planetary Sciences including the evolution of planet Mars and its satellites, Diemos and Phobos (Rosenblatt et al. 2016; Hyodo et al. 2017a, b; Canup and Salmon 2018).

In this chapter I am attempting a chronological development of the idea that a large body (mars-mass or larger) impacting on an earth-mass body can lead to the origin of a system that has the characteristics of the Earth-Moon system. However, this is not intended to be a comprehensive treatment of the topic. Much of the information in this chapter is in the form of direct quotes which I think is the most effective way to communicate the intent of the authors of the articles in journals and chapters in symposium volumes.

6.1 Part I: The GIANT-IMPACT MODEL for the Solution of a Number of Problems in the Earth and Planetary Sciences

The general procedure of the SCIENTIFIC METHOD for solving problems is presented in Chap. 1 and will not be repeated here. But we must start our discussion with a list of facts to be explained by a model for the origin and evolution of the Earth-Moon system, in this case the GIANT-IMPACT HYPOTHESIS (MODEL). Toward the end of this chapter we will examine again the list of facts to be explained in an evaluation of the model.

6.1.1 Items to Be Explained By a Successful Model for the Origin of the Moon and the Earth-Moon System

- The volatile element depletion of the Moon relative to the Earth
- The Potassium Index for solar system bodies
- Volatile element depletion patterns for solar system bodies
- Body density differences between the Earth and Moon

- Oxygen isotope information for Earth, Moon, meteorites, and other bodies in the inner solar system
- Lunar crust and mare rock dates
- The origin of lunar maria
- Temporal and spatial patterns of lunar mare rock distribution
- The asymmetry of lunar mass distribution
- Temporal and spatial patterns of lunar rock magnetization
- Geochemical, mineralogical, and petrological features of the Moon
- Geochemical, mineralogical, and petrological features of the Earth

6.1.2 A Generalized Chronological Development of a Giant-Impact Model for the Origin of the Earth-Moon System

For any major model in the world of science we must attempt to identify the beginning of the concept. Sometimes we are successful and other times we do not go back in history far enough.

6.1.2.1 In the Beginning: Pre-Kona Conference Work on the Giant Impact Concept

In recent times the promoters of the Giant-Impact Model refer to the article by Daly (1946) as the originator of the concept (Baldwin and Wilhelms 1992). Then articles by Hartmann and Davis (1975), Cameron and Ward (1976), Ward and Cameron (1978), and Cameron (1983) got the idea firmly established and the stage was set for a major effort to explain the origin of the Moon and the Origin of the Earth-Moon system by way of the mechanics of large bodies impacting on the primitive Earth.

R. A. Daly (1946) Origin of the Moon and its topography: Proceedings of the American Philosophical Society, v. 90, p. 104–119

From the INTRODUCTION:

> Where there is so much darkness every ray of light is welcome, and it was with sure instinct that Eduard Suess and Thomas Chrowder Chamberlin a half-century ago sought new light from cosmology and from comparison of earth and moon. These masters and others, later on, have been asking a multiple question: what relation have theories of the moon's origin to problems concerning: (1) terrestrial volcanism, igneous action in general; (2) the differentiation of sialic and simatic shells of the outer earth; (3) the theory of isostasy in explanation of the earth's relief – a theory implying mechanical contrast between a strong crust (the lithosphere) and an immediately underlying weak layer or shell (asthenosphere); (4) theories

of mountain-building; (5) the theory of the sealevel figure of the earth; and (6) the existence of continents standing shoulder-high above the floor of the deep ocean?

W. K. Hartmann and D. R. Davis (1975) Satellite-sized planetesimals and lunar origin: Icarus, v. 24, p. 504–515

From p. 512:

> Assuming that a large enough collision occurs after the Earth's core had formed or was forming, the ejected material would be already depleted in iron, as in the fission theory. Advantages of collision over fission are: (1) an energy source is provided to raise the material off the Earth, and (2) the theory is not purely evolutionary, depending on a chance encounter so that it does not require prediction of similar satellites for Mars or other planets.
>
> The material ejected into orbit forms a cloud of hot dust, rapidly depleted in volatiles. As shown by Soter (1971), the particles in such a swarm would interact and rapidly collapse into the equatorial plane, where a satellite could form. The evolution at this point resembles that postulated by Ringwood, except that an energy source is provided that does not necessarily apply to all planets.

From p. 513:

> This model can thus account for the iron depletion, refractory enrichment, and volatile depletion of the Moon, and at the same time account for the Moon's uniqueness; the Moon may have originated by a process that was likely to happen to one out of nine planets.

A. G. W. Cameron and W. R. Ward (1976) The origin of the Moon: Abstracts, Lunar Science Conference VII, p. 120–122

This is the first paragraph, p. 120:

> A key constraint on the origin of the Earth-Moon system is the abnormally large value of the specific angular momentum of the system compared to that of the other planets in the solar system. At an early stage, when the Moon was close to the Earth, most of the angular momentum resided in the spin of the Earth. This spin was presumably imparted by a collision with a major secondary body in the late stages of accumulation of the Earth, with the secondary body adding its mass to the remainder of the protoearth. The collisional velocity must have been close to 11 km/sec, and if the impact parameter was one earth radius, then the mass of the impacting body was comparable to that of Mars. It is probable that the largest accumulative collision should have involved a mass of this order, but the size and location of the impact parameter would have been a matter of chance. It is likely that both bodies would have been differentiated and possibly molten at the time of impact.

W. R. Ward and A. G. W. Cameron (1978) Disc evolution within the Roche limit: Abstracts, Lunar and Planetary Science Conference, p. 1205–1207

This is the first paragraph:

Two recent events have led to a marked interest in the behavior of circumplanetary discs and rings within or near the Roche limit: the surprising discovery of the rings of Uranus (1–3), and the detection of the azimuthal brightness asymmetry of Saturn's rings (4, 5). Circumplanetary discs have also been mentioned in connection with the development and evolution of certain satellite systems (6–10). Our understanding of the behavior of disc system in general has much improved due to the theoretical work of Lynden-Bell and Pringle on accretion discs (11). Cameron has applied some of the ideas to new models of the solar nebula (12) and Goldreich and Tremaine have derived similar evolutionary equations for low mass particulate discs such as Saturn's rings (13). This note concerns the evolution of more massive self-gravitating particulate discs lying interior to the Roche limit. Such a disc might arise either from the tidal decay and break-up of a pre-existing satellite (6–9) or as an impact and/or Roche zone capture by-product of planetary accretion (10, 14).

G. W. Cameron (1983) Origin of the atmospheres of the terrestrial planets: Icarus, v. 56, p. 195–201

From the ABSTRACT, p. 195:

However, not only can gases enter atmospheres; they may also be lost from atmospheres both by adsorption into the planetary interior and by loss into space as a result of collisions with minor and major planetesimals. In this paper a necessarily qualitative discussion is given of the problem of collisions with minor planetesimals, a process called atmospheric cratering or atmospheric erosion, and a discussion is given of atmospheric loss accompanying collision of a planet with a major planetesimal, such as may have produced the Earth's Moon.

COMMENT ON THIS FIRST SET OF QUOTES: The new model was developed to explain a set of key features of the earth-moon system: (1) the iron-poor nature of the Moon relative to the Earth, (2) the volatile depletion of the Moon relative to the Earth, (3) the refractory element enrichment of the Moon relative to the Earth, (4) the abnormally high quantity of angular momentum associated with the earth-moon system (i.e., if all of the angular momentum of the earth-moon system is placed in the Earth alone, then the prograde rotation rate of Earth would by ~5 hours/ day), and (5) THIS FAIRLY IMPROBABLE COLLISIONAL MODEL WOULD MAKE THE EARTH-MOON SYSTEM UNIQUE—UNLIKE ANY OTHER PLANET-SATELLITE SYSTEM.

So the stage is set for an explanation of these features of the earth-moon system by what is now called "THE GIANT IMPACT MODEL FOR THE ORIGIN OF THE MOON AND THE EARTH-MOON SYSTEM."

6.1.2.2 The Kona Conference on the Origin of the Moon (13–16 October 1984) and the Resulting Symposium Volume in 1986

The Kona Conference is the meeting at which the Giant-Impact Model for the origin of the Moon was the overwhelming favorite for a majority of the participants. This " SCIENTIFIC BANDWAGON" was off to a fast start.

Hartmann, W. K., Phillips, R. J., and Taylor, G. J. eds., 1986, ORIGIN OF THE MOON: Lunar and Planetary Institute (Houston), 781 p

From the PREFACE (of the conference volume) (p. vii–viii) by W. H. Hartmann on the history of the most recent version of the GIM:

> Perhaps the most sophisticated new theorizing was that of A. E. Ringwood, who postulated that very hot silicate vapor that was spun off or somehow thrown out of the Earth could condense and accrete into a body resembling the Moon. In 1975 W. K. Hartmann and D. R. Davis and in 1976 A. G. W. Cameron and W. R. Ward proposed that Earth mantle material might have been blown out of the Earth, heated, pulverized, and/or vaporized, and might have thus provided the basic material for forming the Moon. A few papers and a major book by Ringwood followed up some of these ideas but little further work was done in the decade following the last manned lunar landing in 1972.

From the last few sentences of the PREFACE (p. xi) (and good "food for thought"):

> If a giant impact occurred, were circumterrestrial debris predicted as direct crater ejecta, or more by a spin-up of the Earth, yielding an event resembling fission? Or, alternately, can the old ideas of pure fission or pure capture be revived: Or have we missed some alternative – will a wholly new model suddenly emerge from the morass of today's partial models?

The Page Before the PREFACE (Unlabeled Page VI) Has a Few Interesting Comments by at Least Two of the Participants (in Approximately the Same Format as the Page in the Book)

TAYLORS'S AXIOM

> The best models for lunar origin are the testable ones.

TAYLOR'S COROLLARY

> The testable models for lunar origin are wrong– S. Ross Taylor paraphrased by Sean Solomon, at the Conference on the Origin of the Moon, Kona, 1984

COMMENT: To my knowledge there were only two abstracts of Kona Conference papers that had "testable models" as part of the title of the abstract (the italics are for emphasis):

Malcuit, R. J., Winters, R. R., and Mickelson, M. E. (1984b), Directional properties of "circular" maria: Interpretation in the context of a *testable gravitational capture model* for lunar origin: Abstracts Volume, Conference on the Origin of the Moon, Lunar and Planetary Institute, p. 44.

(Note: This was a very extensive poster paper showing the surface morphology of all the "circular" maria, some mare filled craters, and a few other features.)

Malcuit, R. J., Winters, R. R., and Mickelson, M. E. (1984a), A *testable gravitational capture model* for the origin of the Earth's Moon: Abstracts Volume, Conference on the Origin of the Moon, Lunar and Planetary Institute, p. 43. (Note: This was an oral presentation.)

COMMENTS BY HARTMANN et al. (1986, p. xi) ON THE POTENTIAL REVIVAL OF THE CAPTURE MODEL

". . . . can pure capture be revived?"

A Cool Moon

Regardless of where the Moon formed, a cold Moon could not be captured unless it happened via a collisional capture scenario, but a warm Moon yields a different outcome. Winters and Malcuit (1977) demonstrated that a "warm" lunar-like body could be deformed sufficiently to store the energy necessary for capture. The big question was: Could this energy of deformation be dissipated in one tidal oscillation during a close encounter with Earth? In the article by Winters and Malcuit (1977) there was a correction of an equation that was originally derived by Kaula and Harris (1973) for energy storage in a lunar-like body. Peale and Cassen (1978) made a similar correction to that same equation in the Kaula and Harris paper. This revision of the equation made a difference of a factor of two by making the relaxation phase of deformation equal to the tidal energy storage phase. If nearly all of the energy stored in tidal deformation of the moon-like body could be dissipated as thermal energy during one tidal cycle (i.e., during one very close encounter) then gravitational capture of the lunar-like body could occur.

This quantity of energy dissipation would demand a very low Q (specific dissipation factor) for the lunar-like body. Such a low Q factor was not credible until the analysis of Ross and Schubert (1986). Their rationale was that a body like Io (the innermost satellite of Jupiter) gives off about the same quantity of thermal energy that it receives by way of the perijov-apojov tidal oscillations with Jupiter. Thus it must have a Q factor near 1. A low Q value for a lunar-like body encountering an earth-like body is necessary for dissipating the energy for capture within the lunar-like body during one close encounter with Earth [i.e., the lunar-like body must dissipate essentially all of the energy that is necessary to keep the lunar body within the Hill sphere of the Earth-Moon system (about 200 earth radii from Earth)].

A Warm Deformable Earth

The other major misconception for the classical capture model was that most of the energy for lunar capture would be dissipated within the body of the Earth. First of all, very little tidal energy can be stored in an earth-like body by tidal deformation during a close encounter cycle with a lunar-like body. Secondly, a low Q value can never be justified for rock tidal action on an earth-like planet. A Q value near 100 is probably the lower limit (Melchior 1978). Singer (1968, 1970) and other advocates for a capture model theorized that the Earth would be the energy absorber because the Moon was too cold to be sufficiently deformable. Because of this misconception about the energy dissipation capacity of Earth, a capture episode would demand hundreds of close planet-planetoid encounters, a scenario that would be very improbable and would probably result in a collision with Earth because of the solar gravitational perturbations of the lunar body during this extended series of close encounters.

A Response to the Question by Hartmann et al. (1986): YES, PURE CAPTURE CAN BE AND HAS BEEN REVIVED but a few major misconceptions had to be overturned in the process.

A Place of Origin for a Volatile-poor Moon

Still another major hurdle for an acceptable capture model is to find a reasonable PLACE OF ORIGIN for a VOLATILE-POOR lunar body. It is interesting to note that Al Cameron (1972, 1973) suggested that a good place to form such a body is inside the orbit of Mercury. Cameron knew, from his experience analyzing the early history of stars and the sun, that the thermal events associated with the early history of the Sun could cause significant devolatilization of any materials in this area between the Sun and planet Mercury. Since Cameron could not visualize a reasonable method for transferring Luna from its birth place near the Sun to an Earth-like heliocentric orbit, he abandoned his work on the capture model and embarked on a 30-plus year adventure on giant impact studies with a goal of making a lunar-like body in geocentric orbit about an earth-like body.

A Philosophy of Science Issue: Perhaps we as planetary scientists should ponder the extent to which our working lives have been changed by the decision of one person, Al Cameron, to shift his creative abilities from a PROGRADE GRAVITA TIONAL CAPTURE MODEL to a GIANT IMPACT MODEL for the origin of the earth-moon system.

Note: Chap. 7 is devoted to the chronological development of a GRAVITA TIONAL CAPTURE MODEL. I should point out here that many of the ancillary (supporting) concepts for constructing a somewhat convincing Prograde Gravitational Capture Model were not available until about 2009.

6.1.2.3 Post-Kona Conference Papers (1984–1998)

Apparently, a group of investigators convinced NASA SCIENCE ADMINISTRA TORS that the GIANT IMPACT MODEL looked promising. The result was that massive monetary funds were appropriated for very sophisticated numerical simulations of PLANET-PLANETOID COLLISIONS and associated research.

A. G. W. Cameron (1985) Formation of the prelunar accretion disk: Icarus, v. 62, p. 319–327

From the Abstract, p. 319:

> According to the single-impact hypothesis for forming the Moon, the angular momentum needed for the present Earth-Moon system can be imparted to the proto-Earth by a collision with a body having one-tenth of the mass or more.
>
> A successful theory must put considerably more than a lunar mass into orbit, having considerably more angular momentum than is needed to assemble a lunar mass in orbit at 3 Earth radii.
>
> A fairly common characteristic of these cases was the presence of an excess velocity in the collision (above that of a parabolic orbit), implying that the projectile involved in the

collision existed in an Earth-crossing orbit of significant ellipticity. A majority of the mass of the prelunar accretion disk is contributed by the projectile.

From the Discussion section, p. 326–327.

The results of these simulations should not be relied upon for particularly accurate predictions, but rather for exploring the plausibility of a given scenario. My interpretation of the results presented here is that the single massive collision scenario for the formation of the prelunar accretion disk is a plausible scenario and is worth the effort of a more realistic and necessarily more expensive simulation.

W. Benz, W. L. Slattery, and A.G. W. Cameron (1986) The origin of the Moon and the single-impact hypothesis I: Icarus, v. 66, p. 515–535

From the Abstract, p. 515:

Recently the single-impact hypothesis for forming the Moon has gained some favorable attention. We present in this paper a series of three-dimensional numerical simulations of an impact between the protoearth and an object of about 0.1 of its mass.

The amount of angular momentum in the Earth-Moon system thus obtained tends to fall short of the observed amount; this deficiency would be eliminated if the mass of the impactor were somewhat greater than the one assumed here. The scenario for making the Moon from a single-impact event is supported by these simulations.

From the Conclusions, p. 533–534:

From these simulations we conclude that the single-impact hypothesis provides a plausible scenario for making the Moon provided that:

The relative velocity between the impactor and the protoearth is relatively small (less than about 5 km/sec at infinity). If this condition is not fulfilled the impactor is completely dispersed in space.

The impact parameter is not too small. A good approximation is that the center of the impactor should graze the surface of the protoearth. Failure to satisfy this requirement again leads to the complete destruction and spreading out of the impactor and to the accumulation of most of the material onto the protoearth.

The mass ratio of the impactor to the protoearth has to be greater than 1 to 10 (new simulations including iron cores suggest 0.12–0.2) in order to have enough angular momentum in the Earth-Moon system.

W. Benz, W. L. Slattery, and A. G. W. Cameron (1987) The origin of the Moon and the single-impact hypothesis, II: Icarus, v. 71, p. 30–45

From p. 35:

For low mass impactors the collision is almost a grazing one (owing to the angular momentum constraint), therefore the shock is weak, and the impactor does not get completely destroyed but rather "bounces" and is thrown into a very eccentric orbit that makes it collide again with the protoearth. The second collision destroys the impactor

> completely and a great part of it is flung into orbit. The problem, however, is that most of the iron core of the impactor ends in orbit as well, leading to an iron-rich prelunar accretion disk. Large mass impactors, on the other hand, collide with a relatively small impact parameter (again due to the angular momentum constraint)....... The fraction left in orbit is iron poor but its mass is only about a lunar mass or less. Finally, in the intermediate region the collision leads to an iron-poor disk containing well over a lunar mass.

From p. 44:

> We therefore propose that one should consider as prime lunar formation candidates those simulations that lead to about a Moon mass of iron-free material in orbit in the general vicinity of the Roche limit.

From p. 45:

> Since only small relative velocity collisions lead to a state from which a Moon might form, we conclude that the impactor must have had an orbit somewhat similar to that of the Earth in order to collide with such a low velocity. This excludes the possibility that the impactor was perturbed by Jupiter and hit the Earth while on a very eccentric orbit. Since we find the Moon to be mostly formed from material originating from the impactor, the impactor itself must have been formed in a nearby region. This is consistent with the empirical evidence of the similarities between the lunar and terrestrial oxygen isotopic anomalies.

6.1.2.4 Conference on the Origin of the Earth (and Some Important Papers During the Next Decade)

The Conference was held in Berkeley (CA) in December 1988. The resulting CON FERENCE VOLUME was published in 1990:

Newsom, H. E., and Jones, J. J. eds., 1990, Origin of the Earth: Oxford University Press, 378 p

Two paragraphs (2 and 3) from the PREFACE will give us a good idea of the progress on the Giant Impact Model for the origin of the Moon and the Earth-Moon system:

> A strong theme of both the conference and many of this book's papers is the Giant Impact hypothesis. Even the papers that do not specifically address the Giant Impact are influenced by that hypothesis. If the conference had a single conclusion, it was the recognition that the origin of the Moon by giant impact has serious implications for the early history of the Earth. It seems unavoidable that a collision of a Mars-sized body would have vaporized a significant amount of the Earth and would have totally melted the rest. How is it then that there are samples from the Earth's mantle that appear to have undergone minimal processing? This is the paradox that the authors in this book wrestle with but do not solve (at least not to the satisfaction of all).
>
> The Giant Impact hypothesis has been presented by some as a revolution and panacea, similar to the advent of plate tectonics, that can explain the long-standing problems of the angular momentum of the Earth-Moon system, the depletion of iron metal in the Moon, and,

> quite possibly, the difference in chemical composition between the Earth and Moon. As illustrated by the papers in the volume, however, the implications of this theory for the Earth have not led to the unanimous agreement that giant impacts (connected with the origin of the Moon or not) are consistent with all geophysical and geochemical observations. What we need is for all the pieces of the puzzle to fit without serious whittling, carving or shoving. As of now, they don't. We hope that this book will help set the stage for the next generation of investigation and for fitting some pieces of the puzzle together.

COMMENT: The Gravitational Capture Model was represented by one paper at this conference. We had been doing numerical simulations of the capture process for a bit over one year:

Malcuit, R. J., Mehringer, D. M., and Winters, R. R. (1988). NUMERICAL SIMULATION OF THREE-BODY INTERACTIONS WITH AN ENERGY DISSI PATION SUBROUTINE FOR CLOSE ENCOUNTERS BETWEEN TWO BODIES: IMPLICATIONS FOR PLANETARY ACCRETION AND PLANETOID CAPTURE PROCESSES: Abstracts Volume, Conference on the Origin of the Earth, Lunar and Planetary Institute and NASA Johnson Space Center (Houston), p. 52–53. (This paper was presented as a poster paper.)

COMMENT: At this meeting I could demonstrate that it is easier to capture a mars-mass body than a lunar-mass body because a lower displacement Love number is needed for the larger encountering body.

W. Benz, A. G. W. Cameron, and H. J. Melosh (1989) The origin of the Moon and the single-impact hypothesis III: Icarus, v. 91, p. 113–131

From the Abstract, p. 113:

> The debris from the destroyed impacting object tends to form a straight rotating bar which is very effective in transferring angular momentum. If the material near the end of the bar extends well beyond the Roche lobe, it may become unstable against gravitational clumping.

A. G. W. Cameron and W. Benz (1991) The origin of the Moon and the single impact hypothesis IV: Icarus, v. 92, p. 204–216 (see Fig. 6.1 for results of encounter simulation)

From p. 214:

> In all of our simulations, except those in which the Impactor remnant has too much energy and escapes, the collision leaves a largely iron-free orbiting disk. In a few of our simulations we have also left a substantial amount of clumped material in stable orbits. These results pose a problem for understanding the origin of the Moon.

From p. 215:

> We have been asked how probable it is that the Giant Impact would have taken place. There is no surviving evidence for a collision of comparable magnitude with Venus, and Venus seems not to have lost its primordial atmosphere. Thus Giant Impacts are not inevitable. Ringwood (1990) argues that the probability is very small, but he really asks for the

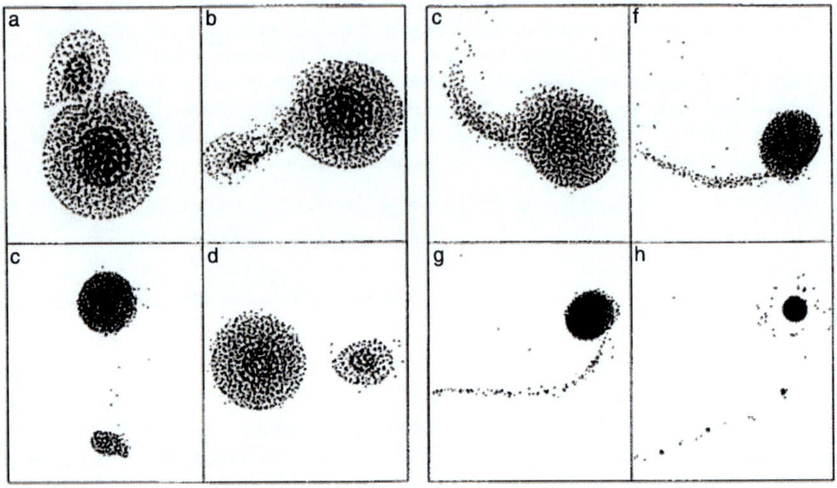

Fig. 6.1 A typical giant-impact sequence of events as of 1991. Iron particles are plotted as solid circles and rock particles are plotted as crosses. In this scenario the impactor separates from the main body and then comes back for a second collision (frame e) and then the resulting debris "tail" forms "clumps". The authors state that "the clumps in h will either pass within the Roche lobe and be tidally destroyed or they will escape". (Diagram from Cameron and Benz (1991, Fig. 1) with permission from Elsevier)

> probability of the Giant Impact should have had in a plane of the collision near that of the ecliptic and that the collision should have been in the prograde direction.

From p. 215:

> We deal with a unique event. In the absence of the Giant Impact, it seems plausible that the Earth would have developed with a massive atmosphere comparable to that possessed by Venus. Just as in the case of Venus, it seems unlikely that life would have developed in such an environment. Thus, there may be a very large selection effect which means that reasoning creatures will speculate about the probability of a Giant Impact only on Earth-like planets which have suffered a Giant Impact and have lost most of their primordial atmosphere. This issue lies well beyond the present reach of science. However, it is useful to raise this issue because in our opinion this possibility invalidates all attempts to estimate the plausibility of the Giant Impact having occurred on our planet.

6.1.2.5 Alistair Cameron is Having Second Thoughts About the GIM

Al Cameron was somewhat satisfied with his (and colleagues) work on the Giant Impact model but there was one outstanding problem that needed a solution. Cameron knew that he needed a mechanism for explaining lunar rock magnetization and that the explanation needed to be associated with a successful model for the origin of the Moon.

A. G. W. Cameron (1992) The giant impact revisited: Abstracts Volume, 23rd Lunar and Planetary Science Conference (Houston), p. 199–200

These are the first two paragraphs of the ABSTRACT:

> During the last six years the author, together with several colleagues, has been carrying out a series of numerical investigations of the Giant Impact theory for the origin of the Moon (1–4). In no sense has this been developed into a complete theory, since all the investigations so far have concentrated on finding conditions in which a disk comprised principally of rock could be placed into orbit about the Earth, within a relatively narrow range of angular momentum constraints, and no attempt has been made to simulate the evolution of this disk to understand the details of the postulated lunar formation process.
>
> In the most recent publication (4) W. Benz and I reported on a series of 41 runs simulating the Giant Impact with a variety of imposed parameters. A general conclusion reached at the end of these calculations was that the outer mantle of the impacted Earth was heated so much that not only would the primitive atmosphere hydrodynamically escape, but so would substantial amounts of vaporized rock. The loss of the rock can also carry away large amounts of angular momentum; the process was therefore termed the Giant Blowoff. This conclusion removed various constraints from our considerations; in particular, any impactor/protoearth ratio greater than about 0.2 and any angular momentum greater than about 1.2 times the present value for the Earth-Moon system can be considered candidate inputs for the Giant Impact.

A. G. W. Cameron (1993) The giant impact produced a precipitated Moon: Abstracts Volume, 24th Lunar and Planetary Science Conference (Houston), p. 245–246

COMMENT: The Balbus-Hawley mechanism was introduced in this paper for possible use in the Giant Impact Model (this is an extended abstract).

From the first paragraph of the ABSTRACT:

> The author's current simulations of Giant Impacts on the protoearth show the development of large hot rock atmospheres. The Balbus-Hawley mechanism will pump mass and angular momentum outwards in the equatorial plane; upon cooling and expansion the rock vapor will condense refractory material beyond the Roche distance where it is available for lunar formation.

From the next to last paragraph of the paper (extended abstract):

> As the vapor flows away from the protoearth it progressively expands and cools and the most refractory components will condense from it, forming small planetesimals. The Balbus-Hawley mechanism will gradually become ineffective as the electrical conductivity decreases, but the remaining noncondensed vapor can probably undergo thermal escape from the protoearth, carrying away some of the angular momentum. The condensed materials, being beyond the Roche distance, are then free to accumulate into the Moon. The Moon thus formed would have a very refractory composition, but it will most probably accumulate some of the material from the impactor, that was left over from the Giant Impact and torqued to several earth radii, but which was never heated too strongly and thus contains less refractory material. The precipitation nature of this lunar formation scenario is reminiscent of some ideas of Ringwood (7).

COMMENT: During the presentation, at which the author of this note was present, Cameron mentioned that this Balbus-Hawley mechanism may eventually help to explain the remanent magnetic signature of lunar rocks. Then Michael Drake, in the audience, stated that the Balbus-Hawley mechanism is related to stellar evolution and in no way is related to planet evolution. To my knowledge, the Balbus-Hawley mechanism was never mentioned by Cameron in future presentations and publications of the Giant Impact Model. Furthermore, the GIM promoters have never suggested any credible mechanism for the generation of a lunar magnetic field at any time in lunar history. The mention of the Balbus-Hawley mechanism was Cameron's attempt to somehow relate the GIM to lunar rock magnetization.

A. G. W. Cameron (1997) The origin of the Moon and the single impact hypothesis V: Icarus, v. 126, p. 126–137

From the first part of the ABSTRACT, p. 126:

> Previous papers in this series have described the smooth particle hydrodynamics (SPH) method, which has been employed to explore the possibility that a major planetary collision may have been responsible for the formation of the Moon. In those simulations the SPH code used particles of equal mass and fixed smoothing lengths; I have found that the results obtained were reliable regarding what happens to the interiors of the colliding planets. Because the particles placed into the surrounding space were isolated rather than overlapping, however, that part of the calculation was unreliable. Then additional cases have been run with 5000 particles in the Protoearth and 5000 in the Impactor, with variable smoothing lengths. Three of the cases had Protoearth/Impactor mass ratios of 5:5, 6:4, and 7:3. The other cases had a mass ratio of 8:2 and a variety of angular momenta. All cases had zero velocity at infinity. In every case the product of the collision became surrounded by evaporated particles of rock vapor, forcing an extended atmosphere; however, relatively little mass extended beyond the Roche lobe. If the Moon formed from a rock disk in orbit around the Earth, then some other mechanism would be needed to transport angular momentum and mass outward in the equatorial plane, so that rock condensates from the hot atmosphere would be precipitated beyond the Roche limit, thus providing material for collection into the Moon.

From the DISCUSSION, p. 137:

> The formation of the Moon as a post-collision consequence of a Giant Impact remains a hypothesis. The previous papers in this series appear to have characterized the internal effects of a Giant Impact on the Protoearth. The present paper has characterized the exterior environment of the Protoearth following a Giant Impact. The most promising direction for future simulations appears to involve Impactor/Protoearth mass ratio around 0.3 to 0.5 but a total mass at the time of the collision substantially less than a present Earth mass, and one would search for conditions in which a lunar-mass body is left in orbit after the collision.

COMMENT: Cameron is becoming concerned about using ever higher mass ratios for the impactor/protoearth tangential collisions that will yield a lunar mass or more in earth orbit that will eventually form a satellite.

6.1.2.6 Now for the Apparent Transfer of the Future Promotion of the Concept of the Giant Impact Model for the Origin of the Moon from Alistair Cameron to Robin Canup

The following passages are from a newspaper article in the New York Times. I am including this in its entirety because of the lucid descriptions of the results of the two different approaches and for the clarity of presentation by the author of the article. In my view, this is a premier example of superb reporting on a major issue in the planetary sciences. Implied in the article is a transfer of PRINCIPAL INVESTIGA TOR status for future calculations from Alistair Cameron to Robin Canup.

J. N. Wilford (1997) ASTRONOMERS R ECALCULATE THE 'WHACK' THAT MADE THE MOON: New York Times, Science Section, p. B12 (29 July) (courtesy of the NYT).

> Before the Apollo lunar explorations, scientists had three competing theories for the origin of the Moon. In the fission theory, the early Moon spun off from Earth, leaving the Pacific Basin as a scar. Another theory conceived of the Moon as a small wandering planet that had passed too close and was captured by Earth's gravity. Or the Moon and Earth might be siblings, both accumulating and evolving as separate planetary objects in the same neighborhood.
>
> Then the analysis of the age and composition of the lunar rocks brought back by Apollo astronauts forced scientists to abandon those theories in favor of a new one: the giant impact theory or, as it is sometimes called, the Big Whack model.
>
> According to this theory, now the accepted wisdom in planetary science, a massive object sideswiped Earth 4.5 billion years ago, in the heavy bombardment of planets and planetary fragments during the solar system's formative period. The collision scattered crustal debris that later coalesced in orbit to form the Moon. Scientists at Harvard University even calculated that the object that collided with Earth must have been as massive as Mars.
>
> Now scientists at the University of Colorado at Boulder think this seriously underestimates how big the object would have to have been to strike Earth with sufficient force to generate the volume of debris required to create the Moon. Their computer modeling indicated that such an object must have been at least 2.5 to 3 times the mass of Mars.
>
> The diameter of Earth is 7,926 miles, almost twice that of Mars and four times that of the Moon. But the mass of Earth is 10 times that of Mars: any object three times as massive as Mars would thus have about one-third the mass of Earth.
>
> Dr. Robin Canup, a Colorado planetary scientist, described the research modeling the Big Whack in an interview last week and is to give a detailed report on Thursday in Cambridge, Mass., at the annual meeting of the American Astronomical Society's Division of Planetary Sciences.
>
> Dr. Canup said the computer model indicated that the collision had probably occurred somewhere between the current orbits of Earth and Mars. A rogue planet and Earth probably struck each other glancing blows. The impact vaporized parts of the other planet as well as the upper layers of Earth's crust and mantle, scattering material into Earth's orbit. The

material spread into a gaseous disk, then formed several small extremely hot moonlets, which eventually coalesced into the single, large Moon seen today.

'Large-scale impacts like this one probably played a crucial role in shaping the solar system,' Dr. Canup said. The Colorado study was conducted in collaboration with the Harvard group, led by Dr. A. G. W. Cameron, although the two reached some different conclusions. The Harvard group, which came up with the estimate of Mars's mass a decade ago, concentrated on the impact event itself and the amount of debris. The Colorado team focused on the other end of the problem: the debris cloud and how it coalesced into the Moon. Dr. Canup said the new analysis showed that only 20 percent to 50 percent of the debris would have been incorporated into the Moon, the rest would have fallen back to Earth. The Harvard research has assumed that nearly all the debris would have wound up in the Moon. But neither study seems to tie up the impact theory with a neat bow. The size of the colliding body proposed by the Colorado research provides enough material for the Moon, but the force of the impact would have left Earth spinning too rapidly. The Harvard model produced the correct initial spin for Earth, but not enough lunar material.

'Our closest celestial neighbor remains a mystery in many ways,' Dr. Canup said.

6.1.2.7 Conference on the Origin of the Earth and Moon (December 1998, Monterey CA)

This conference was organized with the purpose of promoting the accomplishments of the researchers working on the Giant Impact model. What was noteably missing was any mention of a mechanism for causing the remanent magnetism registered in both the ancient lunar (anorthositic) crust and/or the younger lunar basalts. The implication (not stated) was that this grandiose model of the Giant Impact origin for the Earth-Moon system would eventually lead to an explanation of lunar rock magnetization.

R. M. Canup, and K Righter, *eds.*, 2000, Origin of the earth and moon: The University of Arizona Press (Tucson) in collaboration with the Lunar and Planetary Institute (Houston), 555 p

From the ABSTRACT, p. xiii–xv, paragraph 1:

The origin of the Earth-Moon system is one of the longstanding questions in planetary science. The "giant impact hypothesis," which has received increasingly widespread support over the past 15 years, implies that the origins of the Earth and Moon are fundamentally coupled by a collision late in Earth's formation history. This catastrophic event greatly affected the initial thermal, chemical, and dynamical state of the Earth-Moon system, leaving signatures still observable today. The subjects of the Earth and Moon have traditionally been reviewed separately; indeed, the most recent volume addressing both bodies is Ringwood's 1979 work. However, given what is now believed to be their common origin, it is clear that a discussion of the coupled formation and early evolution of both the Earth and Moon is overdue.

COMMENT: There is undoubtedly much enthusiasm for the GIM.

This is paragraph 4:

Compositional constraints on this new scenario have emerged from both petrologic and isotopic studies. First, technical advances in mass spectrometry have allowed measurement of isotopes of tungsten in a variety of terrestrial, lunar, and planetary materials, thus allowing application of the tungsten-hafnium short-lived chronometer to the topic of planetary accretion. Such measurements have provided new constraints on the timing of formation for the Moon and core formation in the Moon and Earth. In particular, they suggest that the Moon formed earlier than the Earth, 50 m.y. after the start of accretion. Second, an explosion of high-pressure experimentation in metal-silicate systems has forced a reevaluation of traditional core formation models for the Earth and Moon. Specifically, the "excess siderophile-element problem" of Earth's primitive upper mantle, long attributed to heterogeneous accretion or disequilibrium processes, can instead be alleviated by high-temperature and high-pressure metal-silicate equilibrium in the early Earth. The presence of terrestrial and lunar magma oceans (the latter long favored by petrologists) is consistent with a hot, early Earth-Moon system. Third, the discovery that comets have distinctly different deuterium-to-hydrogen ratios than terrestrial ocean water has forced consideration of the presence of water during accretion, instead of the previous assumption of dry accretion with later addition of cometary water.

COMMENT: The statement in line 6 is probably considered a mistake by the authors (i.e., the Moon formed earlier than the Earth), but the statement in its present form may turn out to be closer to the truth (my prediction). Also, the last sentence needs to be seriously challenged. The Potassium Index for Solar System objects strongly suggest that there was a gradient of dehydration in the proto-planetary cloud of dust and rocky debris (Taylor 2001). The volatiles then came back from the Asteroid Belt to the terrestrial planet realm at a later time. The Alberede (2009) model suggested that volatile-rich asteroids are the source of much of the water and volatiles that were added to the terrestrial planets as a late veneer about 100 million years (+ or – 50 million years) after the formation of the planet.

E. Kokubo, R. M. Canup, and S. Ida (2000) Lunar accretion from an impact-generated disk, *in* Canup, R. M., and Righter, K., *eds*., Origin of the Earth and Moon, The University of Arizona Press, 555 p

From the ABSTRACT, p. 145:

We review current models for the accumulation of the Moon from an impact-generated debris disk. Such a disk is dynamically distinguished by its substantial mass relative to the Earth and a very centrally condensed radial profile, with a mean orbital radius near the classical Roche limit. In the inner protolunar disk, accretion is inhibited by tidal forces. Typically, a single large moon accretes just outside the Roche limit, at a distance of about 3.5–4.0 × the Earth's radius. A simple relationship between the fraction of the disk mass that is incorporated into the final moon and the initial disk angular momentum has been determined from simulations spanning a wide range of initial conditions, collisional parameterizations, and numerical resolutions. Predicted accretion yields range from 10% to 55%, with most of the remaining material scattered onto the Earth. Recent N-body simulations show the formation of transient gravitational instabilities in the inner disk, leading to rapid disk-spreading rates. These results may, however, be affected by current models' neglect of the thermal state of the disk material. Analysis of the orbital evolution of material due to tidal

> interaction with the Earth suggest that remnants of the initial accretion phase will likely be accumulated by either the largest moon or the Earth leaving a single moon in most cases.

COMMENT: The reader will notice that it is difficult to make a large satellite from the debris of the collision. I note that no one yet has made (1) a lunar-sized body from the debris of the collision and/or (2) a body that has the chemical and physical characteristics of the Earth's Moon. IT JUST MAY BE IMPOSSIBLE!

There was one poster paper on the CAPTURE MODEL at this meeting:

R. J. Malcuit and R. R. Winters (1998) A prograde gravitational capture model for the origin of the Earth-Moon system: Is it compatible with the rock records of the Earth and Moon?: Abstracts Volume, Origin of the Earth and Moon, Lunar and Planetary Institute (Houston), p. 24

COMMENT: There was an opportunity for poster paper presenters and other interested participants at the meeting to give 5 minute oral presentations at essentially an "open microphone" session. Fred Singer gave a brief presentation stating that he thought that the debate on the "origin of the Moon" problem would eventually evolve to an origin by Gravitational Capture. I gave a 5 minute presentation on the "geometry of stable capture zones" and their use in estimating the probability of capture for specified heliocentric orbital configuration of Earth and Luna. If it had not been for Fred and me, the GRAVITATIONAL CAPTURE MODEL would not have been represented!

6.1.2.8 The Robin Canup Era for "Giant Impact Model" Studies

Robin Canup and her research associates have had a major impact in "origin of the Moon" studies. Her main objective was to quantify in greater detail than ever before the various aspects of the GIANT IMPACT MODEL FOR THE ORIGIN OF THE MOON AND THE EARTH-MOON SYSTEM. The early work on the GIM was accomplished by Alistair Cameron. Then there was a transition of the numerical simulation work, as well as a different prospective of the work, to Robin Canup.

R. M. Canup (2004) Origin of the terrestrial planets and the Earth-Moon system: Physics Today, April 2004, p. 56–62

From p. 58:

> Recent models that include more initial objects or a small portion of the nebular gas have found systems with orbits closer to those in our Solar System, although accounting for the nearly circular orbits of Earth and Venus remains an open issue.

From p. 60:

Prior to the Apollo era, three lunar origin hypotheses predominated: capture, fission, and coformation. However, each of these models failed to account for one or more of the major characteristics of the Earth-Moon system. Capturing an independently formed Moon into an Earth-bound orbit does not offer a natural explanation for the lunar iron depletion, and that scenario appears dynamically unlikely. In fission, a rapidly spinning Earth becomes rotationally unstable, causing lunar material to be flung out from the equator. That hypothesis requires Earth-Moon angular momentum to be several times higher than its actual value. Coformation supposes that the Moon grew in Earth orbit from the sweeping up of smaller material from the solar nebula. Although coformation models were successful in producing satellites, they didn't readily explain both the lunar iron deficiency and the Earth-Moon angular momentum, since growth via many small impacts typically delivers little net angular momentum and produces slow planetary rotation.

From p. 62:

Whereas early models proposed that Earth-like planets form through the orderly accretion of nearby small material in the protoplanetary disk, current work instead suggests that solid planets are sculpted by a violent, stochastic final phase of giant impacts.

COMMENT: Perhaps the calculations by Lissauer and Kary (1991) need to be taken more seriously. The stochastic (i.e., oligarchic) body impact model, (mainly a major assumption) does not relate all that well to the real world. Figure 6.2 is from the article by Canup (2004a, Fig. 5).

R. M. Canup (2004b) Simulations of a late lunar-forming impact: Icarus, v. 168, p. 433–456

QUESTIONS ARE ARISING AT THIS POINT IN TIME!

From p. 434:

Despite these computational advances, identifying impacts capable of placing sufficient mass into Earth orbit to yield the Moon while also accounting for the Earth-Moon system mass and angular momentum proved challenging. Since Cameron and Benz (1991), progressively larger impactors relative to the targets were considered in an effort to increase the yield of orbiting material, with Cameron (2000, 2001) considering collisions that all involved impactors containing 30% of the total colliding mass. The type of impact favored by those works involved an impactor with roughly twice the mass of Mars and an impact angular momentum close to that of the current Earth-Moon system.

From p. 453:

The simplest explanation for the Moon's unusual compositional characteristics is that it is the result of an impact that occurred near the very end of terrestrial accretion. We have focused on the "late impact" scenario of Canup and Asphaug (2001), in which a Moon-forming impact occurs when the Earth was >95% accreted.

Another from p. 453:

Results from simulations of the most promising lunar-forming candidates are given in Table 1. The successful impacts – defined as those that produce a sufficiently massive and iron-depleted disk together with a planet-disk system angular momentum $\leq L_{EM}$ involve impactors that contain between 0.11 and 0.14 Earth masses, have a relative velocity at infinity between 0 and 4 km/sec, and an impact parameter between 0.67 and 0.76 . . .

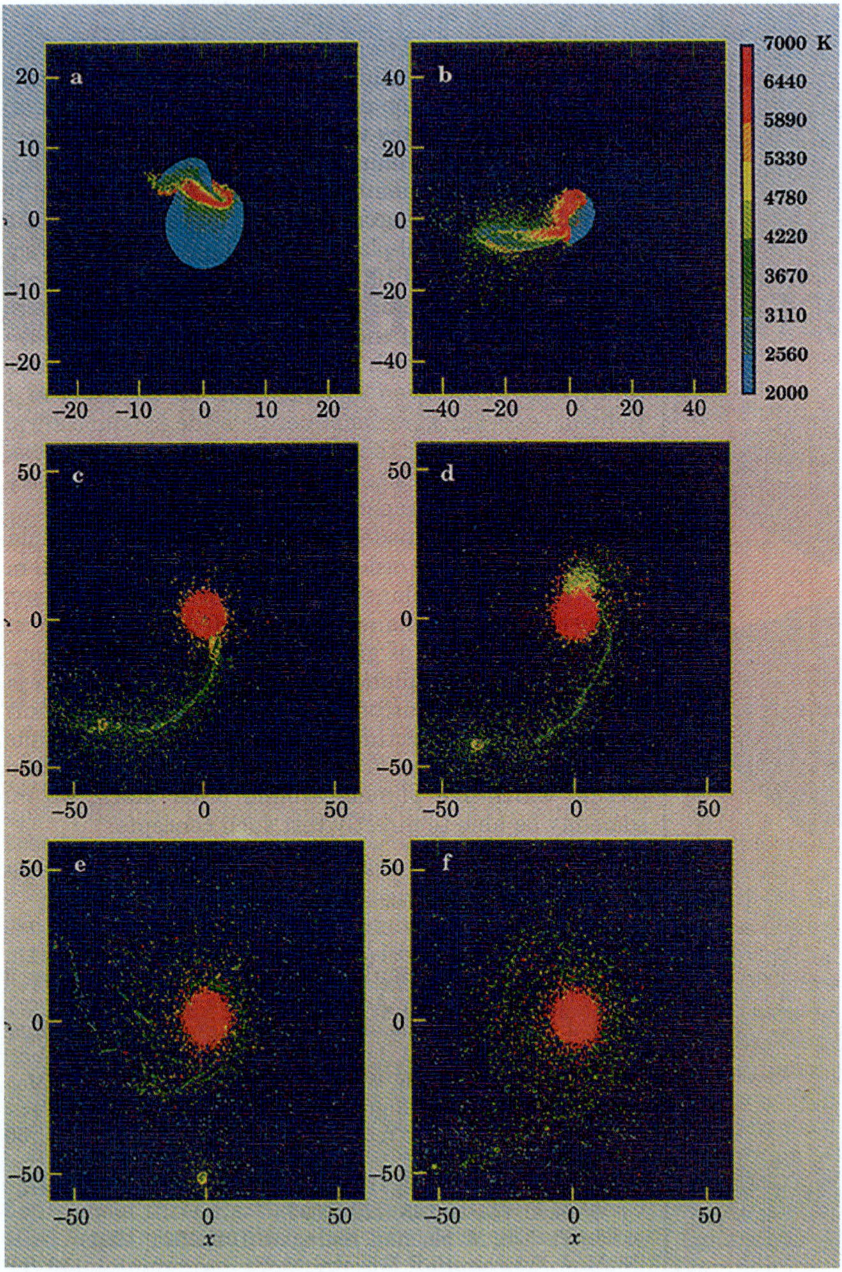

Fig. 6.2 Simulations of a giant impact scenario as of 2004. The simulation is spectacular but the simulation did not result in a satellite with the physical and chemical features of the Moon. View is from the north pole of the solar system. The times for the six frames (**a–f**) are 0.3, 1.4, 4.9, 5.9, 13.5, and 27.0 hours after the collision. The pre-impact mass of the larger body is approximately that of Earth; the mass of the impactor is 1.2 times the mass of Mars. The angular momentum of the system

R. M. Canup (2008a) Lunar forming collisions with pre-impact rotation: Icarus, v. 196, p. 518–538.

From the first paragraph of the text, p. 518:

The leading theory of the Moon's origin is that it formed as a result of the impact of a Mars-sized object with the early Earth (Cameron and Ward 1976). Key strengths of the giant impact theory include its ability to account for the Earth-Moon system angular momentum (which implies a terrestrial day of only about 5 h when the Moon formed close to the Earth), and the Moon's relatively low iron abundance compared to other inner Solar System objects. In addition, dynamical models of the final stages of Earth's accretion suggest that large impacts were common (e.g., Agnor et al. 1999).

R. M. Canup (2008b) Accretion of the earth: Philosophical Transactions of the Royal Society, Series A., v. 366, p. 4061–4075

From the ABSTRACT, p. 4061:

The origin of the Earth and its Moon has been the focus of an enormous body of research. In this paper I review some of the current models of terrestrial planet accretion, and discuss assumptions common to most works that may require re-examination. Density-wave interactions between growing planets and the gas nebula may help to explain the current near-circular orbits of the Earth and Venus, and may result in large-scale radial migration of proto-planetary embryos. Migration would weaken the link between the present locations of the planets and the original provenance of the material that formed them. Fragmentation can potentially lead to faster accretion and could also damp final planet orbital eccentricities. The Moon-forming impact is believed to be the final major event in the Earth's accretion. Successful simulations of lunar-forming impacts involve a differentiated impactor containing between 0.1 and 0.2 Earth masses, and impact angle near 45° and an impact speed within 10 per cent of the Earth's escape velocity. All successful impacts – with or without pre-impact rotation – imply that the Moon formed primarily from material originating from the impactor rather than from the proto-Earth. This must ultimately be reconciled with compositional similarities between the Earth and the Moon.

From the last paragraph of the SUMMARY, p. 4072:

That the final phase in growth of terrestrial planets was dominated by giant impacts has become well accepted in the decades since its first demonstration (Wetherill 1985). As the last major event in the Earth's accretion, the Moon-forming impact is dynamically a relatively well-constrained problem. Successful simulations that can reproduce the current Earth-Moon system invoke impactor sizes and velocities predicted to be common during late stage terrestrial accretion. Models of the subsequent formation of the Moon from the proto-lunar disc are less well developed, and further work is needed to describe the coupled thermal and dynamical evolution of the disc and the Moon's initial thermal state. This should enable improved quantitative compositional and geophysical predictions for an impact-formed Moon, and its expected similarities and dissimilarities with the Earth.

Fig. 6.2 (continued) is near that of the Earth-Moon system. (Diagram from Canup (2004a, Fig. 5) courtesy of the American Institute of Physics)

B. Wood (2011) The formation and differentiation of Earth: Physics Today, v. 64, no. 12, p. 40–45.

From a section titled ORIGIN OF THE MOON: p. 43:

> Our moon is unusual among solar-system satellites because it has a relatively large mass – more than 1% of the mass of the planet it orbits. Theories for the Moon's origin include capture, fission from the parent Earth, co-accretion during planetary growth, and the generation by a colossal impact. All but the last can be excluded on dynamical or compositional grounds. That leaves the current favorite, visualized in figure 1: a collision between the growing Earth and a Mars-sized body toward the end of accretion.[12] Unlike the other theories, such a giant impact could account for the angular momentum of the Earth-Moon system.

COMMENT: Capture from an earth-like heliocentric orbit (i.e., the only way gravitational capture can occur) results in a large major axis geocentric orbit with the angular momentum of a circular orbit of ~30 R_e. The angular momentum of that prograde orbit combined with a prograde earth rotation rate of ~10 hr/day yields the angular momentum of the earth-moon system. THUS, THE LAST STATEMENT OF THE ABOVE PASSAGE IS CHALLENGED!

6.1.2.9 All of this Giant Impact Mania Came to a Head at a Discussion Meeting on the "Origin of the Moon" Sponsored by the Royal Society (London, U. K.). in September 2013 (Proceedings Volume was Published in September 2014)

Most of the scientists that had published articles on the GIM were at this meeting. I would guess that the expectation was that all (or most) of the previously identified problems had been solved to a satisfactory degree by this time. After all, most of the researchers probably thought that the GIM was the only model remaining (a default model) so that the solutions had to flow from that model. Then we have a note in SCIENCE by Dan Clery (listed below). WHAT HAPPENED TO THE GIM??

D. Clery (2013) Impact theory gets whacked: Science, v. 342, p. 183–185

Science journalist, Dan Clery, reported his views that the impact theory got whacked at the Origin of the Moon conference sponsored by the Royal Society in September 2013. The headline of the article was "planetary scientists thought they had explained what made the Moon, but ever better computer models and rock analyses suggest reality was messier than anyone expected". In the article by Clery there is a pictogram showing a lunar-sized planetoid impacting on an earth-mass planet. This pictogram represents an early concept from Hartmann and Davis (1975) that several of these lunar-sized (and smaller) impactors were involved in forming a circumterrestrial debris cloud from which a lunar-sized body would form just beyond the Roche limit of the Earth-Moon system.

D. J. Stevenson, and A. N. Halliday, *eds*. (2014) Origin of the Moon; challenges and prospect: Philosophical transactions of the royal society, A: Mathematical, Physical, and Engineering Sciences, v. 372, No. 2024, Royal Society Publishing (London)

From the PREFACE by D. J. Stevenson, and A. N. Halliday (2014), p. 2–3:

> The late stage collision of the proto-Earth with another planet often called Theia provides the 'least worst' explanation for many of these features. This Giant Impact model has been developed over the years from the early ideas of Reginald Daly [2], to the more comprehensive multiple collision version of Hartmann & Davis [3] and the angular momentum argument of Cameron & Ward [4]. In the 1980's, the powerful smoothed particle hydrodynamics of giant impacts were developed by Benz et al. [5]. And these were followed by more detailed modelling by Canup (reviewed in [6]). In all these computer simulations, the angular momentum (4 above) is generated by the impact itself, Theia striking Earth with a glancing blow. In these simulations, the material forming the Moon is mostly derived from Theia. This feature of successful Giant Impact simulations has been the hardest to reconcile with geochemical data. The fact that the isotopic compositions of silicate Earth and Moon are so similar despite evidence that other objects are different provides evidence that either:
>
> 1. The innermost Solar System from which these two objects formed was not so heterogeneous after all or Theia accreted at a similar heliocentric distance to the Earth [7, 8]. Or
> 2. The atoms of the Moon were derived from the Earth after core formation and the 'traditional' simulations are incorrect [9–11]. New dynamic models have been proposed in which the angular momentum constraint is violated (i.e. the Earth-Moon system began with over twice its current angular momentum) and the excess is extracted by a resonance involving the Sun [10]. Or
> 3. There was isotopic equilibrium between the atoms in the lunar accretion disc and those in the Earth's magma ocean [12].
>
> Determining which of these models is correct is crucial to understanding, not just the formation of the Moon itself, but also the conditions under which terrestrial planets more generally accreted. Each of these suggested resolutions has its difficulties. The first is not readily reconciled with the current accretion models. The second relies on a resonance that may only work for a narrow range of tidal parameters and the third relies on an unlikely high efficiency of mixing (in particular, between the interior of the Earth and the moon-forming disc).
>
> In understanding the Moon's origin and early development, a number of other issues need to be addressed.
>
> First, we do not know the composition of the Moon very well. The GRAIL mission has opened up the opportunity to explore this in detail.
>
> Second, the age of the Moon is poorly defined. It is clearly late (more than 30 Ma) but how late is less certain. The oldest lunar rocks have been redated with more precise modern methods and show no sign of antiquity before about 4.35 Ga – about 200 Ma after the start of the Solar System.
>
> Third, we do not have a clear idea of the manner in which the Moon first developed. It was thought to be a fiery start with a lunar magma ocean, but some of the features of the ages and chemistry of lunar anorthosites might be better explained by more localized magmatism.
>
> Fourth, there is considerable uncertainty about how the Moon became so depleted in volatile elements. Some of this may have arisen from the Giant Impact itself, but it is also possible that early lunar volcanism led to losses.
>
> Last, there is debate about what has happened to the Earth and Moon since the Giant Impact – How did the Moon affect Earth's early evolution and to what extend do the differences in volatiles reflect late additions to the Earth since that time.

This is a Quote (2 Paragraphs) on the "Isotopic Crisis" from the Article by W. K. Hartmann (2014), p. 5 of the Symposium Volume

> In the early era of the giant impact hypothesis [1–3], Clayton et al. [27, 28] were just discovering that objects from different parts of the Solar System have different oxygen isotope ratios, but that the Moon and Earth had virtually identical O isotope ratios, within error bars [29]. As this news spread in the mid-1970s, this fact was seen as strong support for the giant impact model, in which the Moon formed from Earth-like material, not distant material with 'alien isotopes'.
>
> In recent years, new data began to be reported on other elemental isotope ratios. Once again, they were virtually identical between Earth and Moon, relative to error bars, and relative to more distant bodies. One might have assumed that this would be cited as further support that lunar material was related to Earth material and that giant impact was again supported in a first-order sort of way. Instead, the evidence that had once supported the theory seemed to have 'reversed polarity', and now was cited against the theory. As Melosh [30] put it while proposing an 'isotopic crisis', 'Unless the isotopic compositions of the proto-Earth and projectile were nearly identical by some fortuitous coincidence, there should be detectable differences between the isotopic composition of the present Earth and Moon . . .

COMMENT: There is a major difference between the Potassium Index for inner solar system bodies and an index based on Oxygen Isotopes. The author of this book thinks that the Potassium Index yields more logical results than the Oxygen Isotope Index.

This is a Quote from the Article by I. A. Crawford and K. H. Joy (2014), p. 12 of the Symposium Volume

> An important unresolved question is whether the inner Solar System cratering rate has declined monotonically since the formation of the Solar System, or whether there was a bombardment 'cataclysm' between about 3.8 and 4.1 Ga ago characterized by an enhanced rate of impacts (figure 5) [89, 97–99]. Indeed, recent studies of the ages of impact melt samples obtained by the Apollo and Luna missions suggest a very complicated impact history for the Earth-Moon system, with a number of discrete spikes in the impact flux [100]. Clarifying this issue is especially important in an astrobiology context as it defines the impact regime under which life on the Earth became established and the rate at which volatiles and organic materials were delivered to the early Earth [96, 98, 101, 102]. Additionally, as the inner Solar System bombardment history is thought to have been governed, at least in part, by changing tidal resonances in the asteroid belt [97, 103–105] improved constraints on the impact rate will lead to a better understanding of the orbital evolution of the early Solar System.

Figure 6.3 is from Crawford and Joy (2014) and Fig. 6.4 is a plot of some of these same events seen through the lens of a prograde gravitational capture model.

COMMENT: A short note on the end of the viability of the concept of the LATE HEAVY BOMBARDMENT appears in the 25 Jan. 2018 issue of Nature.

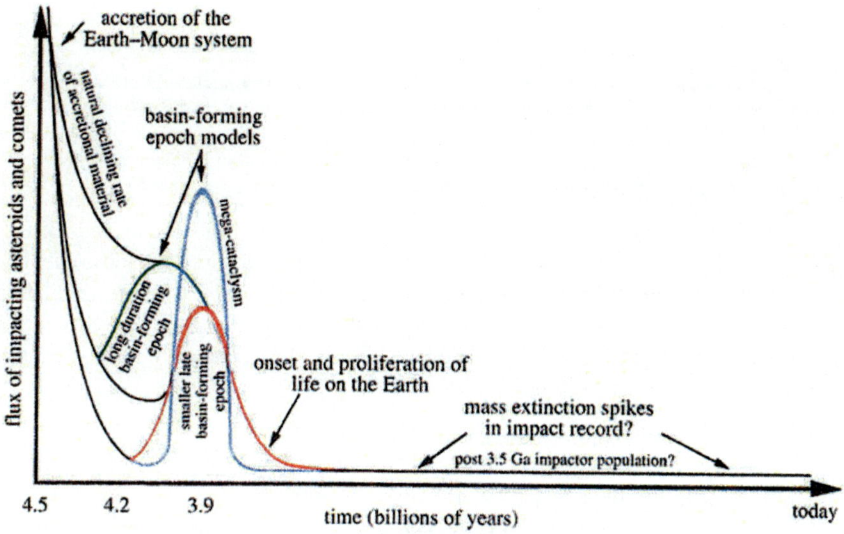

Fig. 6.3 Schematic diagram illustrating a variety of models that may relate to the history of impact cratering on the Moon. Models range from no significant bombardment other than a declining primordial impact flux, a short spike in the impact record at approximately 3.9 Ga, to other more complex models of bombardment. (Diagram from Crawford and Joy (2014, Fig.5) with permission from the Royal Society Publishing)

Fig. 6.4 Schematic diagram of my interpretation of the events represented in Fig. 6.3. The main trend in this illustration is a natural declining rate of space debris following accretion. Then LUNAR CAPTURE occurs at about 3.95 Ga. Then there would be some lunar debris resulting from the tidal disruption event that would impact on both the Moon and Earth as the lunar orbit is undergoing orbit circularization

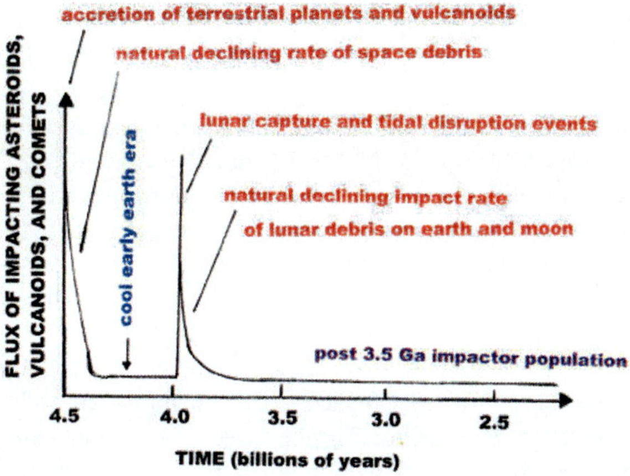

This concept of the LHB has been a ruling paradigm in the planetary science community since 1977. Also be aware that the so-called NICE MODEL was proposed by Gomes et al. (2005) in an article in Nature to explain the LHB.

This is a Quote from A. Mann (2018) Cataclysm's end: Nature, v. 553, p. 395

> One of the original architects of the Nice model, Alessandro Morbidelli of Cote d'Azur Observatory in Nice, admits that the first versions took fine-tuning to get the reshuffling to occur so late. He no longer believes in the LHB, and sees many others in the field trading in the idea of a sudden asteroid deluge for that of a long, declining tail of bombardment. 'My prediction is people will abandon the cataclysm,' he says.

COMMENT: Nonetheless, some big events occurred on the Moon and Earth starting about 3.95 Ga ago. As illustrated on Fig. 6.4, I think that most of these events such as emplacement of the mare lavas, generation of a second phase of lunar magnetism, and destruction of much of the crust of the Earth via subduction into the mantle are the result of PROGRADE GRAVITATIONAL CAPTURE of the Moon.

Now Let us Return to the Clery (2013) Article

Clery (2013) summarized three contrasting scenarios for forming the body of the Moon via a giant impact. The first was the standard model featuring a prograde tangential impact of a mars-mass planetoid on the primitive earth. This standard model has difficulties explaining chemical similarities between Earth and Moon and also has problems explaining the formation of a satellite as large as the Moon. A second model proposed that a massive body impacted on a rapidly rotating primitive earth. This model apparently results in a somewhat better explanation of the chemistry of the two bodies but results in an excess of angular momentum for the Earth-Moon system. A third, recently proposed, model features two equal-sized, half earth mass, bodies colliding to form the Earth-Moon system. These three scenarios are illustrated in Fig. 6.5. Figure 6.6 shows the results of a numerical simulation of a collision of two equal-sized bodies (the third scenario). How the body of the Moon forms from this scenario is not all that clear.

In general, the models summarized in this article are the result of the failure of the "traditional" Giant Impact Model to explain the origin of the Moon. In my opinion, this search for ever more complex models represents the initiation of a "wild goose chase" in which ever more energetic models are proposed to make the Moon as a result of a GIANT impact. After abandonment of the standard model some researchers are considering the effects of an impacting body on a fast-spinning proto-earth, others are working on a model in which two one-half earth mass bodies collide to form a large debris cloud, and now, starting in 2017, we have a SYNESTIA model (Lock and Stewart 2017; Lock et al. 2018). A synestia forms following a high energy, high angular momentum, giant impact on the primitive Earth. In my opinion, a synestia model is the penultimate energy version of the GIM for the origin of the Moon and the Earth-Moon system and will lead to the abandonment of the idea. When that happens, two major paradigms that have guided

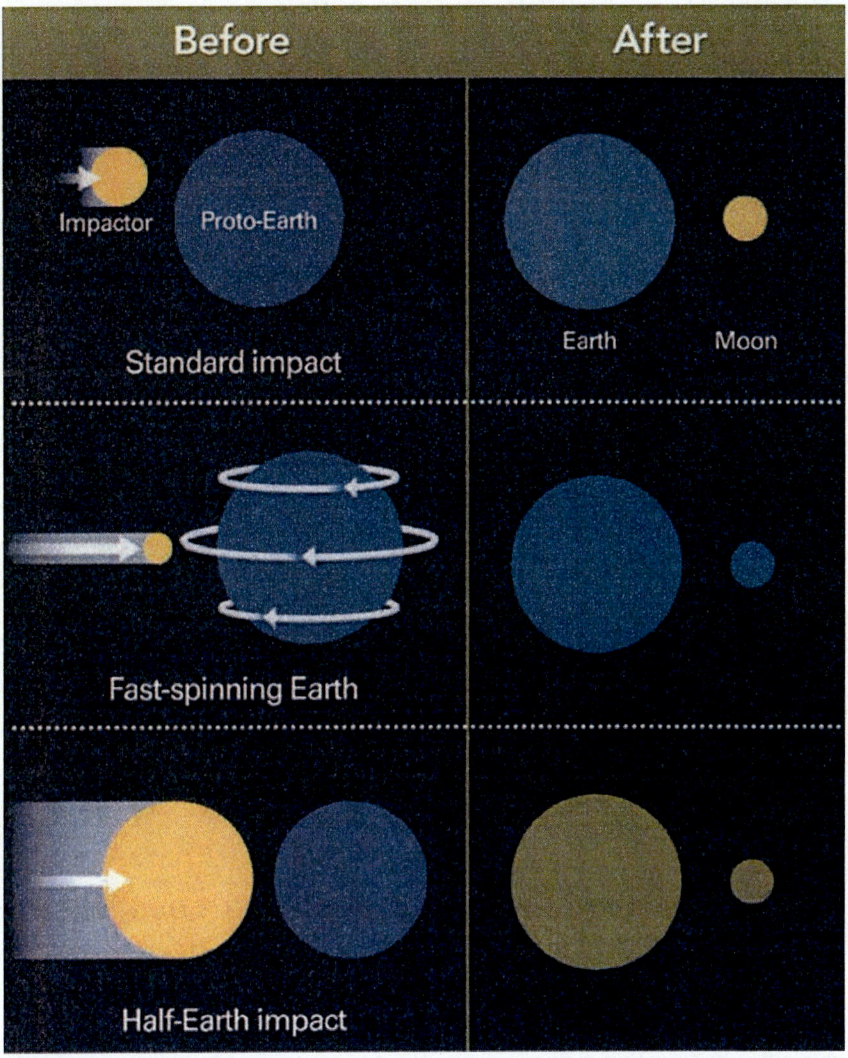

Fig. 6.5 Three attempts to form the Moon via giant impacts. The standard giant impact model (top) has trouble explaining chemical similarities between Earth and the Moon in addition to making a satellite that is as large as the Moon. The intermediate and bottom diagrams, published in 2012, do a better job with the chemistry but result in an excess of angular momentum. (From Clery (2013, p. 198) with permission of AAAS)

research for many investigators in the planetary science community for a span of some 34+ years (i.e., the Giant Impact model and the Synestia model) may need to be abandoned.

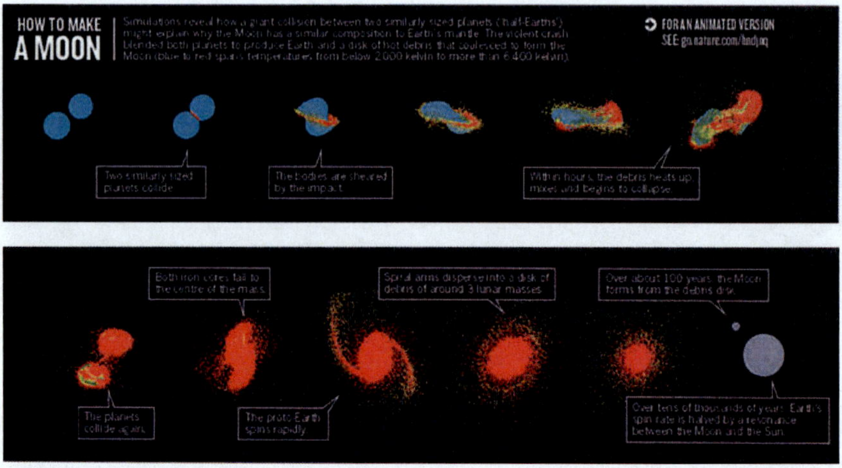

Fig. 6.6 Attempting to make something like the Earth's Moon by smashing together two one-half earth masses. (**a**). First half of the scenario. (**b**). Second half of the scenario. This is the grand scenario for the lower set of pictograms in Fig. 6.5. (From Canup (2013, p. 28–29) with permission from Springer-Nature)

6.1.2.10 Some Post-Symposium Attempts to Improve the GIM

After the symposium sponsored by the Royal Society there were some attempts to form a satellite of lunar composition, using a set of unusual assumptions as illustrated below. But none of these computer models were successful in forming a satellite with the features of the earth's Moon and none of these models resulted in an explanation of lunar rock magnetization.

R. M. Canup (2013) Lunar conspiracies: Nature, v. 504, p. 27–29

The purpose of the paper was to present a model for forming the Earth and Moon by way of a tangential collision between two equal-sized planetoids, each about one-half earth mass (Fig. 6.6).

A quote from the INTRODUCTION of the article, p. 27:

> Current theories on the formation of the Moon owe too much to cosmic coincidences, says Robin Canup. She calls for better models and a mission to Venus

The Synestia Model: An "Enhanced" Stage of the GIM

And now we consider the latest attempt to explain some of the similar chemical isotope ratios between the Moon and Earth. This is the SYNESTIA model (Lock, et al. 2018). A SYNESTIA is defined as "a rapidly spinning donut-shaped mass of vaporized rock". In computer simulations of giant impacts of rotating objects, a synestia can form if the total angular momentum is greater than the co-rotational limit.

From S. J. Lock et al. (2018), The origin of the Moon within a terrestrial Synestia (Plain Language Summary): Journal of Geophysical Research, Planets, v. 123, p. 910

> The favored theory for lunar origin is that a Mars-sized body hit the proto-Earth and injected a disk of material into orbit, out of which the Moon formed. In the traditional Giant Impact Model the Moon forms primarily from the body that hit Earth and is chemically different from Earth. However, Earth and the Moon are observed to be very similar, bringing the traditional model into question. We present a new model that explains the isotopic and chemical compositions of the Moon. In this model, a giant impact, that is more energetic than in the traditional model, drives the Earth into a fast-spinning, vaporized state that extends for tens of thousands of kilometers. Such planetary states are called synestias.

COMMENT: In the author's view, this SYNESTIA model has much more energy than is necessary for forming planets and that there are other, and better, explanations for the similarities and differences of the chemistry of the Earth and Moon. This synestia model appears to be a last, and desperate, attempt to save the concept of a giant impact model. [Note: An artist's rendition of the synestia model is in Fell (2018).]

A few quotes from S. J. Lock et al. (2018) (main article) will give the reader some idea of the "flaws" associated with the ruling paradigm that held center stage for some 34 years (1984–2018)

From p. 911, paragraph 1:

> To date, lunar origin studies have not demonstrated that a single giant impact can explain both the physical and chemical properties of our Moon (Asphaug 2014; Barr 2016).

From p. 911, paragraph 2:

> This scenario, which we refer to as the canonical giant impact, has become the de facto working model for lunar origin. However, studies of the canonical impact and its aftermath have difficulty explaining some key observables of the Earth-Moon system, including: the isotopic similarity between Earth and the Moon; the lunar depletion in moderately volatile elements; the large mass of the Moon; and the present-day lunar inclination.

From p. 911, paragraph 5:

> The MVE depletion of the Moon is a key constraint on lunar origin models. In addition, the volatile depletion of the Moon has been used to argue for a process-based link between giant impacts and MVE loss. Indeed, the lunar depletion has been used to propose that Mercury would be depleted, if it was formed by a giant impact (Peplowski et al. 2011). We must understand the physical processes that led to volatile depletion on the Moon in order to place its data in the context of other bodies in the solar system.

From p. 912, paragraph 1:

> Predicting the final mass of satellites formed by giant impacts is challenging. The methods that are currently used for simulating giant impacts do not include the physics necessary for modeling lunar accretion; therefore, separate calculations of disk evolution are required to infer the mass of the satellite produced by a specific impact.

From p. 912, paragraph 1:

> Given the current results from giant impact calculations and available satellite accretion scaling laws, it is uncertain whether canonical giant impacts can form a sufficiently large moon.

From p. 912, paragraph 3:

> Despite the fact that it has not yet explained major characteristics of the Earth-Moon system, the giant impact hypothesis has not been rejected, primarily due to the lack of another viable mechanism for the origin of the Moon. A range of alternative impact models have been proposed (Canup 2012; Cuk and Stewart 2012; Reufer et al. 2012; Rufu et al. 2017), but each calls upon an additional process or a fortunate coincidence to better explain the Earth-Moon system. Hence, none of these recent variations on an impact origin have gained broad support.

COMMENT: The few sentences from paragraph 3 (above) are especially pertinent to the "History of Science". Is the GIM really the only possible model for the Origin of the Moon? A conclusion from the statement in the paragraph may be that the Giant Impact Model was considered as a "default" model and that any efforts to promote that model were welcome. Now that we know there is another step backward in the "default" process let us consider a PROGRADE GRAVITA TIONAL CAPTURE MODEL, as outlined in detail in Malcuit (2015, Chap. 4), as the "back-step" in the default model process. Also note that the Synestia model has NO explanation for the generation of a lunar magnetic field that might explain lunar rock magnetization.

6.1.2.11 Some Proposed Models for the Generation of a Lunar Magnetic Field Via the GIM

Two important questions for this discussion are: (1) Can the features of lunar rock magnetization tell us anything about the petrologic history of the lunar body? Does an explanation of lunar rock magnetization have any relation to other terrestrial bodies or fragments thereof with remanent magnetic rock signatures?

Earth and planet Mercury are the only terrestrial bodies that have active magnetic fields. The Earth's Moon, Mars, 4 Vesta, the Vestoid meteorites, the Angrite meteorites, and the metallic meteorites all have remanent magnetic signatures of various strengths. A successful explanation for lunar rock magnetization may help to elevate the Earth's Moon to the status of a "Rosetta Stone" of the inner Solar System, a concept suggested by Harold Urey many decades ago (Newell 1973). The Moon is a small body and it is difficult to explain how such a strong field can be generated in a small body without outside assistance. The coefficient of the momentum of inertia is nearly 0.4, that of a homogeneous sphere. Thus, it is very unlikely to have a metallic core. If it did have a small metallic core that core would be useless for use as a lunar dynamo. Planet Mars is ten times more massive than the Earth's Moon and there is substantial evidence that it had an internal dynamo early in its history. The martian magnetic field operated for a period of time between the formation of the martian crust (~4.5 Ga) and the age of the oldest magnetic signature from large impact craters (Arkani-Hamed 2004). There is general agreement that the generative mechanism ceased to operate sometime before 4.0 Ga. The mechanism was probably a core dynamo associated with a low viscosity, iron-rich metallic core or an iron-rich spherical layer surrounding a primitive silicate core as shown in Fig. 1.6a, b. The other small bodies with a remanant magnetic signature could have been magnetized by a similar magnetic field that affected the lunar body if they formed along with the Moon as Vulcanoid planetoids.

A partial list of features to be explained by a successful model of a lunar magnetic field:

- The primitive crust has a magnetic signature that was detected and mapped from orbit and is probably as old as 4.5 Ga
- The mare basalts and breccias have a weaker magnetic signature and many samples have been analyzed
- The strongest magnetization of the mare rocks, ~110 micro tesla, are dated at ~3.95 Ga
- There seems to be an exponential decrease in field strength to at least 3.2 Ga (less than 4 micro tesla
- Some samples of lunar rocks yield dates between 2.5 Ga and 1.0 Ga and have a magnetic signature of around 7 to 3 micro tesla
- A sample with a date at ~4.25 Ga records a magnetic field of ~20 micro tesla

B. P. Weiss and S. M. Tikoo (2014a) Review summary: The lunar dynamo: Science, v. 346, p. 1198)

Figure 6.7 shows a plot of the strength of remanent magnetism in lunar rocks over geologic time.

This is the ABSTRACT from the article:

> The inductive generation of magnetic fields in fluid planetary interiors is known as the dynamo process. Although the Moon today has no global magnetic field, it has been known since the Apollo era that the lunar rocks and crust are magnetized. Until recently, it was unclear whether this magnetization was the product of a core dynamo or fields generated externally to the Moon. New laboratory and spacecraft measurements strongly indicate that much of this magnetization is the product of an ancient core dynamo. The dynamo field persisted for at least 4.25 to 3.56 billion years ago (Ga), with an intensity reaching that of the present Earth. The field then declined by at least an order of magnitude by ~3.3 Ga. The mechanisms for sustaining such an intense and long-lived dynamo are uncertain but may include mechanical stirring by the mantle and core crystallization.

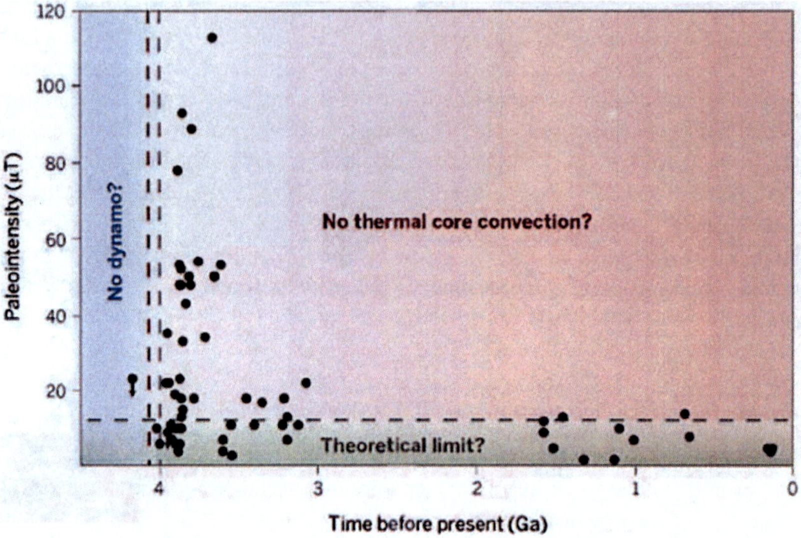

Fig. 6.7 This is the latest compilation of data from the peleointensity and age distribution of magnetized lunar mare materials. Note the "spike" in paleointensity values ~3.95 Ga which is the time of capture in the Prograde Gravitational Capture Model. (Diagram is from Weiss and Tikoo (2014b, Fig. 1) with permission from AAAS)

From p. 1198:

It is unknown whether the Moon has a fully differentiated and melted structure with a metallic core or retains a partially primordial, unmelted interior. The differentiation history of the Moon is manifested by its record of past magnetism (paleomagnetism). Although the Moon today does not have a global magnetic field, the discovery of remanant magnetization in lunar rocks and in the lunar crust demonstrated that there was a substantial lunar surface field billions of years ago. However, the origin, intensity, and lifetime of this field have been uncertain. As a result, it has been unclear whether this magnetization was produced by a dynamo in the Moon's advecting metallic core or by fields generated externally to the Moon. Establishing whether the Moon formed a core dynamo would have major implications for understanding its interior structure, thermal history, and mechanism of formation, as well as for an understanding of the physics of planetary magnetic field generation.

From p. 1198:

It has now been established that a dynamo magnetic field likely existed on the Moon from at least 4.5 billion to 3.56 billion years ago with an intensity similar to that at the surface of Earth today. The field then declined by at least an order of magnitude by 3.3 billion years ago. The early epoch of high field intensities may require an exceptionally energetic power source such a mechanical stirring from mantle precession. The extended history of the dynamo appears to demand long-lived power sources such as mantle precession and core crystallization.

From p. 1198:

The eventual availability of absolutely oriented samples and in situ spacecraft measurements of bedrock should enable the first measurements of the paleo-orientation of lunar magnetic fields. Such directional data could determine the lunar field's geometry and reversal frequency as well as constrain ancient local and global-scale tectonic events.

COMMENT: Note the major "spike" in paleointensity centered on 3.95 Ga in their diagram. Is this spike due to the Late Heavy Bombardment OR due to the DYNAMICS OF THE CAPTURE EPISODE?

From Weiss and Tikoo, (2014b, p. 9)

> The new data make it clear that a dynamo once existed on the Moon. A central remaining mystery is now the nature of the physical mechanism(s) that powered it. The challenge is to find a dynamo-generation mechanism that can account for several outstanding features of the lunar paleointensity record. Foremost is the extremely high paleointensities (average of 77 μT) inferred for the period 3.56 to 3.85 Ga . . .

COMMENT: Was the magnetic field generation continuous?
From the section: "Outlook" in Weiss and Tikoo, p. 9:

> A diversity of geophysical and geochemical evidence now indicates that the Moon formed a ~1 to 4 wt% metallic core. This core was once advecting and generated a dynamo magnetic field. As far back in time as we have paleomagnetic records (4.25 Ga), it appears that the Moon had a dynamo, consistent with an early, partially molten body like that expected after a giant impact origin. The Moon is therefore a highly differentiated object like the terrestrial planets.

COMMENT: In my opinion very few of the statements in the paragraph above are justified from the data and information in the article. These statements appear to be "hype" for the Giant Impact Model for lunar origin! My positive comment about the article is that the patterns of remanent magnetization in the lunar crust and in lunar basalts and breccias need to be addressed. Very few of the multitude of articles on the Giant Impact Model even mention the lunar magnetic field issue because the authors have no adequate explanation for it.

A. L. Scheinberg, D. M. Soderlund, and L. T. Elkins-Tanton (2018) A basal magma ocean dynamo to explain THE early lunar magnetic field: Earth and Planetary Science Letters, v. 492, p. 144

> The source of the ancient lunar magnetic field is an unsolved problem in the Moon's evolution. Theoretical work invoking a core dynamo has been unable to explain the magnitude of the observed field, falling instead one or two orders of magnitude below it. Since surface magnetic field strength is highly sensitive to the depth and size of the dynamo region, we instead hypothesize that the early lunar dynamo was driven by convection in a basal magma ocean formed from the final stages of an early lunar magma ocean; this material is expected to be dense, radioactive, and metalliferous.

COMMENT: The first sentence in the above quote says a good bit: "The source of the ancient lunar magnetic field is an UNSOLVED PROBLEM in the Moon's

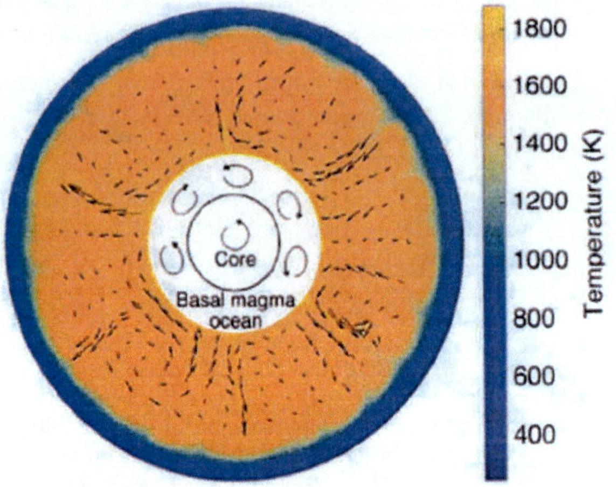

Fig. 6.8 A pictogram of a lunar dynamo generated in a Basal Magma Ocean. An assumption for this model is that the body of the Moon was completely molten early on. If this assumption cannot be justified, then the model will not work. (Diagram from Scheinberg et al. (2018, Fig. 1), with permission from Elsevier)

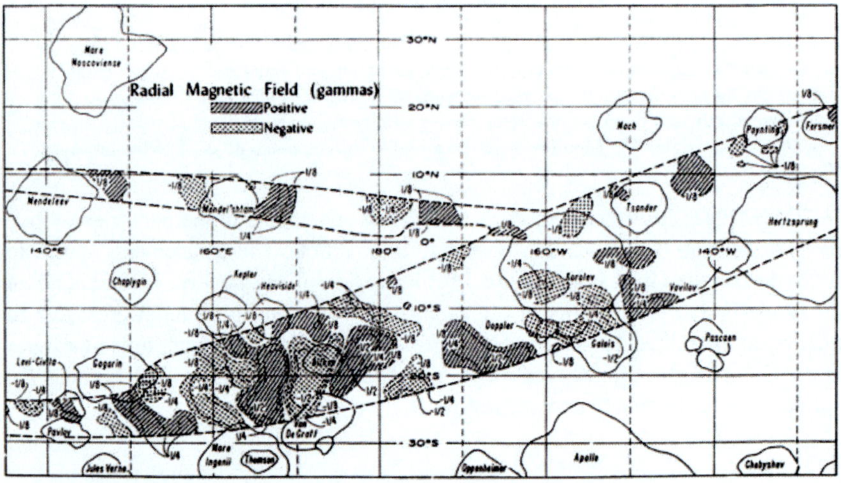

Fig. 6.9 Radial component of the lunar magnetic field mapped by the Apollo 15 and 16 sub-satellites from 135 degrees East to 125 degrees West at altitudes between 65 and 100 km. (From Russell 1980, Fig. 25, curtesy of American Geophysical Union). Note the alternating positive and negative remanent magnetic field patterns on the lunar backside crust. This pattern may relate to the TVI cell magnetic amplifiers shown in Figs. 6.15 and 6.16 in this chapter. The "UP" direction of the amplifiers is the positive direction and the "DOWN" direction of the amplifiers is the negative direction

evolution." In my view the Basal Magma Ocean model, illustrated in Fig. 6.8, needs a power source for running the dynamo.

Figure 6.9 shows the pattern of remanent magnetization imprinted on the primitive crust apparently about 4.5 Ga ago. This information is included in the "facts to be explained by a successful model".

6.2 The Use of Default Models for the Intractable Problems Associated with the Origin of the Moon and the Earth-Moon System

The default model that I am presenting is a combination of a model for the origin of the Moon as an independent Vulcanoid planetoid as well as the origin of the Earth-Moon system via a Prograde Gravitational Capture Model. Intimately associated with the Moon and Earth-Moon system origin models is a model for generating a lunar magnetic field that was proposed by Smoluchowski (1973a, b) during the era of the Apollo lunar landings. My version of his model featuring a series of "shallow shell magnetic amplifiers" is appropriate for explaining the origin of lunar rock magnetization during the two eras of massive "magmatism" associated with the Prograde Gravitational Capture Model.

6.2.1 Some Quotes from R. Smoluchowski (1973a), Lunar tides and magnetism: Nature, v. 242, p. 516–517

> Recent observations and studies of the lunar surface based on Apollo and Explorer flights show that some rocks brought to Earth are magnetized, that there are on the Moon weak, local, fairly randomly oriented magnetic fields but no overall poloidal field, that the magnetic perturbation of the solar wind by the surface of the Moon as observed by lunar orbiters is concentrated at latitudes lower than about 40° (ref. 2) and, finally, that the chemistry of the surface suggests high present and past radioactivity[3]. Here I propose a model, based on these observations, which may lead to the understanding of the origin of lunar magnetism a few billion years ago.
>
> It has been pointed out[1] that the observed lunar magnetism cannot be accounted for either by terrestrial or by solar magnetic fields. The possibility that the Moon once had a sufficiently hot liquid core in which a turbulent convective motion could have produced a magnetic dynamo driven either by a thermal gradient[4, 5] or by precession[6] seems to be marginal for energetic or electromagnetic reasons[7]. The existence of a liquid core is also in seeming conflict with the dynamic stability on the Moon[8]. It is interesting, therefore, to inquire whether a magnetic field could have been generated in an outer liquid shell.
>
> If such an external liquid shell indeed existed then, in principle, a magnetic field could have been generated in it either by thermally driven convection in analogy to what has been suggested for Jupiter[10, 11] or by tidal motions produced by Earth[7].
>
> This admittedly very simplified model leads to the conclusion that the outer layer of the Moon could have been magnetized by local, more or less randomly oriented, magnetic fields produced by tidal currents, that there would be no overall poloidal field and that the magnetization would be observable primarily near the lunar equator where, from equation (1), the tide producing force has its maximum.

Now let us consider a quote by Lipton (2005) that relates to models in general and can be applied to potential default models. A default model for our purposes is defined as "an option that is selected automatically unless an alternative is specified". In other words, a default model can be defined as a model, regardless of complexity, that is tentatively accepted until something better comes along". The main qualification of a default model is that it relates to a number of the features to be explained by a successful model.

6.2.2 P. Lipton (2005) Testing hypotheses: Prediction and prejudice: Science, v. 307, p. 219

> Observations that fit a hypothesis may be made before or after the hypothesis is formulated. Can the difference be relevant to the amount of support that the observations provide for the hypothesis? Philosophers of science and statisticians are both divided on this question, but there is an argument that predictions ought to count more than accommodations, because of the risk of "fudging' that accommodations run and predictions avoid.

And now for the latest attempt to explain a generation mechanism for a lunar magnetic field (an attempt associated with a default model).

6.3 An Attempt to Explain a Shallow-Shell Amplifier Model for the Generation of a Magnetic Field to Cause the Observed Lunar Rock Magnetization

A shallow shell, tidally-driven lunar magnetic dynamo system was first suggested by Smoluchowski (1973a, b) during the Apollo lunar landings era and has been essentially neglected in the recent era of lunar science. That was because there was no strong driving mechanism for the dynamos and there was no suggested mechanism for amplifying a lunar magnetic field that would cause the remanent magnetic signature recorded in lunar rock samples as well as the remanent magnetic signature recorded in the lunar crustal rock patterns. The testable model that I am proposing has two fairly distinct eras for generation of a fairly strong magnetic field. Era I commences a short time after the lunar body accretes as a Vulcanoid Planetoid (a lunar-sized planetoid that we can call LUNA). Era II begins with the dynamic activity associated with a lunar gravitational capture episode.

As stated in the second quote by Scheinberg et al. (2018), the source of lunar magnetism as well a mechanism for generating a sufficiently strong magnetic field are unsolved problems in lunar science. Because of the size and mass of the Moon the choices are very limited. I am pursuing a model proposed by Smoluchowski (1973a, b) because it appears to be very compatible with the PROGADE GRAVI TATIONAL CAPTURE MODEL for the origin of the Earth-Moon system. Smoluchowski (1973a, b) proposed his "THIN LIQUID BASALTIC SHELL DYNAMO" model based on very limited information but the observations and data that he used as the basis of his hypothesis are still valid today. It is interesting to note that his ideas have either been ignored or abandoned because of the lack of an adequate driving force for the "thin liquid basaltic shell dynamo". The other model that has been rejected or not seriously considered by the vast majority of lunar scientists in any way, shape, or form is the PROGRADE GRAVITATIONAL CAPTURE MODEL? Now it appears that a combination of these two ideas can be used as a default model solution for both the "LUNAR MAGNETISM" problem and the "ORIGIN OF THE MOON" problem. Furthermore, several very testable PRE DICTIONS can be made on the basis of this combination of models.

6.3.1 General Predictions from the Model

GENERAL PREDICTION # 1: There should be two very distinct eras of very intense lunar rock magnetization: the first at about 4.5 Ga (soon after the origin of the lunar body) and the second at about 3.95 Ga (the time of the initiation of the Prograde Gravitational Capture episode).

GENERAL PREDICTION # 2: The first era of lunar rock magnetization should be intense but short lived on the order of a million years or perhaps much less. In contrast, the second era of rock magnetization should be initially intense but exponentially decrease to a very low value in about 300 million years.

These predictions are in great contrast to the general assumption of the lunar science community that "Apollo samples of lunar crust indicate a surface field of ~77 micro Tesla that was present between 4.2 and 3.56 Ga and likely continued at a lower magnitude for billions of years" (Scheinberg et al. 2018, p. 144). In my view, the 4.2 Ga beginning date should be pushed back to ~4.5 Ga to be consistent with the magnetic lunar surface rock patterns recorded via orbiting spacecraft as shown in Fig. 6.7. This long timeframe for lunar rock magnetization appears to be an accommodation for the currently very popular GIANT IMPACT MODEL for the origin of the lunar body BUT the GIM proponents have not proposed a mechanism for propelling this continuously operating lunar magnetic field.

6.3.2 Two Major Eras of Lunar Rock Magnetization

ERA I: This era is associated with the magnetization of the primitive lunar anorthositic crust. In the PGCM which is presented in detail in Malcuit (2015, Chap. 4), the lunar body (Luna) forms as a Vulcanoid planetoid (Evans and Tabachnik 1999, 2002) at about 0.15 AU [note also that this general place of origin was suggested by Cameron (1972, 1973)]. Soon after accretion the outer 600 km or so of the lunar body was remelted via electromagnetic induction heating (Sonett et al. 1975) probably during a sequence of FU Orionis outbursts. Differentiation of the magma ocean results in a plagioclase-rich crust formed by "flotation" (Wood et al. 1970;

Gast 1972; and others). As the temperature of the crustal complex falls below the Curie point, the magnitude and direction of the ambient magnetic field gets recorded in ferromagnetic minerals. The solar gravitational tidal amplitude would be about 11 meters (with Luna in a circular orbit at 0.15 AU and a displacement Love number at 0.5). The rotation of the crustal complex then is retarded relative to the lunar interior beneath the LMO and a series of Tidal Vorticity Induction cells (Bostrom 2000) are formed within this magma ocean. A solar magnetic flux is then wrapped and amplified by the rotational action of the electrically conducting magma ocean basalt. The resulting amplifier flux would be several times more intense than the initial "seed" flux of the solar magnetic field. The three-dimensional morphology of the rotating amplifier cells would be like curved, doubly terminated roller bearings. The strongest resulting magnetic field is predicted to be in the lunar equatorial zone of this era and the resulting magnetic flux at the lunar surface would be very weak in the lunar polar areas. There would be plus and minus directions to the flux associated with the rotating amplifier cells (rotating in the retrograde direction relative to the rotation direction of the lunar body). As the LMO cools, the viscosity of the lunar basalts in the LMO increases and the magnetic amplifiers gradually cease to operate.

Figure 6.10 consists of two scale sketches which relate to the first era of lunar rock magnetization.

I should note here that there could be several sibling Vulcanoid Planetoids formed at a similar time and place as Luna (Evans and Tabachnik 1999, 2002). Each of these bodies would experience the FU Ori electromagnetic heating and magma ocean formation. The details of magma ocean formation and of the magnetic amplifier system would depend on the size of the body, the distance from the sun, the rotation rate of the body after accretion, and perhaps other factors. These minor details could be important because these Vulcanoid planetoids could be the source region for "volcanic asteroids" like Vesta, Angra, and other asteroids and meteorites of similar composition.

ERA II: The era commences with a successful capture episode. It is necessary to dissipate 1–2 E28 joules of energy in the body of Luna to capture it by purely gravitational processes. With the initial encounter of a capture scenario at 1.43 earth radii the lunar body would undergo a tidal amplitude (radial distortion) of about 400 km and within the first few years the lunar body would experience tidal amplitudes of around 200 km. The initial encounter would deposit enough energy to remelt a zone of lunar basalt about 50–100 km thick in the LMO region. The maximum tidal amplitude for the earth-like body is about 20 km and there would be several perigee passages that would raise rock tides of around 10 km amplitude. After the initial irregular dynamics of the capture process, the lunar orbit stabilizes to a somewhat regular apogee-perigee rhythmic motion. During this somewhat hectic era, tidal vorticity induction cells (Bostrom 2000) are formed again in the LMO. For this second era the "seed" magnetic field is Earth's magnetic field and this weak field, which is strongest during perigee passages, is amplified by the rotating amplifier cells operating, again in the retrograde direction, in the LMO. The time scale for orbit circularization is about 300 million years to about 20% eccentricity and about 600 million years to about 10% eccentricity. The circularization of the

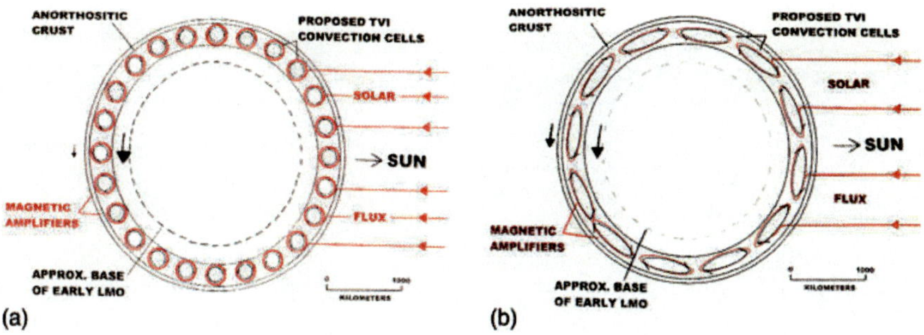

Fig. 6.10 Diagrammatic sketch of the first era of lunar magnetism. (**a**). The early phase when the differential rotation is the most rapid. (**b**). The later phase when the differential rotation has decreased. Note: The timescale for operation of this era is brief, probably in centuries to millennia of time. Also note the direction of motion of the magnetic amplifiers is in the retrograde (clockwise) direction when viewed from the lunar north pole. (From Malcuit (2015, Figs. 4.4 and 4.5) with permission from Springer)

lunar orbit is an exponential function and the resulting decay of the retrograde motion of the tapered roller bearing cells is predicted to be an exponential. This second era of lunar magnetism is illustrated in Fig. 6.11. A more complete treatment of the post-capture orbital evolution can be found in Malcuit (2015, Chap. 4).

The author realizes that this capture model is complex but simple capture models have not been successful. For a capture model there has to be a reasonable place of origin for a volatile-poor and iron-poor lunar body. Forming lunar-like bodies inside the orbit of Mercury (as first suggested by Cameron 1972, 1973) necessitates a transfer scheme for moving Luna and sibling planetoids through the Mercury and Venus orbital zones to reach Earth orbit. Many of these planetoids would not survive this trip and would be casualties of collision, fragmentation, or other fates. A lunar-like body can only be successfully captured from an earth-like heliocentric orbit if the orbit of the planetoid is within + or – 3% of the eccentricity of the earth-like planet's orbit. Although this capture model appears complex, it may relate fairly directly to lunar rock magnetization patterns and such patterns may be the key tests of the PROGRADE GRAVITATIONAL CAPTURE MODEL. The capture model is a default model and it need not be considered seriously until the other models are laid to rest. The somewhat traditional GIANT IMPACT MODEL (in the driver's seat for 30 plus years) does not relate very favorably to lunar rock magnetization patterns. The recently proposed SENESTIA model for lunar origin (Lock and Stewart 2017; Lock et al. 2018) is much more energetic than the traditional GIM. But it is not clear how the SENESTIA model relates to the features of the lunar body and to the patterns of lunar rock magnetization.

Figure 6.12 shows a proposed time scale for the major events for the PGCM that incorporates a shallow-shell magnetic amplifier model for generation of a lunar magnetic field for two distinct eras of lunar history.

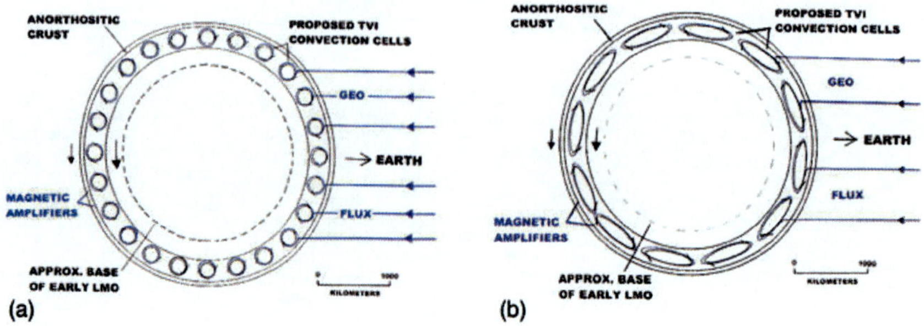

Fig. 6.11 Diagrammatic sketch of the second era of lunar magnetism. (**a**). The early phase when the differential rotation is most rapid. (**b**). The later phase when the differential rotation has decreased. Note: The timescale for operation of this era is several 100 s of Ma. (From Malcuit (2015, Fig. 4.33) with permission from Springer)

6.3.3 More Detailed Predictions from the Model

Listed below are more specific predictions from the PROGRDE GRAVITATIONAL CAPTURE MODEL (PGCM) for patterns of remanent magnetism recorded in the primitive lunar crust and in lunar rock samples.

PREDICTION I: The strongest remanent magnetism should be recorded in the primitive crust in the equatorial zone of the lunar body at the time when it was a Vulcanoid Planetoid. The polar zones of the planetoid should record a weak remanent magnetism that may relate to the solar magnetic flux at the time.

PREDICTION II: There should be positive and negative directions of remanent magnetization recorded in the lunar crust. These plus and minus directions should relate to the toroidal magnetic field of the amplifier cells. These two predictions for the remanent crustal magnetization are consistent with the maps showing the patterns of remanent magnetization like the one shown in Fig. 6.13.

PREDICTION III: There should be very little magnetic activity recorded between about 4.4 Ga and 3.95 Ga because the amplifier cells have no motive force. The exception to this is that close encounters with Mercury and/or Venus could cause some remelting in the LMO and a temporary rejuvenation of amplifier cells. The "stray" field, in this case, would be a weak solar field. The data point at about 4.25 Ga on Fig. 6.7 might be explained by this type of mechanism.

PREDICTION IV: The strongest remanent magnetic signature from mare rocks should be in rocks of about 3.95 Ga: *i.e.*, the time of lunar capture. There should be an exponential decrease in remanent magnetization from the time of capture to about 3.35 Ga (about 600 million years after capture). Such a pattern is consistent with the information compiled by Weiss and Tikoo (2014a, b) and shown in Fig. 6.7.

Possible explanation of data points from lunar samples dated after 3.0 Ga: Elevated lunar tidal amplitudes would be associated with Planet Orbit – Lunar Orbit resonance episodes shown in Fig. 6.12. Such episodes would probably result

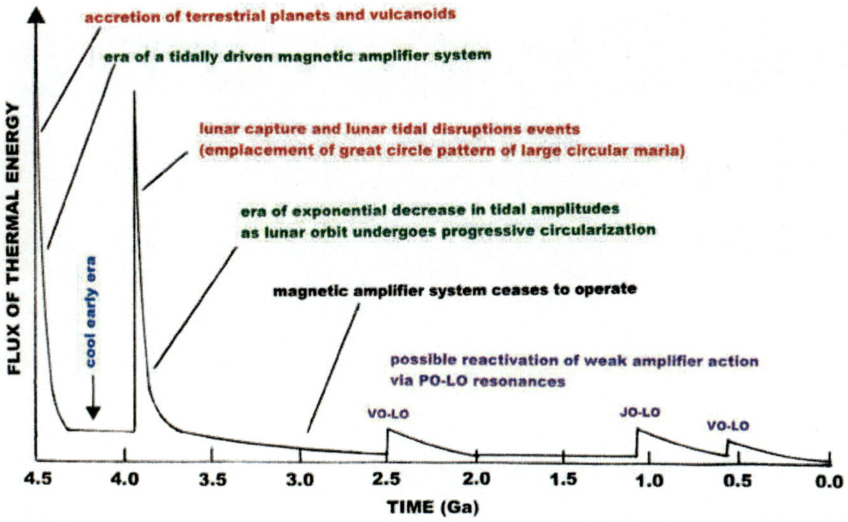

Fig. 6.12 Diagram showing some of the major events for a model featuring generation of a lunar magnetic field at two specific eras in lunar history via a shallow-shell magnetic amplifier system. Shown also are three additional eras of potentially high tidal amplitudes that could be associated with two Venus Orbit – Lunar Orbit resonances and one Jupiter Orbit – Lunar Orbit resonance as explained in Malcuit (2015, Chap. 8). These three PO-LO resonances may relate to young lava flows on the lunar surface that have been recognized by low crater counts (see Schultz and Spudis 1983). Also, the ~20 micro-Tesla data point at 4.25 Ga may relate to a close non-capture encounter or two with planets Venus and/or Mercury. Such close encounters could reactivate the amplifier system if the LMO viscosity is within a certain range of values

in some lunar magmatism and young lava flows (age determination based on crater counts) have been located at various locations on the lunar surface (Schultz and Spudis 1983). Whether there would be a sufficiently strong magnetic flux to be recorded by the cooling basalt is an open question.

6.3.4 Summary for Sects. 6.1, 6.2, and 6.3

A list of items to be explained by a successful model for the origin of the moon and the earth-moon system (a repeat of the list in Sect. 6.1):

- the anhydrous nature of lunar rocks and minerals
- The Potassium Index for solar system bodies
- Volatile element depletion patterns for solar system bodies
- Body density differences between the Earth and Moon
- Oxygen isotope information for Earth, Moon, meteorites, and other bodies in the inner solar system

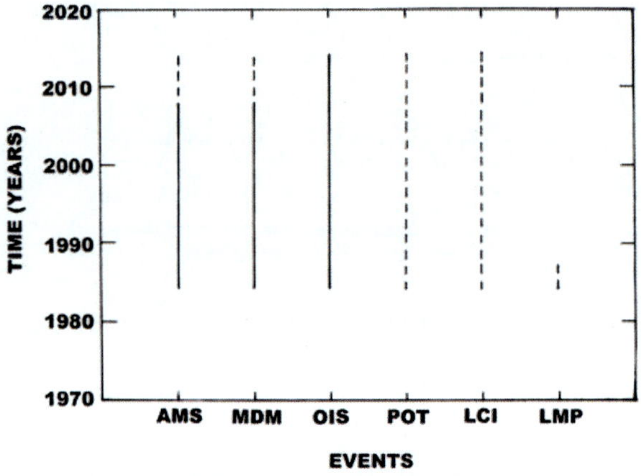

Fig. 6.13 Graphical summary issues associated with the development of the Giant Impact Model from 1984 to Present. SOLID LINE means that the issue is fairly well settled for the mainline (traditional) model of the GIM. DASHED LINE means that the item is debatable (i.e., there is no general agreement); LACK OF A LINE means that there is no reasonable explanation of the issue at this time. Abbreviations: AMS angular momentum of earth-moon system, MDM mass and density of the moon, OIS oxygen isotope similarities of the earth and moon, POT place of origin of Theia, LCI lunar core issue, LMP lunar rock magnetization patterns

- Lunar crust and mare rock dates
- The origin of lunar maria
- Temporal and spatial patterns of lunar mare rock distribution
- The asymmetry of lunar mass distribution
- TEMPORAL AND SPATIAL PATTERNS OF LUNAR ROCK MAGNETIZATION
- Geochemical, mineralogical, and petrological features of the Moon
- Geochemical, mineralogical, and petrological features of the Earth

As the reader has probably noticed, the GIM appears to offer solutions to very few of the items on the list of "facts to be explained by a successful model" and has no good hint of a reasonable explanation for the problem of lunar rock magnetization.

There are two ways to end this discussion of LUNAR ORIGIN MODELS:

1. Admit that the results of the GIANT IMPACT MODEL (in all of its guises) do not offer a reasonable solution to the origin of the Moon and then yield to another default model, and there is at least one remaining, the PROGRADE GRAVITA TIONAL CAPTURE MODEL.
2. Continue research efforts on some version of the Giant Impact Model (like the SYNESTIA model) with the hope of finding a useful solution to the lunar origin problem.

Figure 6.13 is an attempt at a graphical summary for the development of the Giant Impact Model from the time of the Kona Conference in 1984 to the Present.

The information in the time-scale chart is not very favorable to the Giant Impact model. The only favorable argument that has not changed is the one that led to the initiation of the concept of a giant impact: i.e., the nearly identical oxygen isotope ratios of the Earth and Moon. As the chart shows, all of the other arguments that seemed to be in support of the GIM now are not very convincing.

6.3.5 *The Prograde Gravitational Capture Model (PGCM) as a Default Model*

I will admit that this PGCM is complex but it does relate to most of the items on the list of "facts to be explained by a successful model". I really do not want to attempt to convert GIM enthusiasts to the PGCM camp. If you volunteer, fine; if you want to continue pursuing the GIM as your favorite default model, fine.

As Bill Hartmann (1986) stated in the Preface for the book on the Kona Conference, "will a wholly new model suddenly emerge from the morass of today's partial models?" Such is the progress of science: DEFAULT MODEL after DEFAULT MODEL!

6.4 Part II. Building on Success. Extension of the Giant-Impact Model for the Solution of Other Problems in the Planetary Sciences

Judging from recent reports like the Royal Society Volume (Stevenson and Halliday 2014) as well as the articles by Canup (2013), Elliott (2013), and Stewart (2013), the supporters of the GIM have not been all that convincing that they have a reasonable solution for THE ORIGIN OF THE MOON AND THE EARTH-MOON SYSTEM. The Synestia model brings much more energy and angular momentum to the system but, as yet, not many reasonable solutions. Furthermore, the success of the model for explaining other problems in the history of the Solar System as well as the history of development of individual planets and planet-satellite systems has not been very successful. The various authors, however, have been successful in the identification of significant problem areas in the planetary sciences. Again, the approach of this critique is to use many direct quotes from relevant articles and then to comment on the concepts presented in the articles.

These explanations are in order of distance from the sun: Mercury, Venus, Mars, Asteroids, Uranus, Neptune.

6.4.1 On the Origin of Planet Mercury

6.4.1.1 W. Benz, W. L. Slattery, and A. G. W. Cameron (1998) Collisional stripping of mercury's mantle: Icarus, v. 74, p. 516–528

From the Abstract, p. 516:

> We investigated the conditions under which a giant collision between a hypothetical proto-Mercury and a planet one-sixth it mass would result in a loss of most of the silicate mantle of the planet, leaving behind an iron-rich planet and thus explaining the anomalously high density of Mercury. . . . . We show that a head-on collision at 20 km/sec and an off-axis (impact parameter equal to half the radius of proto-Mercury) collision at 35 km/sec are about equivalent as far as damage to proto-Mercury is concerned. . . . . Preliminary estimates show that most of the ejected mass is probably removed from Mercury-crossing orbits. If this turns out to be true, a giant collision is a plausible explanation for the strange density of Mercury.

6.4.1.2 W. Benz, A. Anic, J. Horner, J. A. Whitby (2007) The origin of mercury: Space Science Reviews, v. 132, p. 189–202

From p. 201:

> We have confirmed, using SPH models with a significantly higher resolution than previous efforts that a giant impact is capable of removing a large fraction of the silicate mantle from a roughly chondritic proto-Mercury. The size and velocity of the impactor were chosen to be consistent with predictions of planetary formation and growth, and a plausible Mercury can be obtained for several assumptions about initial temperatures and impact parameter.

This quote is from p. 202:

> Confirming the collisional origin of the anomalous density of Mercury would go a long way toward establishing the current model of planetary formation through collisions which predicts giant impacts to happen during the late stages of planetary accretion. Hence. Small Mercury has the potential to become a Rosetta stone for the modern theory of planet formation!

COMMENT: Malcuit (2015) presents a more "conventional" model for making a high density planet at the orbital radius of planet Mercury. The model features a super-thermal phase of the X-Wind model of Shu et al. (2001). The thermal action vaporizes metallic components of the protosolar cloud and propels the metallic vapor out to about 0.4 AU. The vapor then condenses as it cools to form an "iron line" (ring). This iron-rich condensate along with high-temperature silicate minerals accrete to form an iron-nickel enriched planet with a high specific gravity. In my view, metal-rich planetoids accreting to form Mercury can be explained as the result of early, high temperature X-Wind activity. Using the X-Wind and Iron Line concepts there is no need for a giant impact episode for the formation of planet Mercury as suggested by Benz et al. (1998) and Benz et al. (2007). Although the chemistry and other features of planet Mercury are very interesting and important, I think that the Earth's Moon is a better choice as a "Rosetta Stone" of the inner Solar System.

6.4.2 *Origin and Evolution of Planet Venus Via A Giant Impact Scenario*

6.4.2.1 J. H. Davies (2008) Did a mega-collision dry Venus interior?: Earth and Planetary Science Letters. v. 268, p. 376–383

From the Abstract, p. 376:

> To overcome the above difficulties in explaining a dry Venus interior, a new hypothesis is proposed that Venus formed by a near head-on collision of two large planetary embryos, as might be expected from favoured oligarchic planetary accretion. Such a collision would be sufficiently large to melt totally and briefly vapourise a significant portion of both bodies. This would allow much of the released water to react rapidly with iron. Depending upon the reaction hydrogen is either expected to escape to space or enter the core. Oxygen would form FeO, most of which would enter the core, together with other iron reaction products. Most everything else not caught in the hydrodynamic escape driven by any hydrogen stream would be gravitationally retained by the final body.

A quote from the "conclusion", p. 381:

> I have pointed out the current view of relatively quiescent formation of Venus is probably inconsistent with its much drier interior then Earth. In contrast I propose that Venus suffered a maga-collision during its formation. This allowed water to be removed by reaction with iron. Big collisions are not rare, but expected. This hypothesis would suggest that Earth was strange in having only 'small' collisions; while Venus and Mercury and possibly Mars suffered larger collisions in their evolution. The suggested result of these large collisions has included Mercury losing virtually all of its mantle (Benz et al. 1988); and Venus becoming dry. Mars is a smaller body and would not have suffered as large a collision as Venus and hence it probably was able to retain more water in its interior.

COMMENT: Davies (2008) suggests that the dry condition of Venus (e.g., Baines et al. 2013) can be explained as a result of a head-on collision of two roughly equal-sized bodies. The intent of this paper by Davies is to advance the "band-wagon" concept that large collisional scenarios can be used to explain many Solar System phenomena better than any other explanations. I think that a RETROGRADE GRAVITATIONAL CAPTURE SCENARIO, first proposed by Singer (1970) and presented in significant detail by Malcuit (2015, Chap. 6) explains the condition of planet Venus much better than the giant impact explanation of Davies (2008). In the model of Malcuit (2015) a 0.5 moon-mass Vulcanoid planetoid, named Adonis, is captured into a retrograde orbit about 300 Ma after Solar System formation. After circularization of the post-capture elliptical orbit (about 100 Ma), the satellite undergoes a lengthy circular orbit evolution. Over about 3.5 Ga the circular orbit gradually evolves into an ever smaller orbit raising ever higher amplitude rock and ocean tides on the planet. Tidal energy dissipation in the broadly defined equatorial zone gradually increases to the point where ocean water begins to evaporate to the Venusian atmosphere. The thermal regime of the crust-upper mantle system increases exponentially as the radius of the satellite orbit decreases and planetary rotation rate decreases. Eventually the orbit of Adonis decreases so that it enters the Roche limit for a "solid" body (at about 1.6 venus radii), breaks up in orbit, and then falls to the surface through the dense carbon dioxide atmosphere that formed from the sequential degassing of the Venusian mantle. In this model we are left with a dense carbon dioxide atmosphere causing a super "greenhouse" effect, a fairly recent global resurfacing event, a planet rotating very slowly in the retrograde direction, and a planet with a very low obliquity. In summary, I think that this Retrograde Gravitational Capture Model explains much more about planet Venus than does the explanation of the giant impact scenario of Davies (2008).

6.4.2.2 From R. M. Canup (2013) Lunar conspiracies: Nature, v. 504, p. 27

> Current theories on the formation of the Moon owe too much to cosmic coincidences, says Robin Canup. She calls for better models and a mission to Venus

COMMENT: I agree that we should return to Venus. A testable RETROGRADE GRAVITATIONAL CAPTURE MODEL with predictions about features of the planet are in Malcuit (2015, Chap. 6).

6.4.2.3 C. Gillman, G. J., Golabek, and P. J. Tackley (2016) Evolution of a single large impact on the coupled atmosphere-interior evolution of Venus: Icarus, v. 268, p. 295

From the Abstract:

> We investigate the effect of a single large impact either during the Late Veneer or Late Heavy Bombardment on the evolution of the mantle and atmosphere of Venus.

COMMENT: This study builds on the Giant Impact paradigm. It does confront issues on the early history of Venus but the authors admit that it is difficult to relate their model to the present condition of Venus (i.e., the Global Resurfacing Event). Although this article by Gillman et al. (2016) is more recent and more detailed on the effects of giant impacts on a venus-like terrestrial planet, I think that the RETROGRADE GRAVITATIONAL CAPTURE scenario as outlined in Sect. 6.4.2.1 may be a much more complete explanation for the history of planet Venus than any giant impact scenario that has been proposed at this time in the history of planetary science.

6.4.3 *Origin of the Small Satellites of Mars*

6.4.3.1 R. A. Craddock (2011) Are Phobos and Diemos the result of a giant impact: Icarus, v. 211, p. 1150–1161

From the Abstract, p. 1150:

> Despite many efforts an adequate theory describing the origin of Pbobos and Deimos has not been realized. In recent years a number of separate observations suggest the possibility that the martian satellites may have been the result of a giant impact.

From the Abstract, p, 1150:

> The low mass of Phobos and Deimos is explained by the possibility that they are composed of loosely aggregated material from the accretion disk, which also implies that they do not contain any volatile elements. Their orbital eccentricity and inclinations, which are the most difficult parameters to explain easily with the various capture scenarios, are the natural result of accretion from a circum-planetary disk.

From the section on "satellite formation", p. 1155:

> It has become widely accepted that the Earth's moon formed from the material inserted into orbit during a collision with a Mars-sized object. However, determining how this material actually goes on to form the Moon remains unclear.

6.4.3.2 Rosenblatt et al. (2016) Accretion of Phobos and Deimos from an extended disc stirred by transient moon: Nature Geoscience, v. 9, p. 581–583

From page 582:

Our results provide, for the first time, a working dynamical framework for the formation of two satellites orbiting Mars from an accretion disc generated by a giant collision that occurred in the early history of the planet, such as the one thought to have formed the Borealis basin.

COMMENT: I agree with Craddock (2011) that the small satellites of Mars are not easy to explain but I do not think that a giant impact scenario was involved in their formation. My explanation for the origin of the satellites of planet Mars comes under the heading of a "drop-off" model. I agree with Craddock (2011) that Phobos and Deimos are composed of loosely bound material of probably carbonaceous chrondrite origin. My favored scenario for the origin of Phobos and Deimos, which can be eventually calculated in detail, features a close encounter of an asteroid of carbonaceous chondrite material. During a close, fly-by encounter, within the weightlessness limit of a Mars-planetoid encounter system, asteroidal material is lifted off the sub-Mars region of the surface of the asteroid and is inserted into marocentric orbits. Most of the material that was tidally disrupted from the asteroid would either (1) impact onto the surface of Mars or (2) return to the surface of the asteroid. But some disrupted material could remain in both stable and unstable marocentric orbits. The asteroidal material in the unstable orbits would eventually (perhaps thousands, millions, and billions of years later) impact onto the martian surface. But a few orbiting bodies could attain stable or quasi-stable prograde orbits and eventually evolve into orbits like those of Phobos and Deimos with the aid of atmospheric drag in a primitive martian atmosphere.

Since the origin of Phobos and Diemos is an open question, the models of Craddock (2011) and the more recent models by Rosenblatt and Charnoz (2012) and Rosenblatt et al. (2016) deserve serious consideration. [See Chap. 3 for an extended discussion of the origin of the martian satellites as well as a discussion of the model by Rosenblatt et al. (2016).]

6.4.4 The Origin of Satellites of Asteroids Via Large Impacts Between Asteroids

D. D. Durda, W. F. Bootke, B. L. Enke, W. J. Merline, E. Asphaug, D. R. Richardson, and Z. M. Leinhardt (2004) THE FORMATION OF ASTEROID SATELLITES IN LARGE IMPACTS: RESULTS FROM NUMERICAL SIMULATIONS: Icarus, v. 170, p. 234–257.

From the Abstract, p. 243:

> We present the results of 161 numerical simulations of impacts into 100 km diameter asteroids, examining debris trajectories to search for the formation of bound satellite systems.

A quote from p. 243–244:

> Understanding how asteroid satellites form is important because:
>
> 1. they hold important clues to both the past and present collisional environments of the main asteroid belt;
> 2. models of their formation may provide constraints on internal structures of asteroids beyond those possible from observations of satellite orbital properties alone;
> 3. they represent numerous small-scale potential analogues for the early, large impacts believed responsible for the formation of the Earth-Moon and Pluto-Charon systems.

COMMENT: There is no need for large impacts to form satellites of asteroids as suggested by Durda et al. (2004). Several satellites of asteroids exist but they can be explained by somewhat "run-of-mill" collisions between asteroids.

6.4.5 Origin of the Excessive Tilt Angle of Planet Uranus

Some GIM enthusiasts would like to extend the main features of the model to the far reaches of the Solar System. The details of how the current satellite system of Uranus would fit into this scheme are not specified. A non-GIM model is presented in the following section as an alternative to the GIM scenario.

6.4.5.1 W. L. Slattery, W. Benz, and A. G. W. Cameron (1992) Giant impacts on a primitive Uranus: Icarus, v. 99, p. 167–174

From the Abstract, p. 167:

> Using smooth particle hydrodynamics we have carried out a series of simulations of collisions between a model of a primitive Uranus and impactors with masses ranging from 1.0 to 3.0 M_e. . . . . All simulations reported here have been carried out assuming a relative velocity at infinity between the planets of 5 km/sec. . . . . From these simulations we conclude that there is a fairly large range of giant impacts that could have produced the present period and inclination of the spin axis to the plane of the ecliptic and there is a subset of these that could have deposited suitable material in orbit from which the regular satellites of Uranus could have been assembled.

COMMENT: There is no need for a giant impact of a 1–3 earth-mass impactor (100–300 moon-mass) to explain the very high obliquity of planet Uranus. Kobo-Oka and Nakazawa (1995) developed an analytical model for explaining the 90-plus degree obliquity of Uranus by way of the orbital evolution of a retrograde satellite of about 16–20 moon-masses tidally interacting with planet Uranus (one earth mass = 81 moon masses). Their model simply needs a "front end". They need to start with a satellite like Neptune's Triton (but with a mass of about 20 times the Earth's Moon) that is in a retrograde orbit. One way to get a large satellite into a retrograde orbit is to capture it from a heliocentric orbit. The capture could be the result of a collisional capture (Goldreich et al. 1989) or a purely gravitational capture as suggested by Malcuit et al. (1992).

For either type of capture the logical source for an icy planetoid is the Kuiper Belt. Malcuit et al. (1992) found that capture of a Triton-mass satellite into a prograde orbit around Neptune was very improbable because of the very limited size of the prograde stable capture zones. In contrast, they found that retrograde stable capture zones were fairly large for planetoids encountering Neptune from a heliocentric orbit that is either slightly larger or slightly smaller than the orbit of Neptune.

Although we have not yet defined the geometry of retrograde stable capture zones for Uranus, we expect them to be very similar in morphology to those for planet Neptune and a Triton-mass planetoid. The bottom line is that a giant impact origin for the excessively large obliquity of planet Uranus may be physically possible but unnecessary. Retrograde capture of a 20 moon-mass planetoid, originating in the Kuiper Belt, and the subsequent very lengthy tidal evolution of the orbit will yield a more reasonable result than bringing in a 2–3 earth mass body from the Kuiper Belt.

6.4.5.2 A Non-GIM Alternative for the High Obliquity Condition of Planet Uranus

The following is a sequence of events that is suggested to result in the 98 degree obliquity of planet Uranus.

1. Retrograde capture of a two mars-mass planetoid from a Uranus-like heliocentric orbit (the planetoid originated as a Kuiper Belt Object).
2. Post-capture orbit circularization due to energy dissipation in the planetoid (a gassy-icy planet cannot dissipate much energy during close encounters). Orbit circularization occurs rapidly when the planetoid is in a retrograde orbit.
3. Circular orbit of somewhere between 20 and 30 uranus radii is obtained in 500 Ma or less (maybe much less).
4. The orbits of the regular satellites of Uranus are disturbed during periur passages. Both the regular satellites and the plane of the captured planetoid are probably near the common plane of the planets of the solar system.
5. Now the planetoid at 30 R_U with a period of about 58 earth days is in a circular retrograde orbit. This configuration gives about 533 months/Uranus year. The planet is rotating in the prograde direction with a period of about 10 hours. The regular satellites are assumed to be in the equatorial zone of the planet and they were probably heavily disturbed during the orbit circularization sequence.
6. The increase in the obliquity of Uranus begins and continues as the orbital radius of the planetoid decreases. Angular momentum of the planetoid orbit is slowly transferred to the tilting of planet Uranus and its associated disc of satellites and other orbiting debris. This is the model proposed by Kobo-Oka and Nakazawa (1995).
7. The planetoid orbit remains near the plane of the planets. Eventually as the planetoid orbit decreases in radius, the planetoid gravitationally interacts with the regular satellites of Uranus.
8. As the orbital radius of the planetoid decreases to near the Roche limit for a solid body, this gravitational interaction increases greatly.
9. Then the planetoid breaks up into three major parts and all three interact with the natural satellites and with each other.
10. Most parts of the disrupted planetoid fall onto Uranus but some may interact with the disc of regular satellites and are incorporated into the highly inclined plane of the regular satellites.
11. After about 4 Ga (or 3 Ga or 2 Ga) we end up with what we have today: (a) a planet with very high obliquity (about 98 degrees), (b) a disturbed satellite system, and (c) a phantom planetoid (i.e., one that no longer exists but one that could have caused the changes to the planet-satellite system).

The following pictograms relate to the Uranus tilting problem and are meant as an aid to help explain the Kobo-Oka and Nakazawa (1995) model. Figure 6.14 shows a generalized view of orbital conditions for gravitational capture and Fig. 6.15 is an overview of the orbital circulation sequence. Figures 6.16, 6.17, 6.18, 6.19 and 6.20 show five stages in the tilting operation for planet Uranus that result in the 98 degree obliquity of the planet as suggested by Kobo-Oka and Nakazawa (1995).

These are some of the conditions following orbit circularization. Uranus and its satellite system are rotating in the prograde direction with some small initial obliquity. The two mars-mass planetoid is in a near circular orbit and orbiting in the retrograde direction. This large planetoid orbit is near the plane of the planets. The sidereal period (month) of this large satellite is about 58 earth days. The Uranus year

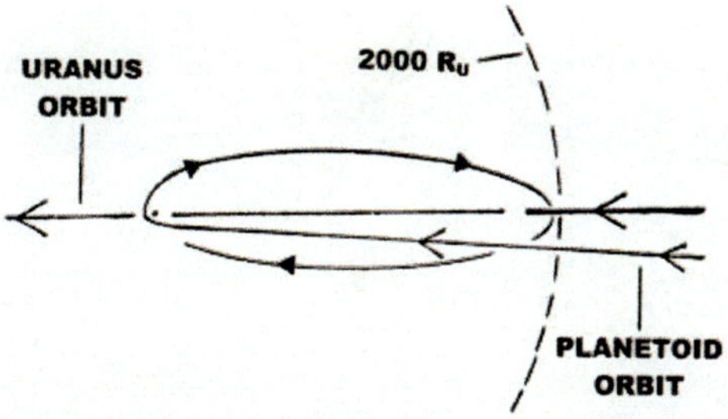

Fig. 6.14 Two dimensional view of capture of a planetoid into a retrograde orbit around Uranus. View from north pole of solar system. The sphere of influence (the Hill sphere) is located at about 2600 R_U. To be securely captured the captured body should have a major axes not exceeding 2000 R_U. The source of the body of the planetoid could be from the space between Uranus and Neptune or from the Kuiper Belt

Fig. 6.15 Overview of orbit circularization sequence after capture. Large orbit has major axis of about 2000 uranus radii (2600 R_U is the maximum dimension). Small circular orbit is about 60 uranus radii for major axis (30 uranus radii from Uranus)

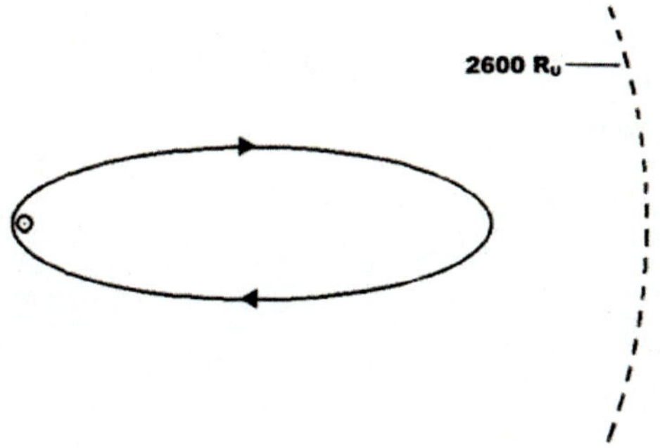

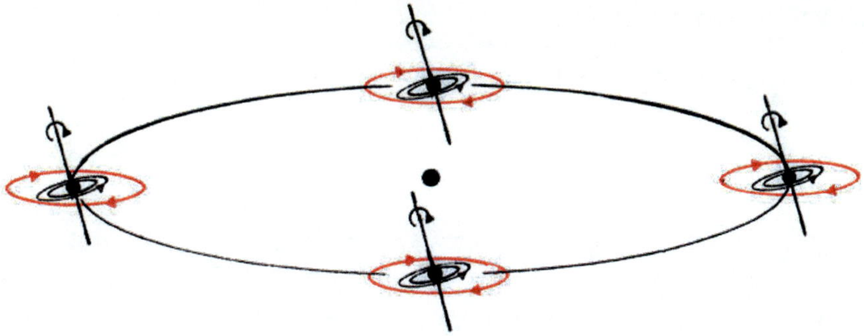

Fig. 6.16 Following orbit circularization the retrograde orbit of Discordia is in about a 30 uranus radii circular orbit and the obliquity of the prograde rotating Uranus and its satellite system is about 15 degrees

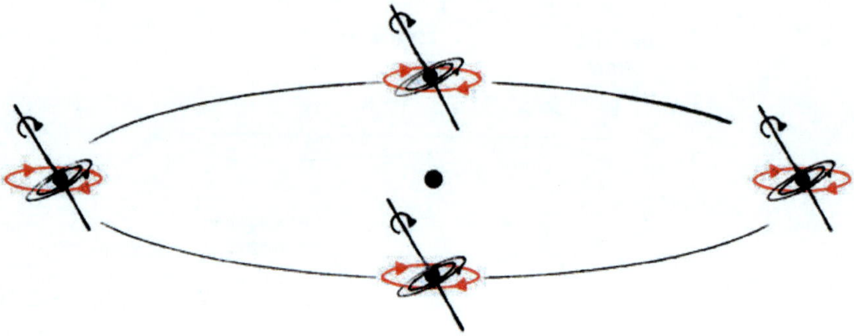

Fig. 6.17 The retrograde orbit of Discordia is now at about 20 uranus radii and the obliquity of prograde rotating Uranus and its satellites is at about 30 degrees

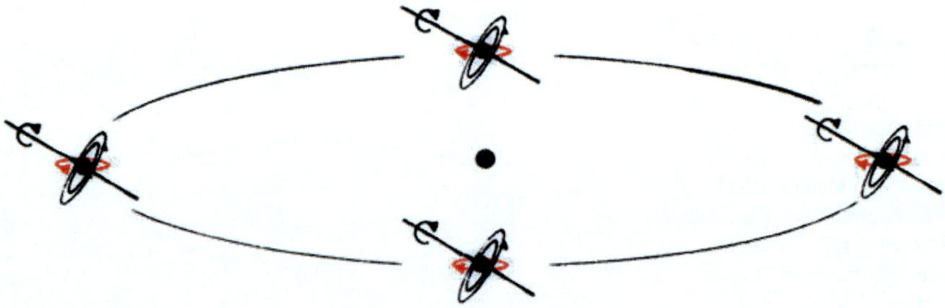

Fig. 6.18 The retrograde orbit of Discordia is now at about 10 uranus radii and the obliquity of the prograde rotating Uranus and it satellites is about 60 degrees

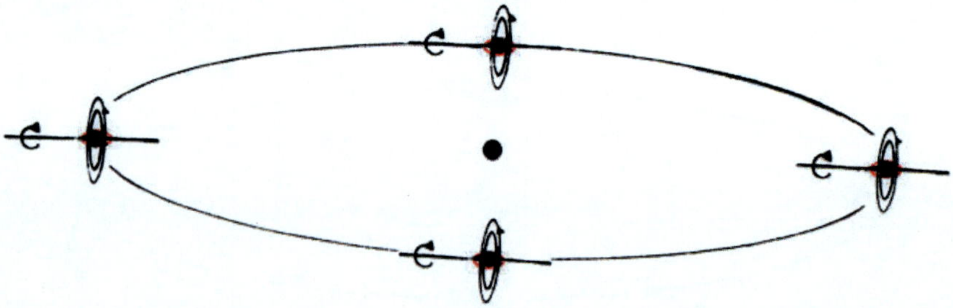

Fig. 6.19 The retrograde orbit of Discordia is now at about 5 uranus radii and the obliquity of the prograde rotating Uranus and its satellites is about 90 degrees

is about 84 years. Thus there are about 533 months/Uranus year with the large satellite at 30 Uranus radii. (Note: Let us name the phantom satellite DISCORDIA, the name of a Roman goddess of discord.)

In Fig. 6.19 the large retrograde satellite is still torqueing the system and the final obliquity will be about 98 degrees. During the circular orbit evolution of Discordia

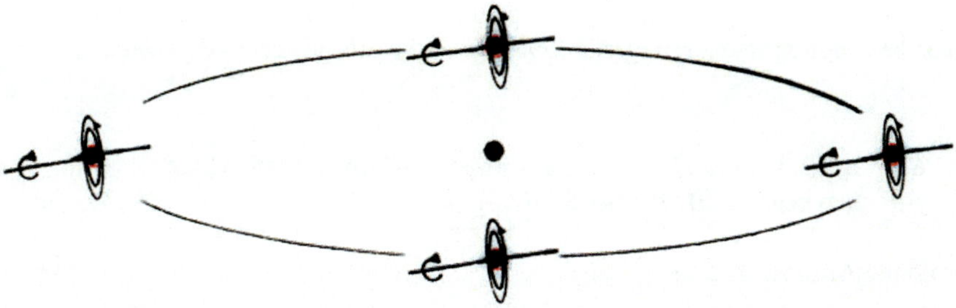

Fig. 6.20 The retrograde orbit of Discordia is now at about 2.5 uranus radii and the obliquity of the prograde (now retrograde) rotating Uranus and its satellite system is about 98 degrees

the satellite system is being severely affected by gravitational interactions with the captured satellite. All of the natural satellite orbits would be affected in some way (*i.e.*, collisions and perturbations resulting in collisions with Uranus; perhaps perturbations resulting in escape from the system).

Figure 6.20 shows the Uranus-Discordia system NEAR THE FINAL SCENE. Uranus and its highly perturbed satellite system is torqued to a 98 degree obliquity. The orbit of Discordia is within the Roche limit for a solid body and has broken up into 3 or more parts which undergo mutual destruction by collisions, close encounters, and escapes. All bodies that avoid escape from the system will eventually fall to the surface of Uranus mainly in the retrograde direction and mainly via atmospheric drag.

NOTE: Once the tidally disrupted debris from the captured satellite settles to the surface, then we are left with Uranus and its satellite system that we see today. The planet is rotating in the retrograde direction only because it was torqued into that conditions (i.e., the obliquity is over 90 degrees) by the now PHANTOM retrograde satellite, Discordia.

This again is, as Fred Singer (1970) stated for planet Venus, "the smile of the Cheshire Cat".

6.4.6 *Origin of the Pluto-Charon System*

The origin of Pluto has been an unsolved problem since the 1930's. In 1978 another complication was added to the puzzle, the retrograde satellite Charon. The orbit of Pluto is closer to the Sun than that of Neptune for part of its heliocentric orbit but the two bodies are protected from collision by an orbital resonance. I am presenting two models for the origin of the Pluto-Charon system. The first is a large impact model recently proposed by Canup (2005). The second is an escape model that my colleagues and I are considering. The original idea was proposed by Lyttleton

(1936) and has been modified by our group. It is still an escape model but Triton must be captured into a retrograde orbit before the escape encounter occurs.

6.4.6.1 R. M. Canup (2005) A giant impact origin of Pluto-Charon: Science. v. 307. P. 546–550

From the Abstract, p. 546:

> Pluto and its moon, Charon, are the most prominent members of the Kuiper belt, and their existence holds clues to outer solar system formation processes. Here hydrodynamic simulations are used to demonstrate that the formation of Pluto-Charon by means of a large collision is quite plausible. I show that such an impact probably produced an intact Charon, although it is possible that a disk of material orbited Pluto from which Charon later accumulated. These findings suggest that collisions between 1000-kilometer-class objects occurred in the early inner Kuiper belt.

From p. 550:

> The tendency for oblique low-velocity collisions between similarly sized objects to produce substantial amounts of material in bound orbit suggests that the impact generation of satellites is a common outcome of late-stage accretion, with the Earth-Moon ($q = 0.01$ and $J = 0.115$) and Pluto-Charon offering examples of the potential range of q and J in systems produced by such events.

NOTE: q is the mass ratio of the final planet-satellite system; J is the mass ratio of the impactor and the target planet.

6.4.6.2 R. A. Lyttleton (1936) On the possible results of an encounter of Pluto with the Neptunian system: Monthly Notices, Royal Astronomical Society, v. 97, p. 108–115

COMMENT: Raymond Lyttleton (1936), just 6 years after the discovery of Pluto, suggested that Pluto may be an escaped satellite of Neptune. An additional complication for any such explanation that was added in 1978 is the presence of the large satellite, Charon, and the charonoids (very small satellites of Pluto). Malcuit et al. (2018) have developed a model for the origin of Pluto as an escaped satellite of Neptune in which Charon is formed in the process of a very close gravitational encounter between a captured retrograde satellite, Triton, in a highly elliptical orbit, and an original satellite of Neptune in a circular prograde orbit. In this model the mass of Triton is substantially larger than that of proto-Pluto (the prograde satellite of Neptune). Proto-Pluto then has a very close (grazing) gravitational encounter with Triton that is well within the Weightlessness Limit of the Triton--proto-Pluto system. During the encounter, ice is "peeled off" the sub-Triton portion of proto-Pluto and goes into a retrograde orbit around planetoid Pluto (and the Pluto-Charon system is formed). The small charonoid satellites of Pluto (Styx, Nix, Kerberos, and Hydra), in this model, are simply masses of ice that attained stable or quasi-stable orbits relative to the gravitational influence of Pluto and Charon and are still in orbit around Pluto. THUS, AGAIN, A GIANT IMPACT MAY NOT BE NEEDED TO EXPLAIN THE PLUTO-CHARON-CHARONOID SYSTEM!

6.4.6.3 Modification and Expansion of the Concept of Raymond Lyttleton that Pluto is an Escaped Satellite of Neptune

The following is a series of sketches that I think satisfies Lyttleton's speculation and these sketches can be the basis for a series of 5-body numerical simulation calculations. An enterprising physics/computer science student perhaps can use this qualitative model as the basis for numerical simulations of a close encounter featuring a major ice peeling operation to form the body of Charon and results in the eventual escape of the Pluto-Charon system into a heliocentric orbit with the characteristics of the heliocentric orbit of Pluto. Figure 6.21 shows the heliocentric orbital geometry for the outer planets and Pluto. Figure 6.22 illustrates the heliocentric conditions necessary for a successful retrograde capture scenario for a triton-mass planetoid. Figures 6.23, 6.24, and 6.25 are sketches illustrating one possible orbital setting for the speculation of Lyttleton (1936) that Pluto is an escaped satellite of Neptune.

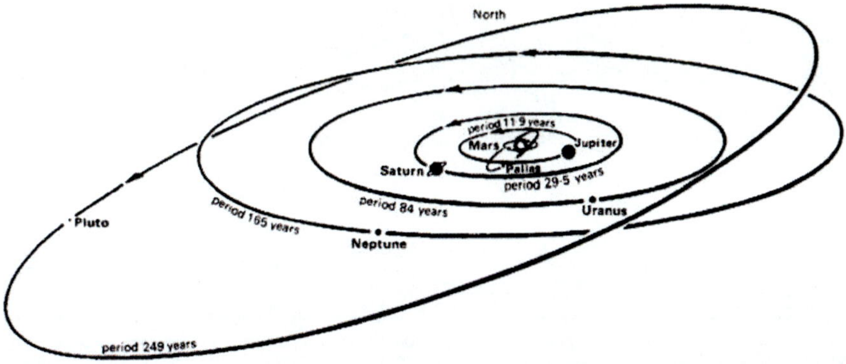

Fig. 6.21 Overview of the orbits of the planets in the Solar System (with Pluto included as a planet). Note that the orbit of Pluto is nearer to the Sun than is Neptune for a portion of Pluto's trip around the Sun. The orbits do not intersect because of the inclination of Pluto's orbit. All of the planets move in the prograde direction (counter-clockwise) around the Sun. View is from the north pole of the Solar System. (From Kopal (1973, Fig.1) courtesy of Oxford Univ. Press)

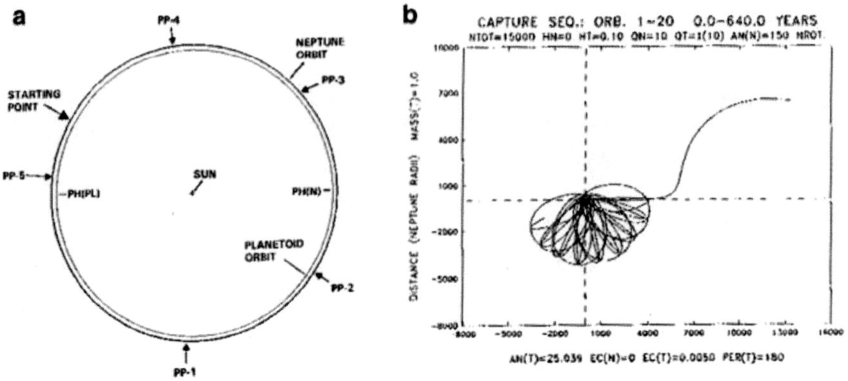

Fig. 6.22 Three-body numerical simulation of retrograde capture of a triton-mass planetoid by Neptune from an orbit that is inside the orbit of Neptune. The Neptune year is about 165 earth years in time. Sufficient energy is dissipated during the first encounter to insert Triton into a stable neptocentric orbit of about 4000 neptune radii. The irregularities in the orbital motion are due to solar perturbations. (**a**). Heliocentric orientation for the encounter. The simulations begins at the STARTING POINT; about 53 earth years later, the first encounter [perinept passage one (PP-1)] occurs. Four subsequent perinept passages are shown on this diagram. (**b**). Neptocentric plot for the encounter showing the morphology of the first 20 orbits. The elapsed time from the STARTING POINT until after the 20th orbit is 640 earth years. (Note: There will be more information on the Neptune-Triton system in Chap. 8. These diagrams are simply to demonstrate that tidal capture of a Triton-mass body by planet Neptune is physically possible)

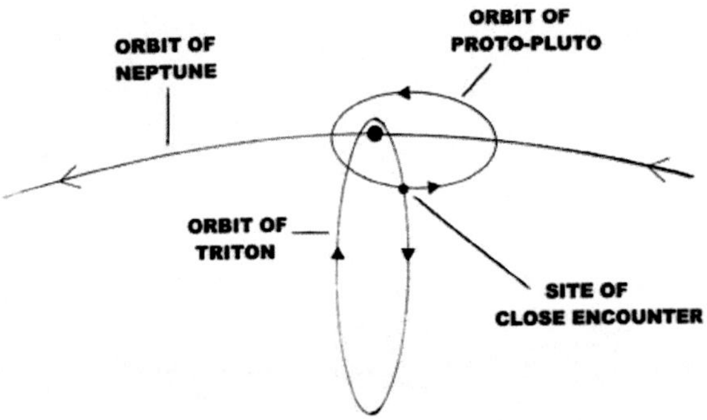

Fig. 6.23 Close-up of the orbital scene for this model for the origin of the Pluto-Charon system. Orbit of planet Neptune is prograde. Orbit of Proto-Pluto (an original satellite of Neptune) in orbiting Neptune in the prograde direction. Triton (a captured Kuiper Belt Object that is significantly more massive than proto-Pluto) is orbiting Neptune in the retrograde direction. The two satellites have slightly different orbital inclinations but the orbits can intersect. The site of a very close encounter is labelled. View is from the north pole of the Solar System

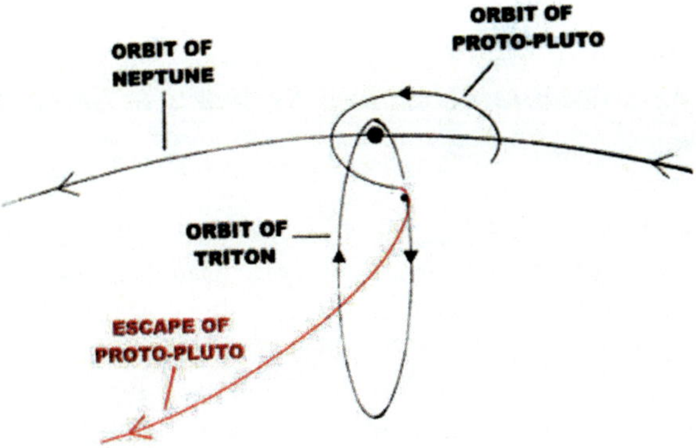

Fig. 6.24 Diagram illustrating the probably trajectory of proto-Pluto resulting from a very close encounter with the larger body, Triton. The encounter is in the retrograde direction (clockwise) about Triton but after the encounter the heliocentric (sun-centered) orbit of proto-Pluto becomes prograde. The orbit of Triton would be altered significantly but would remain retrograde. See Fig. 6.25 for the "transformation" of proto-Pluto to the Pluto-Charon System. View is from the north pole of the Solar System

Fig. 6.25 Close-up of the proposed very close encounter which can be slightly grazing. As the smaller body enters the Weightlessness Limit of the system, a mass of ice is dislodged by tidal processes from the Triton side of proto-Pluto and eventually goes into retrograde (clockwise) orbit around Pluto (i.e., proto-Pluto loses mass and becomes Pluto). The dislodged mass of ice attains a spherical shape due to self-gravity and it now Pluto's satellite, Charon. The charonoids are much smaller masses of ice that attain stable orbits and are small satellites of Pluto. Some of these small satellites have been named (Styx, Nix, Kerberos, Hydra). View is from the north pole of the Solar System

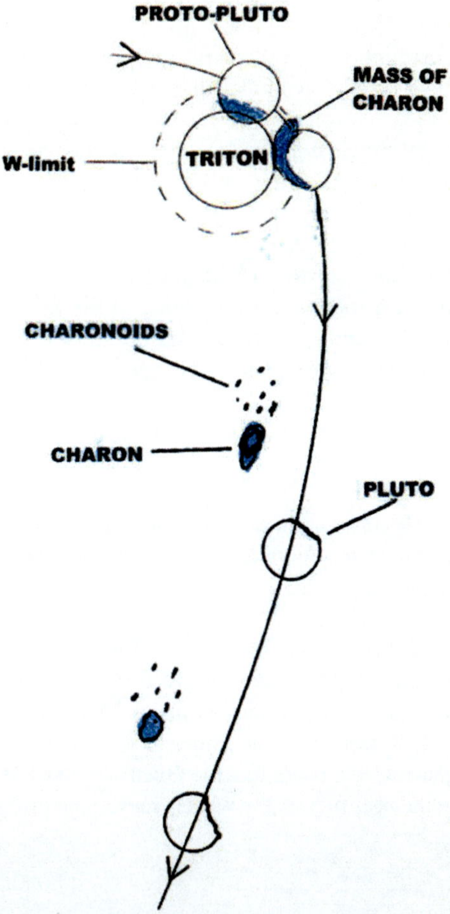

6.4.7 *Discussion of the Paradigm of Giant Impacts for the Solution of Unsolved Problems in Solar system Science*

The Question is: ARE THERE OTHER AND BETTER SOLUTIONS than GIANT IMPACT solutions for the Solar System issues discussed in this section of this book? I think that for each of these issues there are better solutions that do not involve giant impact scenarios. There is no doubt that large impacts occurred early in the history of the Solar System, but the results of these large impacts may be much different than the results stated in the above scenarios!

6.5 Summary

In this chapter we examined the results of the Giant Impact Paradigm for solving the problem of (1) the origin of Moon and the associated problem of the origin of the Earth-Moon system and (2) other outstanding problems in solar system science. There is no doubt that large impacts (i.e., about lunar mass and smaller) occurred in the early history of the solar system but have any of these large impacts resulted in reasonable solutions to solar system problems?

In my view the ultimate test for any successful explanation for the origin of the Moon and the origin and evolution of the Earth-Moon system is an explanation of lunar rock magnetization. There appear to be two different eras involved in the magnetization of lunar rocks. The first is the magnetic mechanism that imprinted a magnetic signature in the primitive crust – a crustal complex that crystallized soon after the formation of the body of the Moon about 4.5 Ga ago. The second era of magnetization is recorded in the lunar mare basalts and associated breccias. This second era started ~3.95 Ga ago and the magnetizing field exponentially decayed over about 300 Ma. In general, the Prograde Gravitational Capture Model explains these two eras of lunar rock magnetization and the proposed mechanisms for generating the two different eras of magnetism can be modelled and tested in detail in the future by interested scientists.

The promotors of the Giant Impact Model for the origin of the Moon and the origin of the Earth-Moon system have not yet identified a viable mechanism for generating a lunar magnetic field of appropriate strength within the body of the Moon.

In the second part of this chapter we examined some of the Giant Impact explanations for other unsolved, or partly solved, problems ranging from the origin of the high iron content of planet Mercury to the origin of the Pluto-Charon system. In each case, the Giant Impact explanation is less impressive than other solutions. In general, it appears that the Giant Impact "bandwagon" is slowing considerably and in the near future "the wheels may come off". There is no doubt that there have been

Giant Impacts (about lunar-mass bodies and smaller) on the terrestrial planets and larger bodies have impacted the outer planets but these impacts have not yet been demonstrated to result in any reasonable solutions of solar system problems.

References Cited

Agnor, C. B., Canup, R. M., & Levison, H. F. (1999). On the character and consequences of large impacts in the late stage of terrestrial planet formation. *Icarus, 142*, 219–237.

Alberede, F. (2009). Volatile accretion history of the terrestrial planets and dynamic implications. *Nature, 461*, 1227–1233.

Arkani-Hamed, J. (2004). Timing of the martian core dynamo. *Journal of Geophysical Research, 109*, 12. https://doi.org/10.1029/2003JE002195.

Asphaug, E. (2014). Impact origin of the moon? *Annual Review of Earth and Planetary Sciences, 42*, 551–578.

Baines, K. H., Atreya, S. K., Bullock, M. A., Grinspoon, D. H., Mahaffy, P., Russell, C. T., Schubert, G., & Zahnle, K. (2013). The atmospheres of the terrestrial planets: Clues to the origins and early evolution of Venus, Earth, and Mars. In S. J. Mackwell et al. (Eds.), *Comparative climatology of the terrestrial planets* (pp. 137–160). Tucson: University of Arizona Press.

Baldwin, R. B., & Wilhelms, D. (1992). Historical review of a long-overlooked paper by R. A. Daly concerning the origin and early history of the Moon. *Journal of Geophysical Research, 97*(E3), 3837–3843.

Barr, A. C. (2016). On the origin of Earth's moon. *Journal of Geophysical Research, Planets, 121*, 1573–1601.

Benz, W., Slattery, W. L., & Cameron, A. G. W. (1986). The origin of the Moon and the single-impact hypothesis, I. *Icarus, 66*, 515–535.

Benz, W., Slattery, W. L., & Cameron, A. G. W. (1987). The origin of the Moon and the single-impact hypothesis, II. *Icarus, 71*, 30–45.

Benz, W., Slattery, W. L., & Cameron, A. G. W. (1988). Collisional stripping of Mercury's mantle. *Icarus, 74*, 516–528.

Benz, W., Cameron, A. G. W., & Melosh, H. J. (1989). The origin of the Moon and the single-impact hypothesis, III. *Icarus, 91*, 113–131.

Benz, W., Slattery, W. L., & Cameron, A. G. W. (1998). Collisional stripping of mercury's mantle. *Icarus, 74*, 516–528.

Benz, W., Anic, A., Horner, J., & Whitby, J. A. (2007). The origin of Mercury. *Space Science Reviews, 132*, 189–202.

Bostrom, R. C. (2000). *Tectonic consequences of earth's rotation* (p. 266). Oxford: Oxford University Press.

Cameron, A. G. W. (1972). Orbital eccentricity of Mercury and the origin of the Moon. *Nature, 240*, 299–300.

Cameron, A. G. W. (1973). Properties of the solar nebula and the origin of the Moon. *The Moon, 7*, 377–383.

Cameron, A. G. W. (1983). Origin of the atmospheres of the terrestrial planets. *Icarus, 56*, 195–201.

Cameron, A. G. W. (1985). Formation of the prelunar accretion disk. *Icarus, 62*, 319–327.

Cameron, A. G. W. (1992). *The giant impact revisited: Abstracts volume, 23rd lunar and planetary science conference* (pp. 199–200). Houston: Lunar and Planetary Institute.

Cameron, A. G. W. (1993). *The giant impact produced a precipitated moon: Abstracts volume, 24th lunar and planetary science conference* (pp. 245–246). Houston: Lunar and Planetary Institute.

Cameron, A. G. W. (1997). The origin of the Moon and the single-impact hypothesis V. *Icarus, 126*, 126–137.

Cameron, A. G. W. (2000). Higher resolution simulations of the giant impact. In R. M. Canup & K. Righter (Eds.), *Origin of the Earth and Moon* (pp. 133–144). Tucson: University of Arizona Press.

Cameron, A. G. W. (2001). From interstellar gas to the Earth-Moon system. *Meteoritics and Planetary Science, 36*, 9–22.

Cameron, A. G. W., & Benz, W. (1991). The origin of the Moon and the single impact hypothesis, IV. *Icarus, 92*, 204–216.

Cameron, A. G. W., & Ward, W. R. (1976). *The origin of the Moon: Abstracts, Lunar Science Conference VII*, pp. 120–122.

Canup, R. M. (2004a). Origin of terrestrial planets and the Earth-Moon system. *Physics Today, 57*, 56–62.

Canup, R. M. (2004b). Simulations of a late lunar-forming impact. *Icarus, 168*, 433–456.

Canup, R. M. (2005). A giant impact origin of Pluto-Charon. *Science, 307*, 546–550.

Canup, R. M. (2008a). Lunar-forming collisions with pre-impact rotation. *Icarus, 196*, 518–538.

Canup, R. M. (2008b). Accretion of the Earth. *Philosophical Transactions of the Royal Society A, 366*, 6061–4075.

Canup, R. M. (2012). Forming a Moon with an earth-like composition via a giant impact. *Science, 338*, 1052–1055.

Canup, R. M. (2013). Lunar conspiracies. *Nature, 504*, 27–29.

Canup, R. M., & Asphaug, E. (2001). Origin of the Moon in a giant impact near the end of Earth's formation. *Nature, 412*, 708–712.

Canup, R. M., & Righter, K. (Eds.). (2000). *Origin of the Earth and Moon* (p. 555). Tucson: University of Arizona Press.

Canup, R., & Salmon, J. (2018). Origin of Phobos and Deimos by the impact of a Vesta-to-Ceres sized body with Mars. *Science Advances, 4*, eaar6887, (6 p.).

Clery, D. (2013). Impact theory gets whacked. *Science, 342*, 183–185.

Craddock, R. A. (2011). Are Phobos and Deimos the result of a giant impact? *Icarus, 211*, 1150–1161.

Crawford, I. A., & Joy, K. H. (2014). Lunar exploration: Opening a window into the history and evolution of the inner solar system. *Philosophical Transactions of the Royal Society A, 372*, 20130315, (21 p.).

Cuk, M., & Stewart, S. T. (2012). Making the Moon from a fast-spinning Earth: A giant impact followed by resonant despinning. *Science, 338*, 1047–1052.

Daly, R. A. (1946). Origin of the Moon and its topography. *Proceedings of the American Philosophical Society, 90*(2), 104–119.

Davies, J. H. (2008). Did a mega-collision dry Venus' interior? *Earth and Planetary Science Letters, 268*, 376–383.

Durda, D. D., Bottke, W. F., Jr., Enke, B. L., Merline, W. J., Asphaug, E., & Leinhardt, Z. M. (2004). The formation of asteroid satellites in large impacts: Results from numerical simulations. *Icarus, 170*, 234–257.

Elliot, T. (2013). Forum: Planetary science – Shadows cast on moon's origin. *Nature, 504*, 90–91.

Evans, N. W., & Tabachnik. S. (1999). Possible long-lived asteroid belts in the inner solar system. *Nature, 399*, 41–43.

Evans, N. W., & Tabachnik, S. (2002). Structure of possible long-lived asteroid belts. *Monthly Notices of the Royal Astronomical Society, 333*, L1-5.

Fell, A. (2018). How the Moon formed inside a vaporized earth synestia. *Science and Technology*, (Feb. 28), p. 3.

Gast, P. W. (1972). The chemical composition and structure of the Moon. *The Moon, 5*, 121–148.

Gillman, C., Golabek, G. J., & Tackley, P. J. (2016). Effect of a single large impact on the coupled atmosphere-interior evolution of Venus. *Icarus, 268*, 295–312.

Goldreich, P., Murray, N., Longaretti, P. Y., & Banfield, D. (1989). Neptune's story. *Science, 245*, 500–504.

Gomes, R., Levison, H. F., Tsiganis, K., & Morbidelli, A. (2005). Origin of the cataclysmic Late Heavy Bombardment period of the terrestrial planets. *Nature, 435*, 466–469.

Hartmann, W. K. (1986). Preface. In Hartmann, W. K., Phillips, R. J., & Taylor, G. J. (Eds.). *Origin of the Moon* (pp. vii–xi). Houston: Lunar and Planetary Institute.

Hartmann, W. K. (2014). The giant impact hypothesis: Past, present (and future?). *Philosophical Transactions of the Royal Society A, 372*, 20140289.

Hartmann, W. K., & Davis, D. R. (1975). Satellite-sized planetesimals and lunar origin. *Icarus, 24*, 504–515.

Hartmann, W. K., Phillips, R. J., & Taylor, G. J. (Eds.). (1986). *Origin of the Moon* (p. 781). Houston: Lunar and Planetary Institute.

Hyodo, R., Genda, H., Charnoz, S., & Rosenblatt, P. (2017a). On the impact origin of Phobos and Diemos. I. Thermodynamic and physical aspects. *The Astrophysical Journal, 845*, 125, (8 p.).. https://doi.org/10.3847/1538-4357/aa81c4.

Hyodo, R., Rosenblatt, P., Genda, H., & Charnoz, S. (2017b). On the impact origin of Phobos and Deimos: II. True polar wander and disk evolution. *The Astrophysical Journal, 851*, 122, (9 p.).. https://doi.org/10.3847/1538-4357/aa9984.

Kaula, W. M., & Harris, A. W. (1973). Dynamically plausible hypotheses of lunar origin. *Nature, 245*, 367–369.

Kobo-Oka, T., & Nakazawa, K. (1995). Gradual increase in the obliquity of Uranus due to tidal interaction with a hypothetical retrograde satellite. *Icarus, 114*, 21–32.

Kokubo, E., Canup, R. M., & Ida, S. (2000). Lunar accretion from an impact-generated disk. In R. M. Canup & K. Righter (Eds.), *Origin of the Earth and Moon* (pp. 145–163). Tucson: The University of Arizona Press.

Kopal, Z. (1973). *The solar system* (152 p). Oxford, London/New York: Oxford University Press.

Lipton, P. (2005). Testing hypotheses: Prediction and prejudice. *Science, 307*, 219–221.

Lissauer, J. J., & Kary, D. M. (1991). The origin of the systematic component of planetary rotation. 1. Planet on a circular orbit. *Icarus, 94*, 126–159.

Lock, S. J., & Stewart, S. T. (2017). The structure of terrestrial bodies: Impact heating, corotation limits, and synestias. *Journal of Geophysical Research, Planets, 122*, 950–982. https://doi.org/10.1002/2016JE005239.

Lock, S. J., Stewart, S. T., Petaev, M. I., Leinhardt, Z., Mace, M. T., Jacobsen, S. B., & Cuk, M. (2018). The origin of the Moon within a terrestrial synestia. *Journal of Geophysical Research, Planets, 123*, 910–951. https://doi.org/10.1002/2017JE005333.

Lyttleton, R. A. (1936). On the possible results of an encounter of Pluto with the Neptunian system. *Monthly Notices of the Royal Astronomical Society, 97*, 108–115.

Malcuit, R. J. (2015). *The twin sister planets, Venus and Earth: Why are they so different?* (p. 401). Cham: Springer International Publishing.

Malcuit, R. J., & Winters, R. R. (1998). *A prograde gravitational capture model for the origin of the Earth-Moon system: Is it compatible with the rock records of the Earth and Moon?* Abstracts volume, Conference on the Origin of the Earth and Moon, Lunar and Planetary Institute, p. 24.

Malcuit, R. J., Winters, R. R., & Mickelson, M. E. (1984a). *A testable gravitational capture model for the origin of the Earth's Moon.* Abstracts volume, Conference on the Origin of the Moon, Lunar and Planetary Institute, p. 43.

Malcuit, R. J., Winters, R. R., & Mickelson, M. E. (1984b). *Directional properties of "circular" Maria: Interpretation in the context of a testable gravitational capture model for lunar origin.* Abstracts volume, Conference of the Origin of the Moon, Lunar and Planetary Institute, p. 44.

Malcuit, R. J., Mehringer, D. M., & Winters, R. R. (1988). *Numerical simulation of three-body interactions with an energy dissipation subroutine for close encounters between two bodies: Implications for planetary accretion and planetoid capture processes.* Abstracts volume, Conference on the Origin of the Earth, Lunar and Planetary Institute and NASA Johnson Space Center, pp. 52–53.

Malcuit, R. J., Mehringer, D. M., & Winters, R. R. (1992). Numerical simulation of tidal capture of a triton-mass body by Neptune: Dimensions of a two-dimensional retrograde stable capture zone. *Transactions. American Geophysical Union, 73*(14)., Spring Meeting Supplement,, 191.
Malcuit, R. J., Winters, R. R., & Doty, S. D. (2018). *A new approach to an old (1936) suggestion for the origin of Pluto as an escaped satellite of Neptune: Geological Society of America*, Abstracts with Programs, *50*(6). https://doi.org/10.1130/abs/2018AM-31888.
Mann, A. (2018). Cataclysm's end. *Nature, 553*, 393–395.
Melchior, P. J. (1978). *The tides of planet Earth* (p. 609). New York: Pergamon Press.
Newell, H. E. (1973). Harold Urey and the Moon. *The Moon, 7*, 1–5.
Newsom, H. E., & Jones, J. H. (Eds.). (1990). *Origin of the Earth* (p. 375). New York: Oxford University Press.
Peale, S. J., & Cassen, P. (1978). Contributions of tidal dissipation to lunar thermal history. *Icarus, 36*, 245–269.
Peplowski, P. N., Evans, L. G., Hauck, S. A., McCoy, T. J., Boynton, W. V., Gillis-Davis, J. J., et al. (2011). Radioactive elements on Mercury's surface from MESSENGER: Implications for the planet's formation and evolution. *Science, 333*, 1850–1852.
Reufer, A., Meier, M. M., Benz, W., & Wieler, R. (2012). A hit-and-run giant impact scenario. *Icarus, 221*, 296–299.
Ringwood, A. E. (1979). *Origin of the Earth and Moon* (p. 295). New York: Springer-Verlag
Ringwood, A. E. (1990). Earliest history of the Earth-Moon system. In H. E. Newsom & J. H. Jones (Eds.), *Origin of the Earth* (pp. 101–134). New York: Oxford University Press.
Rosenblatt, P., & Charnoz, S. (2012). On the formation of the martian moons from a circum-martian accretion disk. *Icarus, 221*, 806–815.
Rosenblatt, P., Charnoz, S., Dunseath, K. M., Terao-Dunseath, M., Trinh, A., Hyodo, R., Genda, H., & Toupin, S. (2016). Accretion of Phobos and Diemos in an extended debris disc stirred by transient moons. *Nature Geoscience, 9*, 581–583.
Ross, M., & Schubert, G. (1986). Tidal dissipation in a viscoelastic planet: Proceedings of the 16th Lunar and Planetary Science conference. *Journal of Geophysical Research, 91*, D447–D452.
Rufu, R., Aharonson, O., & Perets, H. B. (2017). A multiple-impact origin for the Moon. *Nature Geoscience, 10*, 89–94.
Russell, C. T. (1980). Planetary magnetism. *Reviews of Geophysics and Space Physics, 18*, 77–106.
Scheinberg, A. L., Soderlund, D. M., & Elkins-Tanton, L. T. (2018). A basal magma ocean dynamo to explain the early lunar magnetic field. *Earth and Planetary Science Letters, 492*, 144–151.
Schultz, P. H., & Spudis, P. D. (1983). Beginning and end of lunar mare volcanism. *Nature, 302*, 233–236.
Shu, F. H., Shang, H., Gounelle, M., Glassgold, A. E., & Lee, T. (2001). The origin of chondrules and refractory inclusions in chondritic meteorites. *The Astrophysical Journal, 548*, 1029–1050.
Singer, S. F. (1968). The origin of the Moon and geophysical consequences. *Geophysical Journal of the Royal Astronomical Society, 15*, 205–226.
Singer, S. F. (1970). How did Venus loose its angular momentum? *Science, 170*, 1196–1198.
Slattery, W. L., Benz, W., & Cameron, A. G. W. (1992). Giant impacts on a primitive Uranus. *Icarus, 99*, 167–174.
Smoluchowski, R. (1973a). Lunar tides and magnetism. *Nature, 242*, 516–517.
Smoluchowski, R. (1973b). Magnetism of the Moon. *The Moon, 7*, 127–131.
Sonett, C. P., Colburn, D. S., & Schwartz, K. (1975). Formation of the lunar crust: An electrical source of heating. *Icarus, 24*, 231–255.
Soter, S. (1971). The dust belts of Mars. Cornell CRSR-492 (unpublished).
Stevenson, D. J., & Halliday, A. N. (2014). The origin of the Moon. *Philosophical Transactions of the Royal Society A, 372*, 20140289.
Stewart, S. T. (2013). Weak links mar lunar model *in* Forum: Planetary science – Shadows cast on Moon's origin. *Nature, 504*, 90–91.
Taylor, S. R. (2001). *Solar system evolution: A new perspective* (p. 460). Cambridge: Cambridge University Press.

Ward, W. R., & Cameron, A. G. W. (1978). *Disc evolution within the Roche limit.* Abstracts, Lunar and Planetary Science Conference IX, pp. 1205–1207.
Weiss, B. P., & Tikoo, S. M. (2014a). The lunar dynamo (review summary). *Science, 346*, 1198.
Weiss, B. P., & Tikoo, S. M. (2014b). The lunar dynamo. *Science, 346*, 1246753, (10 p.).
Wetherill, G. W. (1985). Occurrence of giant impacts during the growth of the terrestrial planets. *Science, 228*, 877–879, (p. 21).
Wilford, J. N. (1997). *Astronomers recalculate the 'whack' that made the Moon.* New York Times, Science Section, Tuesday, 29 July 1997, p. B12.
Winters, R. R., & Malcuit, R. J. (1977). The lunar capture hypothesis revisited. *The Moon, 17*, 353–358.
Wood, B. (2011). The formation and differentiation of Earth. *Physics Today, 64*(12), 40–45.
Wood, J. A., Dickey, J. S. Jr., Marvin, U. B., & Powell, B. N. (1970). *Lunar anorthosites and a geophysical model of the moon.* Proceedings of the Apollo 11 Lunar Science Conference, Lunar Science Institute, Houston, *1*, 965–988.

Some Numbered References from Quotes from Smoluchowski (1973a) as They Appear in the Reference Section of the Paper

[1]Sonett, C. P., & Runcorn, S. K. (1971). *Comments on Astrophys. and Space Phys., 3*, 49.
[2]Sonett, C. P., & Mihalov, J. D., (1971). *J. of Geophys. Res., 77*, 588.
[3]Wood, J. A., Dickey, J. S., Marvin, U. B., & Powell, B. N. (1972). *Proc. Apollo 11 Lunar Science Conference, 1*, 965.
[4]Elsasser, W. M. (1939). *Phys. Rev., 55*, 489.
[5]Bullard, E. C. (1954). *The Earth as a planet.* In G. P. Kuiper (Eds). Chicago: University of Chicago Press.
[6]Malkus, W. V. R. (1968). *Science, 160*, 259.
[7]Smoluchowski, R. (1973). *The Moon, 6*, 48.
[8]Lunar Science Institute. (1972). *Post-Apollo lunar science.* Houston, Texas: Lunar Science Institute.
[9]Gast, P. W. (1972). *The Moon, 5*, 121.
[10]Hide, R. 1967. *Magnetism and cosmos*, 378. Edinburgh: Oliver and Boyd.
[11]Smoluchowski, R. (1972). *Phys. Earth Planet. Inter., 6.*

Chapter 7
A History of Satellite Capture Studies As Experienced by the Author: A Chronology of Events that Eventually led to a Somewhat Comprehensive Gravitational Satellite Capture Model

> Of the other alternatives, it is perhaps just possible that the moon was originally an independent planet, though it is much less massive than any existing planet.—From, Jeffreys (1929), p. 37.

> The basic geochemical model for the structure of the Moon proposed by Anderson (1973), in which the Moon is formed by differentiation of the calcium, aluminum, titanium-rich inclusions in the Allende meteorite, is accepted, and the conditions for formation of this Moon within the solar system models of Cameron and Pine (1973) are discussed. The basic material condenses while iron remains in the gaseous phase, which places the formation of the Moon slightly inside the orbit of Mercury .—From Cameron (1973a), p. 377

> The genetic kinship between the moon and the differentiated meteorites, together with the fact that these meteorites were already differentiated 4.6 G.y. ago, lends strong support to models of the formation of the moon by capture of such bodies.—From Clayton and Meyeda (1975), p. 1768

To understand the origin and evolution of planet Mars we must first try to understand the origin and evolution of Earth, Moon, and the Earth-Moon system. Then we must endeavor to understand the origin and evolution of planets Mercury, Venus, the asteroids, as well as the gassy/icy planets: Jupiter, Saturn, Uranus, and Neptune. And we cannot forget planetoid Pluto and its relatively large satellite, Charon. Some Information on Earth and Moon is in Chap. 6. Some of the other planets/planetoids have captured satellites that are extant (e.g., Triton) or had interactions with satellites that no longer exist (phantom satellites) (e.g., Adonis of the Venus-Adonis system).

This chapter is a narrative of the author's odyssey into the sea of planetoid capture studies when it was as popular as was Continental Drift in the 1915–1930 era (e.g., Wegener 1924) and as was the Milankovitch Model for the Pleistocene Glaciations in the 1920–1950 era (e.g., Milankovitch 1941). So let us begin and see where it takes us.

The relationship of capture studies to the evolution of planet Mars and to the potential geoforming of planet Mars was presented in Chaps. 4 and 5. In this chapter

R. Malcuit, *Geoforming Mars*, https://doi.org/10.1007/978-3-030-58876-2_7

I intend to involve most of the remainder of the Solar System in the capture study endeavor. Some planets and planetoids have been directly affected by the capture process and still have their captured bodies in orbit about the planet. Examples are Earth with the Moon in a prograde orbit and Neptune with its satellite, Triton, in a retrograde orbit. Other planets show signs of the former existence of large satellites (i.e., phantom satellites). Examples are planets Venus and Uranus.

7.1 My Time at Kent State University

My interest in the "Origin of the Moon" project began during the Fall of 1968. Dr. Richard Heimlich (Kent State University) was involved in a Bighorn Mountains Project and invited me to do a Master of Science Thesis on some aspect of the project. The mutually agreed upon project was on "PETROGENESIS OF GNEISS AND ASSOCIATED ROCKS, SOUTHERN BIGHORN MOUNTAINS, WYOMING". It was mainly a petrology, petrography, and zircon morphology project.

Since the gneisses were among the oldest rocks known on Earth at that time (Archean in age), I immediately got involved in the literature on the Primitive Earth and one of the main authors of articles was Dr. Preston Cloud (University of California, Santa Barbara). He published an article in Science in 1968 summarizing what was known about the primitive Earth and Preston thought that maybe the Moon was involved in some of the action which helped to destroy the older geologic rock record before about 3.6 Ga. In short, Cloud (1968) speculated that the Moon was captured by the Earth and that the dynamics of the capture process could explain the "high temperature events on the Moon" as well as the absence of older rocks on Earth. This speculation got me to thinking about the early history of both the Moon and the Earth and I have been working on various aspects of the interaction between the two ever since.

During the Spring of 1970 I was finishing my Master of Science Project and as I was walking along the edge of the KSU campus on a Full Moon night I just stopped and stared at the Moon. I noticed a pattern on the front face of the Moon that shows up on every full-face lunar map and chart – that the three largest maria are on a straight line and that they decrease in size from left to right. I noted the image in my mind but I did not think about any origin scheme until about 18 months later. I had a Master of Science thesis to finish!

7.2 My Time at Michigan State University

I started my Ph. D work at MSU in September 1970. I did coursework during the school year and during the Spring Quarter I collected a suite of samples (and made thin-sections) of igneous and metamorphic rocks in the Grenville Province (Ontario, Canada) as preliminary work on a Ph. D. project. Then on August

Fig. 7.1 Overview of the western limb of the Moon. Zond 8 photograph of the Mare Orientale region

21, 1971, I had a "day of discovery". I was examining a recently acquired lunar globe in my office and discovered (really rediscovered) a pattern. I did realize that this was the same pattern as when I looked at the Moon on the edge of the Kent State campus in Spring 1970. I noticed that the four largest circular maria on the Moon (only three are visible when viewed from Earth) were located very nearly along a straight line and that there were features to the lunar west and lunar east that were located on that same "great-circle pattern" as shown in Figs. 7.1, 7.2, and 7.3. I was sure that others had noticed that pattern because it was so obvious to me. I was very familiar with both the Geology Library as well as the Physics/Astronomy Library and I did find one really good article by Stuart-Alexander and Howard (1970) titled "Lunar maria and circular basins: A review". The authors described

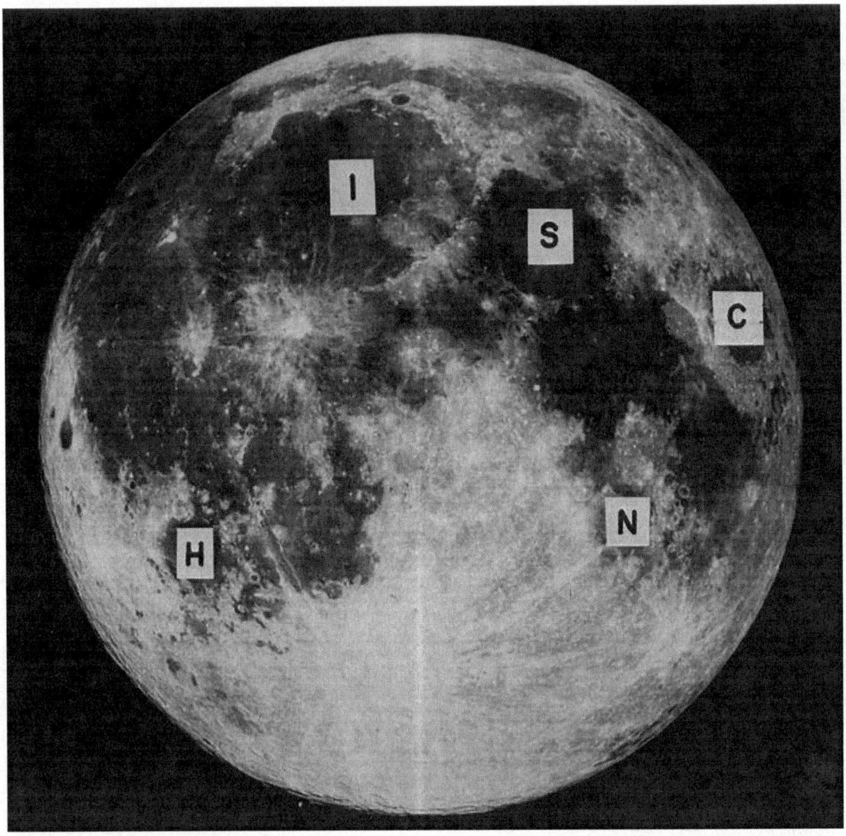

Fig. 7.2 Overview of the front face of the Moon. (Hale Observatory photo). *I* Mare Imbrium, *S* Mare Serenitatis, *C* Mare Crisium, *H* Mare Humorum, *N* Mare Nectaris

an equatorial belt along which several lunar basins were located but they did not point out the pattern that my cursory survey of the lunar globe revealed. The question in my mind was: HOW COULD SUCH A PATTERN OF LARGE CIRCULAR MARIA FORM?

One of my early thoughts was that the maria were circular lava lakes. The conventional model for the circular maria was that the lava intruded from below and filled a circular depression made by a solid impacting body. I was wondering what would happen if a spheroid of lunar basalt impacted the lunar surface from above. This action would certainly form a lava lake. The next question was: IS THERE A WAY TO GET A SPHERICAL BODY OF LUNAR BASALT ABOVE THE LUNAR SURFACE? The analogy that came to mind was a paddle ball with an elastic connector to the paddle. If there was liquid in the ball and it had some "weak spots" on the backside, then the outward force would cause the liquid to pass through the weak spots on the outer surface of the ball. If the ball had gravity of its own, then

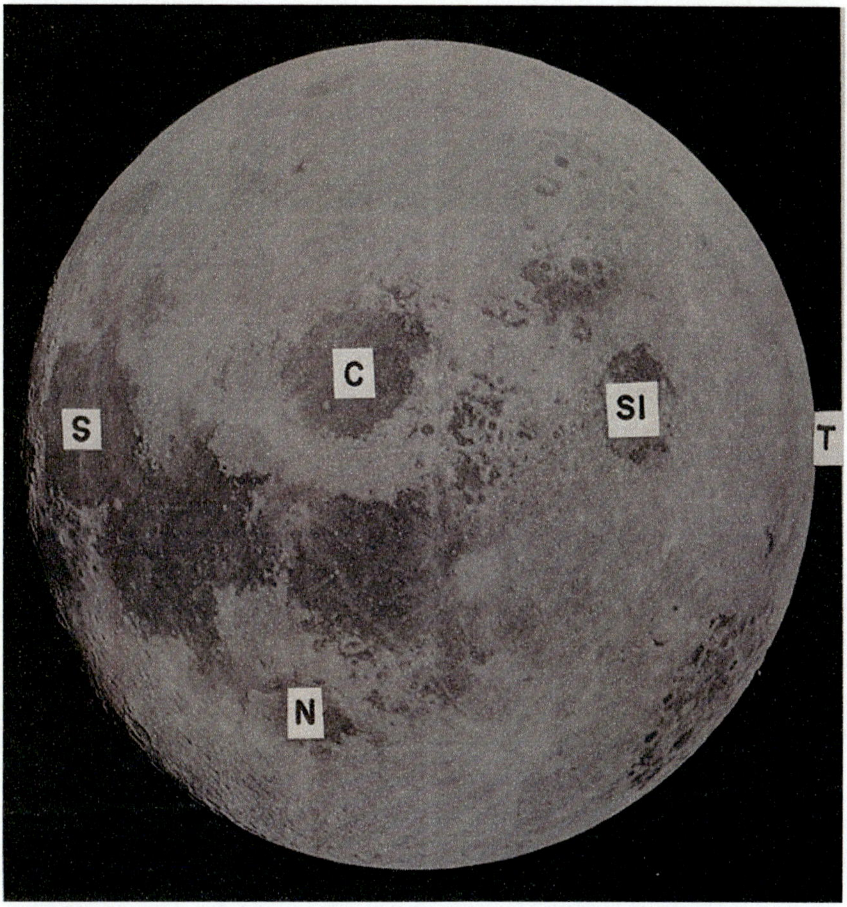

Fig. 7.3 Overview of the eastern limb of the Moon. (Apollo photograph 11-44-6667). *S* Mare Serenitatis, *C* Mare Crisium, *SI* Mare Smythii, *T* Crater Tsiolkovsky, *N* Mare Nectaris

some of the liquid that exited the ball would fall back onto the surface of the ball. Then I got to thinking about a paper by Cloud (1968) in which he proposed that a "close approach of the Moon to the Earth" may have been associated with lunar capture and then I was off and running with what I thought was a new idea.

I wrote a short paper with some lunar photographs and sketches about this "great-circle pattern of large circular maria" and what I thought could happen during a close approach between the Moon and Earth in which the Moon came close enough to the Earth to cause partial tidal disruption of the lunar surface so that lunar basaltic lava could be necked off a lava column and then fall back onto the lunar surface. I then circulated copies of this paper in the departmental coffee room. I got some positive responses and questions from graduate school colleagues as well as paleontology, geology, and geophysics professors. Certainly no one had heard of such an idea before and the model not only related to the circular pattern of individual circular maria but it also related to a great-circle pattern of large circular maria in which the four largest circular maria on the moon decreased in diameter from lunar west to lunar east. Tom Vogel, my advisor, could see that I was very interested in pursuing the idea and asked me to check these ideas out with someone in the Physics/ Astronomy Department.

Gary Byerly, a graduate student colleague, had taken a course recently in Solar System Physics with Tom Stoeckley and he thought that Dr. Stoeckley would be the professor to check with. Gary made an appointment with Dr. Stoeckley and we went over for a visit. Gary had worked with some numerical simulation programs during the Solar System Physics course and he and Dr. Stoeckley thought that this one particular n-body program could be used to test this tidal disruption idea. I explained my generalized model about how this great-circle pattern may be related to a very close fly-by encounter of the Moon to the Earth. In order to get anything lifted from the surface or extracted from the interior, the smaller body, the Moon, would have to undergo a weightless condition in the near surface area. In Dr. Stoeckley's judgment the gravitational force on the front side of the Moon would be sufficient to lift particles, large and small, as well as spheroids of basalt above the lunar surface and they would subsequently impact at a low speed onto the lunar surface. Gary got the program implemented right there in Dr. Stoeckley's office area and on the second run we got a particle to lift off before it fell back onto the lunar surface (*i. e.*, the Weightlessness Limit was a viable concept).

We then got a copy of the program operating on the CDC 6500 at the Computer Center at MSU. I spent a good bit of time in the departmental calculator room working out the trigonometry for launch sites for particles. I did hundreds of simulations and each one required punch cards to be processed and then the cards were put in the queue for computing the results and printing out a long list of numbers and a crude trajectory diagram.

The concept of the Late Heavy Bombardment (Tera et al. 1974) had not yet been developed but it was known that much activity occurred on the lunar surface between 3.6 and 3.8 Ga. Mutch (1970, 1972) promoted the idea that the maria were formed by solid body impact [*i. e.*, bodies impacting at hyper-velocities (speeds) onto the lunar surface over a time period of several hundred million years]. In great contrast, my tidal disruption model strongly suggested that the launch phase would be about 30 minutes long and the main part of the fall-back phase would be about 90 minutes of time. Small bodies of debris could be impacting on the Moon and Earth long after the tidal disruption but the larger bodies that would form large circular lakes of lava on the lunar surface would fall back within that 90 minute timeframe.

During the Fall Quarter we had a special lecturer at MSU by the name of Harold Urey. Urey was one of the original capture proponents and he was giving a talk on "The Chemistry and Origin of the Moon" and he showed some slides of the circular maria. After the talk and the normal Q and A session, I was in a small group discussion and I asked Harold if he ever noticed that the 4 largest circular maria are on a straight line on the lunar surface and that they were very close to a lunar great circle. He said that he had not noticed that before. Then I asked him if he ever considered what could happen during a close approach between Earth and Moon that might be related to capture – like a partial tidal disruption of the Moon in which lunar basalt was pulled from the interior by earth gravity and then deposited on the lunar surface as a string of lava lakes. Harold's response was to the effect: "Son, that just sounds too catastrophic". And that was the end of the conversation and the end of the discussion session.

Tom Vogel, my advisor, noticed that there was to be a special symposium over the holiday break on "The Origin of the Moon" sponsored by the American Association for the Advancement of Science (AAAS) in Philadelphia as part of the annual meeting. He thought that I should consider going to that event. I got registered and went to the meeting. It seemed like all the big names were there: Carl Sagan, John Wells, Keith Runcorn, Harold Masursky, Hannes Alfven, Don Wise, William Kaula, Fred Singer, and a whole host of others. Throughout the meeting there was some mention of a close approach of the Moon to the Earth but no mention of tidal disruption. Fred Singer gave a presentation on his version of lunar capture but, again, there was no mention of tidal disruption.

In those days the AGU Spring Meeting was always in Washington, D. C. in early Spring and my advisor thought that I should submit an abstract on this new idea. I did and it was accepted for an oral presentation in a session on "lunar surface features". The title of the presentation was: "Lunar Maria and Mascons: 'Hypervolcanic' flyby 'scars'" (Malcuit 1972). The chair of the session was E. T. C. Chao from the U. S. Geological Survey. Dr. Chao was not pleased with my tidal disruption model and with the idea that the maria and mascons could be formed by a tidal disruption and fall-back operation. He also thought that the model was "too catastrophic" and stated that as chair of the session. But his main criticism was that mare basalts are not all one age but apparently were extruding onto the lunar surface for 100's of millions of years. This was a valid criticism and it was a weak point of a simple "fly-by" model. The tidal disruption must be somehow related to the capture process so that the Moon would return for additional close encounters.

7.2.1 A Brief Discussion of the Importance of the Weightlessness Limit

Now for a brief digression to explain some aspects of the Weightlessness Limit (Malcuit et al. 1975) which is similar in nature to the "Roche limit for a solid body" (Jeffreys 1947; Aggarwal and Oberbeck 1974; Holsapple and Michel 2006, 2008). The main difference is that the Weightlessness Limit can be very useful for analyzing the effects of close encounters from either a heliocentric orbit or a highly elliptical geocentric orbit. The Roche limit for a solid body is mainly useful for a solid body in a circular planetocentric orbit (an orbit of very low eccentricity). The

Roche limit for a solid body is about 1.6 planet radii. This is the limit at which a somewhat solid body would begin to disintegrate if it were in a circular orbit about a planet. The classical Roche limit is at about 2.89 planet radii and it is the distance at which a fluid body would begin to disintegrate by necking material off at the sub-planet point as well as the anti-sub-planet point. Thus, there is quite a difference between the classical Roche limit and the Roche limit for a solid body.

I was interested in analyzing the effects of planet-planetoid encounters from either a hyperbolic orbit or from a planetocentric orbit of high eccentricity. If the encountering planetoid is solid, then only loose material on the surface at the sub-planet point can be lifted off the surface during the time-frame of an encounter. However, if there is magma available in a sub-surface magma chamber of a planetoid, like a layer of "magma ocean", then the magma could be extracted from the subsurface magma chamber and subsequently could either escape from the planetoid or be transported along the encounter plane to another location on the surface of the planetoid.

Figure 7.4 shows some features of the Weightlessness Limit as it relates to close planetoid-planet encounters.

Equation 7.1 is for the W-limit for the sub-earth point.

$$\frac{GM_e}{r^2} = \frac{GM_e}{(r + R_m)^2} - \frac{GM_m}{R_m^{\ 2}}$$

where G is the gravitational constant, M_e is the mass of the Earth, M_m is the mass of the Moon, r is the distance of separation of centers of Earth and Moon, and R_m is the radius of the Moon.

The first term in the equation is the force on the body of the Moon toward the Earth; the second term is the force on a particle at the sub-earth point on the moon toward the Earth; the third term is the force on the particle by the gravity of the Moon. If a particle along the earth-moon center line is inside the W-limit, it will be lifted off the surface of the Moon and move toward the Earth. If the particle along the earth-moon center line is beyond the W-limit, then it will remain at its normal position on the lunar surface.

Equation 7.2 is for the weightlessness limit for a particle located at the anti-sub-earth point on the lunar surface.

$$\frac{GM_e}{r^2} = \frac{GM_e}{(r + R_m)^2} + \frac{GM_m}{R_m^{\ 2}}$$

The first term is as defined above; the second term is for the force on a particle at the anti-sub-earth point on the Moon by the Earth; the third term is the force on the particle by the gravity of the Moon.

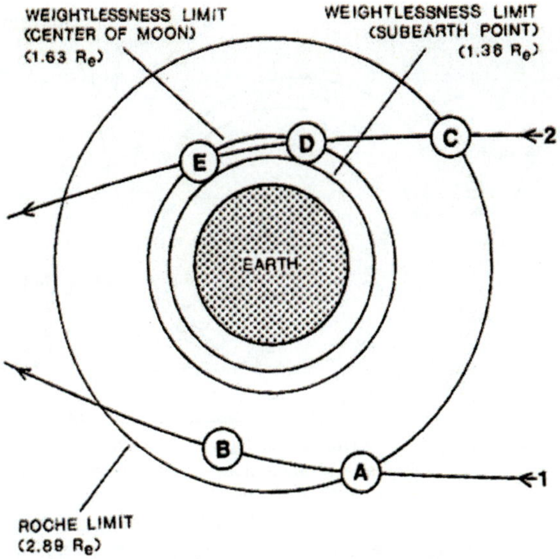

Fig. 7.4 Scale Sketch of the weightlessness limit for the case of the Moon encountering the Earth with neither body being deformed. Note that the classical Roche limit is located at a much greater distance from Earth than the weightlessness limit. Scenario 1: The lunar body enters the classical Roche limit at point A and the perigee point is at point B. A particle at the lunar surface would *not* become weightless under this condition and the lunar body would then exit the classical Roche limit when it reaches a distance of 2.89 earth radii. Scenario 2: The lunar body passes through the classical Roche limit at 2.89 earth radii (point *C*). At point *D* the lunar body passes through the W-limit for a particle at the sub-earth point on the lunar body. From point *D* to point *E* any loose bodies or particles on the lunar surface at or near the sub-earth point, would be lifted off the lunar surface and inserted into sub-orbital trajectories that would either return the particle(s) to the lunar body or the particle(s) would eventually collide with the Earth, the details depending on the position of the lunar body at the time of gravitational liftoff. (From Malcuit (2015, Fig. 5.5) with permission from Springer)

As shown in Fig. 7.4, the W-limit for a particle at the sub-earth point on a spherical (non-deformed) Moon is at 1.36 R_e and the distance between the centers of the Earth and Moon is 1.63 R_e. The W-limit for a particle at the anti-sub-earth point (which is not indicated on the diagram) occurs when the center-to-center distance of the two bodies is at 1.36 R_e. Thus, with a lunar radius at 0.27 R_e weightlessness would occur at the anti-sub-earth point on the lunar surface but the perigee passage would be a near-grazing encounter.

The position of the W-limit for both the sub-earth point and the anti-sub-earth point are displaced to a greater distance from the Earth as the lunar body is gravitationally deformed during an encounter. For more discussion of the scenarios involving deformable bodies see Malcuit (2015, p. 123–126). I note here that it is this major shift in the W-limits that make the process of tidal disruption feasible during close encounters of bodies like the Earth and Moon.

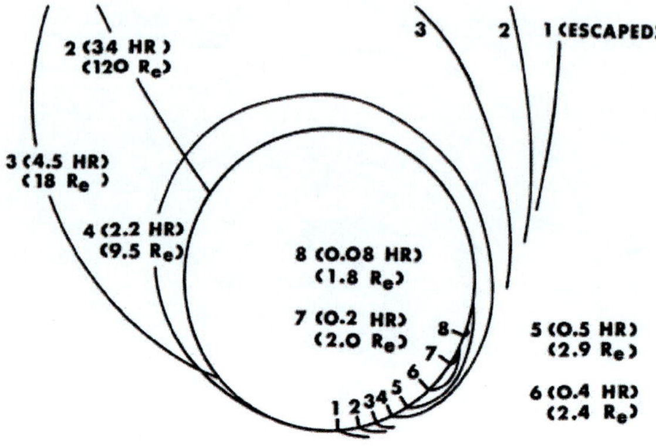

Fig. 7.5 Diagram showing the results of numerical simulations of trajectories of particles launched from specific positions at the sub-earth point on the lunar surface during an encounter that passes within the weightlessness limit of the Earth-Moon system . The launch positions, the flight times, as well as the impact positions are shown. The most favorable location for extracting lunar basalt to neck-off into spheroids appears to be the geometric location of Oceanus Procellarum. Note that the lunar body is not rotating during the timeframe of these simulations. (From Malcuit et al. (1975, Fig. 5) with permission from Springer)

After returning to East Lansing from the AGU meeting, Dr. Aureal Cross, a paleobotanist and one of the coffee room group, asked me to give a lecture to his Historical Geology class on this new idea for the origin of the major features of the Moon. I suggested that we invite other interested students and faculty to the talk to see what they thought of the idea and we did make it an open session.

Then I returned to the work of finishing the numerical simulations some of which are illustrated in Figs. 7.5 and 7.6. I finished my thesis project in the summer of 1972 and started teaching at Denison University during the Fall Semester of that year. I gave an oral presentation (with MSU colleagues) at the Spring AGU meeting in 1973 titled "Computer simulation model of a lunar flyby encounter with Earth" (Malcuit et al. 1973). I got some questions but no serious disagreements because the particles whose trajectories were simulated could be either solid particles or liquid spheroids.

Then I gave another oral presentation at the Spring AGU meeting in 1974 entitled "Lunar maria emplacement via the flyby encounter model" (Malcuit et al. 1974). William Muehlberger (Univ. of Texas) was the chair of the session and he did not like the idea. He was quite sure that the mare basins were excavated via solid body impact and he really did not want to consider any other model.

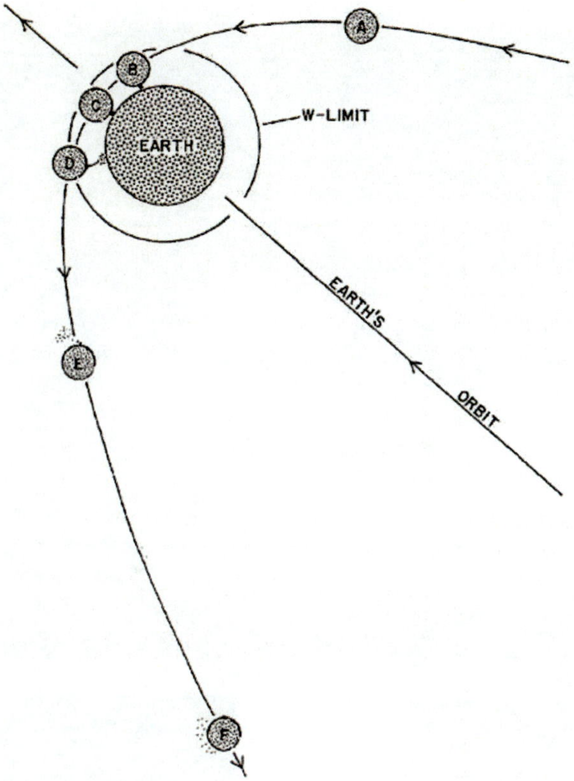

Fig. 7.6 General setting for a TIDAL DISRUPTION SCENARIO. This was the author's early concept of what could happen during a close encounter within the Weightlessness Limit of the Earth-Moon system . Tidal disruption can occur only if magma is available in the lunar mantle and if the encounter is well within the W-limit of the system. Unknown to me at that time was that a warm lunar-like body could not dissipate the energy for capture because it could not have the appropriate viscosity value for dissipating the energy for capture. Thus such an encounter could not be the capture encounter but must be some subsequent encounter soon after capture. Note: The detailed series of pictograms in Figs. 7.9, 7.10, 7.11, and 7.12 are based on this close encounter concept. Pictograms a-f in Figs. 7.9 and 7.10 relate to stages A-F on this scale diagram

7.3 My Early Days at Denison University

When I came to Denison, I got Ron Winters and Mike Mickelson (both in the Physics/Astronomy Department) interested in the "origin of the Moon by capture" project. In 1975 we gave a paper, presented by Mickelson entitled "Accretion-driven and encounter-driven dynamos as sources of lunar rock magnetism" (Mickelson et al. 1975). The session was on the subject of the Lunar Magnetic Field and we got a good response because the models did relate to some aspects of the patterns of lunar rock magnetization. And we did not need to bring in the "catastrophic" nature of

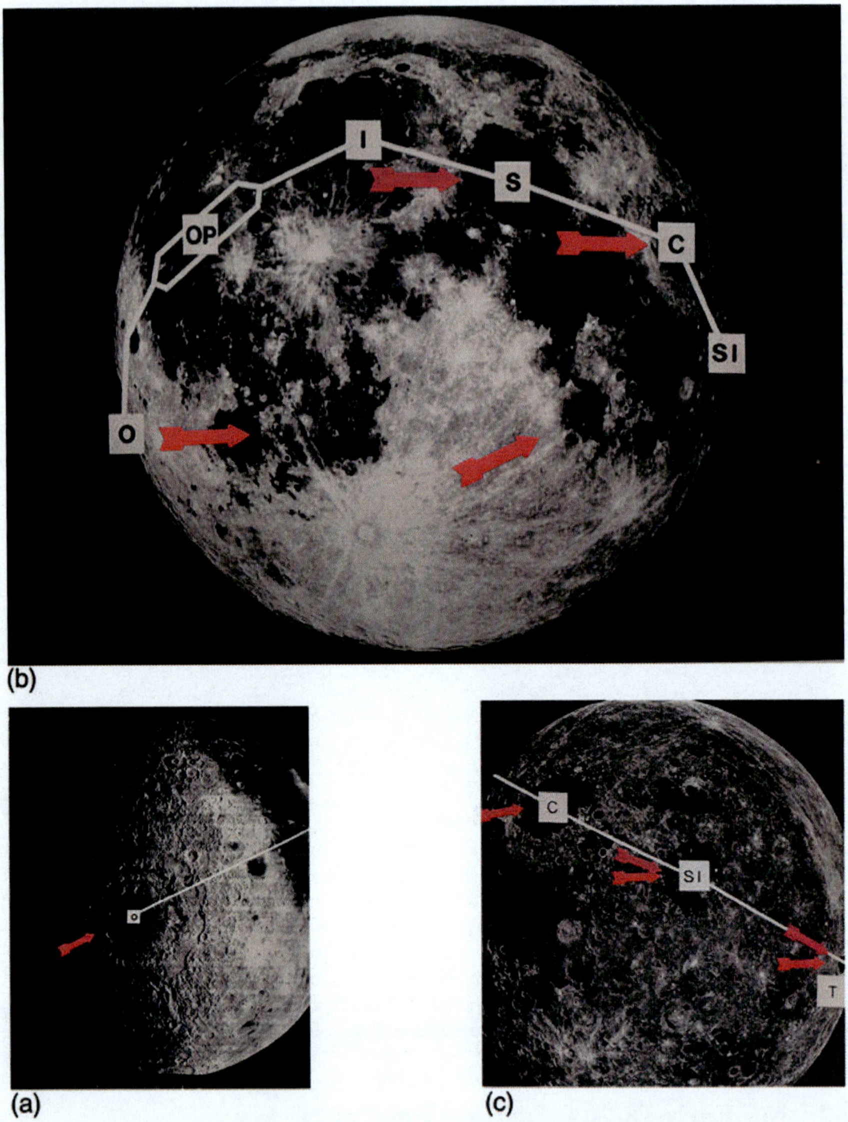

Fig. 7.7 Three lunar photographs that define a primary great-circle pattern of features as well as those that define a secondary great-circle pattern. (**a**). Western limb of the Moon: *O* Mare Orientale; the white line connects features on the primary great-circle pattern (Lunar orbiter photo IV-187 M). (**b**). Front face of the Moon: *O* Mare Orientale, *OP* Oceanus Procellarum, *I* Imbrium, *S* Mare Serenitatis, *C* Mare Crisium, *SI* Mare Smythii. Features on a secondary Great-Circle Pattern are marked only with a *red arrow*: on *left* is Mare Humorum and on the *right* is Mare Nectaris. Mare Moscoviense is on the lunar backside. (Hale Observatory photo). (**c**). Eastern limb of the Moon: *C* Mare Crisium, *SI* Mare Smythii, *T* Crater Tsiolkovsky. (Apollo photo 8-14-2485)

capture and tidal disruption of the lunar body. Then at the 1976 Spring AGU meeting we gave a presentation (given by Winters) entitled "The lunar capture hypothesis: A post-Apollo evaluation" (Winters et al. 1976). In this presentation we demonstrated that the energy for capture could be stored (but not necessarily dissipated, because we did not understand the Q factor relationship at that time) during a single very close, but non-collisional, encounter. These energy storage concepts were published in Winters and Malcuit (1977).

During the Summer of 1976 (August) I gave an oral presentation at the 25th International Geological Congress in Sydney, Australia. The talk was titled "Geological evidence supporting the lunar capture hypothesis: The great-circle pattern of large circular maria and a major thermal episode on Earth" (Malcuit et al. 1976). Carol Hodges (USGS) was the moderator and Bill Muehlberger (Univ. of Texas) was in the audience. Needless to say, Bill did not like the idea when he heard me at the 1974 AGU meeting and he still did not like it. But others in the audience thought that the tidal disruption model fit fairly well for explaining this great-circle pattern of large circular maria, a pattern that most of the audience had not noticed before. But we still could not demonstrate that the Moon was capturable (and it would be another decade before we could demonstrate that it could be captured). The pattern of large circular maria and the interpreted directional properties are shown in Fig. 7.7. As a historical note and some facts not known to me at that time, this great-circle pattern was first reported by Julius Franz (1912, 1913) (a German astronomer) and the Soviet lunar scientists emphasized that relationship after the lunar backside was photographed in 1959 (Lipskii 1965; Lipskii et al. 1966). I did not catch up with that information until 1977 (Lipskii and Rodionova 1977). (There will be more on this subject later.)

Before the 25th IGC meeting I participated in a 10-day field trip to the Flinders Range in South Australia. I had been interested in the problem of the Late Precambrian Glacial Deposits since graduate school and I thought that this may be a good time to check out some of these deposits. In general, there are two episodes in Earth History that I have been keenly interested in throughout my geological career: (1) a major thermal episode on Earth between 3.9 and 3.6 Ga and the associated major events on the Moon (and how these phenomena may be related to the origin of the Moon by gravitational capture), and (2) the Late Precambrian Glacial Deposits which appear to represent alternating "icehouse" and "greenhouse" events on Earth (and how this second set of events may be related to the long-term evolution of the Earth-Moon system).

Getting back to lunar science, I submitted an Abstract for the 8th Lunar Science Conference (Malcuit et al. 1977) titled "Is the Moon a captured body?" The abstract was accepted for publication, but the conference organizers apparently did not consider it important enough for a presentation because of space requirements for several sessions dedicated to the analysis of Mars. Anyhow, in that three-page abstract there is a timescale chart titled "Possible time-scale for earth-moon interactions" which has a section for EARTH EVENTS, and another section for LUNAR EVENTS, and another section for a CAPTURE SEQUENCE of events as shown in Fig. 7.8. This was the first time that such a series of events was related to a gravitational capture scenario. Although there was significant circumstantial

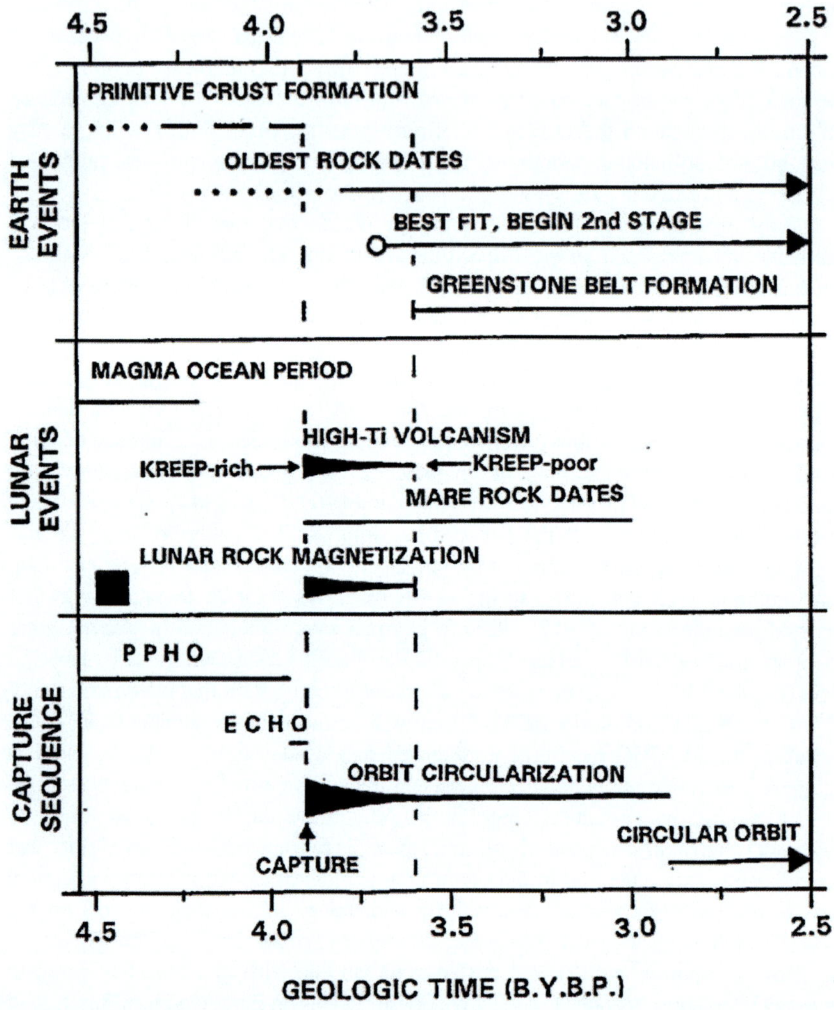

Fig. 7.8 Earth history, Lunar history, and the gravitational capture scenario. Diagram showing the possible relationships of a TIDAL CAPTURE MODEL and a number of terrestrial and lunar events. History of the diagram: First appearance in preliminary form was in Malcuit et al. (1977); a more advanced form was in Malcuit et al. (1992). And now the current form. Sources of data for EARTH EVENTS, oldest rock dates (Black et al. 1971; Goldich and Hedge 1974; Baadsgaard et al. 1976; Windley 1984): Pb-Pb age of Swaziland Sequence (Saager and Koppel 1976); best fit for second stage of lead isotope evolution (Stacey and Kramers 1975; Albarede and Juteau 1984). More recent information on EARTH EVENTS (Valley et al. 2002; Valley 2006; Harrison 2009; Mueller et al. 2011; Bell et al. 2011).. Sources of information for LUNAR EVENTS: lunar-rock magnetization (Cisowski and Fuller 1983; Cisowski et al. 1983, Runcorn 1983); "pre-mare volcanism" (Ryder and Taylor 1976; Taylor 1982); mare-rock dates (Taylor 1982; Ryder 1990). Capture-sequence symbols: PPHO Planet-Perturbed Heliocentric Orbit, ECHO Earth-Crossing Heliocentric Orbit. (From Malcuit (2015, Fig. 5.75) with permission from Springer)

evidence in support of the capture model, we still could not demonstrate that the Moon was capturable.

Another interesting experience occurred during Spring Semester 1977, when my Denison colleagues and I were invited by the Ohio State University Physics/Astronomy Department to give a seminar on our Lunar Capture ideas. I led off with a summary of the geological and lunar evidence with the hard evidence being "A GREAT-CIRCLE PATTERN OF LARGE CIRCULAR MARIA" and the numerical simulations that I did at MSU to show the trajectories of particles and spheroids of lunar basalt and how some of them would impact on the Moon. After my part of the presentation, there was a time for questions. The first question, and one that surfaced at a number of other presentations, is: "What happens to the Moon when it enters the Roche limit?" And, of course, the questioner had his own answer: "The Moon would either blow-up or disintegrate when it gets within 2.89 earth radii, the classical Roche limit." Someone else in the room called to the attention of the astronomer that we are dealing with rocky bodies during such an encounter and the body would need to come much closer than the classical Roche limit before anything would happen. Then there was general agreement that that limit for tidal disruption would be the Weightlessness Limit, which is also referred to in more recent years as the "Roche Limit for a Solid Body" (Aggarwal and Oberbeck 1974; Holsapple and Michel 2006, 2008). Then Ron Winters and Mike Mickelson illustrated that sufficient energy could be stored in the Moon for its capture and that we needed a very low Q value (specific dissipation factor) for capture.

During the Fall Semester 1976, I got a copy of the numerical simulation program from Tom Stoeckley that I used for my thesis work. The program could handle all nine planets in the Solar System but there was no energy dissipation subroutine. Ron Winters got the program going on our fairly primitive Digital Corporation Computer. We were only interested in doing co-planar simulations. We concentrated on encounters of a lunar-mass body with an earth-mass body in an earth-like heliocentric orbit. For many of the simulations, the lunar-mass body made one pass and then went back into heliocentric orbit. But for some encounter configurations, the Moon would orbit the Earth for 2, 3, 4, or more orbits before going back into heliocentric orbit and some simulations resulted in orbital collisions. These simulations were of interest but without an energy-dissipation subroutine, and some justification for its operation, the program was not very useful for capture studies.

In 1977 I became aware of the history of discovery of a great-circle pattern of large circular maria. I could not believe that such a pattern was not recognized by other observers. I became aware of the history of discovery via an article by Lipskii and Rodionova (1977) titled "Antipodes on the Moon" published in a proceedings volume of a Soviet-American Conference on "Cosmochemistry of the Moon and the Planets". They pointed out that the first publication that mentioned this great circle pattern was by Julian Franz (1912) in a book (Der Mond) as well as in another book by Franz (1913). Then the Soviet planetary scientists that got the first photos of the lunar backside stated that the great-circle pattern that Franz reported continued onto the lunar backside. This information was reported in a short article by Lipskii et al. (1965) in Sky and Telescope and in an article in Russian in 1966 titled (in English) "Current problems of the morphology of the Moon's surface" (Lipskii et al. 1966).

(The article was translated to English at a later date.) Robert Dietz referred to this line of circular maria; the article was titled "The meteoritic impact origin of the Moon's surface features" (Dietz 1946). But none of the authors suggested an origin of the pattern which deviated from solid body impact.

During the Summer of 1980 I presented an oral paper titled "The Late Proterozoic Glaciations: Possible product of the evolution of the Earth-Moon system" (Malcuit and Winters 1980) at the 26th International Geological Congress in Paris and then went on to the Island of Islay in Scotland to study some of the Late Precambrian glacial deposits along the north shore of the island.

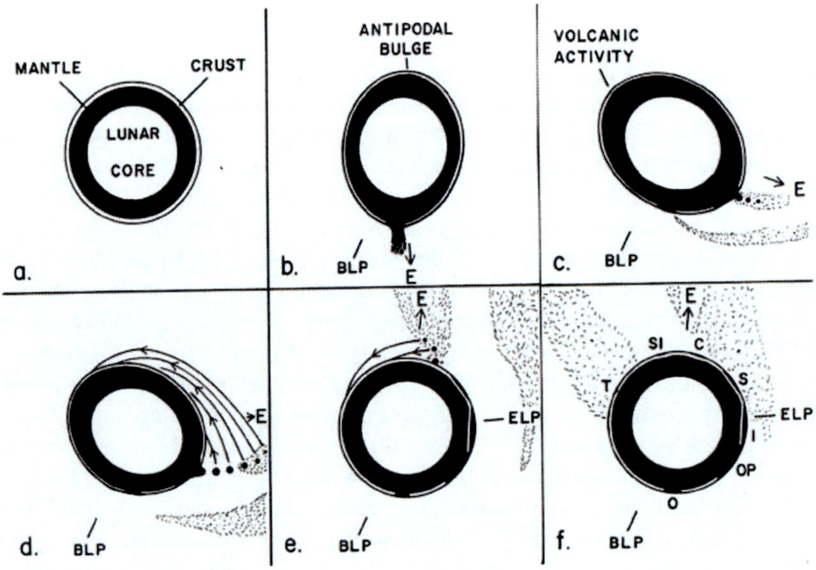

Fig. 7.9 Cross-sectional views of a lunar-like body during a tidal disruption sequence of events. View is from the lunar north pole. (**a**). Initial conditions before the encounter; the body is essentially undeformed. (**b**). The lunar-like body enters the W-limit at the *BLP* (=begin launch phase) position; the *E arrow* points toward Earth. Magma in the first disruption zone is loaded with volatiles which cause a very explosive eruption. This feature is interpreted as the Orientale disruption zone. (**c**). At this stage the lunar-like body is well within the W-limit and a second disruption zone (in the Oceanus Procellarum region) also erupts very explosively. Gas-rich volcanism also occurs in the antipodal bulge region and this area can also experience a weaker version of weightlessness during this very close encounter. (**d**). The lunar-like body now exits the W-limit and this is the end of the Launch Phase of this very close encounter. (**e**). *ELP* end launch phase. At this position the lunar-like body is well beyond the W-limit and the spheroids of basalt and other debris are falling back onto the surface of the lunar body. (**f**). The lunar-like body is moving away from the Earth at this point but material is still falling back onto its surface; the positions of several features in Fig. 7.6 are shown on this diagram as well as the two launch centers; most of the action occurs within about 2 hr. of time. (From Malcuit (2015, Fig. 5.46) with permission from Springer)

For the Lunar Origin Project during the early 1980's I concentrated on drafting illustrations on how the tidal disruption sequence of events could occur in both cross-section and surface views as illustrated in Figs. 7.9 and 7.10. I started with a simple six-element cross-sectional view of a coplanar tidal disruption model and then recast that sequence of events into a surface view from the lunar north pole. But this simple model did not explain the major lunar enigmatic features to be explained which were the locations and directional properties of Mare Humorum, Mare Nubium, Mare Nectaris, and Mare Moscoviense. These features were off the Great Circle Pattern and they needed an explanation via a single tidally disruptive encounter. After some serious sketching, I came to the tentative conclusion that this second great-circle pattern (also pointed out by Runcorn (1983) in a paper in NATURE) could be explained via an initial eruptive phase of the Oceanus Procellarum disruption center from which lunar material, some of which was in the form of basaltic spheroids, was both deflected by a crustal plate and "gas-jetted" to the lunar south

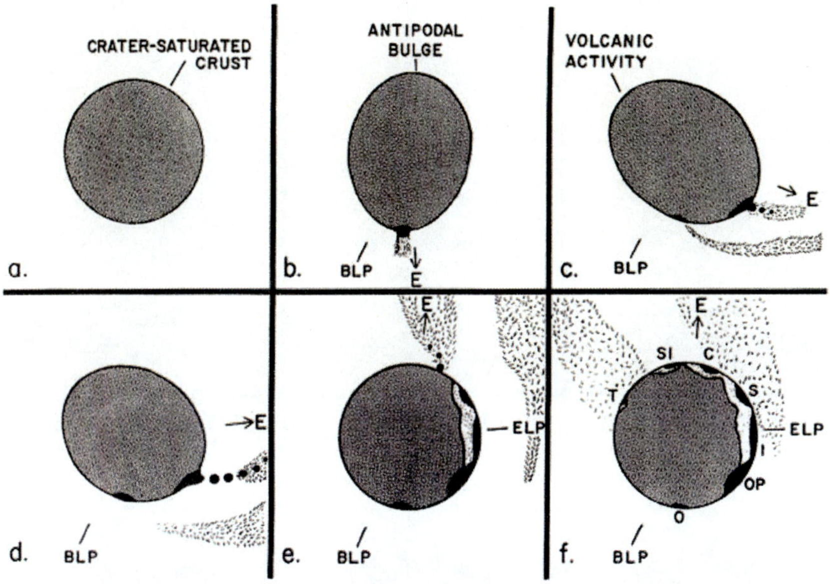

Fig. 7.10 View is from the lunar north pole. (**a**). Crater-saturated anorthositic crust. (**b**). Lunar body enters the W-limit at *BLP* (begin launch phase) position. *Arrow* with E points toward the Earth. First disruption center is an interpretation of the Orientale supra-volcanic form. (**c**). Initial eruptions of the Oceanus Procellarum disruption zone. A few spheroids of basalt are necking off a lava column from this second disruption zone and there is volcanic activity in the antipodal region (the Mare Ingenii region). (**d**). This is the apex of the tidal disruption episode and the lunar body is departing from the W-limit. (**e**). The fall-back phase is well underway and some of the spheroids of lunar basalt have impacted obliquely onto the lunar surface. *ELP* end launch phase. (**f**). At this stage all of the large spheroids have impacted on the lunar surface but much lunar-borne material is in the lunar environment. Most of this multi-sized material will eventually impact on the lunar surface. (From Malcuit (2015, Fig. 5.52) with permission from Springer)

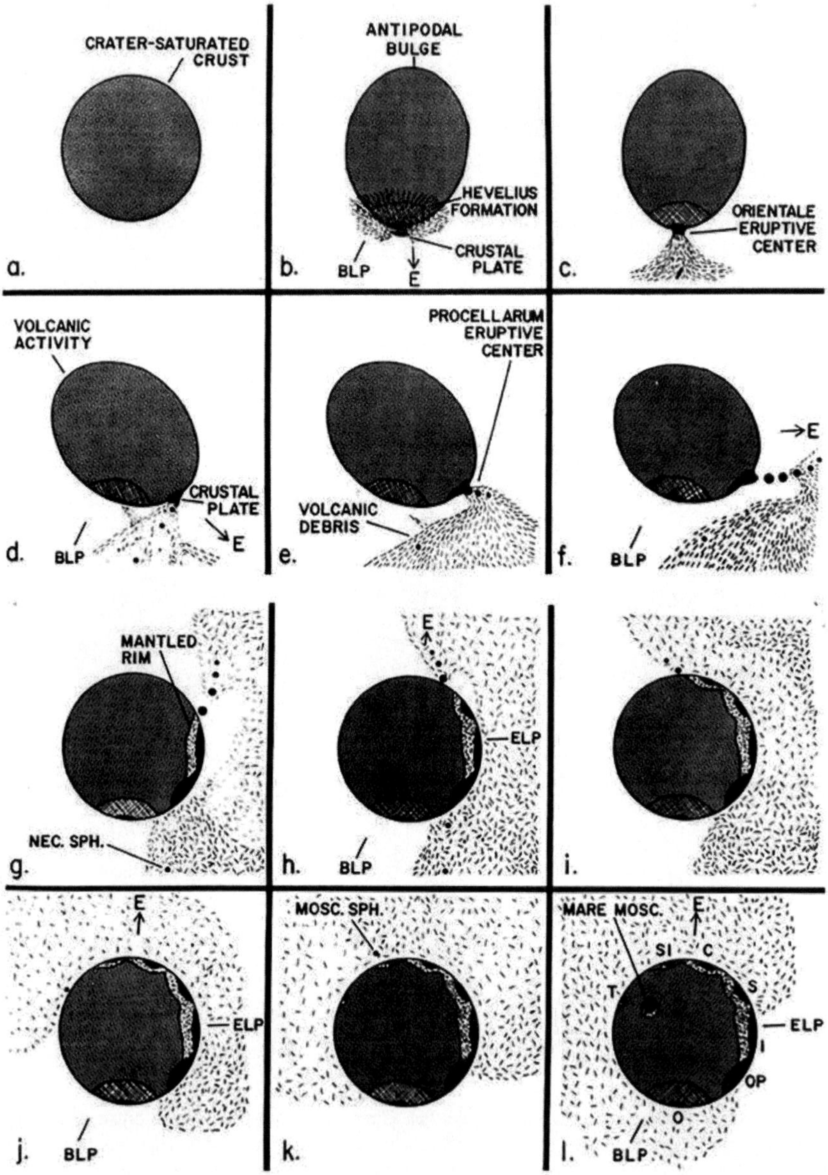

Fig. 7.11 View of activity from the lunar north pole. (**a**). Crater-saturated anorthositic crust. (**b**). The gas-propelled eruption of the Orientale Supra-Eruption center which lead to the development of the Hevelius Formation. The arrow with *E* points toward Earth. A crustal plate temporarily deflects some material back to the lunar surface. *BLP* = begin launch phase. (**c**). Apex of activity associated with the Orientale Supra-Volcanic Center. Most of this material will go to the Earth. The crustal plate is clearly visible and would go to the Earth. (**d**). The initial eruption phase of the Oceanus Procellarum Supra-volcanic Center. A crustal plate deflects material both backward from the normal gravitational flow as well as off the encounter plane. In this scene some spheroids are deflected to the south of the encounter plane and some will fall back in the lunar southern hemisphere and some

and west. This material would then fall back in the lunar "highlands" area to the south of the encounter plane and some would cross over the encounter plane to impact in the lunar northern hemisphere.

During the Summer of 1982 I drafted two sets of 12-element diagrams (Fig. 7.11 and 7.12): one was for a tidal disruption sequence of events showing the evolution of lunar surface features in the northern hemisphere and the second was for the evolution of lunar surface features in the southern hemisphere. With these visual aids I could now explain the directional properties of Mare Humorum, Mare Nubium, Mare Nectaris, and Mare Moscoviense. I should note here that Mare Moscoviense is not all that easy to explain because it has an impact vector that is oriented perpendicular to the encounter plane. But when one follows the material ejected from the initial Oceanus Procellarum launch center, this unusual orientation can be explained.

Since I mentioned Keith Runcorn, I should emphasize the very positive and encouraging interactions that I had with Keith. I met Keith Runcorn in 1972 when he stopped by the Geology Department at MSU. He knew that I was working on a lunar project by way of my AGU presentation and he was interested in my ideas on this great-circle pattern of large circular maria. I would see Keith at AGU and GSA meetings from time to time and he introduced me to Harold Masursky (USGS) who was also interested in the features associated with this great-circle pattern of large circular maria and associated mascons. In 1983 Keith borrowed my idea of a great-circle pattern of large circular maria and tried to apply it to the generative mechanism of the mare-age lunar magnetic field. This article in NATURE was titled "Lunar magnetism, polar displacements and primeval satellites in the Earth-Moon system" (Runcorn 1983). In this article Runcorn pointed out that in addition to the primary great-circle pattern there is a secondary great-circle pattern on which Mare Humorum, Mare Nubium, Mare Nectaris, and Mare Moscovience are located. I was very pleased to see this article because it indicated that someone else had recognized the secondary great-circle pattern and was attempting to relate those patterns of large circular maria to the problem of lunar rock magnetization. About the time of the publication of Runcorn's article in NATURE, I had just finished my set of scale sketches on how

Fig. 7.11 (continued) in the lunar northern hemisphere. (**e**). Normal gravitational flow is occurring from the Procellarum Supra-Volcanic Center. A spheroid is visible in the volcanic debris. (**f**). The termination of the Procellarum Eruptive Center. Note the stack of spheroids from this center as well as the spheroid with the other debris. (**g**). A few of the basaltic spheroids have impacted obliquely onto the lunar surface and are causing "tsunami" wave action, mantled rims, etc., as a result of their impact. Note the position of the Nectaris spheroid that will impact in the lunar southern hemisphere. (**h**). Spheroids are falling back onto the lunar surface along a great-circle pattern. *ELP* = end launch phase position. (**i**). Fall-back phase continues. (**j**). All major spheroids have impacted on the lunar surface. Note the position of the Tsiolkovsky spheroid (not yet impacted). (**k**). All major spheroids have impacted onto the Great-Circle Pattern. The Moscoviense Spheroid is moving from the lunar Southern Hemisphere into the lunar Northern Hemisphere. (**l**). The Moscoviense Spheroid has impacted to form an elliptical lake on the lunar surface with a impact vector about 90° to the trend of the Great Circle of Large Circular Maria. (From Malcuit (2015, Fig. 5.53) with permission from Springer)

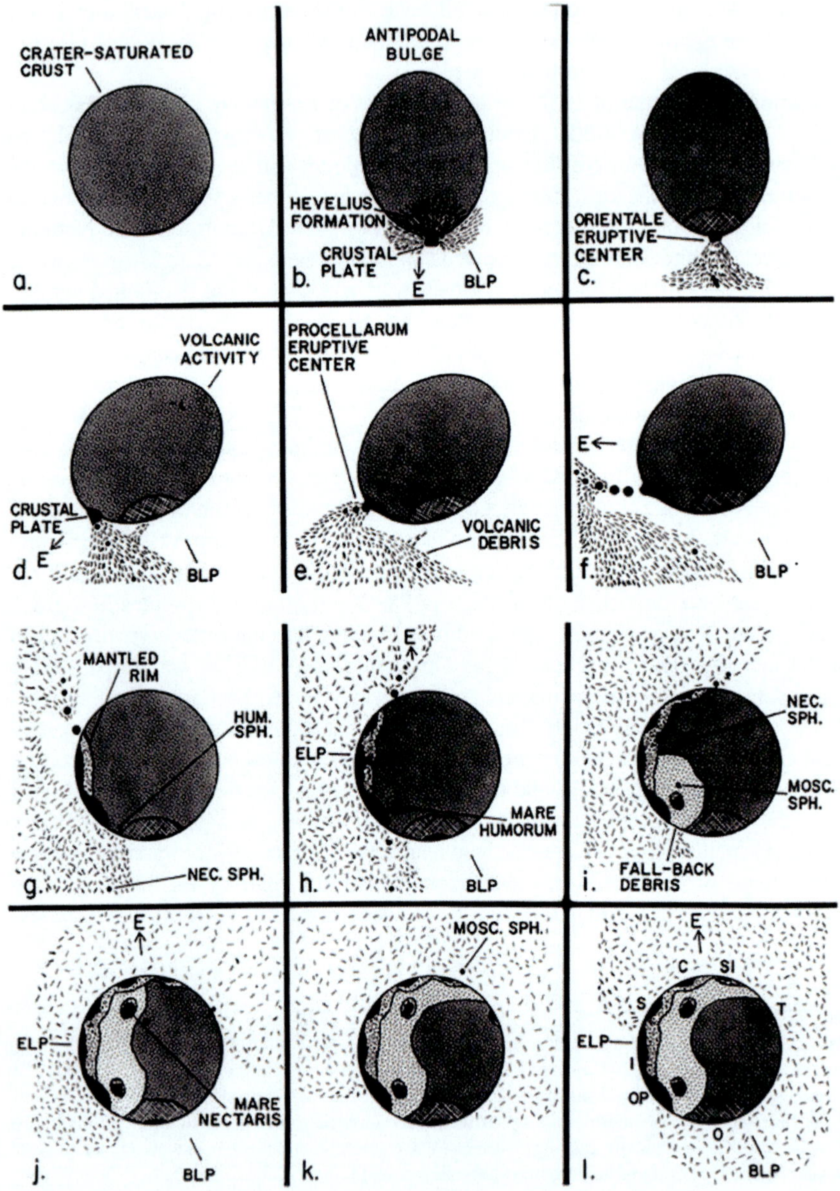

Fig. 7.12 View of activity from the Lunar South Pole. Note that the action is in the opposite direction to that in Fig. 7.10. (**a**). Crater-saturated anorthositic crust. (**b**). Lunar body enter the W-limit and a gas-charged eruption occurs. A crustal plate deflects material back onto the lunar surface to form the Hevelius formation. *BLP* begin launch phase. *Arrow* with *E* points toward Earth. (**c**). Orientale Supra-Eruption center is in full bloom. Most of this material goes to the Earth. A crustal plate is visible and goes to the Earth. (**d**). The Procellarum Supra-Eruption center opens with a very explosive eruption that propels material, including spheroids of basalt, both backward relative to the gravitational influence and into the Lunar Southern Hemisphere. Note the volcanic activity in the antipodal bulge area. Note that the anti-sub-earth region is only slightly within the

this secondary great-circle pattern might relate to the tidal disruption sequence of events associated with gravitational capture (see Figs. 7.11 and 7.12).

By the Fall of 1983 I had a some good sketches on how a tidal disruption sequence of events could generate a great-circle pattern of large circular maria on the lunar surface as well as a more poorly defined second great-circle pattern. But I still did not know when in this tidal capture sequence of events that tidal disruption would occur. It would be four years in the future before we would have any numerical simulations of capture. We knew that the energy for capture could be stored by tidal deformation during a close encounter if the displacement Love number was sufficiently high but we could not justify a sufficiently low Q (specific dissipation factor) for the lunar body to facilitate capture during the initial encounter of a stable capture scenario (i.e., the CAPTURE ENCOUNTER). Thus, the lunar body must be cool enough to have a low Q value but immediately following the capture encounter the lunar body must be warm enough (after dissipating somewhere between 1 and 2 times E28 joules) to undergo tidal disruption during a close encounter that was well within the W-limit. Over 20 years passed before I had a reasonable answer to that dilemma.

During the Summer of 1984 I participated in the 27th International Geological Congress in Moscow (USSR). I went on a 10-day pre-meeting field trip to Kazakhstan to study rock units ranging from Late Precambrian through Upper Ordovician. The only Late Precambrian Glacial Deposits were outwash gravels but the lowermost Cambrian sequences contained sharply defined phosphorite units which represented some unusual atmospheric and oceanographic geochemical events at the transition of the Precambrian to the Cambrian. This sequence that we were studying was one of the few in the world where there is continuous deposition through the Precambrian to Cambrian transition. Before coming to the meeting I had

Fig. 7.12 (continued) W-limit but this leads to a gas-charged eruption in the Mare Ingenii region. (**e**). The Procellarum supra-eruption center is settling down to a normal gravitational necking process to produce basaltic spheroids. (**f**). The lunar body passes out through the W-limit. This is the end of the Launch Phase. Note that a spheroid is mixed in with the small debris. (**g**). The fall-back phase is well underway and some spheroids of lunar basalt have impacted to form "tsunami" waves of basalt to flow over the surface around the impact sites to form mantled rims. The Humorum spheroid is about to impact and the Nectaris spheroid is visible in the lunar debris field. (**h**). Regular fall-back activity continues. The Humorum spheroid has impacted and the Nectaris spheroid is visible. *ELP* end launch phase. (**i**). The Smythii spheroid as well as the Tsiolkovsky spheroid are about to impact along a lunar great circle. The Nectaris spheroid is visible above the lunar surface. The Moscoviense spheroid is visible above the Mare Humorum area. (**j**). The Nectaris spheroid has impacted and the Tsiolkovsky spheroid is about to impact but the Moscoviense spheroid is visible above the lunar surface in the Mare Nectaris area. (**k**). All of the major spheroids are down but the Moscoviense spheroid. It will impact on the lunar northern hemisphere (as shown in Fig. 7.10). (**l**). The bulk of the Fall-Back Phase is over but much small lunar debris is returning to the lunar surface. Note that Mare Humorum and Mare Nectaris are located off the great-circle pattern and are in the lunar Southern Hemisphere. These two features along with Mare Moscoviense are on a secondary great circle in this model. (From Malcuit (2015, Fig. 5.54) with permission from Springer) read an article titled "Contributions of tidal dissipation to lunar thermal history" (Peale and Cassen 1978). In this article they identified a Jupiter Orbit – Lunar Orbit resonance that would occur when the lunar orbit is at 53.4 earth radii. Most reasonable estimates would have the lunar orbit at that distance around the Precambrian-Cambrian transition. I was wondering on the outcrop how the lunar tides may influence the depositional and geochemical events that we were observing and whether or not it had an effect on the alternating "icehouse" and "greenhouse" events recorded in the late Precambrian rock sequences.

Then after returning to Moscow State University for the meetings, I presented an oral paper to the Planetology Section titled "A testable gravitational capture model for the origin of the Earth-Moon system" (Malcuit et al. 1984a). The talk was translated instantaneously into Russian and Italian. I got a very favorable response from the audience. There were several questions from the audience especially on the structure of lunar maria and my explanation of the mascons. In the audience was Zhana Rodionova, coauthor of the 1977 article on "Antipodes on the Moon" (Lipskii and Rodianova 1977). Zhana introduced herself after the talk and we had a very interesting discussion on patterns of lunar features and especially "A GREAT-CIRCLE PATTERN OF LARGE CIRCULAR MARIA". She had read my article (Malcuit et al. 1975) so she understood my genetic model for emplacement of the lunar maria and mascons.

7.4 The Kona Conference on the Origin of the Moon

The time has come in this narrative for a discussion of the Conference on the Origin of the Moon at Kona, Hawaii, in October 1984. Many lunar and planetary scientists were there. I presented two papers for my group (Malcuit et al. 1984a, b). The poster paper was titled "Directional properties of 'circular' maria: Interpretation in the context of a testable gravitational capture model for lunar origin". Keith Runcorn, Bruce Conway, and a few other planetary scientists spent considerable time on this poster and they had very good questions but most of the other participants were not really interested. They had something else in mind: THE ORIGIN OF THE MOON BY GIANT IMPACT (and that probably should have been the name of the conference). My second paper was an oral presentation titled "A testable gravitational capture model for the origin of the Earth's Moon". I must admit that we still had no good mechanism for dissipating the energy for capture although we could demonstrate that the energy necessary for capture could be temporarily stored in the lunar body during a close approach of the Moon to the Earth. But again, certain characters in the audience just could not wait to get these capture, co-formation, and fission talks out of the way so the new idea on the origin of the Moon via GIANT IMPACT could be presented on center stage.

After this Conference it looked to me like the gravitational capture model was a vast minority model and a struggling operation. The organizers of the Symposium Volume rejected our submitted manuscript on the great-circle pattern of large circular maria and the testable features of the capture model (testable mainly via directional indicators of maria and mascons). We still did not have a mechanism for DISSIPATING THE ENERGY FOR CAPTURE IN THE MOON and we did not have A PLACE OF ORIGIN FOR THE MOON. On the other hand, the symposium volume organizers, apparently wanted an article on capture that was not controversial, because it did not relate specifically to any lunar features, so they essentially "republished" an article by Singer (1986) with very light modification from Singer (1977) in EARTH SCIENCE REVIEWS in their GIANT IMPACT volume. In Singer's model, the earth-like body was the supposed energy sink for capture and the energy for capture was off by orders of magnitude. In this way the editors of the symposium volume, in my view, avoided an mention of new ideas concerning lunar capture models.

Such is life in the scientific world when a new paradigm model is born. The capture model, especially with a tidal disruption component was TOO CATA STROPHIC. In its place was the GIANT IMPACT MODEL that is about 3000 times more ENERGETIC (= CATASTROPHIC) than our version of a gravitational capture model! An anti-capture quote from Ross and Schubert (1989) is appropriate here and we need some of their introductory words to put this in context:

> In fact, numerical models of the ocean tide in backward evolution calculations suggest that when the semimajor axis becomes less than about 40-50 R_E, oceanic dissipation falls sharply and the Moon remains "stuck" at about 30 R_E (Hansen 1982; Webb 1982). To circumvent the problem, Hansen (1982) proposed a capture origin for the Moon with capture at about 30 R_E! Our results show that such drastic proposals are not required to reconcile formation of the Moon near the Earth with interactive ocean tides inward of about 30 R_E.—From Ross and Schubert (1989)

This article presented a calculation for an orbit evolution following a giant-impact scenario. Apparently the authors temporarily forgot about the catastrophic nature of a giant-impact event. To repeat, a giant impact scenario is about 3000 times more catastrophic than our version of a capture scenario; our scenario does result, after post-capture orbit circularization, in a 30 to 40 earth radii lunar orbit, the details depending on the rate of angular momentum transfer from the rotating Earth to the lunar orbit during the orbit circularization sequence.

During part of the Fall Semester of 1986 I presented an oral paper at the 12th International Sedimentological Congress (Canberra, Australia) titled "Evolution of the Earth-Moon system (3.8 B. Y. to Recent): Implications for tidal ranges over geologic time" in a session on Tidal Deposits (Malcuit 1986). This presentation was based on a calculation of tidal parameters in which the rate of change for earth rotation was constant at 1 millisecond/century. The study of tidal rhythmites was then in its infancy and I got many positive comments on the presentation. After the meeting I was on my second 10-day field excursion to the Flinders Range (I had been there 10 years before). The field trip leaders were different and we examined many outcrops that were not visited during the previous trip. In addition to Late Precambrian Glacial deposits, of which we saw a good many, we got to examine some good outcrops of tidal rhythmites, including the really good exposures of the Elatina-Reynella rhythmites in the Pichi Richi Pass area of South Australia and George Williams (Univ. of Adelaide) was the leader for that segment of the trip. Toward the end of the 10-day field trip we met (as scheduled) Gene and Carolyn Shoemaker (USGS) to examine a series of outcrops in the Bunyeroo Formation with a layer of impact debris from the recently mapped Acraman impact structure which was about 250 km to the west of us. A few days later, Gene gave an afternoon seminar for the Geoscience Department (University of Adelaide) in the afternoon on his and Carolyn's work on the impact structures in Western Australia and in the evening I gave a talk to the Adelaide Astronomical Society on "The Great-Circle Pattern of Large Circular Maria and Implications for the History of the Earth-Moon System".

7.5 The Era of Serious Capture Studies at Denison University

In 1985 a very talented freshman by the name of David Mehringer enrolled at Denison University. David was interested in Physics, Astronomy, and Computer Science. Eventually Winters got David introduced to the capture idea and by 1986 Mehringer was doing some physics and numerical simulation work for us on the post-capture orbit evolution after making certain assumptions of the h and Q values of the Earth and Moon. Also in 1986, there was a paper published by Ross and Schubert in the Journal of Geophysical Research illustrating that with intermediate viscosity conditions a body like the Moon could have a very low "Q" value. This was super important for us because we could not make any more progress until we could justify a very low "Q" value for the Moon. Ross and Schubert (1986) also pointed out that IO, the innermost satellite of Jupiter, appears to be emitting about as much energy via volcanism as it generates via tidal interaction with Jupiter. Thus IO must have a "Q" value near 1.

During the Summer of 1987 David Mehringer and Ron Winters developed some ideas on how to insert an energy dissipation subroutine into the planetary orbits computer code that had been setting around for a decade. Then Winters went on leave to do work in nuclear physics at the Oak Ridge National Laboratory. By Fall Semester 1987 Mehringer had a useful version of the code running and David and I were about to witness the first calculation showing that gravitational capture was possible in a coplanar, three-body context. Since the Q value of Earth could not be lower than about 100, most of the energy for capture must be dissipated in the Moon. With a Q value of 1 or 2, the energy for capture could be dissipated in the Moon alone during a sufficiently close encounter. Mehringer did a senior thesis on this project and I started doing numerous simulations of escape sequences, collision sequences, and stable capture sequences attempting to determine under what specific conditions successful (stable orbit) capture could occur.

I gave a presentation for the group (Malcuit et al. 1988a) on the project at the 19th Lunar and Planetary Science Conference and many of the attending scientists were impressed by the demonstration that a lunar-like body (lunar mass and density) could be captured by an earth-like body from a heliocentric orbit into a geocentric orbit with a large major axis. NO ONE HAD DONE THIS BEFORE! THIS WAS A FIRST! After some favorable comments from members of the audience, a question was asked: "WHERE DID THE BODY OF THE MOON COME FROM??????? I knew that I did not have a good answer for that question. All I could say was that we can demonstrate, using reasonable orbital parameters that a lunar-like body can be captured from an earth-like heliocentric orbit into a geocentric orbit with large major axis and high eccentricity. A reasonable answer to the question of "WHERE DID THE MOON COME FROM?" would have to wait another 11 years for a tentative answer and 15 years for a more affirmative answer.

A ten-page paper was published in the Proceedings Volume of the 19th LPSC titled "Numerical simulation of gravitational capture of a lunar-like body by Earth" (Malcuit et al. 1989b). In this paper I referred to Don Anderson's (1972) concept that a body of high temperature condensate material, Luna, may be formed in a highly inclined heliocentric orbit at about 1 AU. This was simply a default explanation for a place of origin.

During the Summer of 1989 I presented an oral paper in the Planetology Section of the 28th International Geological Congress (Washington, D. C.) titled "Numerical simulation of intact planetoid capture: Application to planets Venus and Earth" (Malcuit et al. 1989a). In this paper I demonstrated that retrograde capture of a lunar-like planetoid was physically possible for planet Venus (a process that Fred Singer proposed in a paper in Science but could not demonstrate as physically possible) and that prograde capture of a lunar-like body was physically possible for planet Earth. But I still did not have a good place of origin for the planetoids. Preston Cloud (UCSB) was in the audience. Needless to say, he was very pleased to see that gravitational capture of lunar-like planetoids was physically possible. We had a lengthy discussion on planetary geology and the early history of the Earth and Moon after the session.

As background, this is a historical note on a previous interaction with Preston Cloud. When Graham Ryder and I were doing some work on the anorthosites in the San Gabriel Mountains north of Los Angeles in 1973, we arranged to go to Santa Barbara to visit Professor Cloud. After a tour of a portion of his "Biogeology Clean Laboratory" (in which he and colleagues examined lunar samples for evidence of biological entities), we had lunch with Preston Cloud, Mike Fuller, Stan Aramik, and others. I exposed them to my work on a Great-Circle Pattern of Large Circular Maria and how this pattern of circular lunar maria may be related to the origin of the Moon by gravitational capture. So when I showed the front face of the Moon in my presentation at the Washington (DC) meeting, it was not all that foreign to Preston Cloud.

In August-September 1990, I was involved in the 3rd International Archean Symposium in Perth, Australia, where I presented a poster paper titled "A gravitational capture origin for the Earth-Moon system: Implications for the early history of Earth and Moon" (Malcuit et al. 1990). In this paper I was concentrating on the thermal regime of the Earth during the post-capture orbit circularization sequence; the major part of the orbit circularization takes place in about 300 million years. The Hadean Eon was still a geological mystery to most geologists (at that time there was essentially no rock record for that eon) but a few very old zircon crystals of Hadean age had been reported from the Jack Hills Formation of Western Australia. My paper got considerable positive attention and I predicted in that paper that there should be geochemical evidence for recycled primitive crust in the Archean rock record. A twelve-page article was published in the Proceeding Volume of the meeting (Malcuit et al. 1992). A time-scale of Earth Events, Lunar Events, and Capture Sequence of events (similar to the one in the 1977 LSC abstract) was included in that paper.

During the same week as the Archean Symposium was taking place, the Meteoritical Society was having a meeting at Curtin University. A graduate school colleague, Graham Ryder (Lunar and Planetary Institute), was involved in that meeting. Then on Sunday morning, after the meeting but before the post-meeting field trips, we got together for a while at my hotel to discuss various issues in planetary science.

For the March 1995 LPSC meeting I presented a poster paper titled: Numerical simulation of retrograde gravitational capture of a satellite by Venus: Implications for the thermal history of the planet (Malcuit and Winters 1995)." I also had a short version of prograde capture for the Earth-Moon system for comparison and contrast. Bob Dietz was at the meeting and after studying my poster for a while, he said that this model of retrograde gravitational capture for Venus and prograde gravitational capture for Earth explains all kinds of features concerning the history of the two planets. Bob then said that he was going to get some of the NASA people to come and discuss the merits of my paper. He returned with a few characters but he found that they were much more interested in papers favoring the GIANT IMPACT MODEL. Needless to say, Bob Dietz was disappointed at their disinterest in the idea of lunar capture. This incident, I think, is another example of how a "band-wagon" ruling paradigm can rob a good bit of science from the pursuit of science.

Another memorable event, mainly because it was a non-event, was an abstract submitted to the LPSC for the March 1996 meeting. The title of the paper was "GEOMETRY OF STABLE CAPTURE ZONES FOR PLANET EARTH AND I MPLICATIONS FOR ESTIMATING THE PROBABILITY OF STABLE GRAV ITATIONAL CAPTURE OF PLANETOIDS FROM HELIOCENTRIC ORBIT" (Malcuit and Winters 1996a). This paper was the result of literally thousands of individual simulations of Earth-Moon encounters and, in my view is one of the most important concepts in planetoid capture studies. The abstract was accepted for publication in the Abstracts Volume but because of the emphasis on the exploration of Mars, the organizers claimed that they had no space on the program for presentation of this paper.

During the Summer of 1996 I gave two oral presentations at the 30th International Geological Congress in Beijing, China (Malcuit and Winters 1996b, c). One was on a prograde capture model for Earth and its compatibility with the rock records of Earth and Moon. The other paper was on capture scenarios for planets Venus and Earth and how those contrasting scenarios relate to the present state of these two "twin sister" planets. I got many good questions both during and after the sessions. I also went on a weekend field trip to a Proterozoic-age anorthosite deposit where the Chinese were actively mining vanadium and other metal deposits in the Chengde area which is northeast of Beijing. The origin of massif-type anorthosites has also been one of my long-standing interests.

7.6 Conference on the Origin of the Earth and Moon

Eventually I worked my way forward to the symposium on the ORIGIN OF THE EARTH AND MOON in Monterey (CA) in early December 1998 where I presented a poster paper on the capture origin of the Moon and how it relates to the rock records of Earth and Moon (Malcuit and Winters 1998). Fred Singer was at the meeting and he spent a good bit of time with my poster. He said that he had a student who wanted to do something similar to this back in the early 1980s but he and his student could not conceive of an effective energy dissipation mechanism in the context of a three-body interaction code. They were thinking that most of the energy for capture would be dissipated in the Earth.

A big hit at this meeting was a paper Kokubo, Canup, and Ida (published in the symposium volume) (Kokubo et al. 2000) in which they showed that a roughly lunar-mass body could be accreted from a geocentric debris cloud. But the impactor was about a 2 to 3 mars-mass planetoid and the debris cloud was about 3 lunar masses. The resulting body, after much "tweeking" was about 0.7 moon mass. They found that it is very difficult to form a lunar-mass body in the vicinity of the Roche limit. (Note: To this day researchers have not been successful in making a lunar-like body with reasonable lunar characteristics from the debris cloud of a simulated collison.)

On the second day of the meeting the poster paper presenters could volunteer to give a five minute presentation on a certain aspect of their posters. Fred Singer gave a short talk on his negative opinion of the GIANT IMPACT MODEL and stated again that some form of gravitational capture looked better to him than giant impact. In his opinion, the giant-impact model was "too catastrophic". A few talks later I gave a short presentation on the GEOMETRY OF STABLE CAPTURE ZONES and how they can be used to estimate the probability of capture. There were no questions from the audience for either of these two talks mainly because very few in the audience were interested in the capture process. For them, I guess, there was no future in pursuing gravitational capture studies.

The year after my retirement from teaching I participated in the 31th International Geological Congress in Rio de Janeiro, Brazil, where I presented two papers. The oral presentation was on Late Precambrian events on Earth and how the Moon may have been involved via a Jupiter Orbit – Lunar Orbit resonance (Malcuit and Winters 2000a). I got a very positive response from a number of geologists working on Late Precambrian rock sequences. The poster paper was titled "Geometry of retrograde stable capture zones for tidal capture of satellites by planet Neptune (Malcuit and Winters 2000b)." In this poster I demonstrated that retrograde capture has a much higher probability of occurrence than prograde capture for planet Neptune and a Triton-mass planetoid.

In 2001 I participated in a meeting sponsored jointly by the Geological Society of London and the Geological Society of America on "Earth System Processes" in Edinburgh (Scotland). I presented a poster paper titled "Neoproterozoic sedimentation and rifting events on Earth: Possible lunar involvement (Malcuit and Winters 2001)." A number of geologists spent a considerable amount of time examining and discussing my display of information and numerical simulation of orbits, tidal amplitudes, and tidal ranges.

A major reason that I wanted to participate in this meeting was to go on a pre-meeting field trip to the Garvellach Islands north of the Isle of Islay to study very good exposures of Late Proterozoic sequences. Just a few weeks before the scheduled field trip, it was cancelled because of MAD COW DISEASE. What a disappointment!

7.7 An Era of Important Events for Gravitational Capture Studies

Even after 100 Myr, some 80% of our Vulcanoid orbits still have eccentricities <0.2 and inclinations $<10^0$. It is this evidence, together with the low rate of attrition of their numbers, that suggests that they can continue for times of the order of the age of the Solar System — Evans and Tabachnik (1999, p. 42).

7.7.1 Vulcanoid Planetoids

I retired from my teaching duties after the Spring Semester 1999. At that point in time I still had no reasonable explanation for "WHERE THE MOON CAME FROM" and, as far as I could see at that time, maybe I should terminate my pursuit of gravitational capture studies. Then in the Spring of 1999, during my weekly perusal of the journal NATURE, I came across an article by Evans and Tabachnik (1999) titled "Possible long-lived asteroid belts in the inner Solar System". In the article they stated that some of the most stable orbits in the terrestrial planet realm are located between 0.1 and 0.2 AU (*i. e.* inside the orbit of Mercury). A second paper was published three years later (Evans and Tabachnik 2002). I was familiar with the very limited literature on the possibility of Vulcanoid planetoids (but no one had ever seen one) and I knew that the Moon was enriched in refractory elements and depleted in volatile elements like potassium. My question was: Why not form Luna and sibling planetoids in this region of space where the Vulcanoids should form? And I entertained the thought that maybe we humans and our animal friends with eyes have been looking at a fairly large Vulcanoid planetoid (named Luna) for a long time! Figure 7.13 shows the general setting of the Vulcanoid Zone relative to the Sun and the orbit of planet Mercury. In 2012 I discovered (rediscovered) two papers by by Al Cameron (1972, 1973a, b) in which he suggested that the chemistry of the lunar body was compatible with an origin between Mercury and the Sun (see the second quote at the beginning of this chapter). Cameron then abandoned the capture model because he could not visualize, at that time, a transfer process for getting the Moon to an earth-like heliocentric orbit from which it could be captured. Soon after publishing those papers, Cameron shifted his attention to another model and he eventually became a major advocate for the GIANT IMPACT MODEL (Hartmann and Davis 1975; Cameron and Ward 1976).

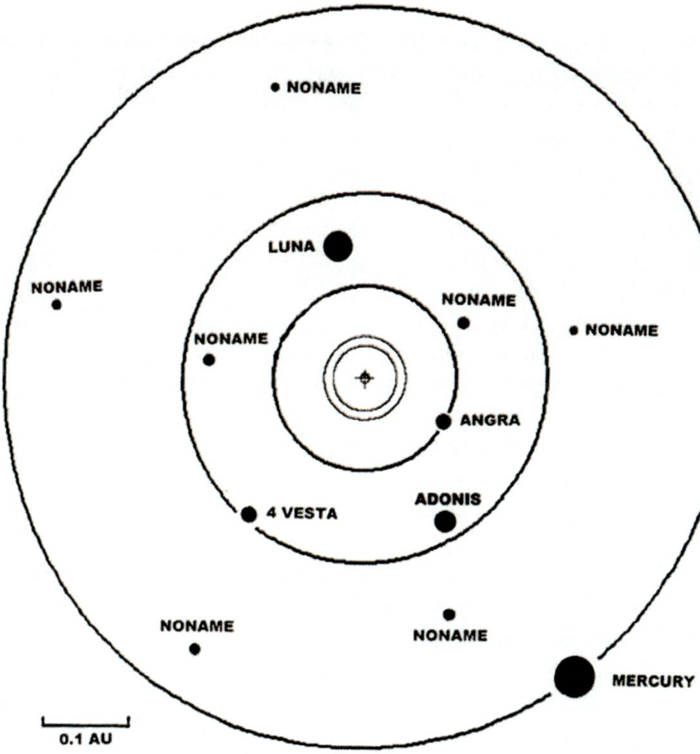

Fig. 7.13 Scale diagram illustrating a possible population of Vulcanoid planetoids in quasi-stable orbits as proposed by Evans and Tabachnik (1999). Luna is the name for the planetoid that becomes the Earth's Moon; Adonis is the name for the planetoid that gets captured by planet Venus in a retrograde direction; Vesta and Angra are the names of volcanic (differentiated) asteroids that ended up, in whole or in part, in the asteroid zone. Many other basaltic bodies could have populated this zone before they got destroyed by collision or got gravitationally perturbed into the region of the terrestrial planets or into the Sun. The small zone next to the Sun is the position of the Reconnection Ring in the X-Wind model of Shu et al. (2001); this is the zone in which Calcium-Aluminum Inclusions, the chemical precursors of the Vulcanoid planetoids in this model, are formed via the X-Wind model.

7.7.2 The X-Wind Model

I got introduced to the X-Wind model at a Gordon Conference on "The Origins of Solar Systems" in New Hampshire in 1993. Frank Shu gave a presentation on the model and how it might relate to the formation of CAIs and chondrules. This X-Wind model was a kind of extension of the T-Tauri model of star formation. At an earlier time the term, T-Tauri, conveyed certain connotations regarding the thermal regime of a nascent star but it is now used as a very general term associated with star formation. The X-Wind phenomena cannot be observed directly because of the dense "cloud cover" around young stars but certain aspects of the X-Wind model can be simulated in special laboratories. The formation of magnetic field lines and their breaking and reforming give off very energetic thermal pulses to their environment. This action would be in the mid-plane of the proto-sun as it is accumulating mass from the proto-solar cloud and commencing its thermonuclear reactor. The high temperature pulses are centered on 0.06 AU (the small doughnut-shaped figure near the Sun in Fig. 7.13); in short, the chemical, radiological, environmental, and thermal environments are conducive to the formation of CAIs.

A few papers on the X-Wind model were published in the 1990s and a key paper was published in "The Sun in Time" volume (Shu et al. 2001). Ross Taylor (2001) in his book titled "SOLAR SYSTEM EVOLUTION: A NEW PERSPECTIVE' favored the X-Wind model for the formation of CAIs. John Wood (2004) did a review of models for CAI and chrondrule formation and concluded that an X-Wind-like model looks good for CAI formation but not for chrondrule formation. It is now pretty well established that the origin of these two entities, CAIs and chrondrules, may overlap slightly in time (Brearley and Jones 1998) and that the thermal and dynamical environments of origin are significantly different. Thus, there seems to be a general agreement that the X-Wind model is good for CAIs but not for chrondrules. A recent article by Salmeron and Ireland (2012) looks favorable for the formation of chondrules in a "Disk-Wind" environment; the Disk-Wind is less energetic and slightly later in time than the X-Wind.

7.7.3 The Cool Early Earth Model

The interpretation of Valley et al. (2002) that the oxygen isotope ratios of Hadean-age zircon crystals suggest that the early Earth could have had ocean water on the surface was an important boost to the Gravitational Capture Model. The oldest zircons that show low surface rock temperature signatures are ~4.4 Ga (see Fig. 7.14). There is a major change in zircon oxygen isotope signatures at ~4.0 Ga and the abundance of datable supra-crustal zircons decreases considerably in the era 4.0 to 3.6 Ga. I date the Capture Episode at 3.95 Ga [which is the traditional maximum of the Late Heavy Bombardment of the Moon (Wasserburg et al. 1977). In my interpretation of the Cool Early Earth model, the Earth was moon-less for about 600 million years with a rotation rate of about 10 hr/day. The Sun should have been about 20% less powerful in that era so that would help keep the surface somewhat cooler. Although the mantle would still be quite warm and undergoing some thermal convection, the surface would probably be affected mainly by a vertical tectonic regime as proposed by Bedard (2006). Atmospheric circulation would be vigorous with a rotation rate of 10 hr/day and geochemical weathering processes along with the hydrologic cycle should result in significant sedimentary activity during that era. Figure 7.14 also shows a visual representation of the relationship of the Gravitational Capture model some important events in the Cool Early Earth model.

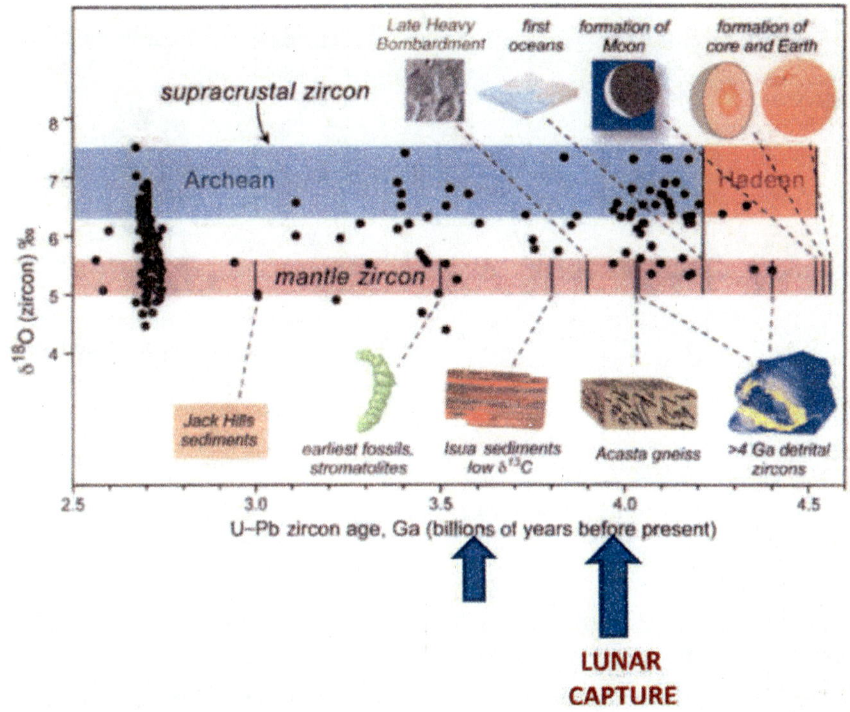

Fig. 7.14 Plot of Crystallization Age vs. Oxygen Isotope Ratios for zircons from the first 2.1 Ga of Earth history. The time of lunar capture and the duration of the main part of the orbit circularization sequence is indicated by the *arrows*. (Diagram adapted from Valley, (2006, Fig. 2) with permission from Mineralogical Society of America)

Although Preston Cloud named this era the Hadean Eon, he named it because of the lack of a recognizable rock record, a rock record that he thought was continuously destroyed by thermal processes until about 3.8 Ga. I am sure that Preston Cloud would be pleased with the Cool Early Earth idea because the very high rock tides resulting from the capture process is what helps to destroy the primitive rock record at 3.95 Ga. As time goes on, there is more and more evidence accumulating in support of the concept of a Cool Early Earth.

7.7.4 The Concept of Recycling the Hadean-age Primitive Crust

Armstrong (1981, 1991) suggested that there was isotopic evidence that an all-enclosing primitive crust existed in the earliest history of planet Earth and that this crust was eventually recycled into the mantle. Later work by Amelin (2005) and

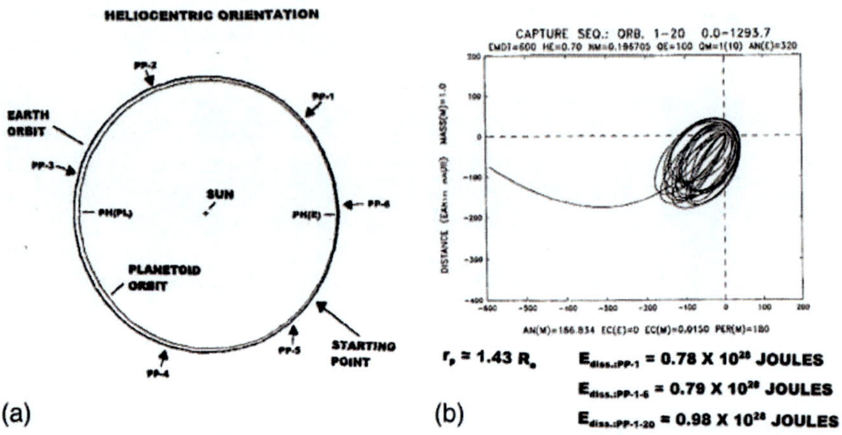

Fig. 7.15 A stable prograde capture scenario. a. Heliocentric orientation showing positions of the first six perigee passages. b. Geocentric plot showing the first 20 orbits (~3.5 years). Note that energy dissipation during the initial encounter was not sufficient for stable capture but energy dissipation during the 18th pass secured the capture. (From Malcuit (2015, Fig. 5.8) with permission from Springer)

Zeh et al. (2008) and others have confirmed this concept. A few quotes will help us understand the process in more detail:

> A final and perhaps most intriguing question is the mechanism whereby the signatures of extreme differentiation in the mantle-crust system during the first 500 million years were almost completely erased. Although there is some evidence for early differentiation in the oldest rocks, the average $^{176}Hf/^{177}Hf$ evolution curve defined by most mantle-derived rocks over the rest of Earth's history diverges from the chondritic curve at about 4 billion years, which is when the rock record begins (10).—From Amelin (2005), p. 1915.

> The existence of a depleted mantle reservoir in the Limpopo's hinderland is reflected by the ~3.6 Ga zircon population, which shows epsilon $Hf_{3.6Ga}$ between -4.6 and +3.2. In a global context, our data suggest that a long-lived, mafic Hadean protocrust with some tonalite-trondhjemite-granodiorite constituents was destroyed and partly recycled at the Hadean/Archean transition, perhaps due to the onset of modern-style plate tectonics .—From Zeh et al. (2008), p. 5304

My explanation of this episode of recycling of the primitive crust into the mantle is that it is associated with the very high rock tides caused by the process of GRA VITATIONAL LUNAR CAPTURE. Figures 7.15, 7.16, 7.17, 7.18, and 7.19 illustrate some of the major features of the simulated capture. (For additional information go to Malcuit (2015, Chap. 5).

7.7.4.1 A Typical Post-capture Orbit Circularization Sequence

An added bonus is that the mare-age lunar magnetic field can be generated simultaneously by rock tidal action on the lunar body during the orbit circularization

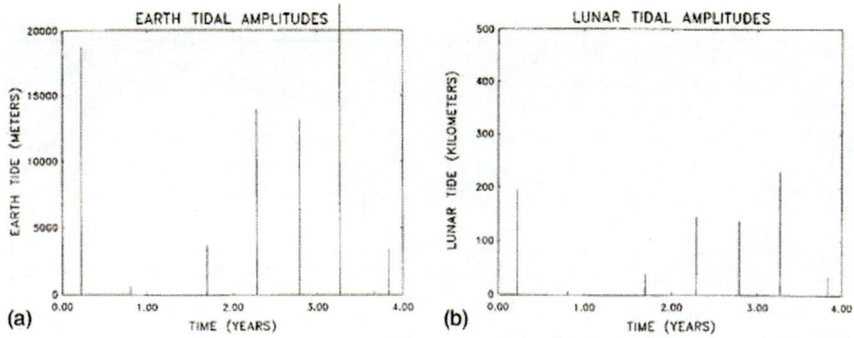

Fig. 7.16 Plots of tidal amplitudes for both bodies for this scenario. (**a**). Tidal amplitudes for Earth for the first 4 years of this stable capture scenario. Note that an encounter at about 3.25 years is a closer encounter than the capture encounter. (**b**). Tidal amplitudes on the lunar-like planetoid for the first 4 years of this scenario. The very close encounter at about 3.25 years raises rock tides higher than the capture encounter. This is the encounter than can permit tidal disruption of the lunar-like body if magma is available in the lunar upper mantle. This is perigee passage 18 (PP-18). (From Malcuit (2015, Fig. 5.9) with permission from Springer)

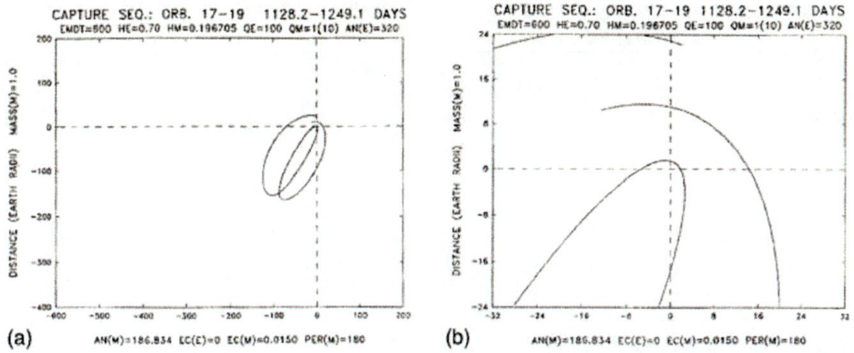

Fig. 7.17 Geocentric orbital setting for the 18th perigee passage. (**a**). Geocentric plot showing orbits 17, 18, and 19. (**b**). Close-up view of PP-17, PP-18, and PP-19. Note that PP-18 is much closer than the other two. (From Malcuit (2015, Fig. 5.10) with permission from Springer)

sequence shown in Fig. 7.20. The timing is the same for each of the major operations: *i. e.*, 3.95–3.6 Ga. Figure 7.21 shows sketches of a map view and equatorial cross-section of the primitive earth to illustrate a possible recycling mechanism.

7.7.4.2 Evolution of the Circular Lunar Orbit

Under normal conditions (i.e., without gravitational influence from neighboring planets) the lunar orbit would progressively increase in orbital radius as the Earth

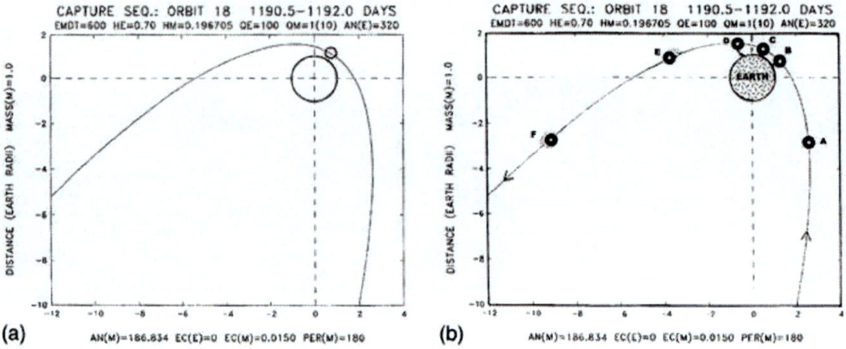

Fig. 7.18 General setting for a TIDAL DISRUPTION SCENARIO. (**a**). Numerical simulation of PP-18. (**b**). The author's concept of the broad view of what could happen during a close, tidally disruptive encounter which is well within the W-limit. Note that this figure is similar to Fig. 7.6. The concept is the same (tidal disruption during a very close encounter); the difference is that this orbital trajectory is part of a stable capture sequence and Fig. 7.6 was simply a very close encounter, not necessarily associated with a capture sequence of events. (From Malcuit (2015, Fig. 5.11) with permission from Springer)

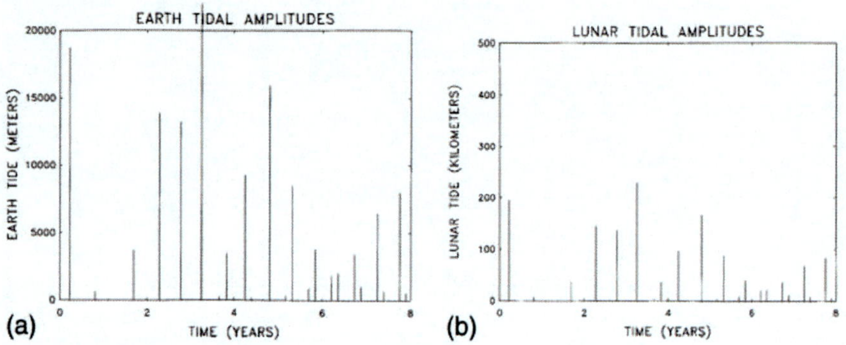

Fig. 7.19 Tidal amplitude plots for the first 8 years of this scenario featuring tidal disruption of the Moon on the 18th perigee passage. (**a**). Earth tidal amplitudes. (**b**). Lunar tidal amplitudes. (From Malcuit (2015, Fig. 5.13) with permission from Springer)

decreases in rotation rate as illustrated in Fig. 7.22. However, there can be irregularities in the process caused by orbital perturbations by Jupiter and Venus. These irregularities are illustrated and discussed by Malcuit (2015, Chap. 8). Figure 7.23 is a plot of the semi-major axis of the lunar orbit vs. geologic time from the time of capture to the present time.

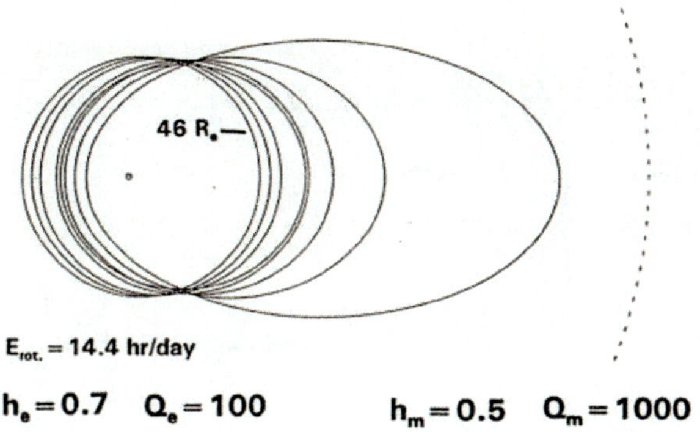

Fig. 7.20 Calculated two-body scenario in which energy is dissipated within body the satellite via radial tides and within the planet via tangential tides. The angular momentum of the lunar orbit is increasing as the rotational angular momentum of Earth is decreasing. The h (displacement Love number) and Q (specific dissipation factor) values of the Earth and Luna control the rate of angular momentum transfer. For this scenario the final rotation rate is 14.4 hr/day and the lunar orbital radius at 10% eccentricity is at 46 earth radii and the two-body timescale for circularization to 10% is about 1.4 Ga. (Note: After adjustments for 3-body orbital perturbations the timescale is shortened to a more realistic number of about 1.0 Ga). (From Malcuit (2015, Fig. 428b) with permission from Springer)

7.7.4.3 Summary of the Tidal and Thermal Effects of the Post-capture Orbit Evolution

In general, the Earth dissipated very little of the energy associated with the capture and subsequent orbit evolution. Although the Earth undergoes high and irregular tidal activity early in the orbit circularization sequence, the surface temperature would increase very little and ocean water, although tidally displaced periodically, would be only mildly affected thermally. By constrast, the Moon would dissipate enough energy to remelt a large portion of the former magma ocean region and a magnetic dynamo can operate again in the lunar magma ocean zone. Figure 7.24 is a plot of tidal amplitudes for the first 25 thousand years for both the Earth and Moon. Note that while the Earth is undergoing tidal amplitudes up to 20 km during the capture encounter, the Moon endures up to 250 km tidal amplitudes. Figure 7.25 shows the thermal regime for the two interacting bodies for the first 25 thousand years after capture. Note that the energy dissipation within the Earth approaches the estimated energy from decay of radioactive elements within a few tens of thousands of years after capture. Thus there is very little excess thermal energy deposited within the earth's mantle during the capture episode. Most of the activity on Earth is the tectonic processing of the crust in the broadly-defined equatorial zone which leads to recycling of a major portion of the primative crust into the earth's mantle. In contrast, the combination of thermal and tidal activity leads to massive melting of the basalts in the lunar mantle. A zone between 50 to 100 km thick is melted in the magma ocean zone during the capture encounter alone. Then more energy is absorbed during each close encounter following the capture encounter.

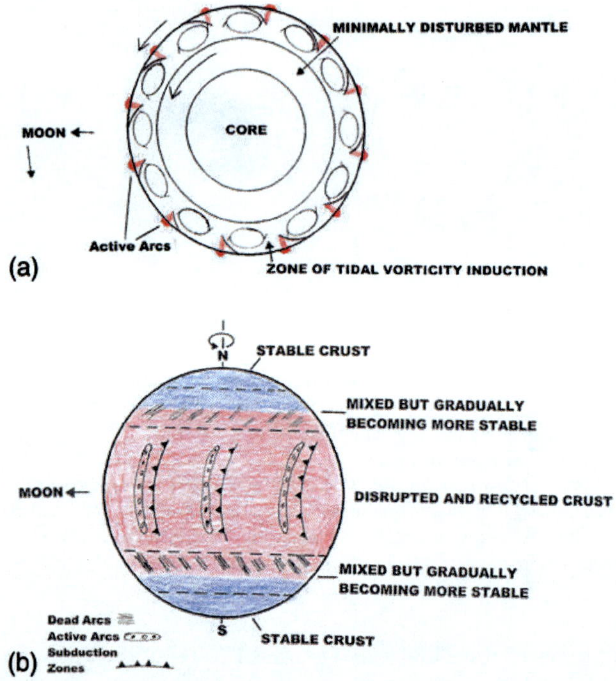

Fig. 7.21 (**a**). Diagrammatic sketch of an equatorial cross-section of Earth as viewed from the north pole. The zone of tidal vorticity induction, the location of subduction zones and volcanic arcs, and the direction of circulation of the TVI cells are shown. (**b**). Diagrammatic sketch of the surface of Earth as viewed from the equatorial plane of the planet. This is for an early stage in the orbit circularization sequence (perhaps about 350 Ka after capture). Note that in time the subduction zone and associated volcanic arc activity decreases and volcanic arc remnants are spalled off as dead (or fossil) volcanic arcs. (From Malcuit (2015, Fig. 5.73) with permission from Springer)

7.7.5 *Summary Diagrams for the Timeframe for Development of the Ancillary Models which are Important for the Tidal Capture Model*

Gravitational Capture has never been a very credible model for the origin of the Moon and the Earth-Moon system mainly because there was little obvious supporting evidence for the model. The main weaknesses were that (a) there was no credible source region for the body of the Moon and (2) there was no credible mechanism for dissipating the energy for transferring the body of the moon from a heliocentric orbit to a geocentric orbit.

Cloud (1968, 1972), however, proposed that the Earth underwent a "major thermal episode" mainly because of the absence of a rock record before the early Archean Eon. He also speculated that the Moon may have been in close proximity to Earth at that time in geologic history. Malcuit et al. (1975) than pointed out that that a

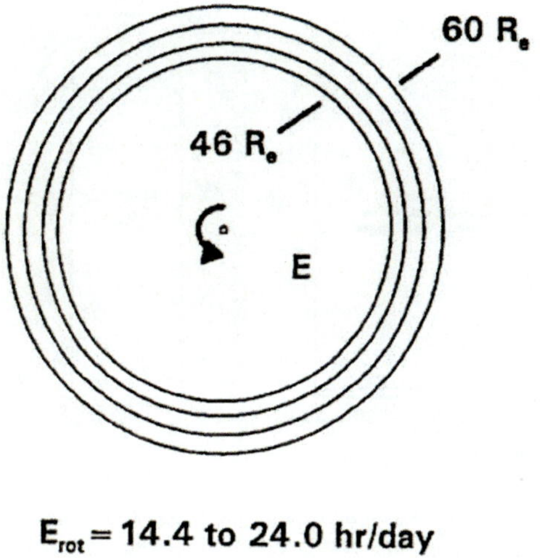

Fig. 7.22 Orbital diagram showing the general sequence of orbital states associated with the expansion of a circular lunar orbit. The first orbital state in this diagram is 46 earth radii (result for the scenario in Fig. 7.20) and this is associated with an earth rotation of 14.4 h/day. Then over about 3.0 billion years the lunar orbit gradually expands to the present 60 earth radii as the Earth rotation rate gradually slows to the present, very comfortable 24 h/day. (From Malcuit (2015, Fig. 9.1) with permission from Springer)

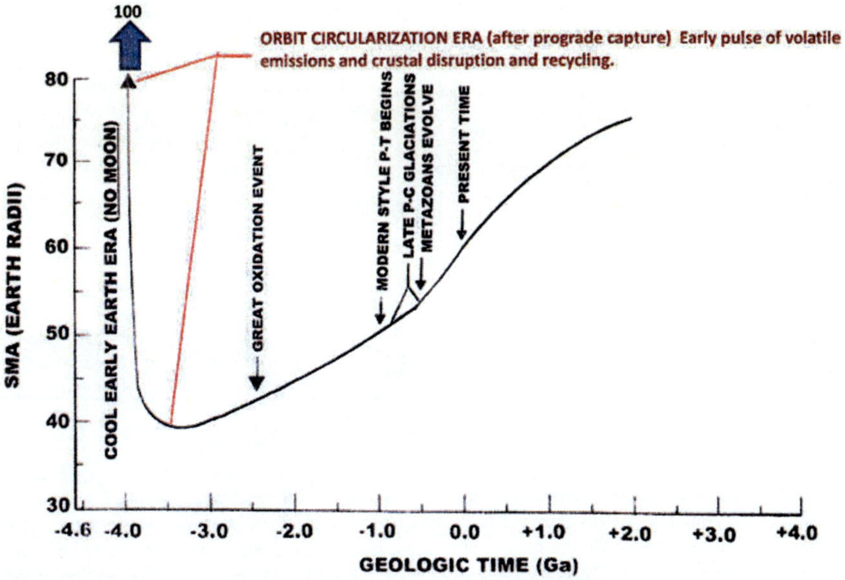

Fig. 7.23 Plot of semi-major axis of the lunar orbit vs. geologic time for a "standard" version of the evolution of the lunar orbit. The assumption of this plot is that once the lunar orbit circularized after the capture episode, the lunar orbit does not deviate much from a circular mode and maintains the present eccentricity of 0.0550. (From Malcuit (2015) Fig. 9.1 with permission from Springer)

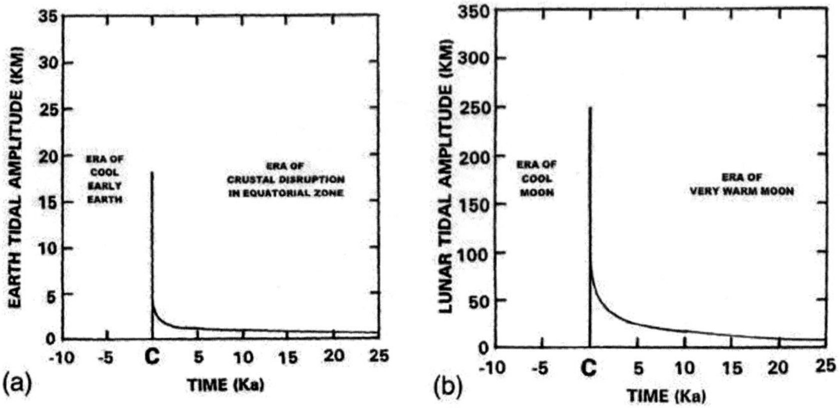

Fig. 7.24 Plot of tidal amplitudes for both Earth and Moon for the first 25 thousand years after capture. The symbol, C, marks the time of capture. (**a**). Earth tidal amplitudes vs. time; note that the maximum tidal amplitude for this normal capture sequence of events is about 20 km. (**b**). Lunar tidal amplitudes vs. time; note that the maximum tidal amplitude for this normal capture sequence is about 250 km

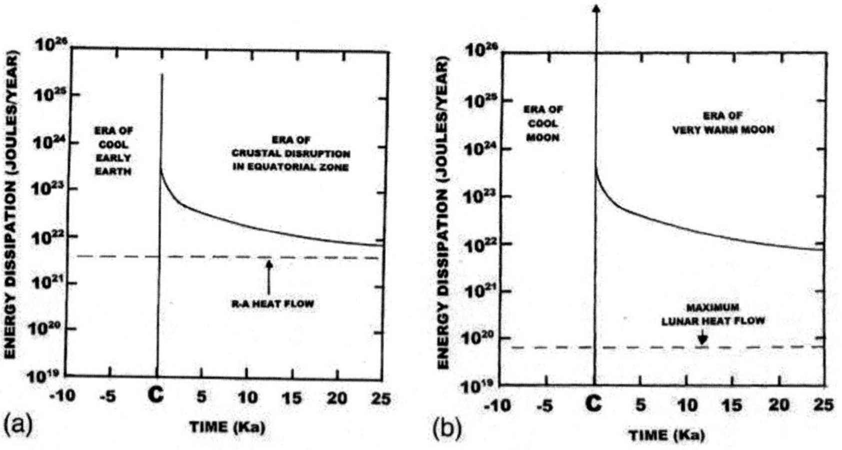

Fig. 7.25 Plot of energy dissipation for both Earth and Moon for the first 25 thousand years after capture. The symbol, C, marks the time of capture. (**a**). Energy dissipation vs time for Earth. (**b**). Energy dissipation vs. time for the lunar body. In general, the Earth stay "cool" during the capture process but there is enough thermal energy dissipated in the Moon to remelt a large portion of the magma ocean zone

Great Circle Pattern of large circular maria may relate to to a very close encounter of the Moon to the Earth about 3.95 Ga ago. Both the "major thermal episode" and the "great circle pattern of large circular maria" can be considered as informed speculations that may or may not lead to some important discovery. Wilson (1998) presented a Scale of Credibility for the pursuit of scientific speculations (endeavors). The scale goes from INTERESTING to SUGGESTIVE to PERSUASIVE to COM PELLING and finally to OBVIOUS.

Figure 7.26b has a horizontal axis listing the various ancillary observations and/or models that could be used as support for the capture model and their development over time. The Great-Circle Pattern of Large Circular Maria (GCP) was the first lunar surface pattern to be interpreted as the signature of the capture process (and ironically it was the last pattern to be explained via the capture sequence of events). In fact, after the numerical simulations were started I did not mention the Great-Circle Pattern in abstracts and publications until after 2006 when I could explain it in some detail. This Great-Circle Pattern I would classify as INTERESTING on the Wilson scale. The Numerical Simulations of Capture (NSC) elevated the credibility to SU GGESTIVE because we then knew that gravitational capture was physically possible if the moon-like body was in a near earth-like heliocentric orbit. The identification of Stable Vulcanoid Orbits (SVO) for planetoids orbiting between the Sun and planet Mercury elevated the gravitational capture idea to a strong SUGGESTIVE category because I now had a reasonable place of origin for Luna, a place of origin also suggested by Cameron (1972, 1973a,b). The X-Wind Model (XWM after a decade or so of existence) was becoming accepted as a source of Calcium-Aluminum Inclusions (CAIs). These high temperature solids are found mainly in chondritic meteorites but are thought to have formed in the "reconnection ring" of the X-Wind

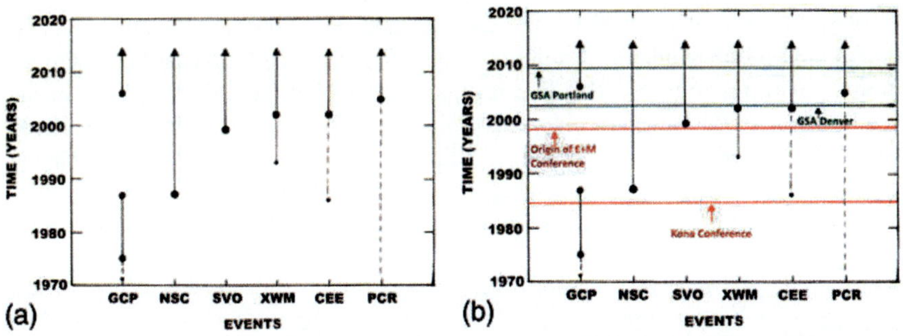

Fig. 7.26 Time chart for development of important supporting models for the Gravitational Capture Model for the origin of the Earth's Moon. GCP great circle pattern, NSC numerical simulations of capture, SVO stable vulcanoid orbits, XWM X-Wind model, CEE cool early earth model, PCR primitive crust recycled. (**a**). Plot of the times of initial publication of the models; dashed lines refer to early mention of the model. (**b**). Times of various presentations at meetings: bottom red line = Kona Conference (Malcuit et al., 1984a, b, c); top red line = Origin of Earth and Moon Conference (Monterey, CA)(Malcuit and Winters, 1998); bottom green line = Geological Society of America National Meeting (Denver, CO) (Malcuit and Winters, 2002); top green line = Geological Society of America National Meeting (Portland, OR) (Malcuit 2009)

Model which is located very near to the nascent Sun centered on about 0.06 AU. It has been known for some time that the chemistry of the Moon is very similar to the chemistry of a composite CAI (*e.g.*, Gast 1972; Taylor, 2001) and that the Moon is similar in chemistry to the Angrite meteorites and the HED group of meteorites (Ruzicka, Snyder, and Taylor 1999, 2001). This information added another notch to the SUGGESTIVE list. Thus the X-Wind Model, in a round-about way, added to the SUGGESTIVE cateogry. The Cool Early Earth (CEE) Model of Valley et al. (2002) moved the Gravitational Capture concept even closer to the PERSUASIVE category. Then in 2006, I finally came up with an explanation for the origin of the Great-Circle Pattern of Large Circular Maria. I found a family of prograde capture similations in which the nth perigee passage was closer to the Earth than the capture encounter. This type of encounter could then be used to explain the Great-Circle Pattern in terms of a partial tidal disruption of the lunar body. One of the best orbital scenarios featured the closest perigee passage on the 18th encounter, thus I call this the 18th Pass Scenario.

Now we come to the category of the Precambrian Crust Recycled (PCR). Back in 1990 I proposed in a poster paper at the 3rd International Archean Symposium that the super high rock tides associated with a stable capture scenario could lead to massive plate tectonic-like action in the equatorial zone of Earth. The idea was published in the Symposium Volume (Malcuit et al. 1992). From 2005 to the present there have been several papers published on the geochemical evidence (mainly on chemical isotope ratios) for a massive recycling of the primitive crust starting ~3.95 Ga and continuing to ~3.60 Ga [e.g., Amelin (2005), Zeh et al. (2010, 2014) and Sobolev et al. (2019)]. It is interesting to note that the second stage of generation of an internal lunar magnetic field has about the same timing. These two extended events, in my view, have a common cause: i.e., lunar capture and the subsequent orbit circularization. The Earth causes tidal vorticity induction circulation in the Moon via lunar rock tides and concomitantly, the Moon causes rock tides in the equatorial zone of Earth that results in tidal vorticity induction cells in the upper mantle of Earth which leads to a combination of crustal subduction and related volcanic arc activity on the Earth as illustrated in Fig. 7.21. The geochemical evidence that relates to the crustal recycling process should elevate the credibility index closer to COMPELLING on the Wilson Scale of Credibility.

7.8 Summary

As stated before, the Prograde Gravitational Capture Model has never been the favored model for the origin of the Moon problem. But when we look at the evidence, the Capture Model may be the only model that relates to the great diversity of geochemical, geophysical, meteoritical, and astronomical evidence that is available in our current era of the history of science. If it is eventually accepted, it will be as a DEFAULT MODEL, but not until all ther reasonable possibilities are exhausted.

References

Aggarwal, H. R., & Oberbeck, V. R. (1974). Roche limit of a solid body. *Tha Astrophysical Journal, 191*, 577–588.

Albarede, F., & Juteau, M. (1984). Unscrambling the lead model ages. *Geochimica et Cosmochimica Acta, 48*, 207–212.

Amelin, Y. A. (2005). A tale of early earth told in zircons. *Science, 310*, 1914–1915.

Anderson, D. L. (1972). The origin of the Moon. *Nature, 239*, 263–265.

Anderson, D. L. (1973). The Moon as a high-temperature condensate. *The Moon, 8*, 33–57.

Armstrong, R. L. (1981). Radiogenic isotopes: The case for crustal recycling on a near-steady-state no-continental-growth Earth. *Philosophical Transactions, Royal Society of London A, 301*, 443–472.

Armstrong, R. L. (1991). The persistent myth of crustal growth. *Australian Journal of Earth Sciences, 38*, 613–630.

Baadsgaard, H., Lambert, R. S. J., & Krupicka, J. (1976). Mineral isotope age relationships in the polymetamorphic Amitsoq gneisses, Godthaab district, West Greenland. *Geochimica et Cosmochimica Acta, 40*, 513–537.

Bedard, J. H. (2006). A catalytic delamination-driven model for coupled genesis of Archaean crust and sub-continental lithospheric mantle. *Geochimica et Cosmochimica Acta, 70*, 1188–1214.

Bell, E. A., Harrison, T. M., McCullach, M. T., & Young, E. D. (2011). Early Archean crustal evolution of the Jack Hills zircon source terrane inferred form Lu-Hf, $^{207}Pb/^{206}Pb$ and delta ^{18}O systematics of Jack Hills zircons. *Geochimica et Cosmochimica Acta, 75*, 4816–4829.

Black, L. P., Gale, N. H., Moorbath, S., Panhurst, R. J., & McGregor, V. R. (1971). Isotopic dating of very early Precambrian amphibolite facies geneisses from the Godthaab district, West Greenland. *Earth and Planetary Science Letters, 12*, 245–259.

Brearley, A. J., & Jones, R. H. (1998). Chapter 3: Chrondritic meteorites. In J. J. Papike (Ed.), *Planetary materials: reviews of mineralogy* (Vol. 36, pp. 1–398). Washington: Mineralogical Society of America.

Cameron, A. G. W. (1972). Orbital eccentricity of Mercury and the origin of the Moon. *Nature, 240*, 299–300.

Cameron, A. G. W. (1973a). Properties of the solar nebula and the origin of the Moon. *The Moon, 7*, 377–383.

Cameron, A. G. W. (1973b). History of the solar system. *Earth-Science Reviews, 9*, 125–137.

Cameron, A. G. W., & Pine, M. R. (1973). Numerical models of the primitive Solar Nebula. *Icarus, 18*, 377–406.

Cameron, A. G. W., & Ward, P. (1976). *Origin of the Moon: Abstract, Lunar Science VII* (pp. 120–122). Lunar Science Institute.

Ciscowski, S. M., & Fuller, M. (1983). *Lunar sample magnetic stratigraphy: Abstracts of the 14th Lunar and Planetary Science Conference* (pp. 115–116). Houston: Lunar and Planetary Institute.

Ciscowski, S. M., Collinson, D. W., Runcorn, S. K., Stephenson, A., & Fuller, M. (1983). A review of lunar paleointensity data and implications for the origin of lunar magnetism. *Journal of Geophysical Research, 88*, A691–A704.

Clayton, R. N., & Meyeda, T. K. (1975). *Genetic relations between the moon and meteorites: Proceedings of the 6th Lunar Science Conference* (pp. 1761–1769). Houston: Lunar Science Institute.

Cloud, P. E., Jr. (1968). Atmospheric and hydrospheric evolution on the primitive Earth. *Science, 160*, 729–736.

Cloud, P. E., Jr. (1972). A working model of the primitive earth. *American Journal of Science, 272*, 537–548.

Dietz, R. S. (1946). The meteoritic impact origin of the Moon's surface features. *Journal of Geology, 54*, 359–375.

Evans, W. N., & Tabachnik, S. (1999). Possible long-lived asteroid belts in the inner solar system. *Nature, 399*, 41–43.

Evans, W. N., & Tabachnik, S. (2002). Structure of possible long-lived asteroid belts. *Monthly Notices of the Royal Astronomical Society, 333*, L1–L5.

Franz, J. (1912). *Der Mond* (2nd ed., p. 120). Leipzig: BG Teubner.

Franz, J. (1913) Nova Acta, Abh. Der Kaiserl. Leop: Deutschen Akademie der naturforscher, v. 99, 1. Halle.

Gast, P. W. (1972). The chemical composition and structure of the Moon. *The Moon, 5*, 121–148.

Goldich, S. S., & Hedge, C. E. (1974). 3,800-Myr granitic gneiss in South-Western Minnesota. *Nature, 252*, 467–468.

Hansen, K. S. (1982). Secular effects of oceanic tidal dissipation on the Moon's orbit and the Earth's rotation. *Reviews of Geophysics and Space Physics, 30*, 457–480.

Harrison, T. M. (2009). The Hadean crust: Evidence from $\geq$ 4 Ga zircons. *Annual Reviews of the Earth and Planetary Sciences, 37*, 479–505.

Hartmann, W. K., & Davis, D. R. (1975). Satellite-sized planetesimals and lunar origin. *Icarus, 24*, 504–515.

Holsapple, K. A., & Michel, P. (2006). Tidal disruptions: A continuum theory for solid bodies. *Icarus, 183*, 331–348.

Holsapple, K. A., & Michel, P. (2008). Tidal disruptions: II. A continuum theory for solid bodies with strength, with applications to the solar system. *Icarus, 193*, 283–301.

Jeffreys, H. (1929). *The Earth, its origin, history and physical constitution* (2nd ed., p. 346). Cambridge: Cambridge University Press.

Jeffreys, H. (1947). The relationship of cohesion to Roche's limit. *Monthly Notices of the Royal Astronomical Society, 107*, 260–272.

Kokubo, E., Canup, R. M., & Ida, S. (2000). Lunar accretion from an impact-generated disk. In R. M. Canup & K. Righter (Eds.), *Origin of the earth and moon* (pp. 145–178). The University of Arizona Press (in collaboration with Lunar and Planetary Institute).

Lipskii, Y. N. (1965). Zond-3 photography of the Moon's far side. *Sky and Telescope, 30*, 338–341.

Lipskii, Y. N., et al. (1966). Current problems of the morphology of the Moon's surface. *Kosmicheskie Issledovaniva, 4*(6), 912–922. (Translated to English at a later date.).

Lipskii, Y. N., & Rodianova, Z. F. (1977). *Antipodes on the Moon, in the Soviet-American Conference on Cosmochemistry of the Moon and the Planets* (pp. 755–761).

Malcuit, R. J. (1972). Lunar maria and mascons: "hypervolcanic" flyby "scars". *Transactions, American Geophysical Union, 53*, 430.

Malcuit, R. J., Byerly, G. R., Vogel, T. A., & Stoeckley, T. R. (1973). Computer simulation model of a lunar flyby encounter with Earth. *Transactions, American Geophysical Union, 54*, 358.

Malcuit, R. J., Byerly, G. R., & Vogel, T. A. (1974). Lunar maria emplacement via the flyby encounter model. *Transactions, American Geophysical Union, 55*, 331.

Malcuit, R. J., Byerly, G. R., Vogel, T. A., & Stoeckley, T. R. (1975). The great-circle pattern of large circular maria: Product of an Earth-Moon encounter. *The Moon, 12*, 55–62.

Malcuit, R. J., Winters, R. R., & Mickelson, M. E. (1976). *Geological evidence supporting the lunar capture hypothesis: The great circle pattern of large circular maria and a major thermal episode on Earth* (Vol. 2, pp. 613–614). Sydney: Abstracts, 25th International Geological Congress.

Malcuit, R. J., Winters, R. R., & Mickelson, M. E. (1977). *Is the Moon a captured body?* (pp. 608–610). Abstracts Volume, Eighth Lunar Science Conference.

Malcuit, R. J., & Winters, R. R. (1980). *The Late Proterozoic glaciations: Possible product of the evolution of the Earth-Moon System* (Vol. 2, p. 600). Paris: Abstracts, 26th International Geological Congres.

Malcuit, R. J., Winters, R. R., & Mickelson, M. E. (1984a). *A testable gravitational capture model for the origin of the Earth-Moon system* (Vol. VIII, p. 296). Moscow, U. S. S. R: Abstracts, 27th International Geological Congress.

Malcuit, R. J., Winters, R. R., & Mickelson, M. E. (1984b). *A testable gravitational capture model for the origin of the Earth's Moon: Abstracts Volume, Conference on the Origin of the Moon* (p. 43). Lunar and Planetary Institute.

Malcuit, R. J., Winters, R. R., and Mickelson, M. E., 1984c, Directional properties of "circular" maria: Interpretation in the context of a testable gravitational capture model for lunar origin: Abstracts Volume, Conference on the Origin of the Moon, Lunar and Planetary Institute, p. 44.

Malcuit, R. J., 1986, Evolution of the Earth-Moon system (3.8 B. Y. to Recent): Implications for tidal ranges over geologic time: Abstracts Volume, 12th International Sedimentological Congress (Canberra), p. 198.

Malcuit, R. J., Mehringer, D. M., & Winters, R. R. (1988a). *Computer simulation of "intact" gravitational capture of a lunar-like body by an earth-like body; Abstracts, Lunar and Planetary Science XIX* (pp. 718–719). Houston: Lunar and Planetary Institute.

Malcuit, R. J., Mehringer, D. M., & Winters, R. R. (1988b). *Numerical simulation of three-body interactions with an energy dissipation subroutine for close encounters between two bodies: Implications for planetary accretion and planetoid capture processes: Abstracts Volume, Conference on the Origin of the Earth* (pp. 52–53). Berkeley: Lunar and Planetary Institute and NASA Johnson Space Center.

Malcuit, R. J., Mehringer, D. M., & Winters, R. R. (1989a). *Numerical simulation of intact capture: Application to planets Venus and Earth: Abstracts, 28th International Geological Congress* (Vol. 12, pp. 355–356). Washington, DC.

Malcuit, R. J., Mehringer, D. M., & Winters, R. R. (1989b). *Numerical simulation of gravitational capture of a lunar-like body by Earth: Proceedings, 19th Lunar and Planetary Science Conference* (pp. 581–591). Houston: Cambridge University Press and Lunar and Planetary Institute.

Malcuit, R. J., Mehringer, D. M., & Winters, R. R. (1990). *A gravitational capture origin for the Earth-Moon system: Implications for the early history of Earth and Moon* (pp. 7–8). Perth: Extended Abstracts Volume, Third International Archaean Symposium.

Malcuit, R. J., Mehringer, D. M., & Winters, R. R. (1992). A gravitational capture origin for the Earth-Moon system: Implications for the early history of Earth and Moon. In J. E. Glover & S. E. Ho (Eds.), *The Archaean: Terrains, Processes and Metallogeny: Geology Department (Key Center) and University Extension* (Vol. 22, pp. 223–235). Perth: The University of Western Australia, Publication.

Malcuit, R. J., & Winters, R. R. (1995). *Numerical simulation of retrograde gravitational capture of a satellite by Venus: Implications for the thermal history of the planet: Abstracts Volume, 26th Lunar and Planetary Science Conference* (pp. 829–830). Lunar and Planetary Institute.

Malcuit, R. J., and Winters, R. R. (1996a) Geometry of stable capture zones for planet Earth and implications for estimating the probability of stable gravitational capture of planetoids from heliocentric orbit: Abstracts Volume, 27th Lunar and Planetary Science Conference, Lunar and Planetary Institute, p. 799–800.

Malcuit, R. J., & Winters, R. R. (1996b). *A prograde gravitational capture model for the origin of the Earth-Moon system: Is it compatible with the geological record of the Earth and Moon: Abstracts Volume* (Vol. 1, p. 6). Beijing: 30th International Geological Congress.

Malcuit, R. J., & Winters, R. R. (1996c). *Numerical simulation of gravitational capture scenarios for planets Venus and Earth: Implications for the contrasting thermal histories of these planets: Abstracts Volume* (Vol. 3, p. 506). Beijing: 30th International Geological Congress.

Malcuit, R. J., & Winters, R. R. (1998). *A prograde gravitational capture model for the origin of the Earth-Moon system: Is it compatible with the rock records of the Earth and Moon?: Abstracts Volume, Origin of the Earth and Moon* (p. 24). Monterey: Lunar and Planetary Institute.

Malcuit, R. J., & Winters, R. R. (2000a). *Neoproterozoic rifting and sedimentation events on Earth: Possible lunar connection: Abstracts*. Rio de Janeiro: 31st International Geological Congress, CD-ROM, Section 9.6.

Malcuit, R. J., & Winters, R. R. (2000b). *Geometry of retrograde stable capture zones for tidal capture of satellites by planet Neptune: Abstracts*. Rio de Janeiro: 31st International Geological Congress, CD-ROM, Section 28.

Malcuit, R. J., & Winters, R. R. (2001). *Neoproterozoic sedimentation and rifting events on Earth: Possible lunar involvement: Programmes with Abstracts* (p. 101). Edinburgh: Joint Meeting on Earth System Processes, Geological Society of America and Geological Society of London.

Malcuit, R. R., & Winters, R. R. (2002). The Cool Early Earth model and the Lunar Capture model: Are they compatible?: Abstracts with Programs, National Meeting of the Geological Society of America (Denver), 34, 6, 273.

Malcuit, R. J. (2009). A prograde planetoid capture episode about 3.95 Ga ago: Is this model compatible with the information from Hadean- and Archean-age zircon crystals?: Geological Society of America. *Abstracts with Programs, 41*(7), 686.

Malcuit, R. J. (2015). *The twin sister planets, Venus and Earth: Why are they so different?* (p. 401). Springer.

Mickelson, M. E., Winters, R. R., & Malcuit, R. J. (1975). Accretion-driven and encounter-driven dynamos as sources of lunar rock magnetism. *Transactions, American Geophysical Union, 56*, 387.

Milankovitch, M. (1941). Kanon der Erdbestrahlung und seine andwendung auf das eiszeitenproblem: Royal Serbian Academy Special Publication 133, 633 p. (English translation published in 1969 by Israel Program for Scientific Translations).

Mueller, P. A., Wooden, J. L., & Nutman, A. P. (2009). *U-Pb ages (3.8–2.7 Ga) and Nd isotope data from the newly identified Eoarchean Nuvvuagittuq supracrustal belt* (Vol. 121, pp. 150–163). Superior Craton: Geological Society of American Bulletin.

Mutch, T. A. (1970). *Geology of the Moon: A stratigraphic view* (p. 324). Princeton: Princeton University Press.

Mutch, T. A. (1972). *Geology of the Moon: A stratigraphic view* (2nd ed., p. 391). Princeton: Princeton University Press.

Peale, S. J., & Cassen, P. (1978). Contributions of tidal dissipation to lunar thermal history. *Icarus, 36*, 245–269.

Ringwood, A. E. (1990). Slab—Mantle interactions: 3. Petrogenesis of intraplate magmas and structure of the upper mantle. *Chemical Geology, 82*, 187–207.

Ross, M., & Schubert, G. (1986). Tidal dissipation in a viscoelastic planet: Proceedings of the 16th Lunar and Planetary Science Conference. *Journal of Geopyhysical Research, 91*, D447–D452.

Ross, M. N., & Schubert, G. (1989). Evolution of the lunar orbit with temperature and frequency-dependent dissipation. *Journal of Geophysical Research, 94*(B7), 9533–9544.

Runcorn, S. K. (1983). Lunar magnetism, polar displacements and primeval satellites in the Earth-Moon system. *Nature, 304*, 589–596.

Ruzicka, A., Snyder, G. A., and Taylor, L. A., (1999), Giant impact and fission hypotheses for the origin of the Moon: A critical review of some geochemical evidence, *in* Snyder, G. A., Neal, C. R., Ernst, W. G., eds., Planetary petrology and geochemistry: Geological Society of America, International Book Series, v. 2, p. 121–134.

Ruzicka, A., Snyder, G. A., & Taylor, L. A. (2001). Comparative geochemistry of basalts from the Moon, Earth, HED asteroid, and Mars: Implications for the origin of the Moon. *Geochimica et Cosmochimica Acta, 65*, 979–997.

Ryder, G. (1990). Lunar samples, lunar accretion and the early bombardment of the Moon. *Transactions, American Geophysical Union, 71*(10), 313–322–323.

Ryder, G., & Taylor, G. J. (1976). *"Pre-mare" volcanism: Abstracts of the 7th Lunar and Planetary Science Conference* (pp. 755–757). Houston: Lunar and Planetary Institute.

Saager, R., & Koppel, V. (1976). Lead isotopes and trace elements from sulfides of Archean greenstone belts in South Africa—A contribution to the knowledge of the oldest known mineralizations. *Economic Geology, 71*, 44–57.

Salmeron, R., & Ireland, T. R. (2012). Formation of chondrules in magnetic winds blowing through the proto-asteroid belt. *Earth and Planetary Science Letters, 327–328*, 61–67.

Shu, F. H., Shang, H., Gounelle, M., Glassgold, A. E., & Lee, T. (2001). The origin of chondrules and refractory inclusions in chondritic meteorites. *Astrophysical Journal, 548*, 1029–1050.

Singer, S. F. (1968). The origin of the Moon and geophysical consequences. *Geophysical Journal, Royal Astronomical Society, 15*, 205–226.

Singer, S. F. (1970a). The origin of the Moon and its consequences. *Transactions, American Geophysical Union, 51*, 637–641.

Singer, S. F. (1970b). How did Venus lose its angular momentum? *Science, 170*, 1196–1198.

Singer, S. F. (1977). The early history of the Earth-Moon system. *Earth Science Reviews, 13*, 171–189.

Singer, S. F. (1986). Origin of the Moon by capture, *in* Hartmann, W. K., Phillips, R. J., and Taylor G. J., eds., Origin of the Moon: Lunar and Planetary Institute, Houston, p. 471–485.

Stacey, J. S., & Kramers, J. D. (1975). Approximation of terrestrial lead isotope evolution by a two-stage model. *Earth and Planetary Science Letters, 26*, 207–221.

Stuart-Alexander, D. E., & Howard, K. A. (1970). Lunar maria and circular basins—A review. *Icarus, 12*, 440–456.

Sobolev, A. V., Asafov, E. V., Gurenko, A. A., Arndt, N. T., Batanova, V. G., Portnyagin, M. V., Garbe-Schonberg, D., Wilson, A. H., & Byerly, G. R. (2019). Deep hydrous mantle reservoir provides evidence for crustal recycling before 3.3 billion years ago. *Nature, 571*, 555–559.

Taylor, S. R. (1982). *Planetary science: A lunar perspective* (p. 481). Houston: Lunar and Planetary Institute.

Taylor, S. R. (2001). *Solar System evolution: A new perspective* (2nd ed., p. 460). Cambridge: Cambridge University Press.

Tera, F., Papanastassiou, D. A., & Wasserburg, G. J. (1974). Isotopic evidence for a terminal lunar cataclysm. *Earth and Planetary Science Letters., 22*, 1–21.

Valley, J. W. (2006). Early Earth. *Elements, 2*, 201–204.

Valley, J. W., Peck, W. H., King, E. M., & Wilde, S. A. (2002). A cool early Earth. *Geology, 30*, 351–354.

Wasserburg, G. J., Papanastassiou, D. A., Tera, F., & Huneke, J. C. (1977). Outline of a lunar chronology. *Philosophical Transactions, Royal Society of London, Series A., 285*, 7–22.

Webb, D. J. (1982). Tides and the evolution of the Earth-Moon system. *Geophysical Journal, Royal Astronomical Society, 70*, 261–271.

Wegener, A. (1924). *The origin of continents and oceans (translated to english from the 3rd german edition by J. G. A., Skerl)* (p. 212). London: Methuen and Company.

Wilson, E. O. (1998). *Consilience: The unity of knowledge* (p. 368). New York: Knoph Doubleday Publishing Co.

Windley, B. F. (1984). *The evolving continents* (2nd ed., p. 399). New York: Wiley.

Winters, R. R., & Malcuit, R. J. (1977). The lunar capture hypothesis revisited. *The Moon, 17*, 353–358.

Winters, R. R., Malcuit, R. J., & Mickelson, M. E. (1976). The lunar capture hypothesis: A post-Apollo evaluation. *Transactions, American Geophysical Union, 57*, 272.

Wood, J. A. (2004). Formation of chrondritic refractory inclusions: The astrophysical setting. *Geochimica et Cosmochimica Acta, 68*, 4007–4021.

Zeh, A., Gerdes, A., Klemd, R., & Barton, J. M. (2008). U-Pb and Lu-Hf isotope record of detrital zircon grains from the Limpopo Belt—Evidence for crustal recycling at the hadean to early-Archean transition. *Geochimica et Cosmochimica Acta, 72*, 5304–5329.

Zeh, A., Gerdes, A., Barton, J. M., Jr., & Klemd, R. (2010). U-Th-Pb and Lu-Hf systematics of zircon from TTG's, leucosomes, anorthosites and quartzites of the Limpopo Belt (South Africa): Constraints for the formation, recycling, and metamorphism of Paleoarchean crust. *Precambrian Research, 179*, 50–68.

Zeh, A., Stern, R. A., & Gerdes, A. (2014). The oldest zircons of Africa – Their U-Pb-Hf-O isotope and trace element systematics and implications for Hadean to Archean crust-mantle evolution. *Precambrian Research, 241*, 203–230.

Chapter 8
Comparative Analysis of the Gravitational Capture Potential for the Terrestrial Planets and Planet Neptune

Among the more perplexing features of the Solar System's morphology is the scarcity of satellites in the inner Solar System: neither Mercury nor Venus seems to have satellites, the Earth has only one, and Mars is orbited by the tiny Phobos and Deimos. This is in sharp contrast with the case of the outer planets which have extensive satellite systems.—From Burns (1973)

Here we explore the possibility that Triton formed on a heliocentric orbit and was subsequently captured by Neptune (4, 5). A number of possible mechanisms might have been responsible for the initial capture of Triton by Neptune. For reasons that will become clear later, we favor the idea that Triton's capture followed its physical collision with a regular satellite of Neptune.—From Goldreich et al. (1989)

The approximate limit to a secondary's (e.g., planet's or moon's) gravitational dominance is given by the extent of its Hill sphere. . . . A test particle located at the boundary of a planet's Hill sphere is subjected to a gravitational force from the planet comparable to the tidal difference between the force of the Sun on the planet and that on the test body. . . . Planetocentric orbits which are stable over long periods of time are those well within the boundary of a planet's Hill sphere; all known natural satellites lie in this region.—From de Pater and Lissauer (2001, p. 24)

The origin of satellites is a big issue in the planetary sciences. There is a place for fission, co-formation, and capture models but the value of Giant Impact Models for the formation of large, dynamically stable satellites has not been demonstrated. The concept of the STABLE CAPTURE ZONE (SCZ) appears to be applicable mainly, but not exclusively, to the gravitational capture model. A very important feature of the SCZ is that the probability of capture can be fairly accurately assessed from the dimensions of the SCZ. When we find satellites associated with exoplanets, especially terrestrial-type exoplanets, determining the dimensions of the SCZs for the exoplanet-satellite pair may help us determine the origin of the satellite. So at least a partial answer to the question in the quote by Burns (1973) (initial quote of this chapter) as well as the question raised by Goldreich et al. (1989) (second quote of this chapter) may be associated with the concept of the STABLE CAPTURE ZONE.

R. Malcuit, *Geoforming Mars*, https://doi.org/10.1007/978-3-030-58876-2_8

Two recent articles emphasize the importance of large satellites to the habitability potential of exoplanets or features of the planet that can be explained by the presence of a large satellite. Piro (2018) pointed out the potential importance of a large satellite associated with a terrestrial exoplanet orbiting a low mass parent star. First of all, he suggests that the presence of a large satellite causes the obliquity (tilt angle) of the planet to become more stable. Secondly, Piro (2018) concludes that if the period of the large satellite is not synchronous with the rotation rate of the planet, significant tidal heating of the planet can result; the combination of energy dissipation in the planet and tangential rock and ocean tidal activity in the surface and near-surface environment of the terrestrial exoplanet could result in tectonic activity on that exoplanet. In general, Piro (2018) suggests that both stabilization of the obliquity and tectonic activity on a terrestrial exoplanet with liquid water on the surface could aid in the development of a biological system on a terrestrial exoplanet.

Shakar et al. (2019) identify and discuss several important features that tend to make a planet habitable. They stress the importance of (1) atmospheric composition and its variation over time, (2) heliocentric orbit stability over time, (3) something like plate tectonics operating over time, (4) magnetic field operation over time, and (5) recycling of surface and near-surface water over time. A major conclusion of Shakar et al. (2019) is that a "stagnant-lid" planet is not favorable for long-term habitability. It is also interesting to note that there is NO mention of a large satellite in this article on exoplanet habitability although a large satellite could be a major aid in overcoming a stagnant-lid condition.

It is the view of the author of this book, as well as Piro (2018), Doglioni and Anderson (2015), Doglioni and Panza (2015), Ward and Brownlee (2000) and others, that a large satellite like the earth's moon is very important for the development of the features listed above and that the presence of a large satellite is probably a necessary, but not sufficient, feature for the development of a habitable terrestrial planet or exoplanet.

Shakar et al. (2019) point out that we should get more detailed information in the near future on exoplanets. The James Webb Space Telescope (JWST) (to be launched this year) should be capable of detecting certain critical chemical components of planetary atmospheres like water and carbon dioxide. Looking farther ahead they project that the Transiting Exoplanet Survey Satellite (TESS) may have the capability of detecting large satellites associated with exoplanets. In my view, these space telescope projects are very important in the search for potentially habitable exoplanets.

An associated major concept in the analysis of the capture potential of planetary satellites is the Hill Sphere (see the third quote above). The approximate extent of the Hill Sphere can be determined by the following equation from de Pater and Lissauer (2001):

$$R_H = \left(\frac{m_2}{3(m_1 + m_2)}\right)^{1/3} a \tag{8.1}$$

Table 8.1 Values for the approximate radial extent of the Hill sphere for circular orbits for the eight major planets in the Solar System

planet	Orbital radius (AU)	Calculated limit	Practical limit
Mercury	0.387	90.98 Rme	85 Rme
Venus	0.723	167.8 Rv	160 Rv
Earth	1.000	236 Re	200 Re
Mars	1.524	321.2 Rma	300 Rma
Jupiter	5.2	747 Rj	700 Rj
Saturn	9.5	1085 Rs	1000 Rs
Uranus	19.2	2748 Ru	2600 Ru
Neptune	30.0	4786 Rn	4500 Rn
Pluto (perihelion)	30.0	8938 Rpl	8500 Rpl
Pluto (aphelion)	50.0	14,928 Rpl	13,000 Rpl

(Note that the dimensions of the limits are given in planetary radii.) Pluto values are the values for perihelion and aphelion

Where m_2 is the mass of the secondary (a satellite in this case) and m_1 is the mass of the primary (a planet in this case), and a is the distance from the sun in astronomical units. Table 8.1 gives the approximate radial extent of the Hill Sphere for the eight major planets in the Solar System. Thus for stable capture, the initial encounter of an encounter sequence that results in stable capture must (1) be within one of the stable capture zones of the planet and (2) the major axis of the orbit of the body following an energy dissipating encounter must not exceed the dimensions of the Hill sphere of the planet.

8.1 Geometry of Stable Capture Zones and their Importance to Planetary Science

A stable capture zone can be defined as a region of parameter space in which the orientation of the orbit of the encountering planetoid is favorable for a stable post-capture orbit if sufficient energy is dissipated for its capture into a planetocentric orbit.

There are three items that are super important for stable capture of satellites by tidal dissipation processes:

1. ORIENTATION – the orbit of the planetoid must be within narrow limits relative to the orbit of the planet (see Figs. 8.1 and 8.2),
2. ENERGY STORAGE – the h (displacement Love number) of the planetoid must be sufficiently high to store the energy for capture in the planetoid by tidal deformation processes during the initial close gravitational encounter of an encounter sequence.

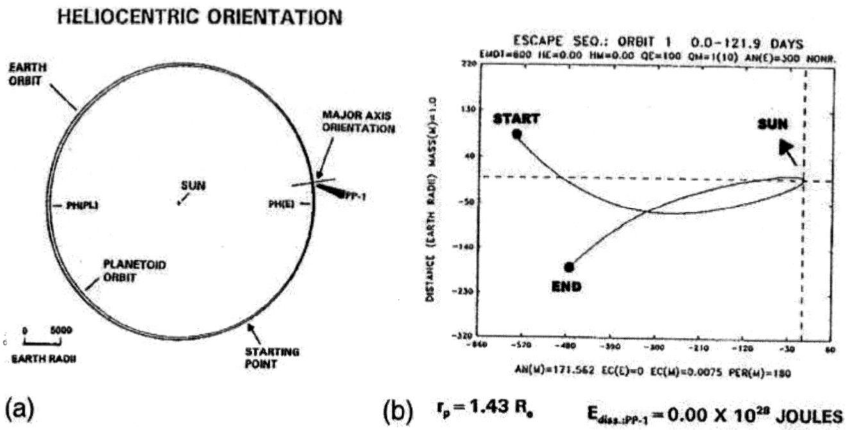

Fig. 8.1 Heliocentric and geocentric orientations that are necessary for stable prograde capture of a lunar-like planetoid from a heliocentric orbit. (**a**) The orientation of the major axis of the first encounter at PP-1 must be nearly perpendicular to a tangent of the heliocentric orbit. (**b**) The geocentric diagram shows such an orientation for the geocentric encounter. Other details of the stable capture conditions are stated in the text

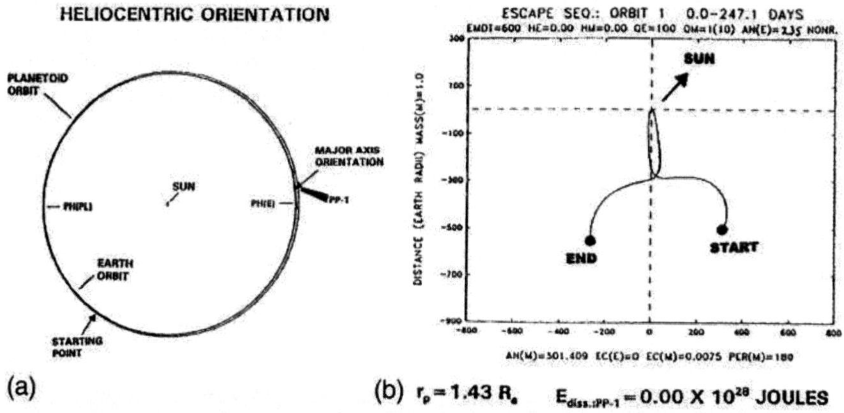

Fig. 8.2 Heliocentric and geocentric orientations that are necessary for stable retrograde capture of a lunar-like planetoid from a heliocentric orbit. (**a**) The orientation of the major axis of the first encounter a PP-1 must be nearly parallel to a tangent of the heliocentric orbit. (**b**) The geocentric diagram shows such an orientation for the geocentric encounter. Other details of the stable retrograde capture conditions are stated in the text

3. ENERGY DISSIPATION – the Q (specific dissipation factor) of the planetoid must be very low (like 1, 2, or 3) so that a large fraction of the stored energy is subsequently dissipated during the timeframe of that close gravitational encounter.

Using Earth and a lunar-mass planetoid as examples, Fig. 8.1 shows the necessary orientation for a stable prograde capture. For capture in the PROGRADE direction, the orientation of the major axis of the planetoid during the first encounter of an encounter sequence must be within a few degrees (about + or – 5 degrees) of a line perpendicular to the tangent of the orbit of the planetoid. Then if sufficient energy is dissipated during the close encounter, the planetoid will be inserted into a stable capture orbit (when there are no significant external disturbances such as gravitational perturbations by neighboring planets).

Again, using Earth and a lunar-mass planetoid as examples, Fig. 8.2 shows the necessary orientation for a stable retrograde capture of a lunar-like planetoid from a heliocentric orbit. For capture in the RETROGRADE direction, the major axis of the planetoid during the first encounter of an encounter sequence must be within a few degrees (about + or – 5 degrees) of a line parallel to the tangent of the orbit of the planet.

The heliocentric orbit eccentricity and inclination of a planet relative to the plane of the solar system are important for assessing the stability of a satellite orbit. Since any stable capture must be close to the plane of the planets, the higher the inclination the lower the probability of stable capture. Since all encounters in this chapter (and in this book) are in the plane of the planets, heliocentric orbit inclination is not a variable. Heliocentric orbit eccentricity of the planet, however, is a factor for captures in the plane of the planets. The higher the heliocentric orbit eccentricity of the planet, the lower the probability of stable capture because the energy to be dissipated for stable capture is increased. Figure 8.3 shows the variations in heliocentric orbit eccentricity and inclination for the planets of the solar system over geologic time as calculated by Laskar (1994). Cursory inspection of the diagrams

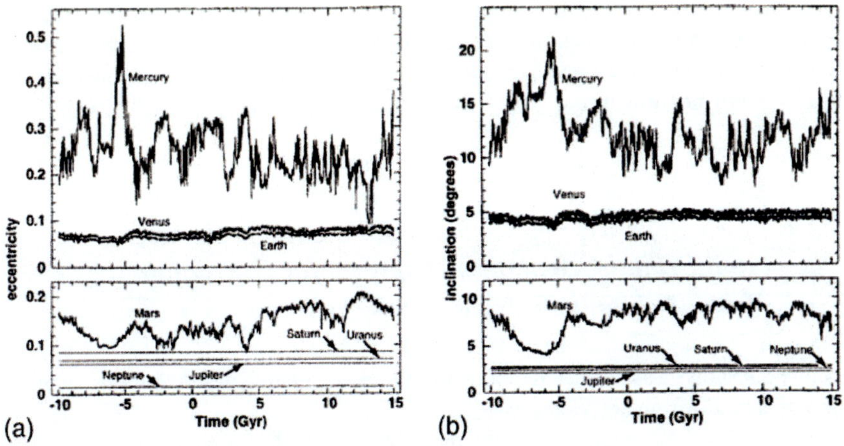

Fig. 8.3 Results of numerical integrations (from Laskar (1994, Fig. 1), courtesy of Springer) of the averaged equations of motion for various Solar System bodies for 10 Ga backward and 15 Ga forward in geologic time. **(a)** Plot of orbital eccentricity vs. geologic time. **(b)** Plot of orbital inclination vs. geologic time

reveals that two terrestrial planets, Mercury and Mars, have significant variations in eccentricity (and inclination) over geologic time and that planets Venus and Earth have very little variation over time.

The purpose of this chapter is to do a survey of a number of planets in our Solar System by defining the geometry of the SCZs for the planet and a planetoid of specific mass and then discussing the implications of this geometry for the stable capture of satellites for that planet. The planets to be analyzed are: Mercury, Venus, Earth, (Mars), and Neptune. (Note: The case for Mars was analyzed in sufficient detail in Chaps. 4 and 5.)

8.2 Planet Mercury

Planet Mercury is very near the Sun. Retrograde capture of a satellite of appropriate size is physically possible from either an outside orbit or an inside orbit if the orbit of Mercury is very nearly circular. Prograde capture is physically possible either from an outside orbit or an inside orbit but the necessary displacement Love numbers are very high. Thus, although capture of a planetoid in a prograde orbit is physically possible (*i.e.*, within an SCZ), it is highly improbable. Collision with some body of appropriate mass with the planetoid as a close encounter is occurring would also aid in the energy dissipation for capture. This can be classified as a collisional capture episode. Such a collisional capture event makes the capture scenario even more improbable.

Furthermore, given the eccentricity and inclination cycles for planet Mercury as calculated by Laskar (1994) and shown in Fig. 8.3, the energy dissipation requirements for capture are increased and the planetocentric orbit stability is decreased. For example, when the heliocentric eccentricity of the planet is increased, the encountering planetoid could escape back into heliocentric orbit after one or a few encounters as the planet moves to a greater distance from the Sun.

8.2.1 Two-Body Representation, Stable Capture Zones, and Orbital Orientation Diagrams for Gravitational Capture Simulations for Mercury

Figure 8.4 shows a scale sketch of a two-body representation of gravitational capture of a planetoid for planet Mercury.

Figure 8.5 shows plots of the geometry of the Stable Capture Zones for a 0.5 moon-mass planetoid encountering Mercury. Note the very limited area of the retrograde SCZs and the near non-existence of the prograde SCZs.

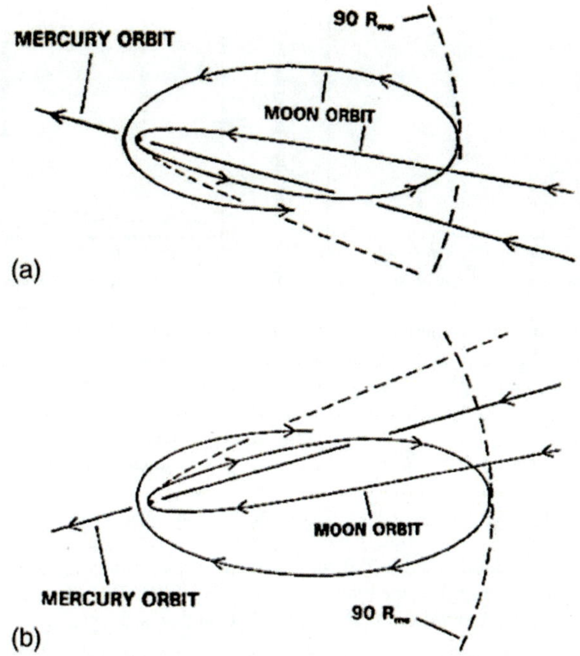

Fig. 8.4 Two-body representation of gravitational capture for planet Mercury. Both the planet and planetoid are in very similar, but not identical, prograde heliocentric orbits. (**a**) The planetoid has a close encounter with the planet in the prograde direction. If no energy is dissipated within the two bodies, then the planetoid departs from the planet along the *dashed line* and attains an orbit very similar to its pre-encounter orbit. If sufficient energy is dissipated in a combination of the two bodies, then the planetoid is inserted into a highly elliptical orbit that stays within the stability limit for the sun-planet-planetoid system [i.e., the Hill sphere for the system (Roy 1965; de Pater and Lissauer 2001)]. (**b**) A planetoid has a close encounter with the planet in the retrograde (clock-wise) direction. If no energy is dissipated within the two interacting bodies during the encounter, then the planetoid departs from the planet along the dashed line and attains an orbit very similar to its pre-encounter orbit. If sufficient energy is dissipated in a combination of the two bodies, then the planetoid is inserted into a highly elliptical orbit that stays within the stability limit for the sun-planet-planetoid system. (View is from the north pole of the solar system)

The plots showing the results of the numerical simulations need some explanation also. Figure 8.6 shows the major features of a non-rotating coordinate system as well as a rotating coordinate system. Most of the plots in this chapter use the non-rotating coordinate system because it is more easily understood. But for certain projects, such as defining the limits of STABLE CAPTURE ZONES, the rotating coordinate system is very important. This special use of the rotating coordinate system will be demonstrated in greater detail in a later section of this chapter featuring numerical simulations associated with the Neptune-Triton system. Figure 8.6 gives an

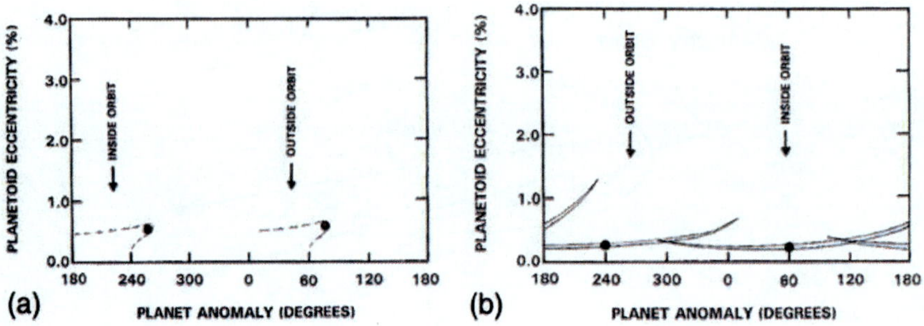

Fig. 8.5 Geometry of stable capture zones for planet Mercury when Mercury is in a circular orbit. (**a**) Stable Capture Zones for a 0.5 moon-mass planetoid encountering planet Mercury in the PROGRADE direction. The black dot positions on the plot require the lowest energy to be dissipated in the planetoid for stable capture. These are the locations in terms of Planet Anomaly and Planetoid Eccentricity that are used in the prograde encounter scenarios in this section of the chapter. (**b**) Stable Capture Zones for a 0.5 moon-mass planetoid encountering planet Mercury in the RETROGRADE direction. The black dot positions on the plot require the lowest energy to be dissipated in the planetoid for stable capture. These are the locations in terms of Planetoid Anomaly and Planetoid Eccentricity that are used in the retrograde encounter scenarios in this section of the chapter. Note: The label INSIDE ORBIT is for planetoids that encounter the planet from an orbit that is slightly smaller than that of the planet; the label OUTSIDE ORBIT is for planetoids that encounter the planet from an orbit that is slightly larger than that of the planet

overview of the features of these two methods of plotting the results. Figure 8.7 gives additional information for understanding the HELIOCENTRIC ORIENTATION plots and the PLANETOCENTRIC plots.

8.2.2 Prograde Encounter and Capture Simulations: Inside Orbit

Figure 8.8a shows the heliocentric setting for an encounter in the prograde direction that could result in capture if sufficient energy is dissipated during a close encounter. The planetoid is approaching Mercury from an orbit that is slightly inside the orbit of Mercury. The STARTING POINT as well as the orientation of the planetoid orbit are shown on the diagram. Since no energy is dissipated during the encounter, the planetoid escapes back into a heliocentric orbit soon after the encounter. The calculated time of this encounter scenario is 68.5 earth days.

Figure 8.9 shows the heliocentric and mercocentric orientations and the positions of the perigee passages for a 3-orbit escape scenario using the same encounter parameters as in Fig. 8.8. Figure 8.9b shows the overall dimensions of the escape scenario. Note that the starting position is the same as in Fig. 8.6. Since significant

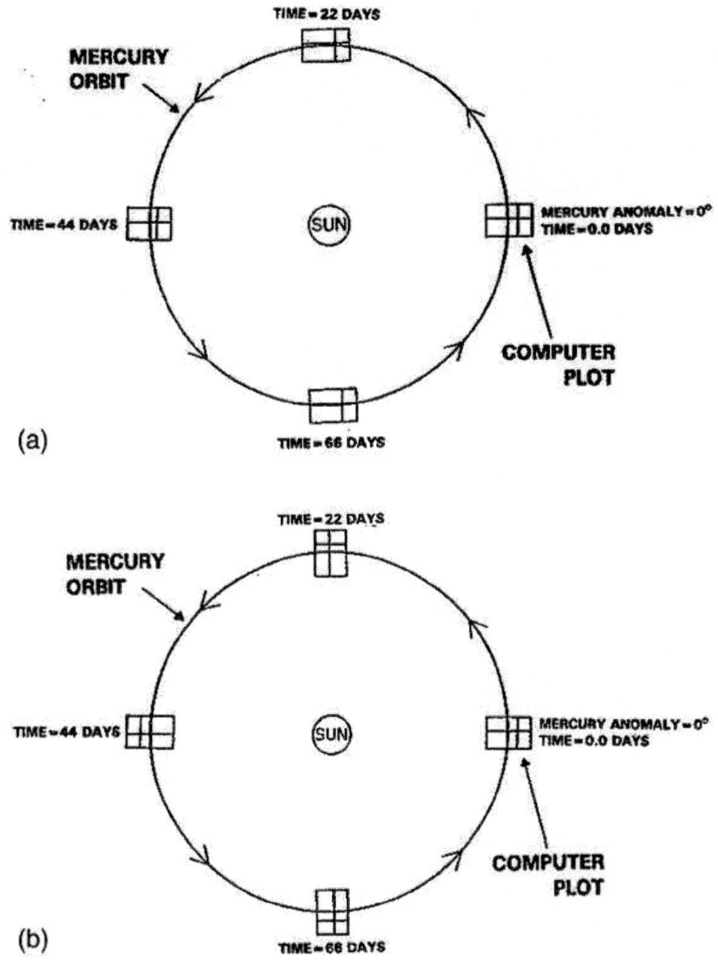

Fig. 8.6 Some orientation information for understanding the computer plots. (**a**) Diagram illustrating a non-rotating coordinate system for plotting the results of the numerical simulations. The computer plot does not rotate as it theoretically revolves around the Sun. Planet Mercury is at the origin of the non-rotating plot. (**b**) Diagram illustrating a rotating coordinate system for plotting the results of the numerical simulations. In this case the computer plot rotates as it theoretically revolves around the Sun. Special symmetry information can be gained from this plot but the plotted information is more difficult to understand. Thus plots in this chapter (unless noted otherwise) use the non-rotating coordinate system and all orbital plots are viewed from the north pole of the solar system. The sidereal year for Mercury is about 88 days

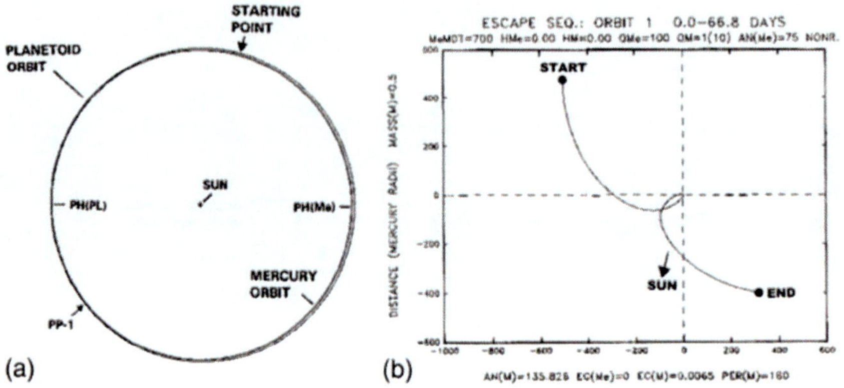

Fig. 8.7 Some terminology for the three-body calculation using planet Mercury and a 0.5 moon-mass planetoid as an example. (**a**) STARTING POINT marks the beginning of the calculation. PP-1 is the position of the first close encounter. PH(Me) is the perihelion of the orbit of Mercury; PH (PL) is the perihelion of the orbit of the 0.5 moon-mass planetoid. In this example the orbit of the planetoid is outside the orbit of Mercury. (**b**) Organization of the information on the planetocentric computer plot is the following: *First title line*: results of simulation (escape, collision, or capture); number of orbits; duration of simulation in earth days. *Second title line*: *MeMDT* initial mercury-planetoid distance of separation in planet radii; *HMe* h for mercury; *HM* h for planetoid; *QMe* Q of Mercury; *QM* Q of planetoid; *AN(Me)* planet anomaly at beginning of calculation. *Bottom line*: *AN (M)* planetoid anomaly at the beginning of the calculation; *EC(Me)* eccentricity of the orbit of mercury; *EC(M)* eccentricity of the orbit of the planetoid before the encounter; PER(M) pericenter radius position for the orbit of the planetoid. *Left side line*: Scale for both axes in planet radii; fractional mass of planetoid relative to mass of earth's moon. The STARTING POINT for the planetoid for this simulation is in the upper left quadrant; the END POINT is in the lower right quadrant. Arrow indicates direction of Sun at the beginning of the simulation. View is from the north pole of the Solar System

energy is dissipated by rock tidal processes with the displacement Love number for the planetoid at 0.54, the planetoid is very near to a captured condition.

Figure 8.10 shows the heliocentric and mercocentric orientations for a successful capture scenario. The STARTING POINT as well as the positions of the first 8 perigee passages shown on the heliocentric plot. Figure 8.10b shows the overall geometry of this calculated successful capture. The total time for this scenario is 655.6 earth days. Figure 8.10c shows the morphology of the orbits at an intermediate scale and Fig. 8.10d shows a close-up of the close encounters. Note that the displacement Love number of the planetoid is 0.56 (just 0.02 more than that of the 3-orbit escape scenario in Fig. 8.9).

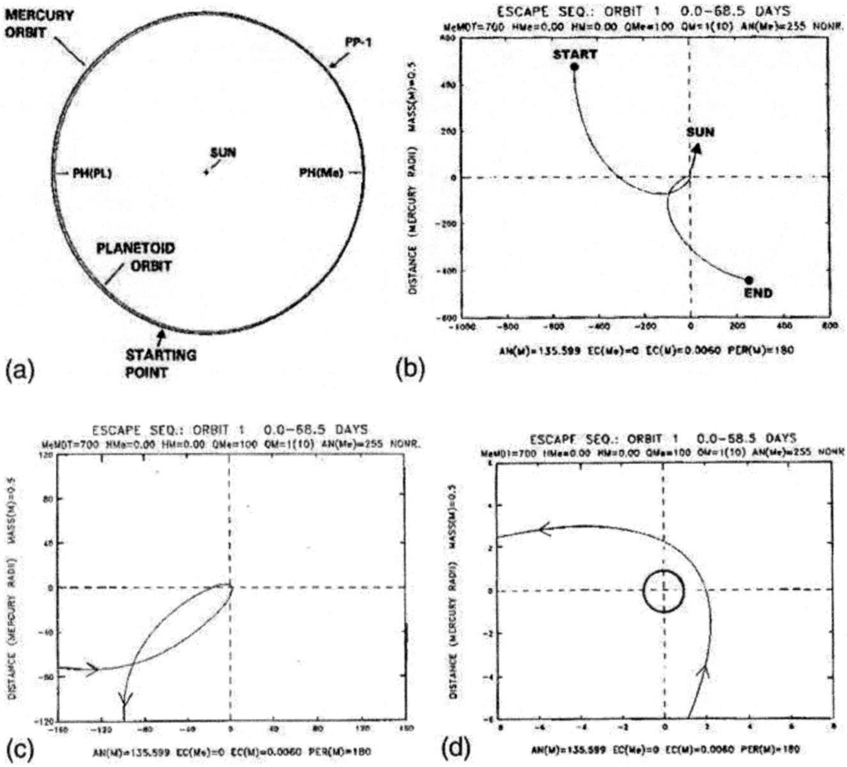

Fig. 8.8 Results of a close encounter in the prograde direction of a 0.5 moon-mass planetoid with planet Mercury in which no energy was dissipated. (**a**) Heliocentric orientation for the encounter. The simulation begins at the STARTING POINT. The close encounter occurs about 40 days later at PP-1 (perigee passage number 1). No energy is dissipated by tidal processes and the planetoid escapes to a heliocentric orbit about 68.5 days after PP-1. (**b**) Mercocentric orientation for the encounter. Mercury is at the origin of the plot. The *arrow* on the plot indicates the position of the Sun at the beginning of the simulation. The simulation begins with the encountering planetoid at 700 mercury radii which is well beyond the boundary of the Hill Sphere (which is at about 90 mercury radii). (**c**) Closer view of the encounter with arrows indicating the direction of motion. (**d**) An even closer view of the encounter

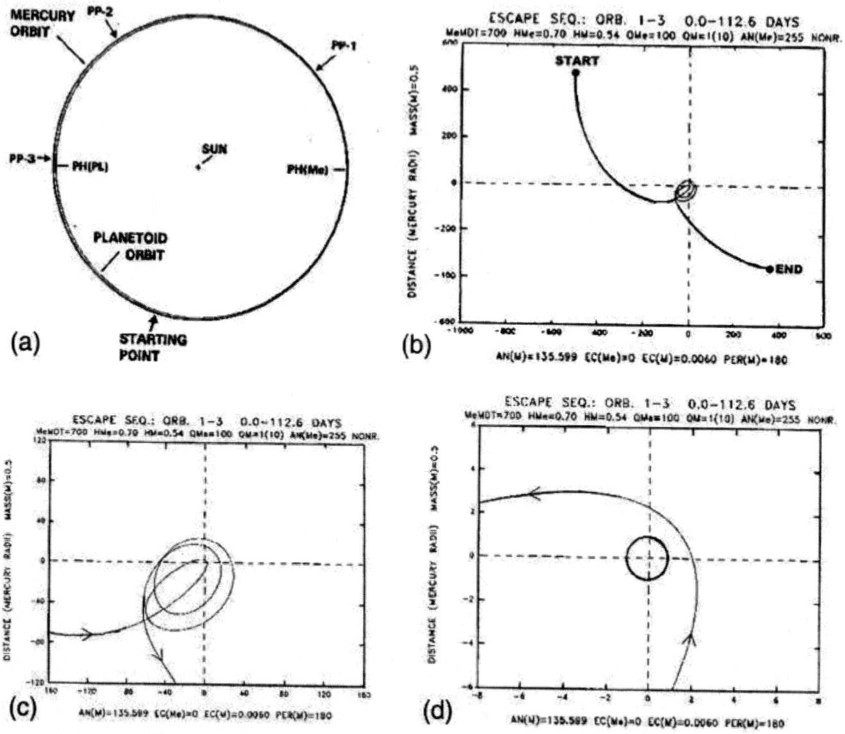

Fig. 8.9 Results of a three orbit escape scenario. (**a**) The heliocentric orientation showing the starting point and the positions of the three close encounters of this encounter sequence. (**b**) The mercocentric orientation shows the pattern of close encounters. Solar gravitational perturbations cause the variation in the orbital path. A significant quantity of energy is dissipated during the first encounter but not enough for stable capture. (**c**) A closer view of the encounters with arrows showing the direction of motion of the encountering body. (**d**) Close-up of the first encounter (which is the only close encounter of this escape scenario)

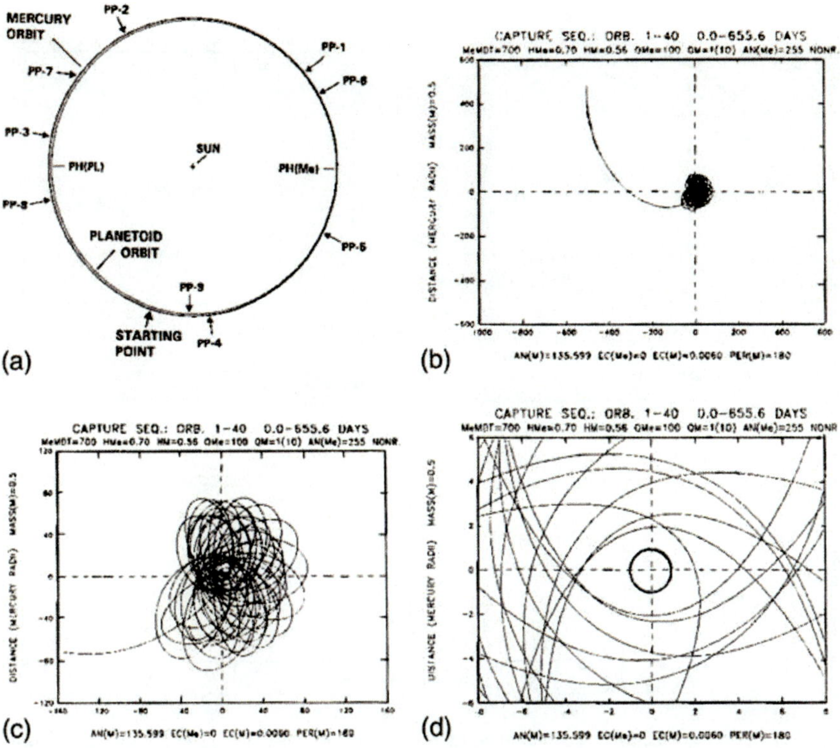

Fig. 8.10 Results of a stable capture scenario. (**a**) Positions of the first nine perimerc passages are shown on the heliocentric orientation. (**b**) Orbital pattern of the first 40 orbits (655.6 earth days) are shown on the merocentric plot. (**c**) A closer view of the orbits for this stable capture scenario. (**d**) Pattern of close encounters relative to planet Mercury. Note that all encounters are in the prograde direction. Tidal energy dissipation during PP-1 = 0.162 E28 joules; tidal energy dissipation for the 40-orbit scenario = 0.190 E28 joules

8.2.3 Prograde Encounter and Capture Simulations: Outside Orbit

Figure 8.11 shows the heliocentric and mercocentric orientations and the STARTING POINT for a 0.5 moon-mass planetoid encountering planet Mercury from a heliocentric orbit that is slightly larger than that of planet Mercury. The morphology of the orbital path of the planetoid, when no energy is dissipated, looks very similar to that for a simulation for an encounter from an orbit that is slightly smaller than that of Mercury. But on close inspection there are some differences.

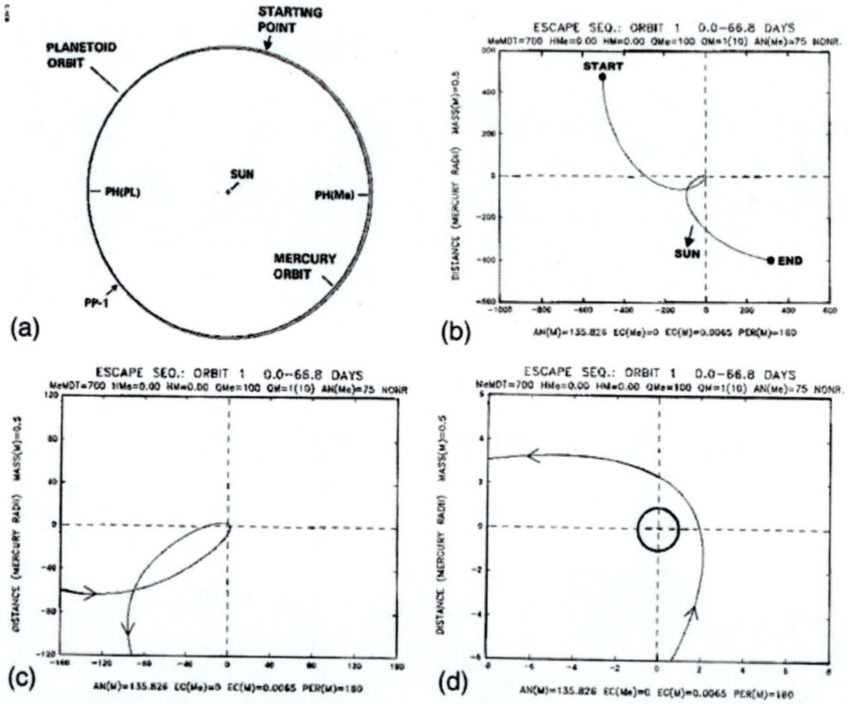

Fig. 8.11 Results of a close encounter in which no energy is dissipated. (**a**) Heliocentric orientation for the encounter. The simulation begins at the STARTING POINT. The close encounter occurs about 40 days later at PP-1. No energy is dissipated by tidal processes and the planetoid escapes into a heliocentric orbit about 66.8 days after the encounter. (**b**) Mercocentric orientation for the encounter. The *arrow* on the plot indicates the position of the Sun at the beginning of the simulation. The simulation begins with a distance of separation of 700 mercury radii. (**c**) Closer view of the encounter with arrows indicating the direction of motion. (**d**) An even closer view of the encounter

Figure 8.12 shows a scenario in which the planetoid makes three perimerc passages and then escapes back into a slightly changed heliocentric orbit. The displacement Love number (h) for the planetoid for this three-orbit escape scenario is 0.46 and an h value of 0.50 is needed for stable capture. Figure 8.13 shows the heliocentric and merocentric orientation for a stable capture from an orbit that is slightly larger than that of a circular orbit of Mercury.

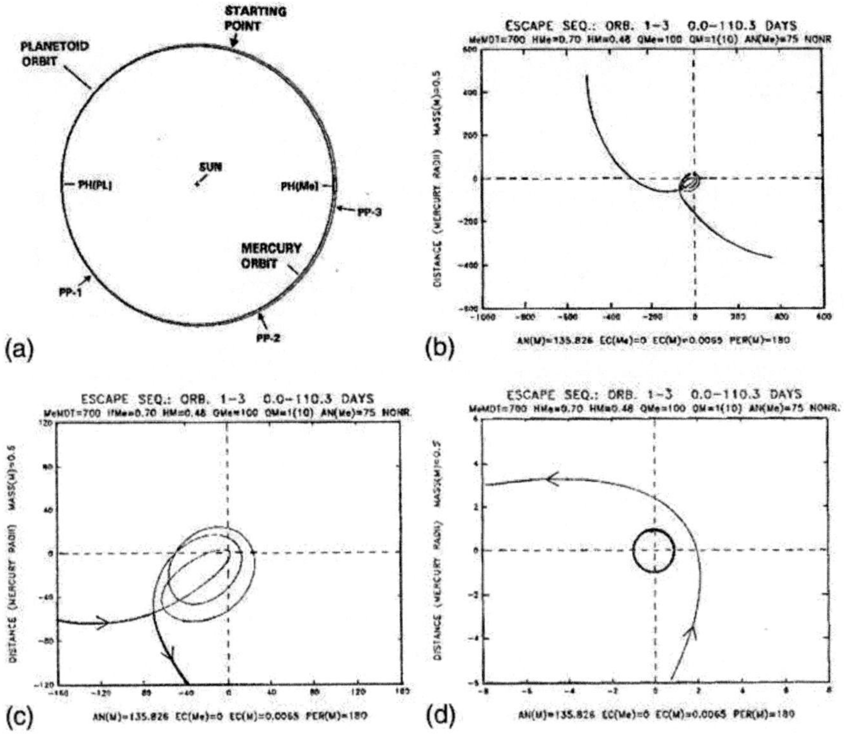

Fig. 8.12 Results of a three orbit escape scenario. (**a**) The heliocentric orientation showing the STARTING POINT and the positions of the three close encounters of this encounter sequence. (**b**) Mercocentric orientation showing the pattern of close encounters. Solar gravitational perturbations cause the variation in the orbital path. A significant quantity of energy is dissipated during the first encounter but not enough for stable capture. (**c**) A closer view of the encounters with arrows showing the direction of motion of the encountering body. (**d**) Close-up of the first encounter (which is the only close encounter in this escape scenario)

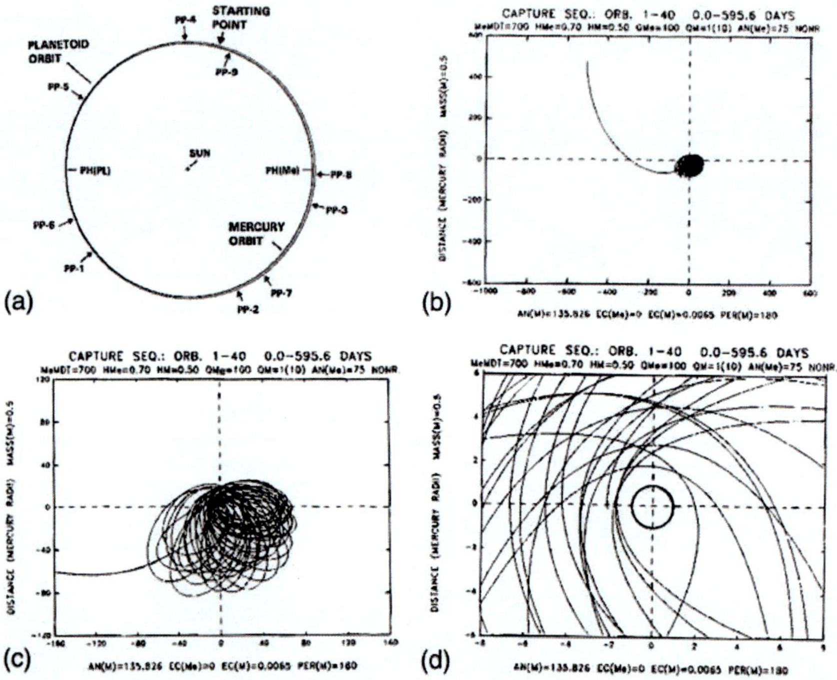

Fig. 8.13 Results of a stable capture scenario. (**a**) Positions of the first eight perimerc passages are shown on the heliocentric orientation. (**b**) Orbital pattern of the first 40 orbits (595.6 days) are shown on the mercocentric plot. (**c**) A closer view of the orbits of this stable capture scenario. (**d**) Pattern of close encounters relative to planet mercury. Tidal energy dissipation during PP-1 = 0.145 E28 joules; tidal energy dissipation for the 40-orbit scenario = 0.228 E28 joules

8.2.4 Retrograde Encounter and Capture Simulations: Inside Orbit

Figures 8.14, 8.15, 8.16, and 8.17 illustrate a series of planetoid encounters with planet Mercury in the RETROGRADE direction from an orbit that is slightly smaller than the orbit of Mercury (i.e., the planetoid orbit is inside the orbit of Mercury). The reader will notice that orbits generated from retrograde encounters have a much different morphology that those associated with prograde encounters. This is a general observation and these differences are obvious for similar planetoid encounters with planets in both the inner and the outer solar system (e.g., the Neptune-Triton system). Much additional information for the encounter scenarios is in the figure captions.

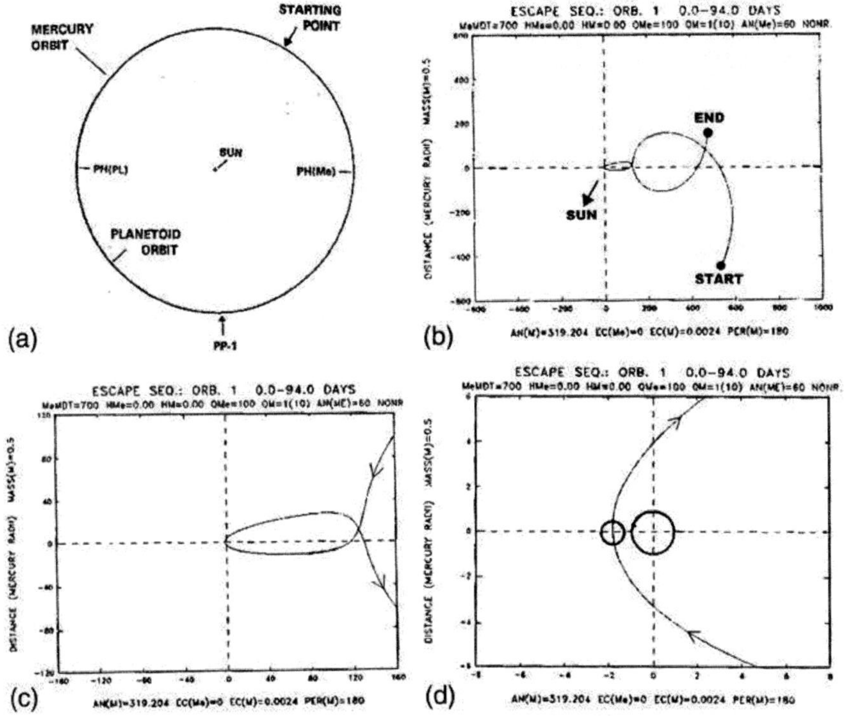

Fig. 8.14 A one-orbit escape scenario. (**a**) Heliocentric orientation showing the starting point for the encounter and the position of the first and only perimerc passage about 51 days after the START position. (**b**) Mercocentric orientation showing the starting point, the position of the sun at the beginning of the encounter, the position of the one and only perimerc passage, and the escape. (**c**) Close-up view of the perimerc passage in the retrograde direction. The arrows indicate the direction of motion. (**d**) Close-up of a scale model of the diameter of the planetoid relative to that of planet mercury. Arrows indicate the direction of movement during the encounter

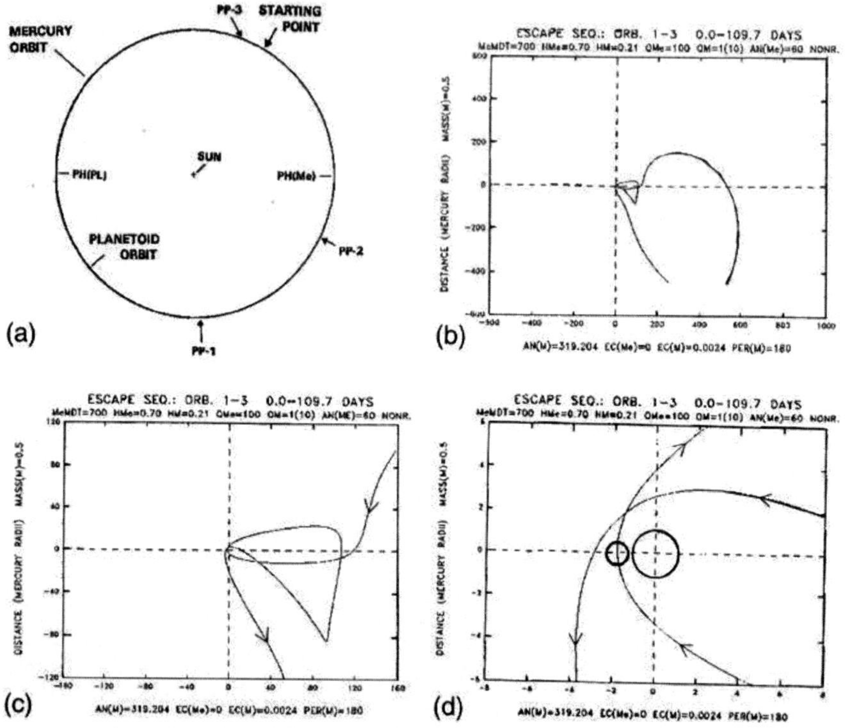

Fig. 8.15 A 3-orbit escape scenario. (**a**) Heliocentric orientation showing the STARTING POINT and the positions of the three perimerc passages. (**b**) Mercocentric plot showing and the geometry of the orbits before the escape to heliocentric orbit. (**c**) Closer view of the geometry of the orbits. Arrows indicate the direction of motion of the planetoid. (**d**) Closeup of two close encounters. Note that the final encounter leads to escape in the prograde (counter-clockwise) direction

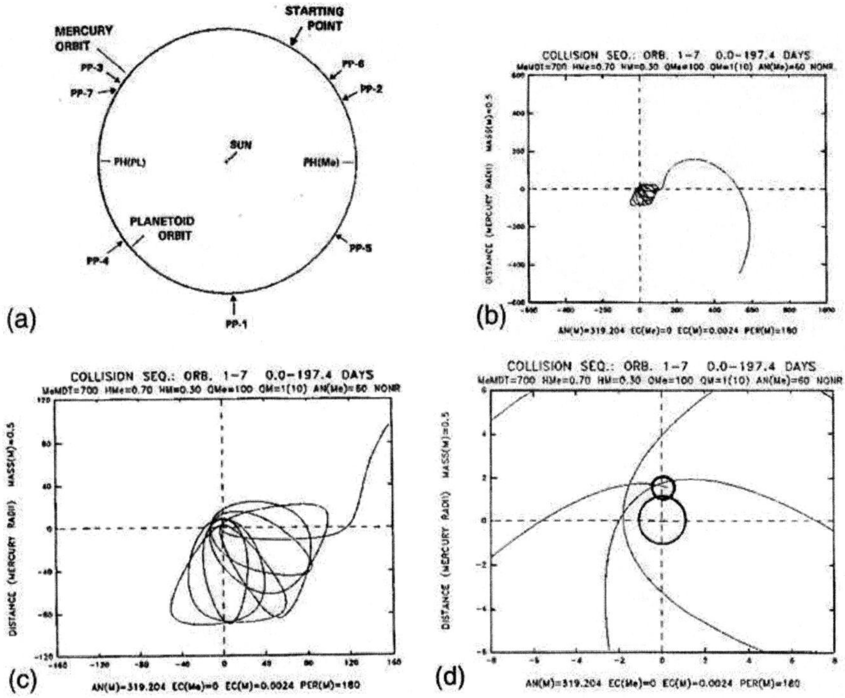

Fig. 8.16 A seven-orbit retrograde collision. (**a**) Heliocentric orientation showing the STARTING POINT and the positions of six perimerc passages and the collision (PP-7). (**b**) Merocentric view of the encounter sequence. (**c**) Closer view of the geometry of the mercocentric orbits; note that all perimerc passages are in the retrograde direction. (**d**) Close-up view of a few close encounters and the collision encounter

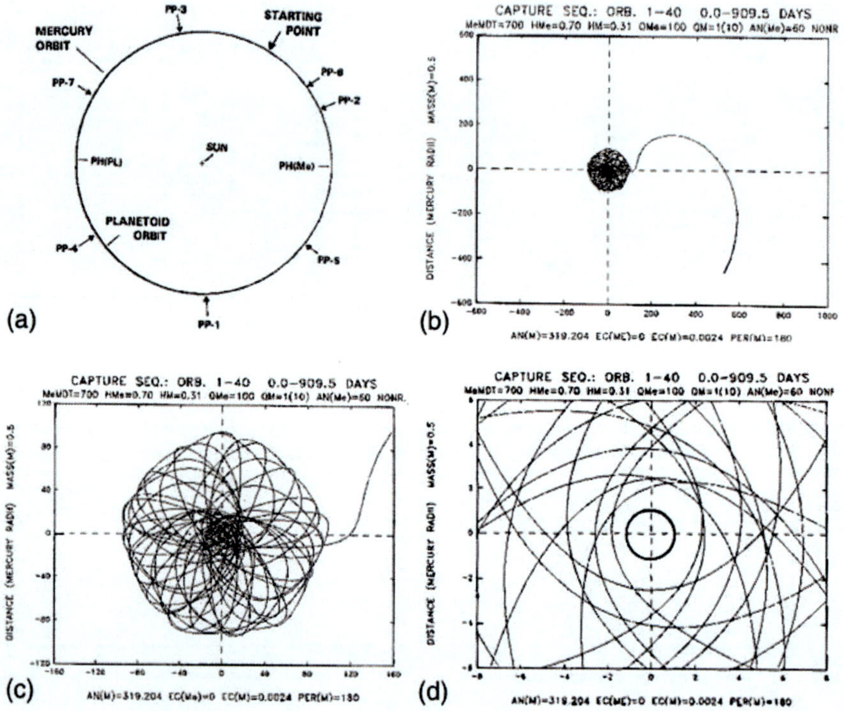

Fig. 8.17 A 40-orbit stable capture scenario. (**a**) Heliocentric orientation for the simulation showing the STARTING POINT and the positions of the first seven perimerc passages. (**b**) Mercocentric orientation of all 40 orbits of the capture scenario. (**c**) Closer view of the geometry of a typical stable capture scenario for mercury. (**d**) Close-up view of the geometry of the orbits relative to the size of planet Mercury. All planetoid orbits are in clockwise (retrograde) direction. Tidal energy dissipation during PP-1 = 0.091 E28 joules; tidal energy dissipation for the 40-orbit scenario = 0.109 E28 joules

8.2.5 *Retrograde Encounter and Capture Simulations: Outside Orbit*

Figures 8.18, 8.19, 8.20, and 8.21 illustrate a series of planetoid encounters with planet Mercury in the RETROGRADE direction from an orbit that is slightly larger than the orbit of Mercury (i.e., the planetoid orbit, in this case, is outside the orbit of Mercury).

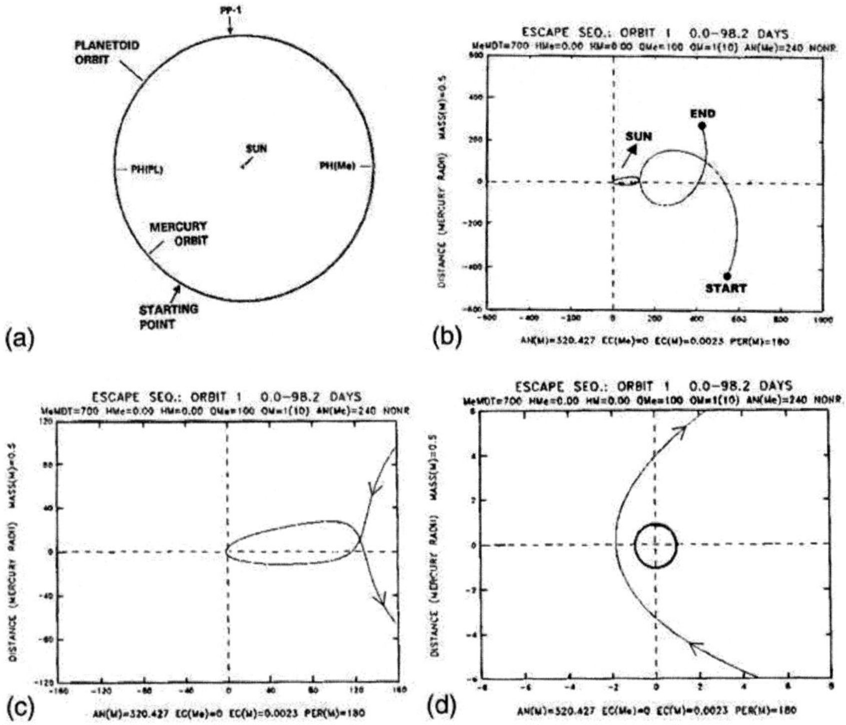

Fig. 8.18 A one-orbit escape scenario. (**a**) Heliocentric orientation showing the STARTING POINT and the position of the one and only perimerc passage about 52 days from the START point. (**b**) Mercocentric orientation showing the starting point, the position of the sun at the beginning of the simulation, the position of the perimerc passage, and the escape to heliocentric orbit. (**c**) Close-up of the perimerc passage in the retrograde direction. The arrows indicate the direction of motion. (**d**) Close-up of the close encounter relative to the position of planet Mercury. Arrows indicate the direction of motion during the encounter

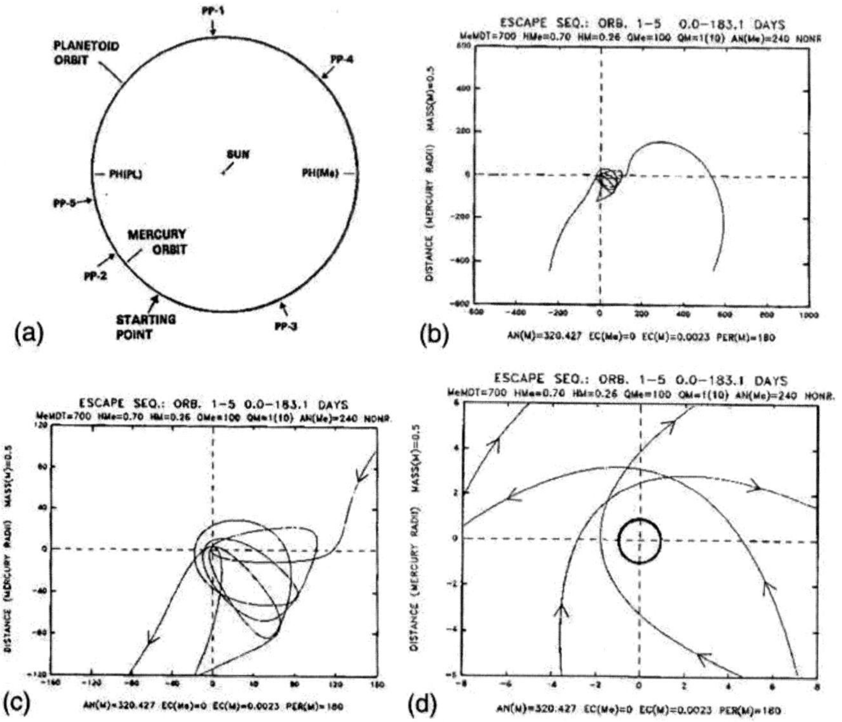

Fig. 8.19 A five-orbit escape scenario. (**a**) Heliocentric orientation showing the STARTING POINT and the positions of the five perimerc passages. (**b**) Mercocentric plot showing the starting position and the geometry of the orbits before the escape to heliocentric orbit. (**c**) Closer view of the geometry of the orbits. Arrows indicate the direction of motion of the planetoid. (**d**) Close-up of four of the perimerc passages relative to the diameter of Mercury. Note that the final encounter leads to escape in the prograde direction

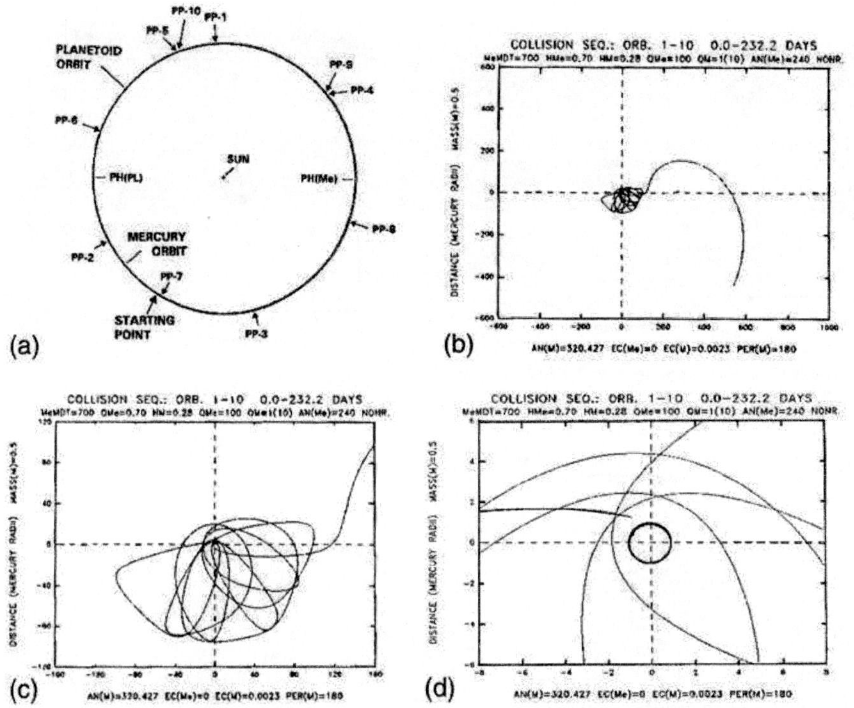

Fig. 8.20 A 10-orbit retrograde orbit collision scenario. (**a**) Heliocentric orientation showing the STARTING POINT, the positions of nine perimerc passages, and the position of the collision encounter (PP-10). (**b**) Mercocentric view of the encounter sequence. (**c**) Closer view of the geometry of the mercocentric orbits; note that all perimerc passages are in the retrograde (clockwise) direction. (**d**) Close-up view of a few close encounters before the collision encounter

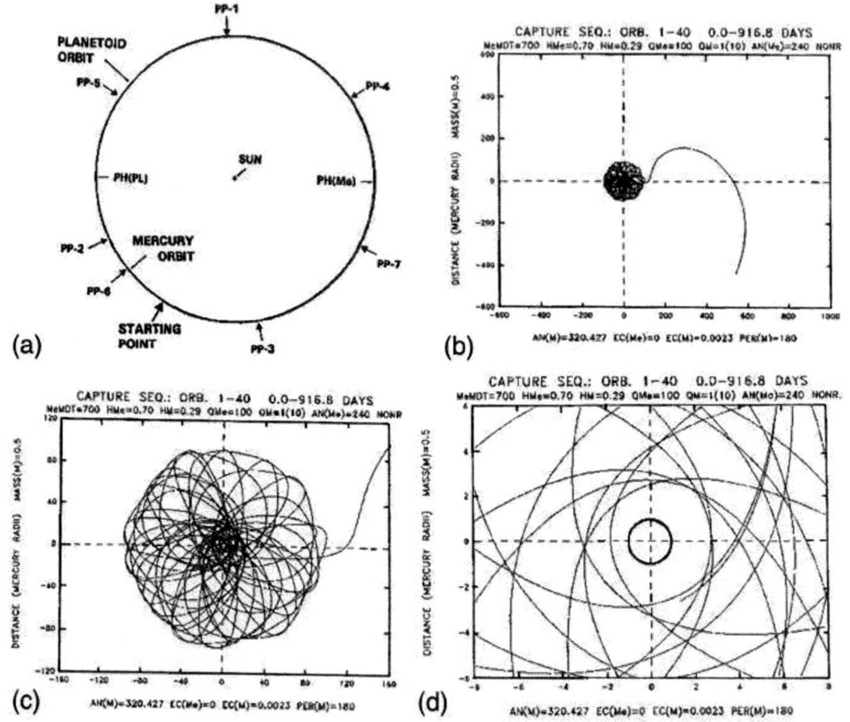

Fig. 8.21 A 40-orbit stable capture scenario. (**a**) Heliocentric orientation for the simulation showing the STARTING POINT and the positions of the first seven perimerc passages. (**b**) Mercocentric orientation for all 40 orbits of the capture scenario. (**c**) Closer view of the geometry of a stable retrograde capture scenario for Mercury. (**d**) Close-up view of the perimerc passages; note that all passages are in the retrograde (clockwise) direction. Tidal energy dissipation during PP-1 = 0.086 E28 joules; energy dissipation for the 40-orbit scenario = 0.094 E28 joules

8.2.6 Summary of the Mercury-Planetoid Simulations

Although stable gravitational capture of a 0.5 moon-mass planetoid is physically possible for Planet Mercury, it is highly improbable. The capture simulations are for a mercury-like planet in a circular orbit. Even with a circular orbit, the geometry of the stable capture zones is very limited. There is no doubt that retrograde capture is favored by the SCZ geometry. As the heliocentric eccentricity increases, so does orbit instability and the useful portion of the SCZs decreases. As inclination of the heliocentric orbit of the mercury-like orbit increases, the stability of the orbit of the captured body decreases and the SCZ decreases in geometry. The displacement Love numbers for the prograde orbit scenarios are beyond the realm of credibility. Although the displacement Love numbers for retrograde capture are much lower than the prograde counterparts, they may not be realistic. The bottom line is that planet Mercury has no natural satellites and the chances of a planet in Mercury's location in another star-planet system would have a very small chance of having a large natural satellite.

Planet Mercury has many impact craters with the largest being the Caloris Basin. A major question is: Was this basin the result of a "straight-in" collision by a body in a heliocentric orbit? Or was it formed via the impact of a planetoid from a mercocentric orbit that was associated with a failed capture scenario of either a Vulcanoid planetoid or a metal-rich body originating near the Iron Line?

8.3 Planet Venus

Planet Venus is a large terrestrial planet and is more distant from the Sun than Mercury. Unlike Mercury, Venus is in a near circular orbit. In general, the potential for gravitational capture is much greater for Venus than for Mercury.

8.3.1 Two-Body Representation, Stable Capture Zones, and Orbital Orientation Diagrams for Gravitational Capture Simulations for Venus

Figure 8.22 shows two-body renditions of both prograde and retrograde gravitational capture scenes.

Fig. 8.22 Two-body diagrammatic sketches of capture for planet Venus and 0.5 moon-mass planetoid. (**a**) Both the planet and planetoid are in very similar, but not identical, prograde heliocentric orbits. The planetoid has a close encounter with the planet. If no energy is dissipated within the two bodies, then the planetoid departs from the planet along the *dashed line* and attains an orbit very similar to its pre-encounter orbit. If sufficient energy is dissipated in a combination of the two bodies, then the planetoid is inserted into a highly elliptical orbit that stays within the stability limit for the sun-planet-planetoid system [i.e., within the Hill sphere for the system (Roy 1965; de Pater and Lissauer 2001)]. (**b**) Scale sketch of an encounter in the retrograde (clock-wise) direction. The explanation is the same for this diagram except that the encounter is on the opposite side of the planet. View is from the north pole of the solar system

Figure 8.23 shows the dimensions of the stable capture zones with the most useful (lower energy) portions shaded in red and the somewhat higher energy zones in blue. Compared to those of planet Mercury (Fig. 8.5) the retrograde SCZs are much wider but cover about the same extent of planet anomaly. The prograde SCZs for Venus are much more extensive and more distinct than those for planet Mercury, but nonetheless, the prograde SCZs for Venus are not very useful for capture via dissipation of energy within the planetoid during one periven passage. Cursory examination of the geometry of all four SCZs suggests that retrograde gravitational capture is strongly favored over prograde gravitational capture.

Figure 8.24 shows sketches for both the non-rotating and the rotating coordinate system for plotting the results of planetoid encounters with planet Venus. All plots for Venus-planetoid encounters will use the non-rotating coordinate system.

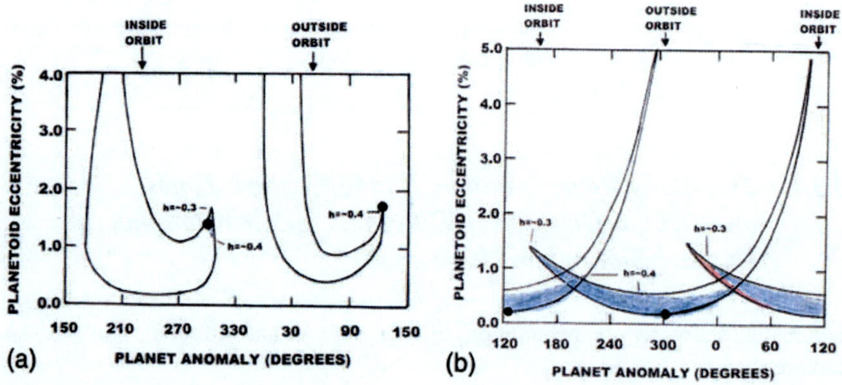

Fig. 8.23 Prograde and retrograde stable capture zones for Venus and a 0.5 moon-mass planetoid. (**a**) Plot of prograde stable capture zones. (**b**) Plot of retrograde stable capture zones. Discussion of the relative probability of stable capture is in the text. The position of the simulations used for both prograde and retrograde encounters are marked with black dots

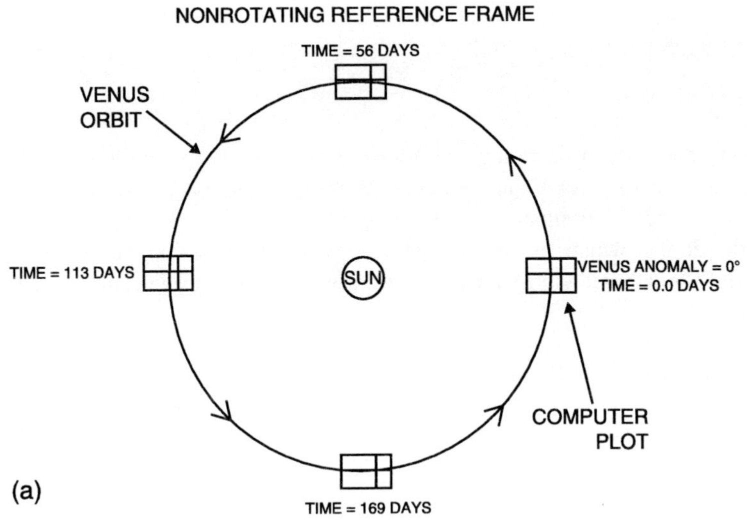

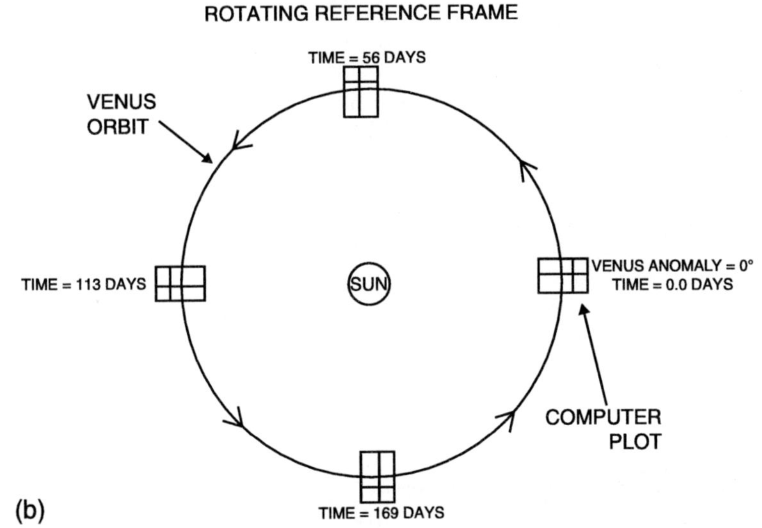

Fig. 8.24 Some orientation information for understanding the computer plots. (**a**) Diagram illustrating a non-rotating coordinate system for plotting the results of the numerical simulations. The computer plot does not rotate as it theoretically revolves around the Sun on a yearly cycle (~226 days). Planet Venus is at the origin of the rotating plot. (**b**) Diagram illustrating a rotating coordinate system for plotting the results of the numerical simulations. In this case the computer plot rotates as it theoretically revolves around the Sun. Special symmetry information can be gained from this type of plot but the plotted information is more difficult to understand. Thus all plots in this chapter (and book) use the non-rotating coordinate system (unless stated otherwise) and all orbital plots are viewed from the north pole of the solar system

8.3.2 *Prograde Encounter and Capture Simulations: Inside Orbit*

Figures 8.25, 8.26, and 8.27 show calculated scenarios for prograde encounters for a 0.5 moon-mass planetoid with planet Venus from a heliocentric orbit that is slightly smaller than the orbit of Venus. Figure 8.25 features a one-orbit escape scenario in which no energy is dissipated in either body; Fig. 8.26 features a four-orbit escape scenario during which nearly enough energy is dissipated in the encountering planetoid for capture; and Fig. 8.27 shows a successful prograde capture scenario.

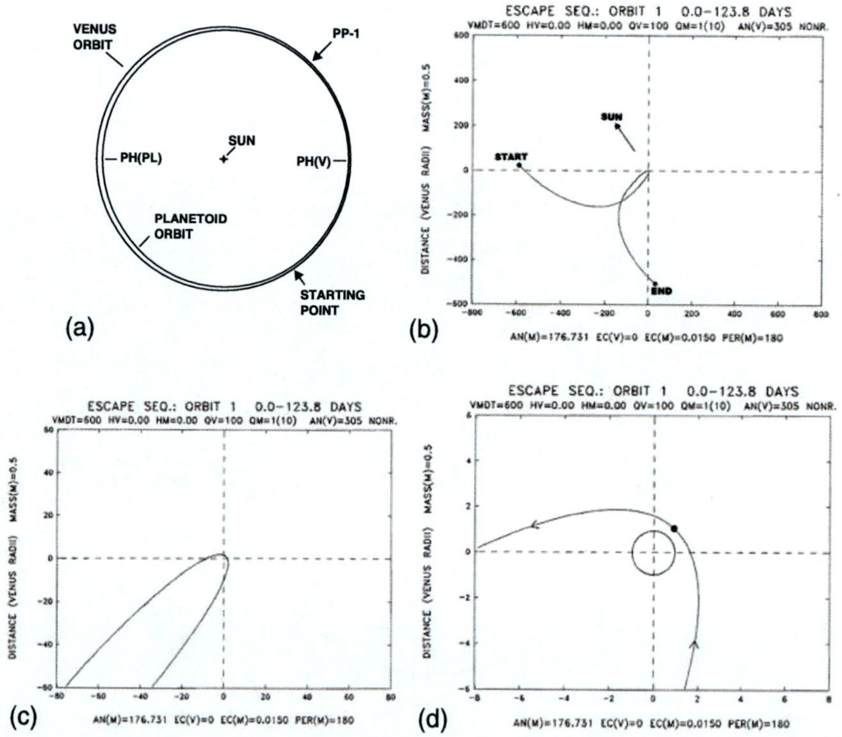

Fig. 8.25 Results of a close encounter in which no energy was dissipated. (**a**) Heliocentric orientation for the encounter. The simulation begins at the STARTING POINT. The close encounter occurs about 50.2 24-hr days later at PP-1 (periven passage number 1). No energy is dissipated by tidal processes and the planetoid escapes to a heliocentric orbit. (**b**) Venocentric orientation for the encounter. Venus is at the origin of the plot. The *arrow* on the plot indicates the position of the Sun at the beginning of the simulation. The simulation begins at 600 venus radii which is well beyond the boundaries of the Hill sphere (which is about 160 venus radii). (**c**) A closer view of the encounter. (**d**) An even closer view showing the relative diameters of the interacting bodies and arrows indicating the direction of motion

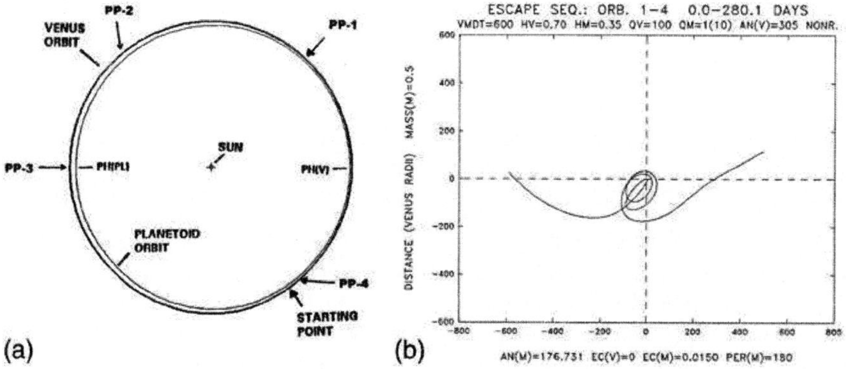

Fig. 8.26 Results of a four-orbit escape scenario. (**a**) Heliocentric orientation showing the STARTING POINT and the positions of the three close encounters of this encounter sequence and a fourth very distant encounter along the escape route. (**b**) Venocentric orientation showing the pattern of close encounters. Solar gravitation perturbations cause the variation in the orbital path of the planetoid. A significant quantity of energy is dissipated during the first encounter but not quite enough for stable capture. View is from the north pole of the solar system

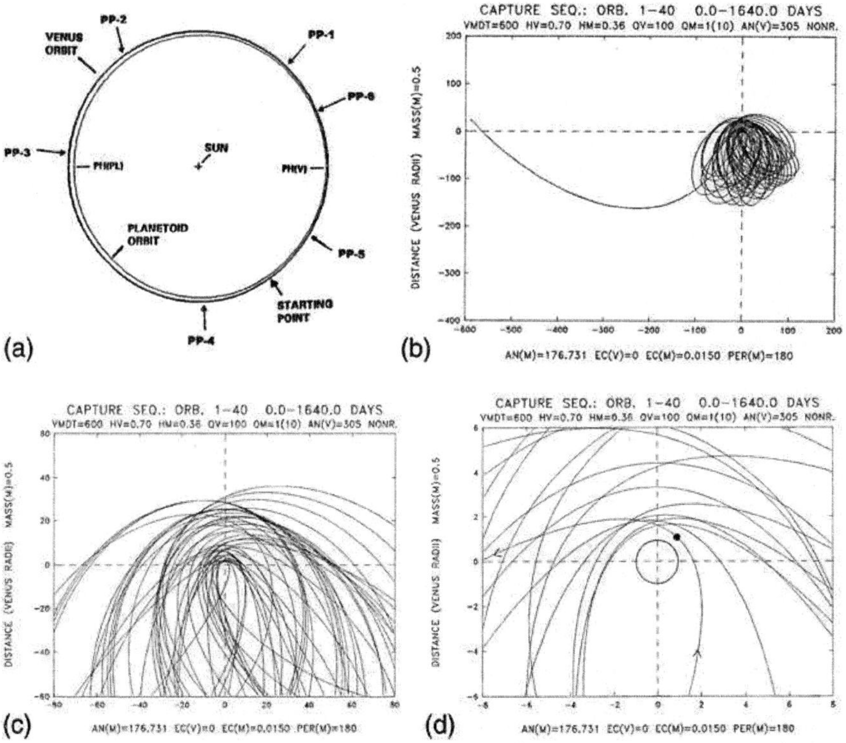

Fig. 8.27 Results of a stable capture scenario. (**a**) Heliocentric orientation showing the STARTING POINT and the positions of the first six periven passages. (**b**) Venocentric plot showing the orbital pattern of the first 40 orbits (1640 earth days). Note that 4.796 E27 joules of energy dissipation is sufficient for capture for this set of orbital parameters; total energy dissipation for the 40 orbit scenario is 5.433 E27 joules. (**c**) A closer view of the orbits for this stable capture scenario. (**d**) Pattern of close encounters relative to diameter of planet venus. Note: All orbits are in the prograde (counter-clockwise) direction and the view is from the north pole of the solar system

8.3.3 *Prograde Encounter and Capture Simulations: Outside Orbit*

Figures 8.28, 8.29, and 8.30 show calculated scenarios for prograde encounters for a 0.5 moon-mass planetoid with planet Venus from a heliocentric orbit that is slightly larger than the orbit of Venus. Figure 8.28 features a one-orbit escape scenario in which no energy is dissipated in either body. Figure 8.29 shows a three-orbit escape scenario during which nearly enough energy is dissipated within the interacting bodies for capture. Figure 8.30 features a successful capture scenario.

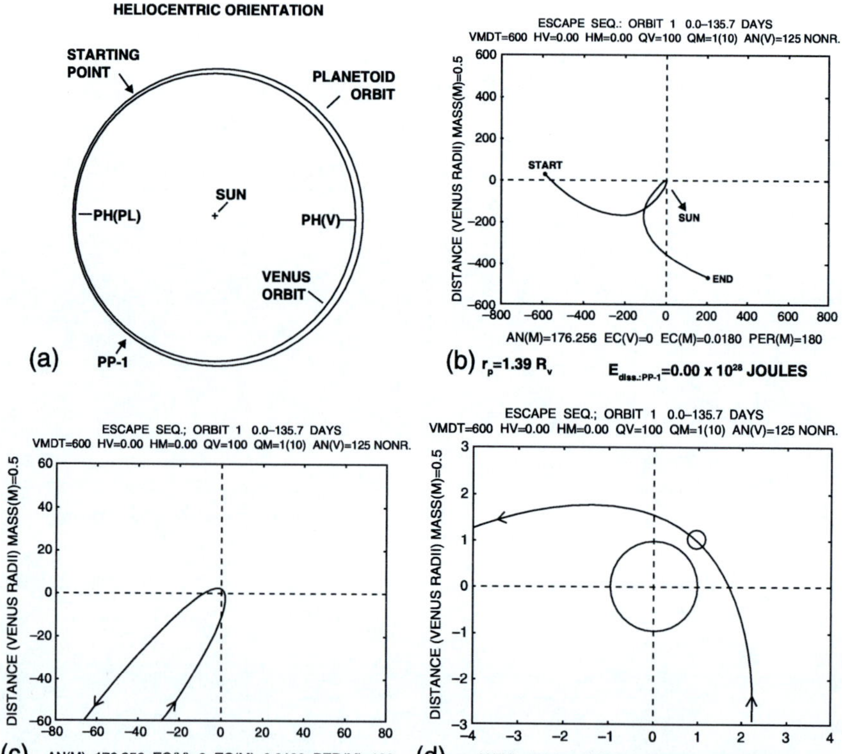

Fig. 8.28 Results of a close encounter in which no energy is dissipated. (**a**) Heliocentric orientation for the encounter. The simulation begins at the STARTING POINT. The close encounter occurs about 75.3 24 hr days later at PP-1. No energy is dissipated by tidal processes and the planetoid escapes to a heliocentric orbit. (**b**) Venocentric orientation for the encounter. Venus is at the origin of the plot. The *arrow* on the plot indicates the position of the Sun at the beginning of the simulation. The simulation begins at 600 venus radii which is well beyond the boundaries of the Hill sphere (which is about 160 venus radii). (**c**) A closer view of the encounter with *arrows* indicating the direction of motion. (**d**) An even closer view showing the relative diameters of the interaction bodies. View is from the north pole of the solar system

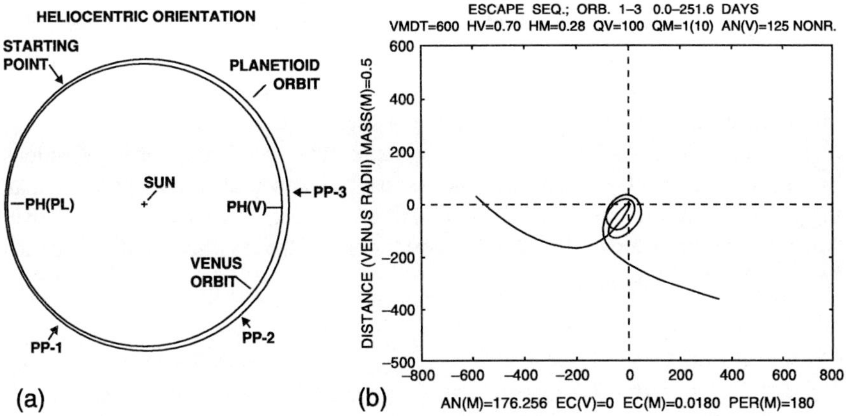

Fig. 8.29 Results of a three-orbit escape scenario. (**a**) Heliocentric orientation showing the STARTING POINT and the positions of the three close encounters of the encounter sequence. (**b**) Venocentric orientation showing the pattern of close encounters. A significant quantity of energy is dissipated during the first encounter but not quite enough for stable capture. View is from the north pole of the solar system

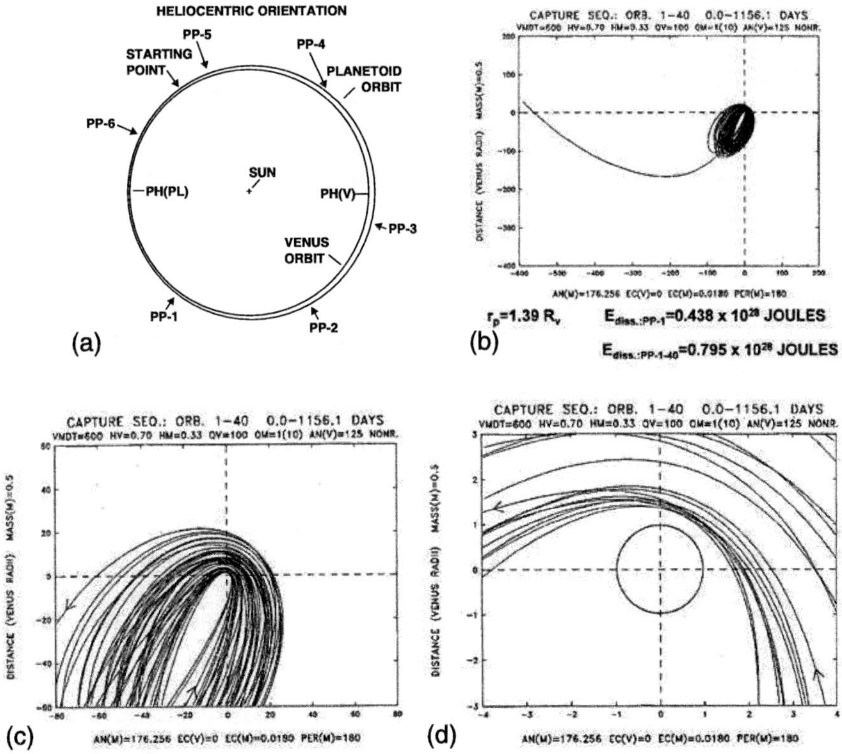

Fig. 8.30 Results of a stable capture scenario. (**a**) Heliocentric orientation showing the position of the STARTING POINT and the positions of the first six periven passages. (**b**) Venocentric diagram showing the pattern of the close encounters. Solar perturbations cause the variation in the orbital path of the planetoid. (**c**) A closer view of the pattern of the close encounters associated with the stable capture scenario. (**d**) A very close view of the encounters relative to the position of Venus. View is from the north pole of the solar system

When we compare and contrast the capture scenarios from an orbit that is slightly inside the orbit of Venus and an orbit that is slightly larger than that of Venus, we find that they have much in common. The major difference is the starting positions for the simulations. The other difference is that the planetoid is moving faster than the planet when it is on the inside orbit and, in contrast, it is moving more slowly when it is on the outside orbit. But the morphology of the resulting orbital scenarios look a good bit the same.

8.3.4 *Retrograde Encounter and Capture Simulations: Inside Orbit*

Figures 8.31, 8.32, 8.33, and 8.34 show the results of calculated scenarios for retrograde encounters for a 0.5 moon-mass planetoid with Venus where the

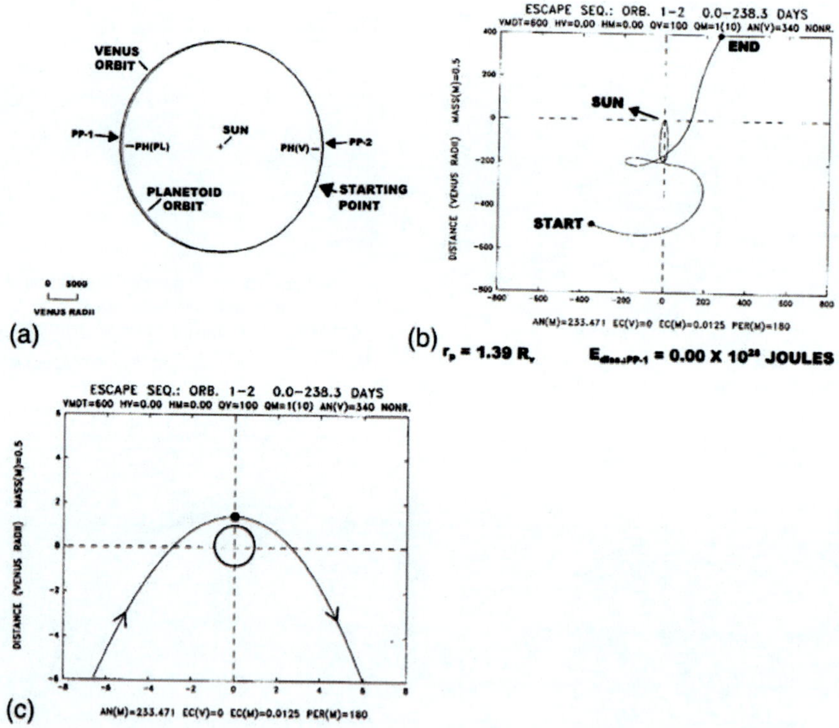

Fig. 8.31 A two-orbit escape scenario. (**a**) Heliocentric orientation showing the STARTING POINT for the encounter and the position of both periven passages (PP-1 and PP-2). (**b**) Venocentric orientation for the numerical simulation. Note that the second encounter is in the prograde direction and then the planetoid escapes. (**c**) A close-up of the initial encounter. Arrows indicate the direction of motion. The surfaces of the two bodies are separated by about 1000 km. View is from the north pole of the solar system

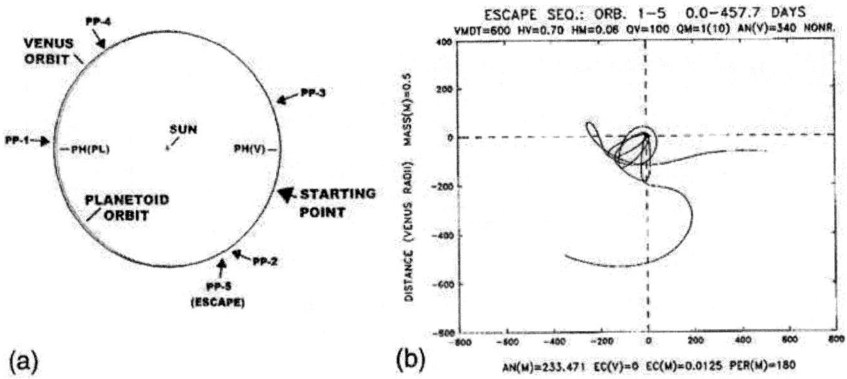

Fig. 8.32 A five-orbit escape scenario. (**a**) Heliocentric orientation showing the positions of the first five periven passages. (**b**) Venocentric plot showing the geometry of the orbits before the escape. Note that on the fifth periven passage the planetoid again encounters the planet in the prograde direction. View is from the north pole of the solar system

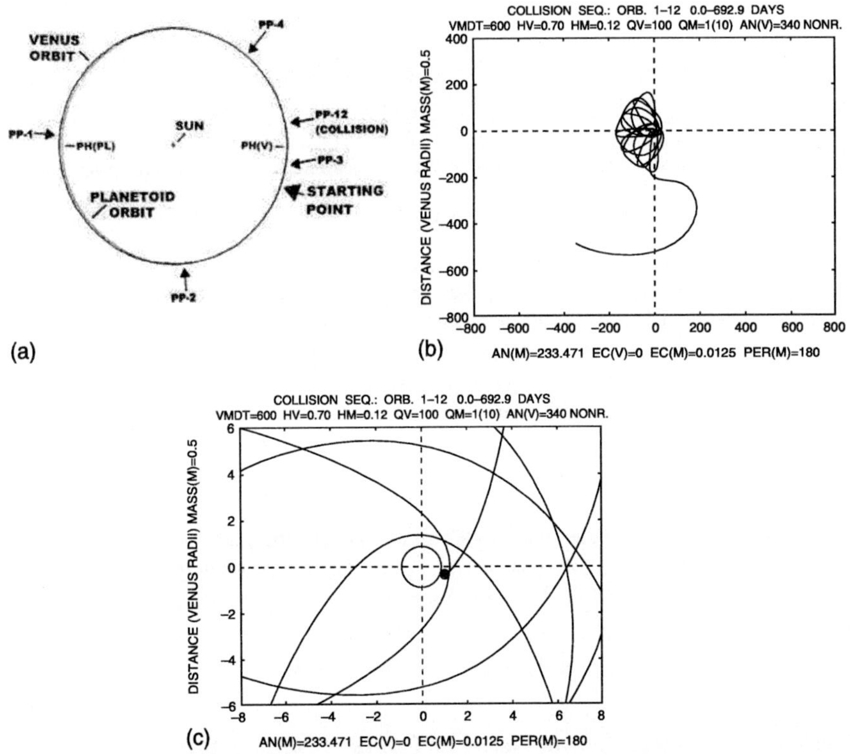

Fig. 8.33 A 12-orbit collision scenario. (**a**) Heliocentric orientation showing the positions of the first four periven passages and that of PP-12. (**b**) Venocentric plot of the simulation. On the twelfth periven passage, the planetoid collides with the planet at a distance of 0.84 venus radii in the retrograde direction. This is a solid collision and the planetoid would not survive. (**c**) Close-up of the collision scene. View is from the north pole of the solar system

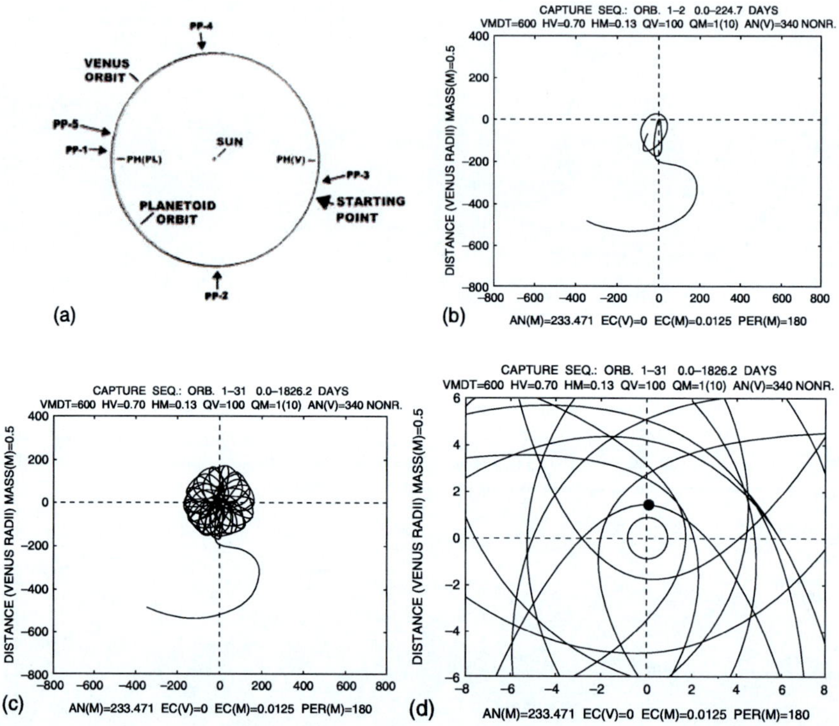

Fig. 8.34 Results of a stable retrograde capture scenario. (**a**) Heliocentric orientation for the simulation showing the positions of the first five periven passages. (**b**) Venocentric orientation of the first 2 orbits for this stable capture scenario. (**c**) Venocentric plot of the first 31 orbits. (**d**) Close-up of the periven passages relative to the surface of Venus. All orbits are in the retrograde (clockwise) direction. View is from the north pole of the solar system

heliocentric orbit of the planetoid is slightly smaller than that of Venus. Figure 8.31 features a somewhat unusual two-orbit escape scenario in which no energy is dissipated in either body; Fig. 8.32 shows a five-orbit escape scenario in which the planetoid escapes in the prograde direction relative to planet Venus; Fig. 8.33 features a 12-orbit collision scenario; and Fig. 8.34 shows the results of a stable capture scenario. The reader will notice that many retrograde scenarios in which there is not enough energy dissipation for capture on the first encounter will result in a grazing or a solid collision. Such results must be considered when estimating the probability of occurrence of prograde vs. retrograde gravitational capture.

8.3.5 Retrograde Encounter and Capture Simulations: Outside Orbit

Figures 8.35, 8.36, 8.37, and 8.38 show the results of calculated scenarios for retrograde encounters featuring a 0.5 moon-mass planetoid encountering planet

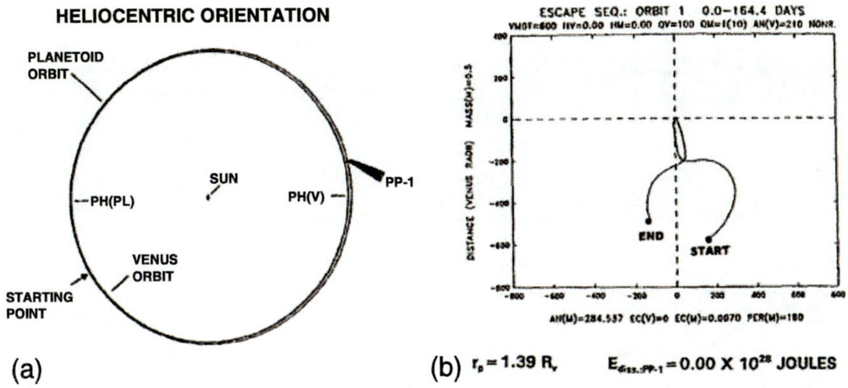

Fig. 8.35 Results of a close encounter in which no energy is dissipated. (**a**) Heliocentric orientation for the encounter. The simulation begins at the STARTING POINT. The close encounter occurs about 103 24 hr days later at PP-1. (**b**) Venocentric plot for the numerical simulation with the START point of the planetoid and the END point marked on the diagram. No energy is dissipated by tidal processes and the planetoid escapes to a heliocentric orbit in about 164.4 days. View is from the north pole of the solar system

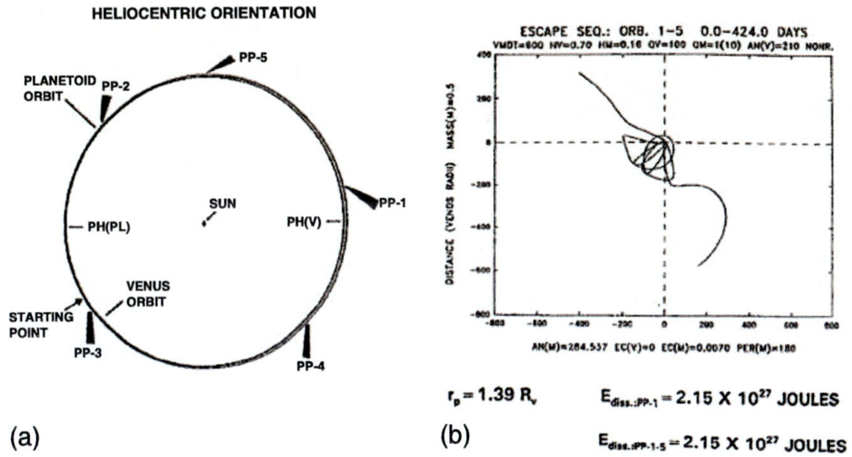

Fig. 8.36 Results of a retrograde five-orbit escape scenario. (**a**) Heliocentric orientation showing the STARTING POINT and the positions of the five close encounters of this encounter sequence. (**b**) Venocentric orientation showing the pattern of close encounters. Note that the fifth encounter is in the prograde orientation and that the escape is in the prograde direction relative to planet venus. View is from the north pole of the solar system

Venus from an orbit that is slightly larger than that of the planet. Figure 8.35 shows the results of a one-orbit escape scenario in which no energy is dissipated in the interacting bodies; Fig. 8.36 shows an encounter sequence that results in an prograde

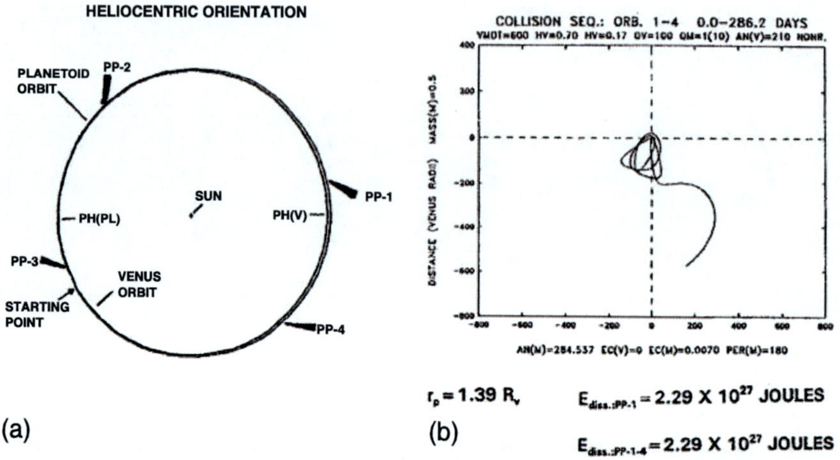

Fig. 8.37 Results of a 4-orbit collision scenario. (**a**) Heliocentric orientation showing the STARTING POINT and the positions of the three non-collision encounters and the collision encounter (PP-4). (**b**) Venocentric orientation showing the pattern of close encounters. All encounters are in the retrograde (clockwise) direction. View is from the north pole of the solar system

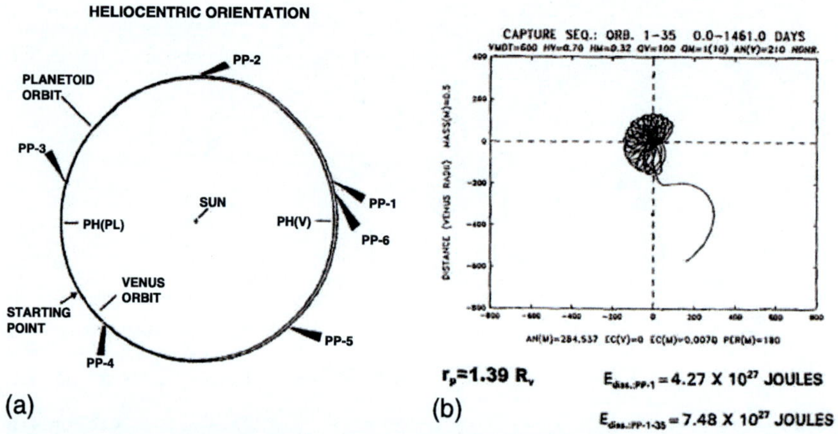

Fig. 8.38 Results of a stable retrograde capture scenario. (**a**) Heliocentric orientation showing the STARTING POINT and the position of the first six periven passages. (**b**) Venocentric plot showing the pattern of the orbits of this scenario. All orbits are in the retrograde direction. View is from the north pole of the solar system

escape to a heliocentric orbit after five periven passages; Fig. 8.37 results in a collision on the fourth encounter; and Fig. 8.38 shows the results of a stable retrograde capture scenario.

8.3.6 *Summary of Venus-Adonis Simulations*

This set of simulations is for our Twin Sister planet. The distance from the Sun is a bit closer for Venus than it is for Earth but the mass of the planet is very similar to that of the Earth. Adonis is one-half the mass of Luna.

We find that prograde capture is so improbable as to make it, for practical purposes, impossible. The useful portion of both prograde SCZs are not much more than pinpoints. In great contrast, the retrograde SCZs for Venus are laterally extensive with significant regions favorable for capture if the planetoid body conditions at the time of a close encounter are in the range to facilitate the required energy dissipation for successful capture.

When exploring exoplanet systems similar to our solar system, we would expect a planet in a similar situation to be either (1) a planet without a satellite, (2) a planet with a retrograde satellite at some stage of orbital evolution, or (3) a planet that has a very slow prograde rotation or some rate of retrograde rotation, a very dense atmosphere, a basaltic surface with massive lava flows, and no satellite.

8.4 Planet Earth

Planet Earth is the largest terrestrial planet and shares many orbital and planetary features with planet Venus. The main difference between these two large terrestrial planets is the atmospheric and hydrospheric conditions on the planets. Because of its larger size and greater distance from the Sun, Earth has a somewhat greater potential for capturing planetoids from near its heliocentric orbit.

8.4.1 *Two-Body Representation, Stable Capture Zones, and Orbital Orientation Diagrams for Gravitational Capture Simulations for Earth*

Figure 8.39 shows two-body scale sketches of both prograde and retrograde gravitational capture scenes.

Figure 8.40 shows the dimensions of the stable capture zones with the most useful portions (lower energy dissipation for capture) shaded in red and the somewhat higher energy zones in blue. Compared with those of planet Venus (Fig. 8.23) the retrograde SCZs are very similar in both Planet Anomaly and Planetoid Eccentricity. The differences are in the details. The retrograde SCZs for Earth have a larger area in the red zone and thus are more effective for gravitational capture processes.

There is a major difference between the useful portions of the prograde SCZs for Venus and Earth. The morphology of the two sets of prograde SCZs is similar, but the most useful area (below $h = 0.3$) is very limited for Venus relative to Earth. This

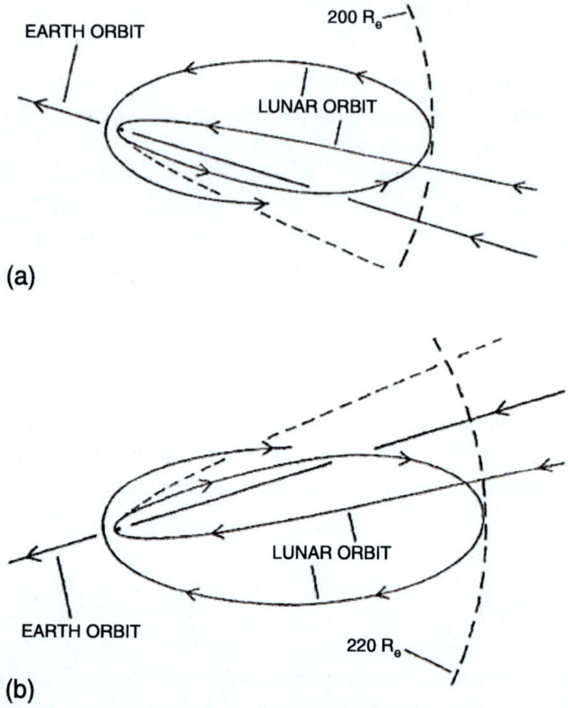

Fig. 8.39 Two-body diagrammatic sketches of capture for planet Earth and a moon-mass planetoid. **(a)** Both the planet and planetoid are in very similar, but not identical, prograde heliocentric orbits. The planetoid has a close encounter with the planet. If no energy is dissipated within the two bodies, then the planetoid departs from the planet along the *dashed line* and attains an orbit very similar to its pre-encounter orbit. If sufficient energy is dissipated in a combination of the two bodies, then the planetoid is inserted into a highly elliptical orbit that remains within the stability limit for the sun-planet planetoid system [i.e., within the Hill sphere for the system (Roy 1965; de Pater and Lissauer 2001)]. **(b)** Scale sketch of an encounter in the retrograde (clock-wise) direction, the explanation is the same for this diagram except that the encounter is on the other side of the planet. View is from the north pole of the solar system

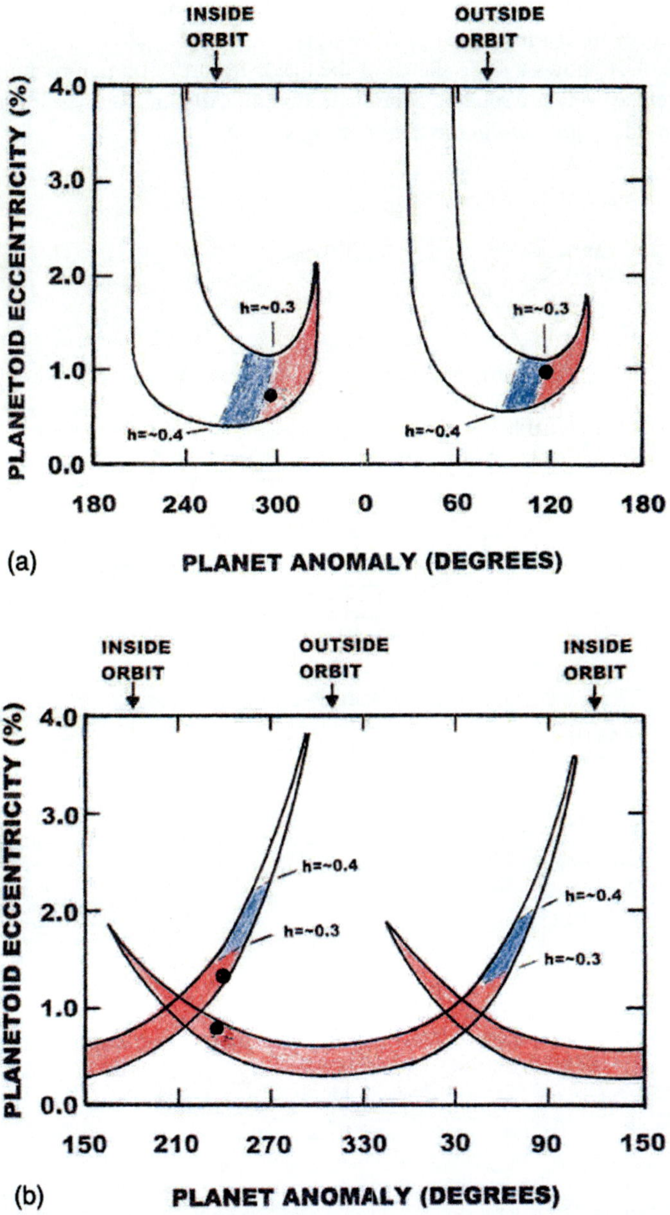

Fig. 8.40 Prograde and retrograde SCZs for Earth and a lunar mass planetoid. (**a**) Plot of prograde stable capture zones. (**b**) Plot of retrograde stable capture zones. Discussion of the relative probability of stable capture of planetoids of appropriate size and mass is in the text. The positions of the four sets of simulations used in figures in this section of the chapter are indicated on the SCZ plots (i.e., the black dots)

minor fact of the geometry of stable capture zones may have had significant consequences for the history of planet Earth!.

Figure 8.41 shows a scale sketch of the major features of a non-rotating coordinate system as well as a rotating coordinate system. All plots for the earth-planetoid system in this section are in the non-rotating system.

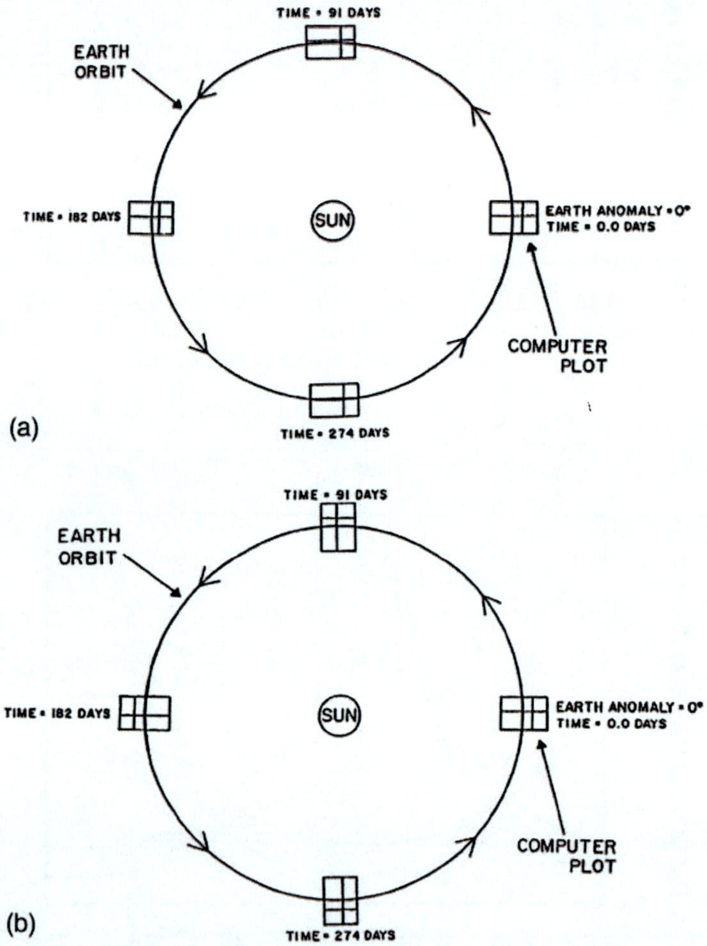

Fig. 8.41 Some orientation information for understanding the computer plots. (**a**) Diagram illustrating a non-rotating coordinate system for plotting the results of the numerical simulations. The computer plot does not rotate as it theoretically revolves around the Sun on a yearly cycle (~365.25 days). Planet Earth is at the origin of the non-rotating plot. (**b**) Diagram illustrating a rotating coordinate system for plotting the results of the numerical simulations. In this case the computer plot rotates as it theoretically revolves around the Sun. Special symmetry information can be gained from this type of rotating plot but the plotted information is more difficult to understand. Thus all plots in this chapter use the non-rotating coordinate system (unless stated otherwise) and all orbital plots are viewed from the north pole of the solar system

8.4.2 *Prograde Encounter and Capture Orientation: Inside Orbit*

Figures 8.42, 8.43, and 8.44 show calculated scenarios for prograde encounters for a lunar-mass planetoid with planet Earth from a heliocentric orbit that is slightly smaller than the orbit of Earth. Figure 8.42 features a one-orbit escape scenario in which no energy is dissipated in either body; Fig. 8.43 features a three-orbit escape scenario during which nearly enough energy is dissipated in the encountering planetoid for capture; and Fig. 8.44 shows a successful prograde capture scenario.

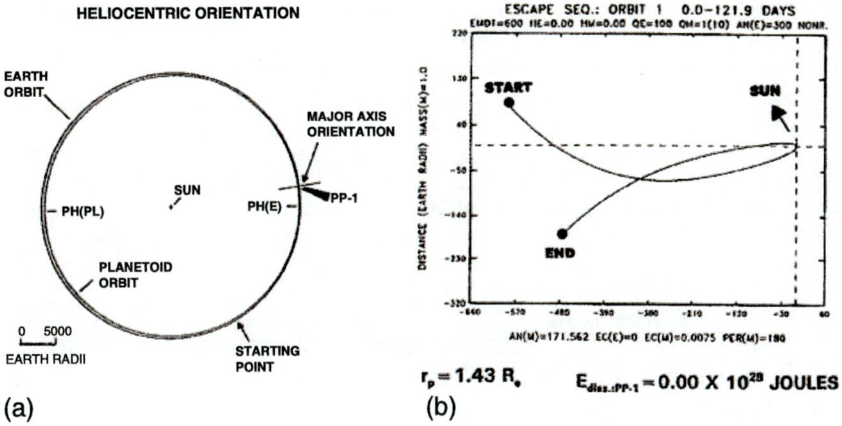

Fig. 8.42 Results of a close encounter in which no energy is dissipated. (**a**) Heliocentric orientation for the encounter. The simulation begins at the STARTING POINT. The close encounter occurs about 66 days later at PP-1 (perigee passage number 1). (**b**) Geocentric orientation for the encounter. Earth is at the origin of the plot. The *arrow* on the plot indicates the position of the Sun at the beginning of the simulation. The simulation begins at 600 earth radii which is well beyond the boundaries of the Hill sphere (which is about 200 earth radii). Stable prograde capture can occur only when the major axis of the orbit is within + or – 3 degrees of a line perpendicular to the orbit of the planet. This orientation characteristic is present in both diagrams above

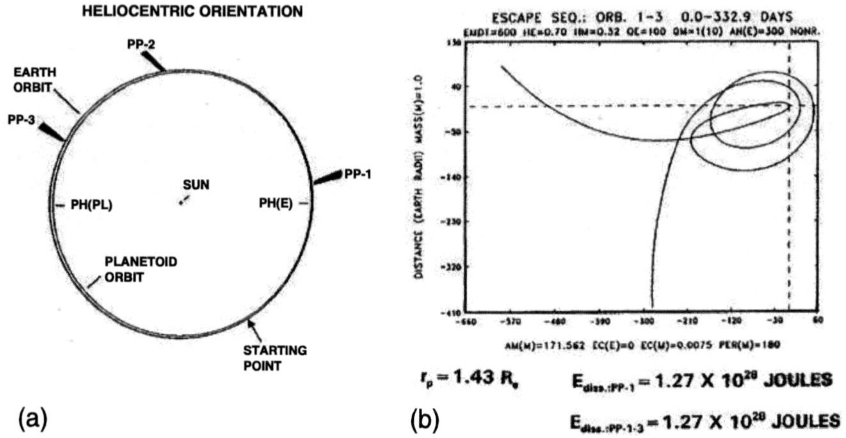

Fig. 8.43 Results of a three orbit escape scenario. (**a**) Heliocentric orientation showing the STARTING POINT and the positions of the three close encounters of this encounter sequence. (**b**) Geocentric plot showing the pattern of close encounters. Solar gravitational perturbations cause the variation in the orbital path of the planetoid. A significant quantity of energy is dissipated during the first encounter but not quite enough for stable capture. View is from the north pole of the solar system

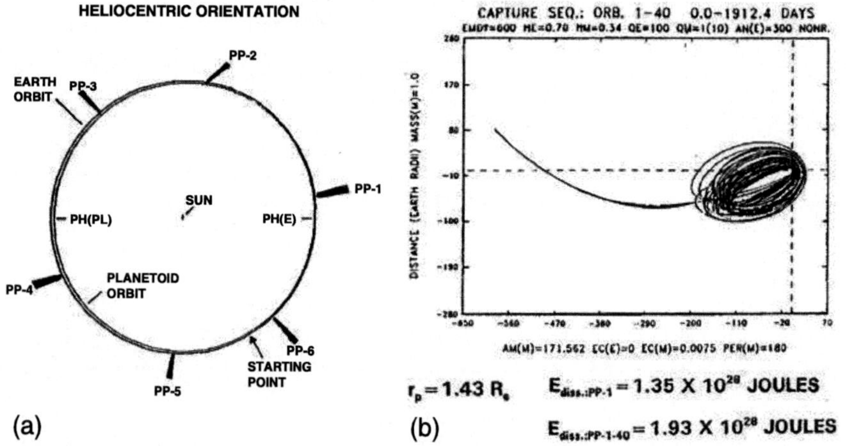

Fig. 8.44 Results of a stable capture scenario. (**a**) Heliocentric orientation showing the STARTING POINT and the positions of the first six perigee passasges. (**b**) Geocentric plot showing the orbital pattern of the first 40 orbits (1912.4 days, ~5.24 years). Note that dissipation of 1.35 E28 joules is sufficient for capture for this set of orbital parameters. View is from the north pole of the solar system

8.4.3 *Prograde Encounter and Capture Simulations: Outside Orbit*

Figures 8.45, 8.46, and 8.47 show calculated scenarios for prograde encounters for a lunar-mass planetoid with planet Earth from a heliocentric orbit that is slightly larger than the orbit of Earth.

Figure 8.45 features a one-orbit escape scenario in which no energy is dissipated in either body; Fig. 8.46 features a three-orbit escape scenario during which nearly enough energy is dissipated in the encountering planetoid for capture; and Fig. 8.47 shows a successful prograde capture scenario.

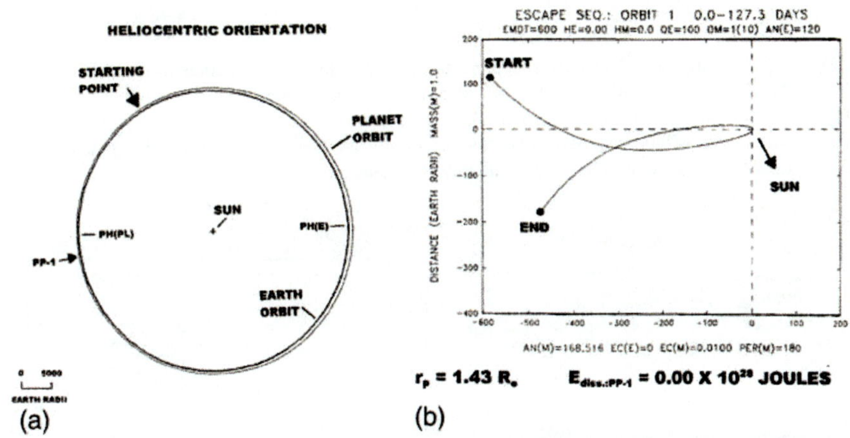

Fig. 8.45 Results of a close encounter in which no energy is dissipated. (**a**) Heliocentric orientation for the encounter. The simulation begins at the STARTING POINT. The close encounter occurs about 67 days later at PP-1 (perigee passage number 1). No energy is dissipated by tidal processes and the planetoid escapes to a heliocentric orbit about 100 days after PP-1. (**b**) Geocentric orientation for the encounter. Earth is at the origin of the plot. The *arrow* on the plot indicates the position of the Sun at the beginning of the simulation. The simulation begins at 600 earth radii which is well beyond the boundaries of the Hill sphere (which is about 200 earth radii). View is from the north pole of the solar system

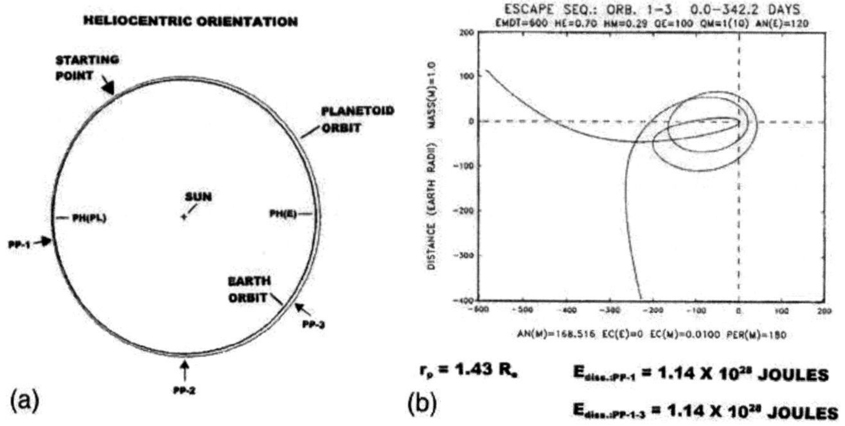

Fig. 8.46 Results of a 3-orbit escape sequence. (**a**) Heliocentric orientation showing the STARTING POINT and the positions of the three close encounters of this encounter sequence. (**b**) Geocentric plot showing the pattern of close encounters. Solar gravitational perturbations cause the variation in the orbital path of the planetoid. A significant quantity of energy is dissipated during the first encounter but not quite enough for stable capture

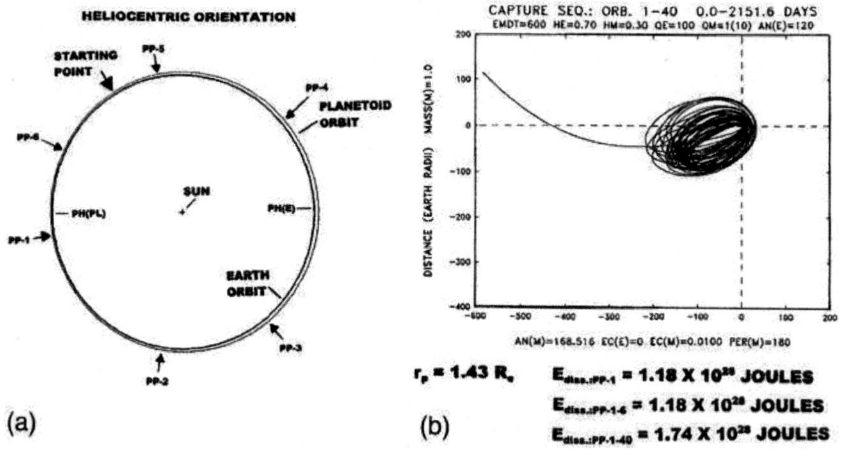

Fig. 8.47 Results of a stable prograde capture scenario. (**a**) Heliocentric orientation showing the STARTING POINT and the positions of the first six perigee passages. (**b**) Geocentric plot showing the orbital patterns of the first 40 orbits (2151.6 days, ~5.89 years). Note that 1.18 E28 joules is sufficient for capture for this set of orbital parameters

8.4.4 *Retrograde Encounter and Capture Simulations: Inside Orbit*

Figures 8.48, 8.49, 8.50, and 8.51 illustrate calculated retrograde scenarios for a lunar-mass planetoid with Earth from a heliocentric orbit that is slightly smaller than that of the planet. Figure 8.48 features a one-obit escape scenario in which no energy is dissipated; Fig. 8.49 features a 22-orbit escape scenario; Fig. 8.50 shows a retrograde encounter scenario in which the energy dissipation during the first encounter is not sufficient for capture and the planetoid has a collision encounter with the planet on the 14th encounter; Fig. 8.51 shows a stable retrograde capture scenario. Note that 0.99 E28 joules of energy dissipation is sufficient for stable capture.

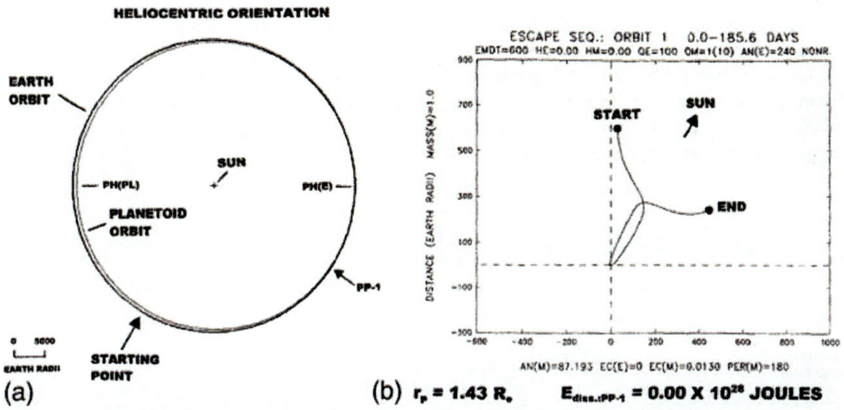

Fig. 8.48 Results of an encounter in which there no energy is dissipated. (**a**) Heliocentric orientation for the encounter. The simulation begins at the STARTING POINT. The close encounter occurs about 87 days later at PP-1 (perigee passage number 1). The simulation is ended after 185.6 days. (**b**) Geocentric plot showing the starting position, ending position, and the arrow points to the position of the Sun at the beginning of the simulation. View is from the north pole of the solar system

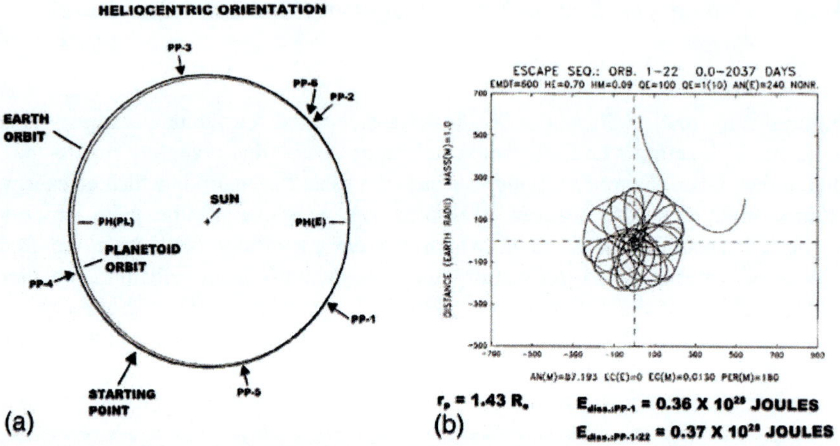

Fig. 8.49 Results of a 22-orbit escape scenario. (**a**) Heliocentric orientation showing the STARTING POINT and the positions of the first 6 perigee passages. (**b**) Geocentric plot showing the orbital patterns of a 22-orbit escape scenario. Note that a very small quantity of energy dissipation is associated with this 22-orbit escape scenario. Orbital resonance action is responsible for this apparent stability which eventually ends with an escape back into heliocentric orbit. View is from the north pole of the solar system

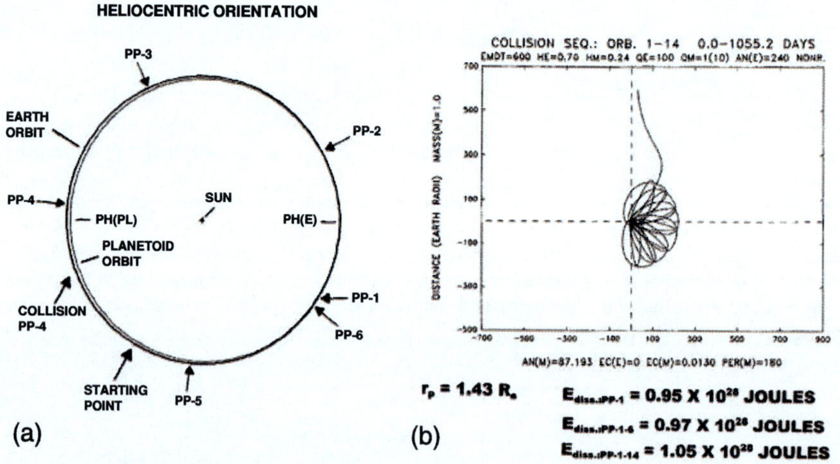

Fig. 8.50 Results of a 14-orbit collision. (**a**) Heliocentric orientation for this collision scenario. The positions of the first six perigee passages as well at the position of the collision (PP-14) are shown on the diagram. (**b**) Geocentric plot showing the orbital pattern for this collision scenario. Note that over 2.5 times more energy dissipation is associated with this collision scenario (not counting the energy associated with the collision itself) than for the 22-orbit escape scenario shown in Fig. 8.49. View is from the north pole of the solar system

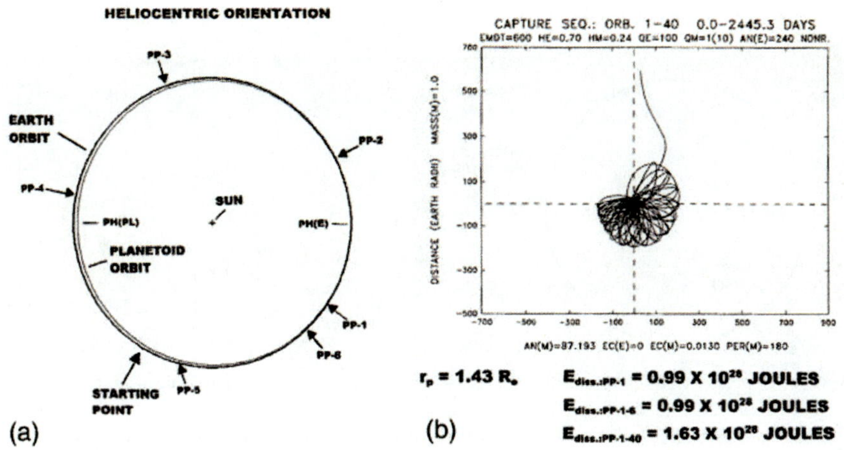

Fig. 8.51 A stable retrograde capture scenario. (**a**) Heliocentric orientation showing the STARTING POINT and the positions of the first 6 perigee passages. (**b**) Geocentric plot showing the pattern of the first 40 orbits (2445.3 day; ~6.70 years). View is from the north pole of the solar system

8.4.5 *Retrograde Encounter and Capture Orientation: Outside Orbit*

Figures 8.52, 8.53, and 8.54 show calculated scenarios of retrograde capture from a heliocentric orbit that is slightly larger than that of the orbit of the earth-like planet. Figure 8.52 shows an encounter scenario in which no energy is dissipated. Figure 8.53 shows a scenario in which some energy is dissipated during the initial encounter but not nearly enough for capture. The morphology of the escape orbit in the case of no energy dissipation (Fig. 8.52) looks very similar to that of Fig. 8.48 but the orientation of the starting points in both the heliocentric plot and the geocentric plots are much different. The result is a 3-orbit escape scenario in which the planetoid goes back into a heliocentric orbit after a prograde encounter with Earth. In Fig. 8.54 there is sufficient energy dissipation for stable retrograde capture.

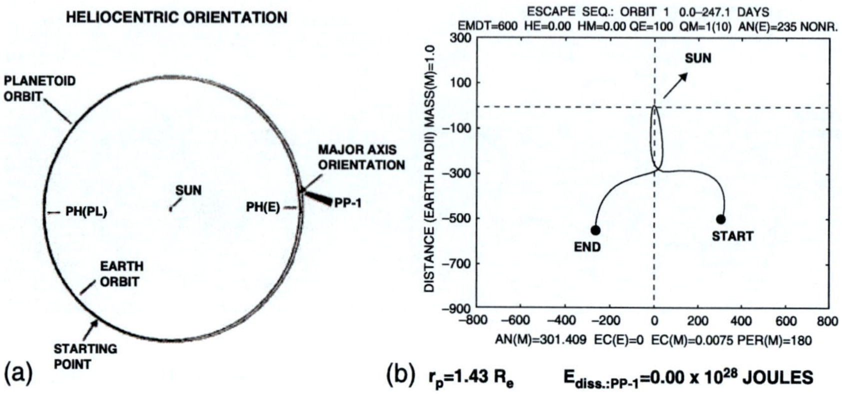

Fig. 8.52 Results of an encounter in which no energy is dissipated. (**a**) Heliocentric orientation showing the STARTING POINT and the position of the close encounter at PP-1. (**b**) Geocentric plot for the encounter. The starting position and ending positions are shown as well as the position of the Sun at the beginning of the simulation. Note that the major axis of the encounter must be within about + or – 3 degrees of a line parallel to the tangent of the orbit of the planet. View is from the north pole of the solar system. Note that the sun angle is very similar for the retrograde encounters in Figs. 8.48b and 8.52b. As shown in Fig. 8.40b, these two orbital scenarios are in different SCZs

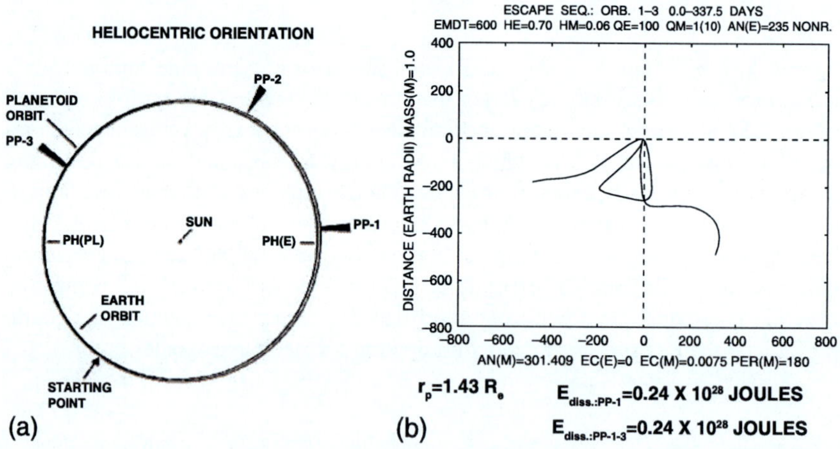

Fig. 8.53 Results of a three-orbit escape scenario. (**a**) Heliocentric orientation for this escape scenario showing the STARTING POINT and the position of the 3 perigee passages. (**b**) Geocentric plot showing the pattern of close encounters. Note that the planetoid escapes in the prograde direction on its way back into a heliocentric orbit. View is from the north pole of the solar system

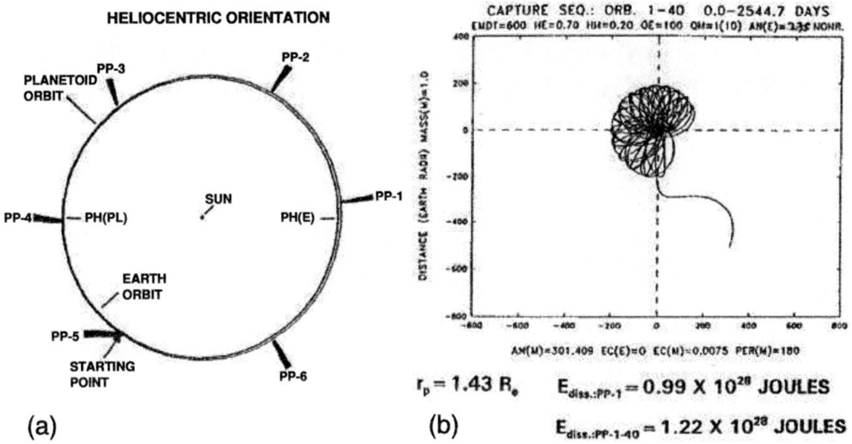

Fig. 8.54 Results of a stable retrograde capture scenario. (**a**) Heliocentric orientation showing the STARTING POINT as well as the positions of the first six perigee passages for this simulation. (**b**) Geocentric plot showing the orbital pattern for this stable capture scenario. View is from the north pole of the solar system

8.4.6 Summary for Earth-Luna Simulations

This set of simulations relates to our planet. The useful part of the prograde SCZs are much more extensive than the counterparts for planet Venus and the upper right portion of the prograde SCZs feature very low encounter speeds that are characteristic of very successful capture scenarios. Nonetheless, the retrograde SCZs for the Earth-Luna encounters are far greater in lateral extent and feature much lower displacement Love numbers for successful capture than in the case for the Venus-Adonis system.

When exploring exoplanet systems somewhat similar to our solar system, we could expect to see (1) a planet with no satellite (a source for candidates for capture is not assured and the probability of capturing a candidate planetoid is not favorable), (2) a planet with a prograde post-accretion rotation rate and with a satellite in prograde orbit (the lower probability option according to the geometry of SCZs) at some stage of development, and (3) a planet with a large satellite in retrograde orbit in some stage of development: for example (a) a planet with a slow prograde rotation rate and a retrograde satellite, (b) a planet with a slow retrograde rotation rate and a satellite in retrograde orbit, or (3) a planet with no satellite but with a massive carbon dioxide atmosphere, a global resurfacing with basalt, and a high surface temperature (i.e., in planet Venus conditions). IN GENERAL, THE CURRENT CONDITION OF THE EARTH-MOON SYSTEM APPEARS TO BE THE LOW PROBABILITY OPTION.

8.5 Note: The Situation for Capture of Planetoids by Planet Mars Was Covered Extensively in Chaps. 4 and 5. It Would Be Redundant to Repeat this Information in this Chapter

Side-Bar on the Use of the Rotating Coordinate System for Determining Boundaries of Stable Capture Zones

This note is to emphasize the utility of the rotating coordinate system for determining the geometry of stable capture zones. In each of the pairs of simulations in Figs. 8.55, 8.56, 8.57, 8.58, 8.59, and 8.60, the left plot is the non-rotating coordinate system and the right plot is the rotating coordinate system. Using only the non-rotating coordinate system, defining the Stable Capture Zones would be very difficult. The process of finding the boundaries of the SCZs would be mainly a "trial-and-error" process. For each change in the initial parameters of an encounter, one would need to do several trial encounters by changing the displacement Love number to see which scenarios would result in stable capture. Geometric clues would be very unreliable.

Using the rotating coordinate system, one can predict, fairly accurately, which initial parameters will result in stable capture scenarios if sufficient energy is dissipated within the interacting bodies. The geometry of the encounter plot, with no energy dissipation, gives a strong indication of which encounters will result in stable capture scenarios.

For prograde encounters in the rotating coordinate system (Figs. 8.55b, 8.56b, 8.57b, and 8.58b) the incoming and outgoing limbs of the encounter must be nearly symmetrical relative to the x-axis, usually within + or – 3 degrees. Then one can be fairly sure that the encounter scenario will result in

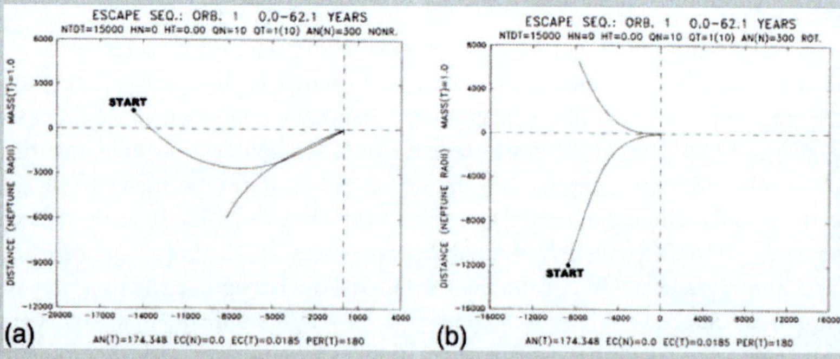

Fig. 8.55 Results of a one-orbit prograde encounter between Neptune and a Triton-mass planetoid from an orbit that is inside the orbit of Neptune. (**a**) Non-rotating coordinate system. (**b**) Rotating coordinate system. Note that this orientation is the signature of stable prograde capture when the planetoid is encountering the planet from an orbit that is inside the orbit of the planet (in this case Neptune). This is the lower limit of the SCZ for this planet anomaly

(continued)

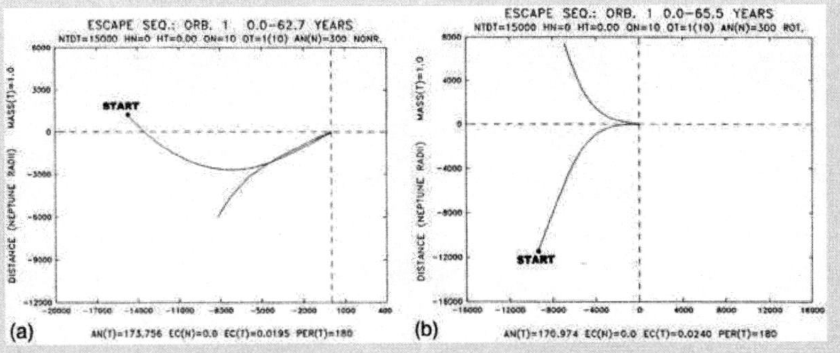

Fig. 8.56 Results of a one-orbit prograde encounter between Neptune and a Triton-mass planetoid. (**a**) Non-rotating coordinate system. (**b**) Rotating coordinate system. This is an intermediate orientation which is well within the SCZ for this planet anomaly. The planetoid is encountering the planet from an orbit that is inside the orbit of the planet

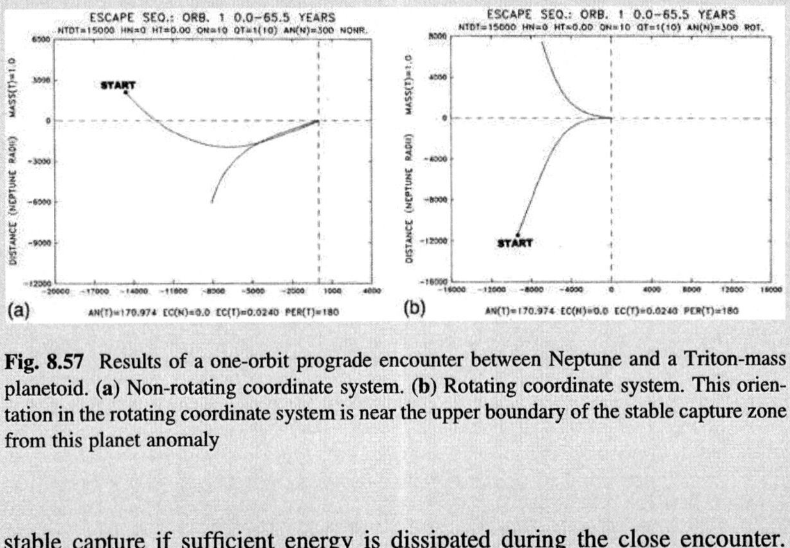

Fig. 8.57 Results of a one-orbit prograde encounter between Neptune and a Triton-mass planetoid. (**a**) Non-rotating coordinate system. (**b**) Rotating coordinate system. This orientation in the rotating coordinate system is near the upper boundary of the stable capture zone from this planet anomaly

stable capture if sufficient energy is dissipated during the close encounter. After a few "trial-and error" runs, one can determine the boundaries of the SCZs and "template" plots can be used to compare and contrast the results of future plots of encounter simulations. This geometric analysis is useful for any planet-planetoid combination (e.g., Neptune-Triton, Earth-Moon, Mars-0.1 moon-mass planetoid, etc.).

(continued)

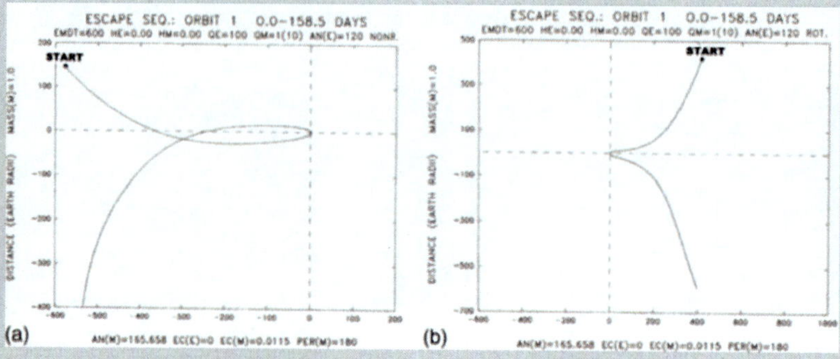

Fig. 8.58 Results of a one-orbit prograde encounter between Earth and a lunar-mass planetoid in which the planetoid orbit is slightly larger than that of the planet (i.e., outside orbit). (**a**) Non-rotating coordinate system. (**b**) Rotating coordinate system. Note that the signature of a planetoid encounter from an orbit that is larger than that of the planet is just the opposite as that for an orbit that is smaller than that of the planet

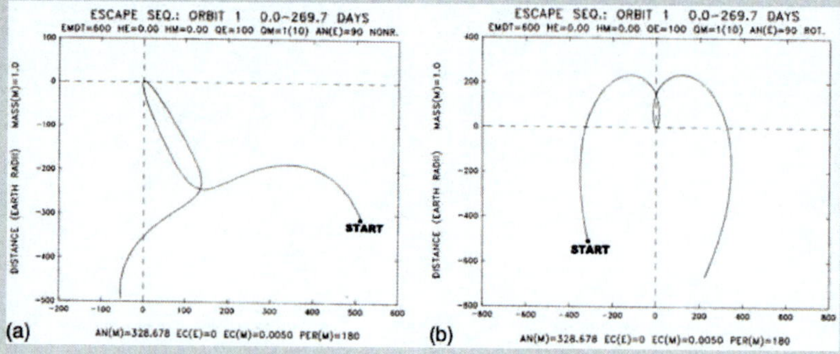

Fig. 8.59 Results of a one-orbit retrograde encounter between Earth and a lunar-mass planetoid in which the planetoid orbit is slightly smaller than that of the planet (i.e., inside orbit). (**a**) Non-rotating coordinate system. (**b**) Rotating coordinate system. Note that the orbital limbs of the close approach are symmetrical about the vertical axis. Also, since the alignment is very close to the vertical axis, this orientation is near the center of the SCZ for this particular planetoid amomaly

For retrograde encounters in the rotating coordinate system (Figs. 8.59b and 8.60b), the geometry of the encounter must be symmetrical about the y-axis, within about + or – 3 degrees. Then encounter scenarios can be run and templates made to help map out the boundaries of the retrograde SCZs. Again, the geometry of the plots is useful for any planet-planetoid combination.

(continued)

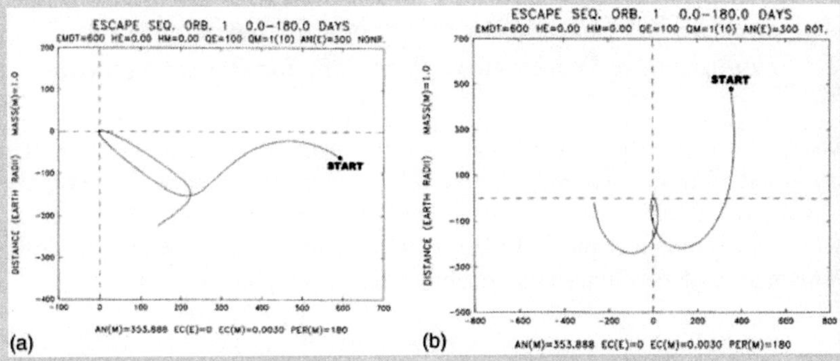

Fig. 8.60 Results of a one-orbit retrograde encounter between Earth and a lunar-mass planetoid in which the planetoid orbit is slightly larger than that of the planet (i.e., outside orbit). (**a**) Non-rotating coordinate system. (**b**) Rotating coordinate system. Note the opposite orientation of the encounter in (**b**) versus Fig. 8.59b. Also note that the symmetry of the orbital limbs for the encounter is displaced to the right. This orientation for this particlular planet anomaly is the lower planetoid eccentricity boundary for the SCZ

8.6 Planet Neptune

Planet Neptune is a large planet near the outer edge of the Solar System. The odd feature for Neptune is that its largest satellite (Triton) is in a retrograde orbit about the planet and the other satellites of Neptune are in prograde orbits. Most concerned planetary scientists agree that Triton is a captured satellite. The capture mechanisms suggested for this retrograde satellite are (1) collisional capture and (2) gravitational capture.

For the collisional capture scenario Triton would have an essentially head-on collision with another body, perhaps a previously existing satellite, as Triton is having a close retrograde encounter with Neptune. The collision, in this case, is the energy sink for capture. Following the collision the orbit of Triton is changed from a near Neptune-like heliocentric orbit to an elliptical neptocentric orbit which in time evolves into the circular retrograde orbit that Triton has today.

For the gravitational capture scenario, the initial encounter of the capture scenario must be close enough (within 1.10 neptune radii) to raise sufficiently high body tides on the encountering body to dissipate sufficient orbital energy to keep the encountering body well within the Hill sphere for Neptune. After a lengthy period of orbit circularization, Triton attains the circular retrograde orbit that it has today. At present the circular retrograde orbit of Triton has a semi-major axis of ~15 neptune radii and an inclination of ~20 degrees.

8.6.1 *Two-Body Representation, Stable Capture Zones, and Orbital Orientation Diagrams for Gravitational Capture of a Triton-Mass Planetoid for Planet Neptune*

Figure 8.61 is a two-body scale sketch of the geometry of a capture scene for planet Neptune and a Triton-mass body. Figure 8.62 depicts the geometry of the prograde and retrograde stable capture zones for the two bodies. Figure 8.63 consists of sketches of the computer plots of the non-rotating and rotating coordinate systems used to analyze plots of planetary encounter and capture interactions.

8.6.2 *Prograde Encounter and Capture Simulations: Inside Orbit*

Figures 8.64, 8.65, 8.66 and 8.67 illustrate a series of calculated scenarios between a triton-mass planetoid and a neptune-mass planet. In this case the Hill sphere is large (~4500 neptune radii) and the planetoid orbits can move through much space. Figure 8.64 illustrates a progrde encounter from an inside orbit in which no energy

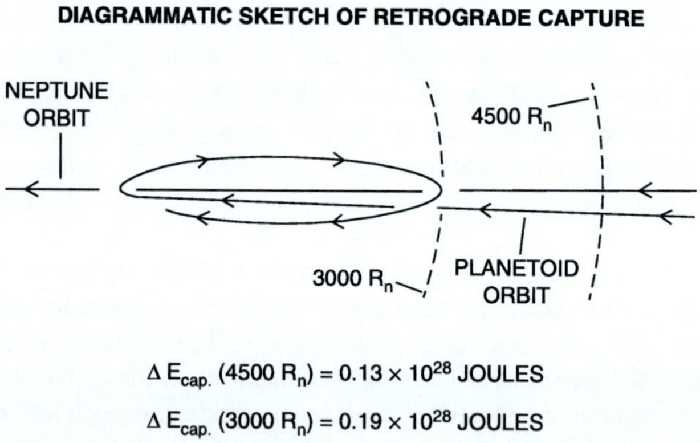

Fig. 8.61 Diagrammatic sketch of two-body capture of a triton-mass body from a Neptune-like heliocentric orbit into a retrograde orbit. Note that if the capture is into a large major axis orbit (e.g., 4500 R_n), then a smaller quantity of energy needs to be dissipated within the interacting bodies. Capture into the smaller orbit (e.g., 3000 R_n) would result in a more stable orbit relative to solar and planetary gravitational perturbations

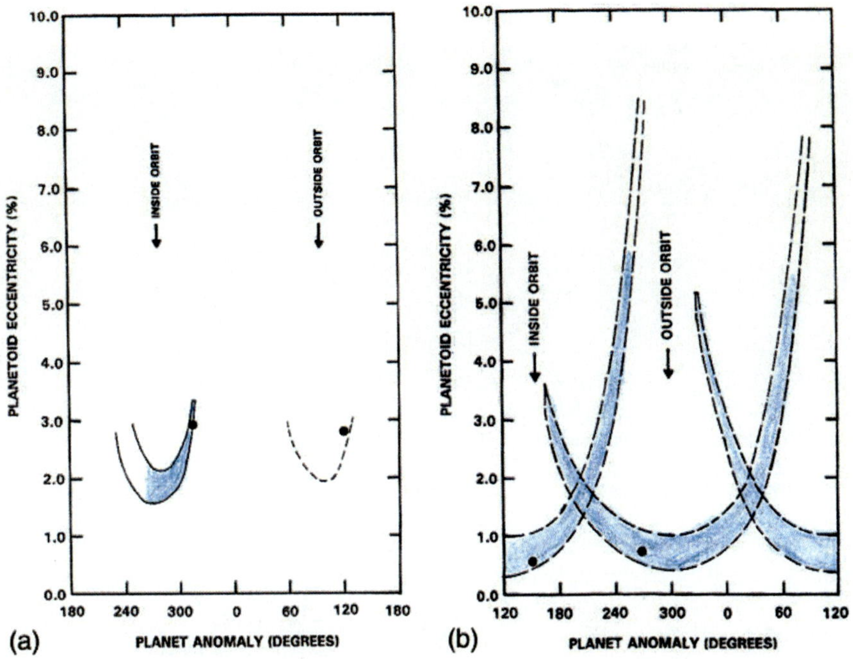

Fig. 8.62 Geometry of prograde and retrograde SCZs for planet Neptune and Triton mass planetoid. (**a**) Note the limited extent of the prograde SCZs. The inside prograde SCZ is limited but defined but the outside SCZ is "phantom" in that a stable capture orientation cannot be attained. (**b**) In contrast, the retrograde SCZs are large, up to 5.0% planetoid eccentricity. However, the most useful part of the retrograde SCZs is between 0.5% and 1.5% for orbits that are inside the orbit of Neptune or outside the orbit or Neptune

is dissipated. There is some energy dissipated in Fig. 8.65 (0.042 E28 joules) which changes the morphology of the orbit in a minor way. Figure 8.66 shows a 5-orbit scenario (e-diss = 0.048 E28 joules) with a solid collision on the fifth encounter. Figure 8.67 illustrates a 3-orbit escape scenario (e-diss = 0.053 E28 joules) and Fig. 8.68 shows a stable prograde capture scenario (e-diss = 0.058 E28 joules on the first encounter and 0.059 E28 after 40 orbits).

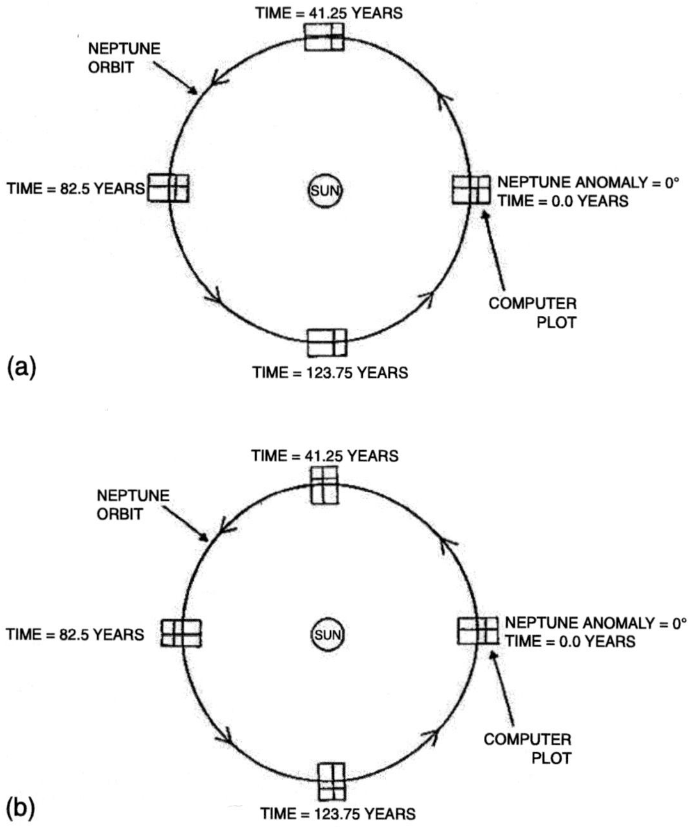

Fig. 8.63 Coordinate systems for plotting the results of interactions between Neptune and Triton. Note that one trip around the sun for Neptune is ~165 earth years. (**a**) Diagram illustrating the non-rotating coordinate system for plotting the results of the numerical simulations. (**b**) Diagram illustrating a rotating coordinate system for plotting the results. In the rotating case the computer plot rotates as it theoretically revolved about the sun. (Note that the importance of the rotating coordinate system was discussed in a sidebar earlier in this chapter)

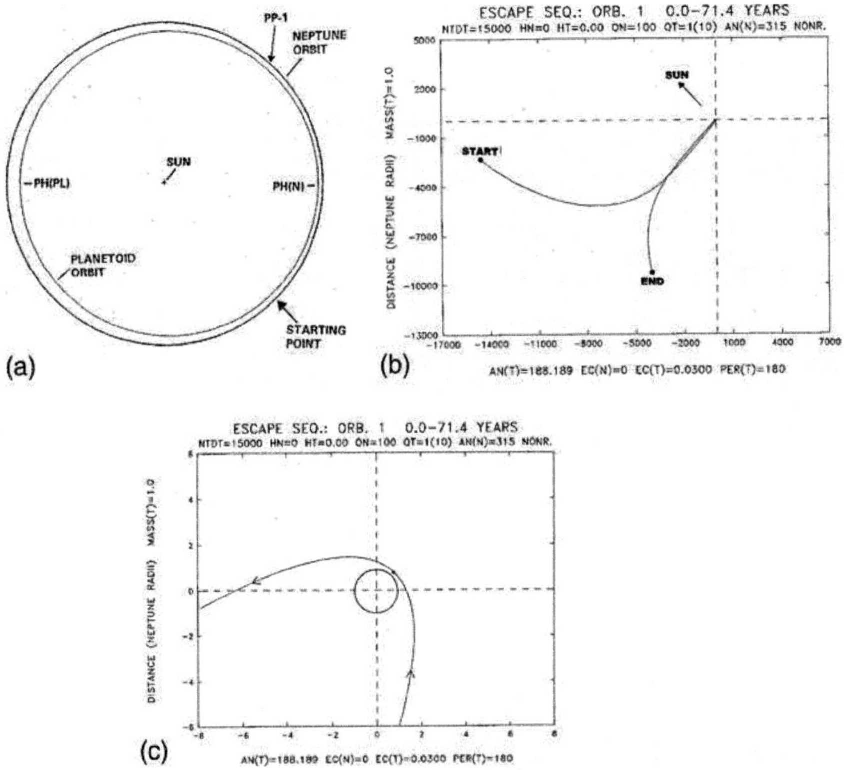

Fig. 8.64 A one-orbit prograde escape scenario. (**a**) Heliocentric orientation showing the STARTING POINT for the simulation and the position of the first perinept passage (about 42.6 earth years). (**b**) Neptocentric orientation for the numerical simulation. The *arrow* points to the position of the sun as the simulation begins. The simulation is terminated after 71.4 earth years. (**c**) Close-up of this initial encounter. The surfaces of the two bodies are separated by ~2000 km but since Neptune has a large gaseous envelope, a minor "collision" would entail a trip through some clouds

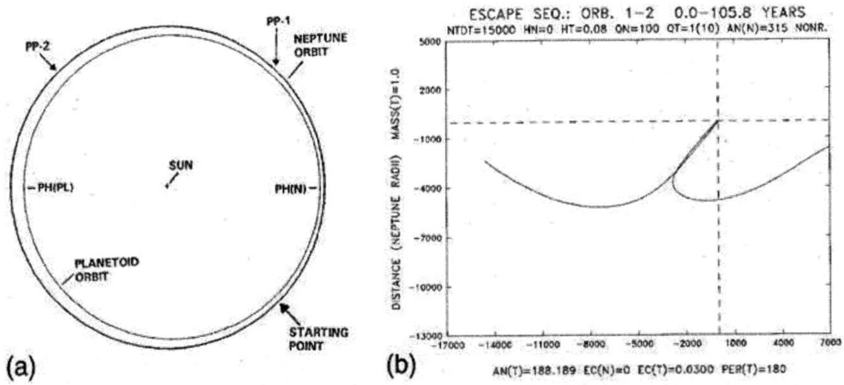

Fig. 8.65 A two-orbit escape scenario. (**a**) Heliocentric orientation showing the STARTING POINT and the positions of the two perigee passages. (**b**) Neptocentric orientation for the simulation. Note that a small quantity of energy dissipation within the triton-like planetoid causes significant change in the morphology of the outgoing limb of the encounter

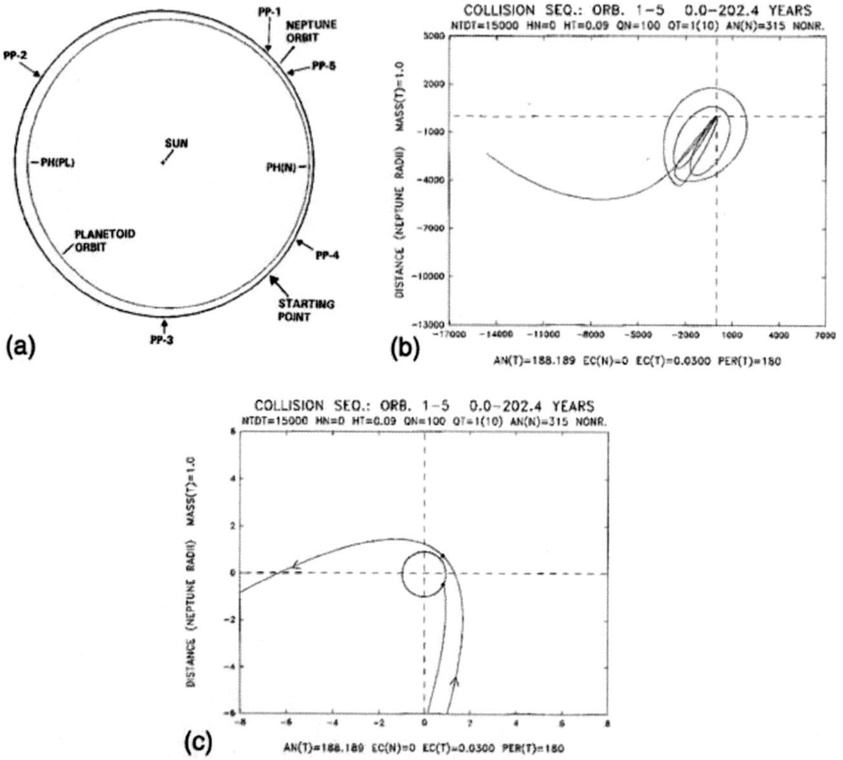

Fig. 8.66 Results of a 5-orbit collision scenario. (**a**) Heliocentric orientation for the scenario. The position of the STARTING POINT, the four non-collision encounters as well as the collision encounter (PP-5) are shown on this diagram. (**b**) Overview of the results of the orbital simulation. The irregularity of the orbital geometry is caused by solar perturbations. (**c**) Close-up view showing perinept passage 1 (the initial encounter of the encounter sequence) and the collision encounter (PP-5). This is a "solid" collision in the prograde direction and the planetoid would lose its identity and be absorbed within the mass of Neptune

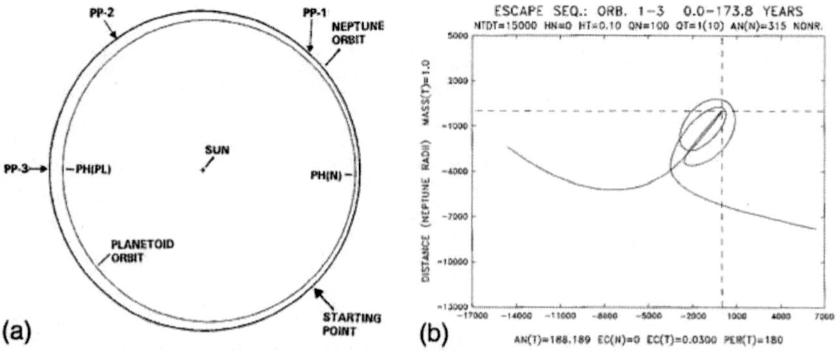

Fig. 8.67 Results of a three-orbit escape scenario. (**a**) Heliocentric orientation for the simulation. The STARTING POINT as well as the positions of the three perinept passages are shown on the diagram. (**b**) After PP-3 the planetoid escapes to a slightly altered heliocentric orbit

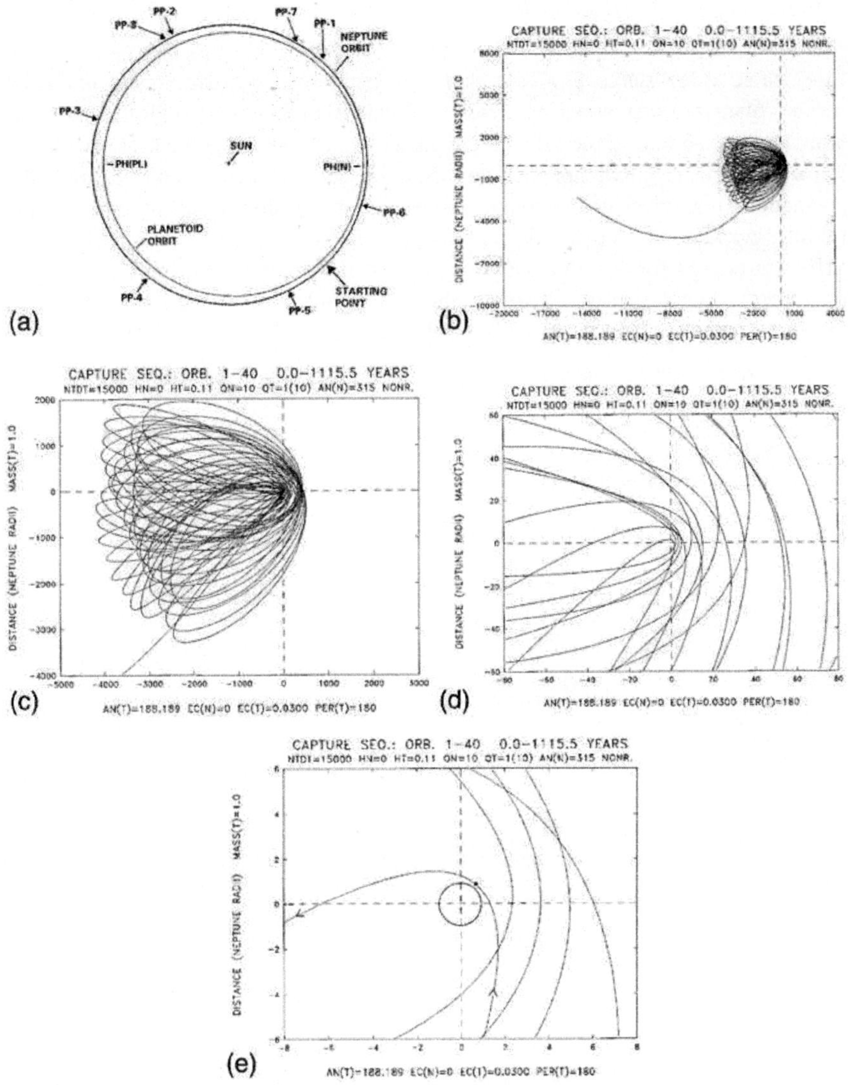

Fig. 8.68 Results of a stable capture scenario. (**a**) Heliocentric orientation for the stable capture scenario. The positions of the STARTING POINT and the first 8 perinept passages are shown on the diagram. Only a small quantity of energy dissipation was needed to change the three-orbit escape scenario into a stable capture scenario. The displacement Love number for the Triton-like planetoid for the five-orbit collision was 0.09, for the 3-orbit escape scenario was 0.10, and for the 40 orbit capture scenario was 0.11. Thus, a small quantity of energy dissipation, in some cases, makes a major change in the outcome of the scenario. (**b**) Overview of the neptocentric orientation for the encounter. (**c**) Closer view of all 40 orbits of this scenario. It is obvious that there is some gravitational resonance action with the sun involved in this simulation. (**d**) Close-up of the pattern of perinept passages. (**e**) Close-up of the initial encounter (with the arrows indicating the direction of motion) and 4 other encounters that are in the view of this small frame. All of the perinept passages in this simulation are in the prograde direction

8.6.3 Prograde Encounter Simulations: Outside Orbit

Since there is no stable prograde capture orientation for a triton-mass planetoid (or any planetoid) encountering a Neptune-mass planet from an orbit that is somewhat larger than that of the orbit of the planet, we can examine the best orientation knowing very well that none of the scenarios will result in stable capture. All encounters, regardless of the quantity of energy dissipation, result in either escape (if little energy is dissipated during the first encounter) or an orbital collision if sufficient energy for a normal capture scenario is dissipated. Figure 8.69 shows a one-orbit escape scenario in a non-rotating coordinate system and Fig. 8.70 shows the same calculation in a rotating coordinate system. In reference to the plot in

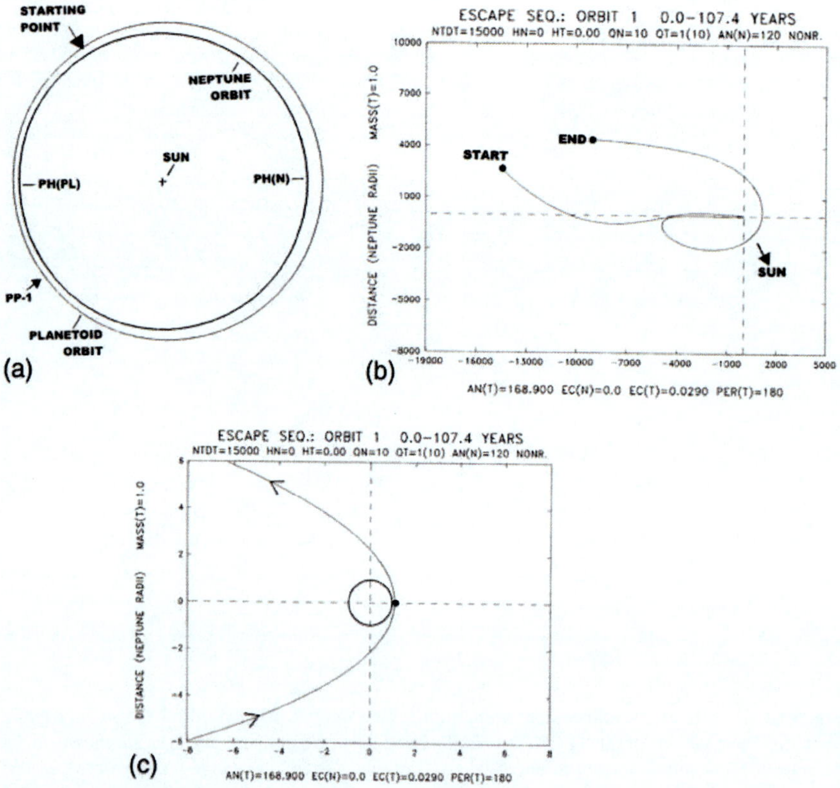

Fig. 8.69 Results of a 1-orbit escape scenario. (**a**) Heliocentric Orientation for the encounter. The STARTING POINT as well as the first perinept passage are shown on the diagram. There is a distant prograde perinept passage associated with this scenario, but it is quite distant from Neptune. (**b**) Neptocentric orientation of the initial encounter with the arrows indicating a prograde orientation. The problem with this encounter orientation is that it is NOT associated with a stable capture zone. This SCZ is undefined so there will be no stable capture orientations

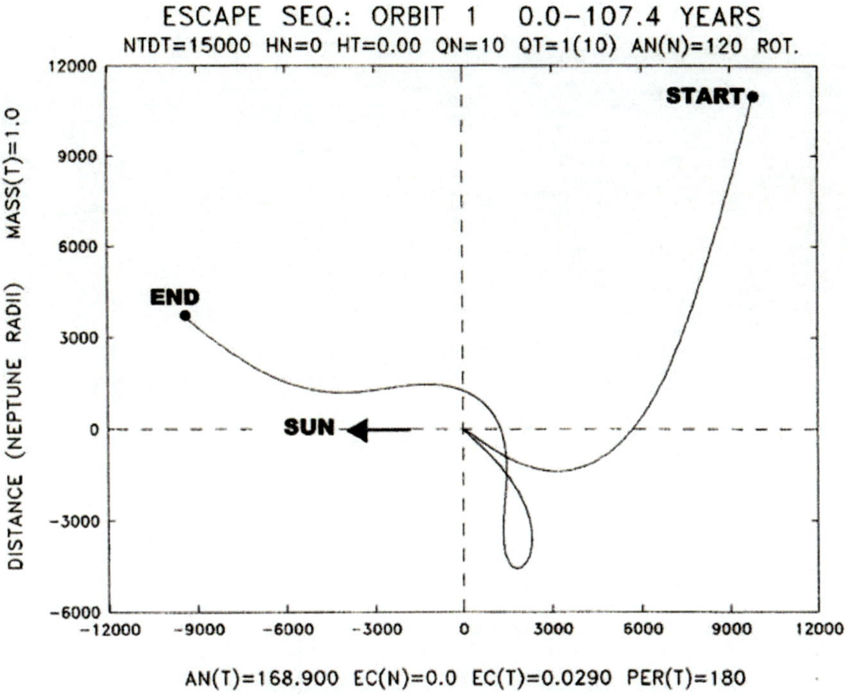

Fig. 8.70 Rotating coordinate system for the Fig. 8.69b The position of the Sun is indicated on the plot. Note that in the rotating coodinate system plot the Sun is always in the same position. The beginning and end of the encounter simulation are marked on the plot. The orientation of the encounter is in the vicinity of a stable orientation for an encounter from an inside orbit in the prograde orientation, but it is off by several degrees. Thus is is predictable that such an encounter will result in either ESCAPE or COLLISION (i.e., STABLE CAPTURE IS NOT AN OPTION)

Fig. 8.70, a stable capture orientation for an encounter from an outside orbit should be symmetrical to a line parallel to the x-axis, within a few degrees, that goes through the origin of the plot. Clearly the plot in Fig. 8.70 is off from that position by some 40 degrees. Figure 8.71 depicts a scenario in which the orbit shifts from prograde to retrograde and finally escapes from the planet into a prograde heliocentric orbit after a prograde encounter with Neptune. In Fig. 8.72 more energy is dissipated in the Triton-like planetoid and the result is a collision in the retrograde direction on the fifth encounter.

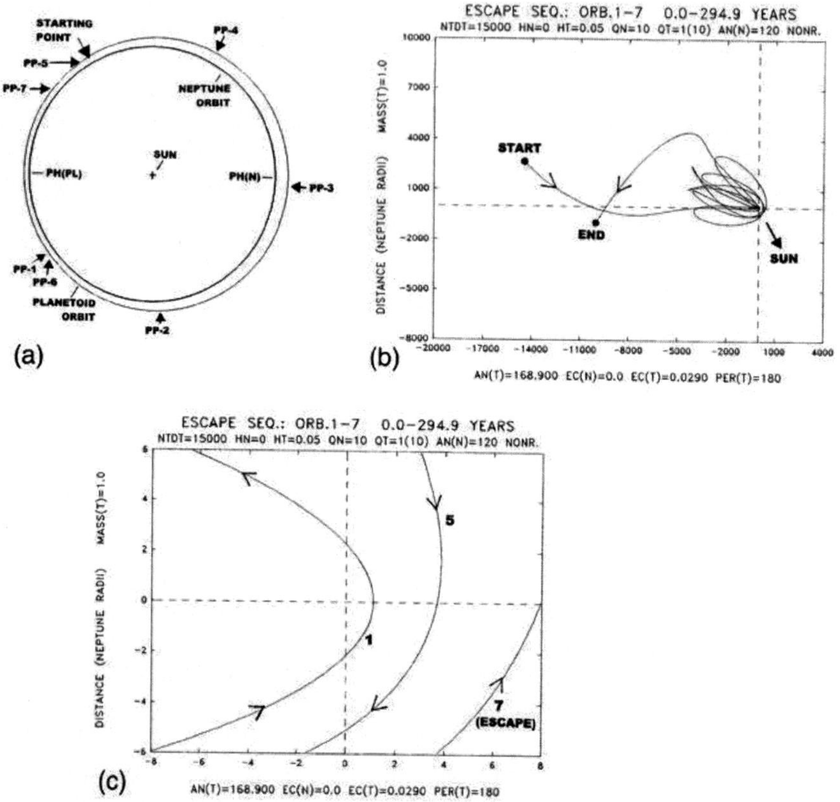

Fig. 8.71 Results of a 7-orbit escape scenario for this orbital orientation that is well outside a stable capture zone (in this case where the SCZ is undefined). (**a**) Heliocentric orientation for this 7 orbit collision scenario. The STARTING POINT as well as the positions of the 7 perinept passages are shown on the diagram. (**b**) Overview of this complex orbital scenario featuring several changes from prograde to retrograde orientation. (**c**) Close-up of the initial perinept passage (PP-1), PP-5 (retrograde), and PP-7 (prograde escape)

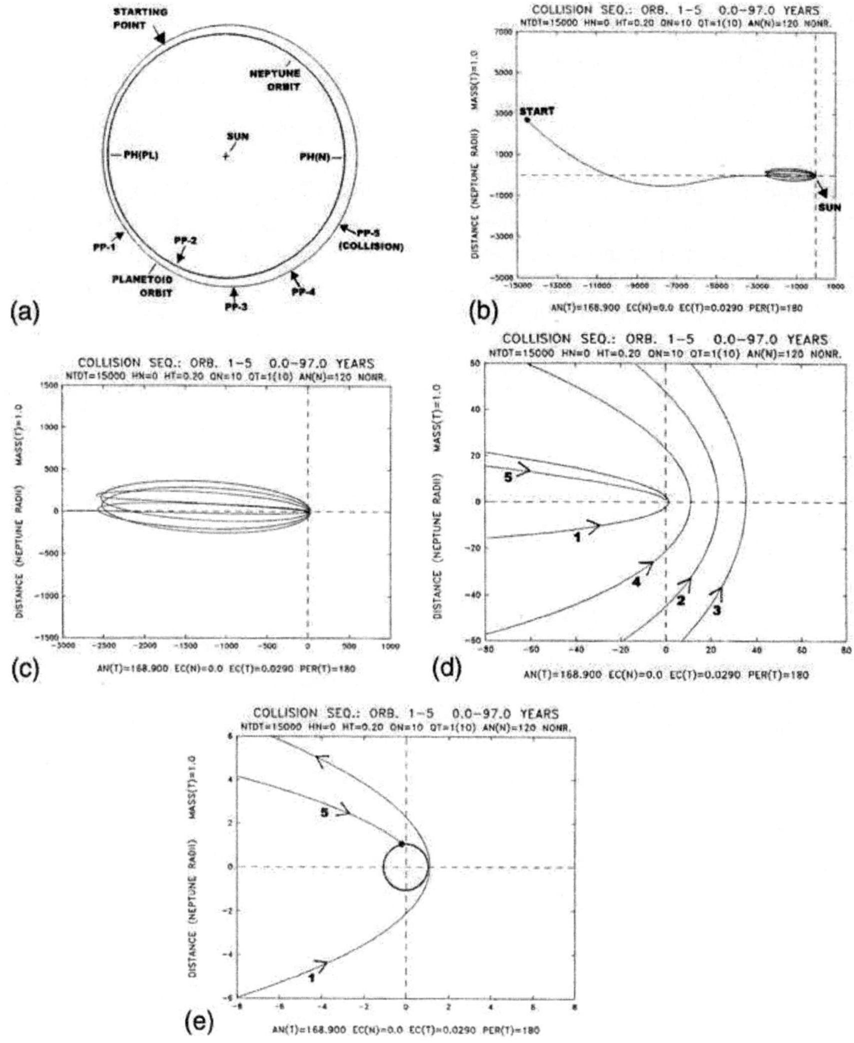

Fig. 8.72 Results of a 5-orbit collision scenario during which there is sufficient energy dissipated for insertion into a "capture" orbit, but the orientation is not conducive to capture. (**a**) Heliocentric orientation showing the STARTING POINT as well as the positions of the 5 perinept passages. Note that all of this action occurs in a little over one-half a Neptune year). (**b**) Neptocentric orientation of the orbital scenario. (**c**) A closer view of this unstable orbit scenario. (**d**) An intermediate view of the pattern of close encounters. Note that all of the encounters are in the prograde direction but the last one, the collision encounter, is notably retrograde. (**e**) Close-up view of the encountering planetoid with planet Neptune. The collision on orbit 5 is a "solid" collision and the planetoid would not be expected to survive. It would be absorbed by planet Neptune

8.6.4 *Retrograde Encounter and Capture Simulations: Inside Orbit*

Figures 8.73, 8.74, 8.75, and 8.76 show a sequence of simulations of retrograde encounters from an orbit that is slightly smaller than the orbit of the planet. This is an encounter orientation that will result in stable retrograde capture if sufficient energy is dissipated in the encountering planetoid. In Fig. 8.73 no energy is dissipated and the planetoid escaped back into heliocentric orbit after one close encounter. In Fig. 8.74 insufficient energy is dissipated for capture and the planetoid escapes back into heliocentric orbit after 11 orbits and it escapes in the prograde direction

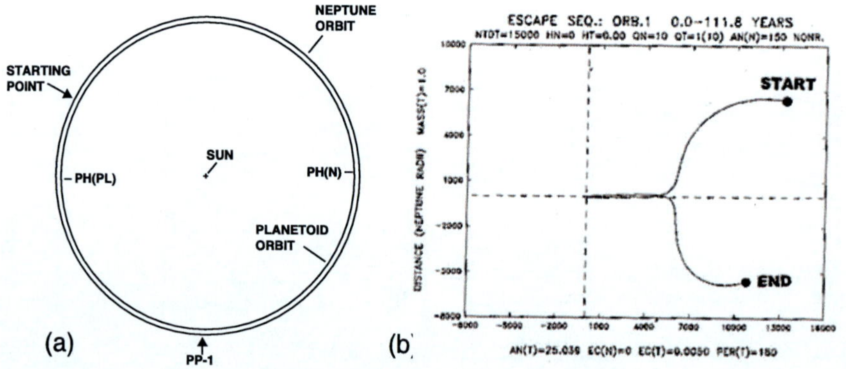

Fig. 8.73 One orbit retrograde escape scenario. (**a**) Heliocentric orientation showing the STARTING POINT and the position of the first perinept passage. (**b**) Neptocentric orientation of the encounter

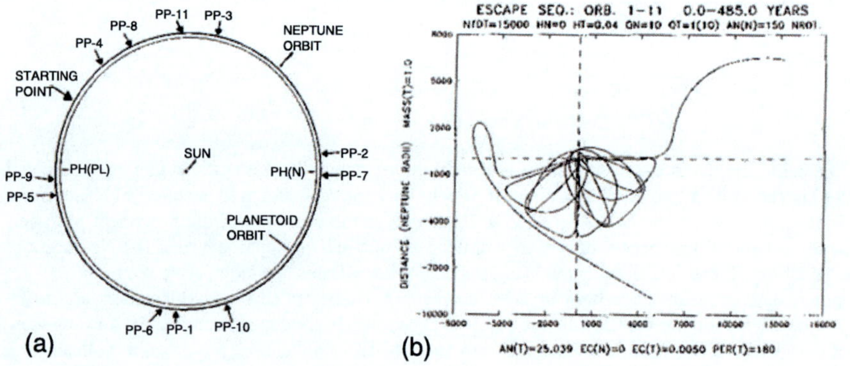

Fig. 8.74 An 11-orbit escape scenario. (**a**) Heliocentric orientation of the orbital scenario showing the STARTING POINT and the positions of the 11 perinept passages. (**b**) Neptocentric plot showing an overview of the orbital scenario. Note that the planetoid escapes back into heliocentric orbit in the prograde direction relative to Neptune

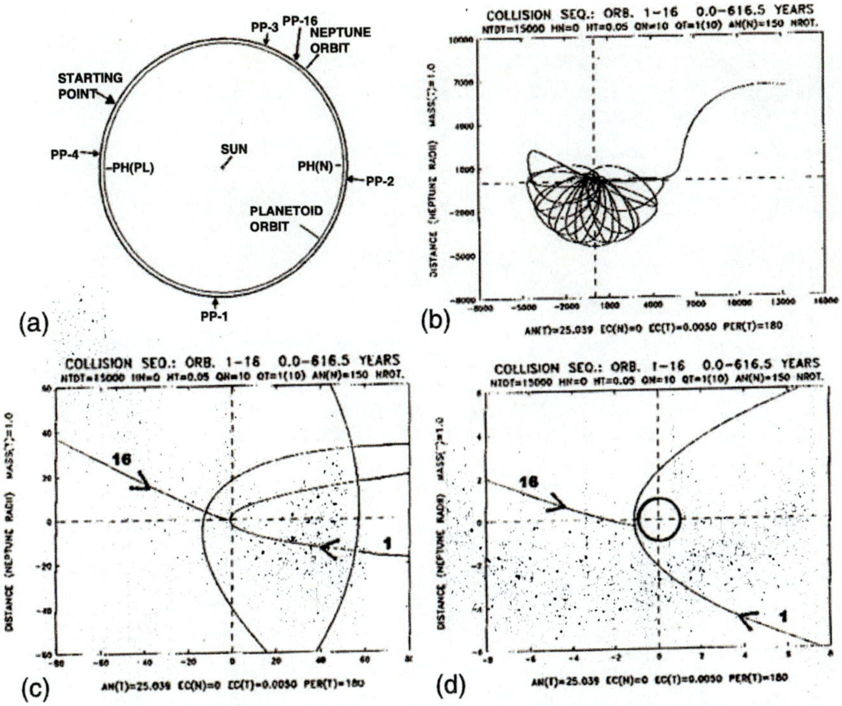

Fig. 8.75 A 16-orbit collision scenario. (**a**) Heliocentric orientation showing the STARTING POINT and the position of first 4 perinept passages as well as the heliocentric position of the collision on PP-16. (**b**) Neptocentric orientation of the orbital scenario. (**c**) A closer view featuring the initial retrograde perinept passage as well as the collision in the prograde direction. (**d**) Neptocentric view of PP-1 and the prograde collision of PP-16. This is nearly a straight-in collision

relative to the planet. Figure 8.75 shows a 16-orbit collision scenario and the collision is in the prograde direction. Figure 8.76 depicts a stable capture scenario of 20 orbits with the h = 0.10 for the planetoid. This is the minimum value of h for capture and the energy dissipated for capture is 0.050 E28 joules. Figure 8.77 shows a stable capture scenario of 40 orbits with an h value of the planetoid at 0.20 and 0.101 E28 joules of energy are dissipated during the first encounter. The major axis soon after capture for the scenario with the lower h value is about 4000 neptune radii and the scenario with the higher h value is about 3000 neptune radii.

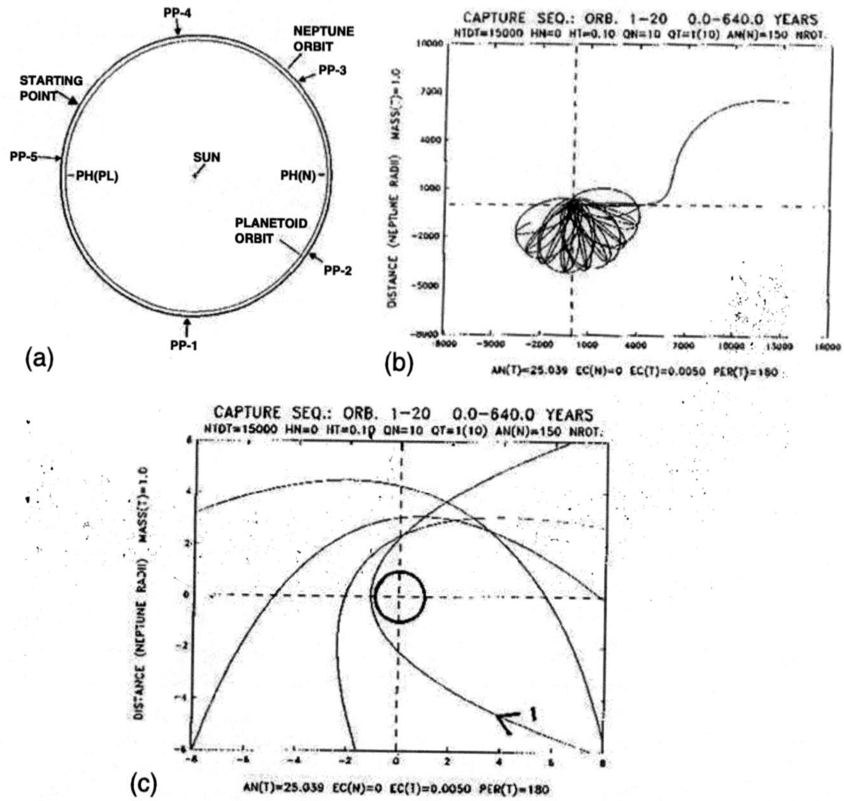

Fig. 8.76 A 20-orbit capture scenario with minimal displacement Love number. (**a**) Heliocentric orientation for this stable capture scenario showing the position of the STARTING POINT as well as the first 5 perinept passages. (**b**) Neptocentric overview of this stable capture scenario. (**c**) Close-up view showing PP-1 and other close encounters associated with this scenario. Note that PP-1 is the only close encounter in this orbit scenario. Also note that the displacement Love number for this scenario is 0.10

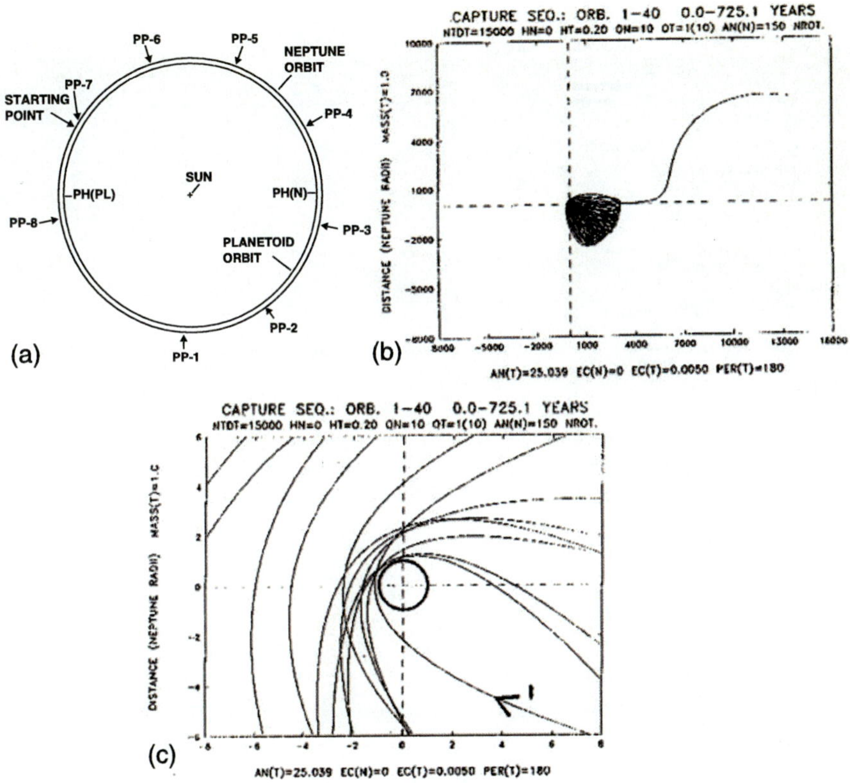

Fig. 8.77 A 40-orbit stable capture scenario with a higher displacement Love number but the same orientation as in the previous scenario. (**a**) Heliocentric orientation showing the position of the STARTING POINT and the first 8 perinept passages. (**b**) Neptocentric orientation showing the orbital evolution of this scenario. (**c**) Close-up of several perinept passages. Note that 2 of the perinept passages are a bit closer to Neptune than PP-1

8.6.5 *Retrograde Encounter and Capture Simulations: Outside Orbit*

Figures 8.78, 8.79, 8.80, and 8.81 demonstrate a series of calculated scenarios with the last one resulting in a stable capture in the retrograde direction from an orbit slightly larger than the orbit of Neptune. Figure 8.78 is a one-orbit escape scenario in which no energy is dissipated. Figure 8.79 is a 6-orbit escape scenario with escape in the prograde direction. Figure 8.80 shows a 16-orbit collision scenario with the collision in the retrograde orientation relative to the center of the planet (h for this

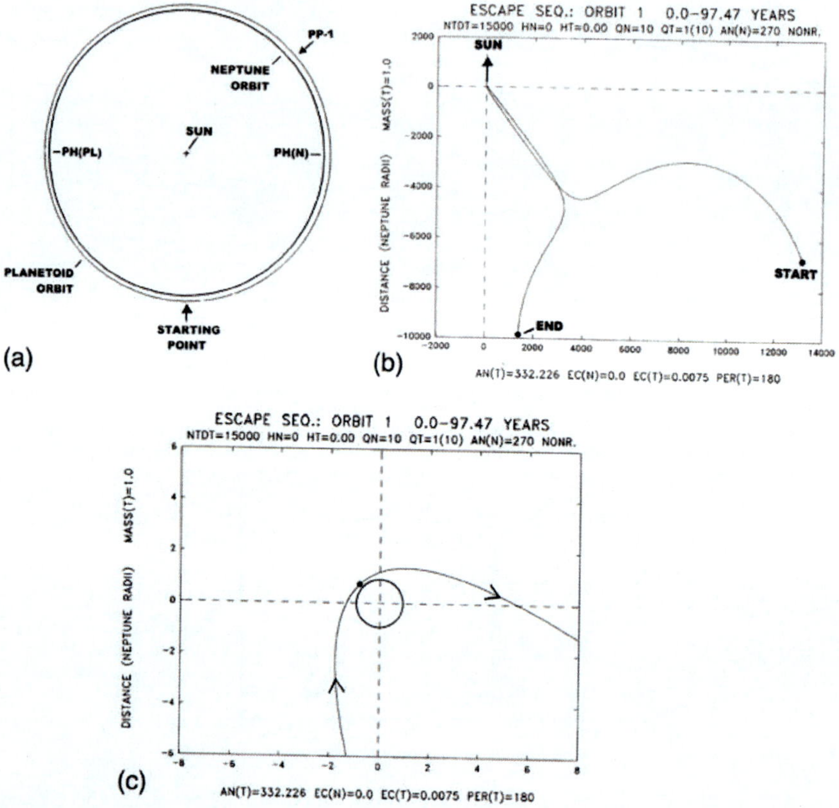

Fig. 8.78 A one-orbit escape scenario. (**a**) Heliocentric orientation for this escape scenario showing the position of the STARTING POINT as well as the position of PP-1. (**b**) A neptocentric plot showing the START position, the position of the SUN at the beginning of the simulation, and the END position for this encounter scenario. (**c**) Close-up of the encounter relative to the surface of Neptune

scenario is 0.16). Figure 8.81 shows a stable capture scenario of 40 orbits in which the planetoid has an h value of 0.18. Note that the major axis of the post-capture orbit is slightly less in Fig. 8.81 than in Fig. 8.80. Energy dissipation in the planetoid for stable capture for this orientation and an h = 0.2 is about the same as for a capture encounter from an inside orbit (~0.10 E28 joules).

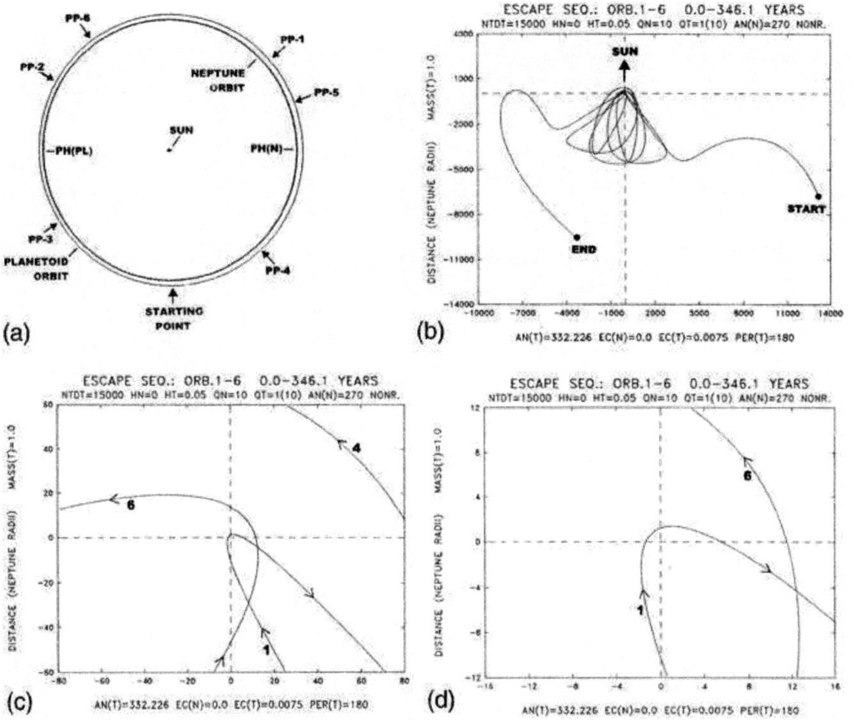

Fig. 8.79 Results of a 6-orbit escape scenario. (**a**) Heliocentric orientation showing the STARTING POINT as well as all 6 perinept passages. (**b**) Close-up of perinept passages 1, 4 and 6. Note that both PP-4 and PP-6 are in the prograde direction. (**c**) Close-up view of PP-1 and PP-6. PP-6 leads to an escape in the prograde direction

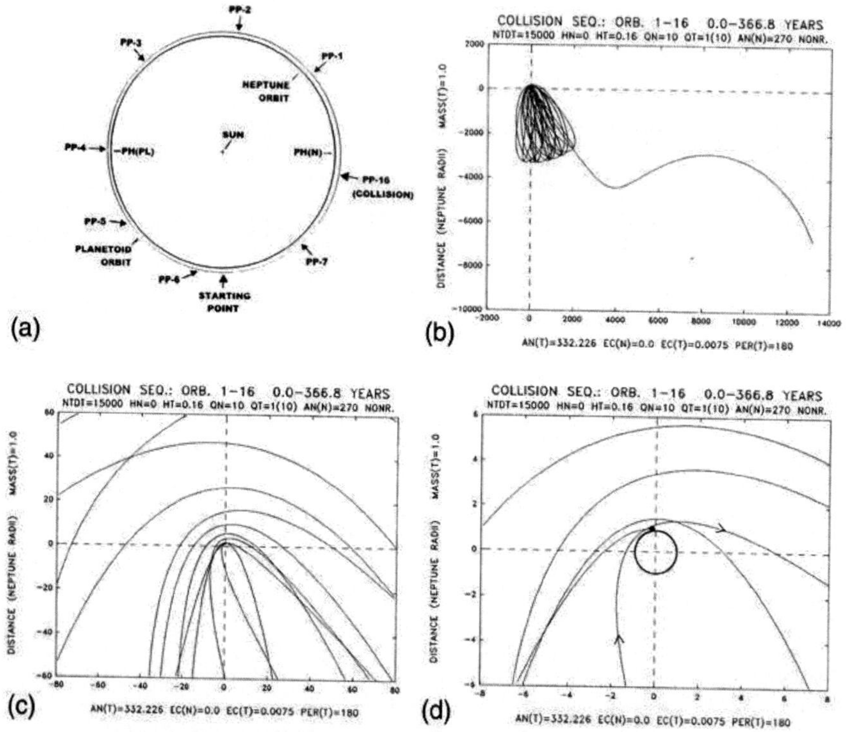

Fig. 8.80 Results of a 16-orbit collision scenario. (**a**) Heliocentric orientation for this collision scenario showing the position of the STARTING POINT as well as the first 7 perinept passages and the position of the collision on PP-16. (**b**) Neptocentric view of the encounter sequence resulting in collision. (**c**) Intermediate view of the series of close encounters before collision. (**d**) View of the collision encounter relative to the planet. In this case the planetoid would "graze" the atmosphere and lose some orbital energy but would continue on its neptocentric orbital path for a while

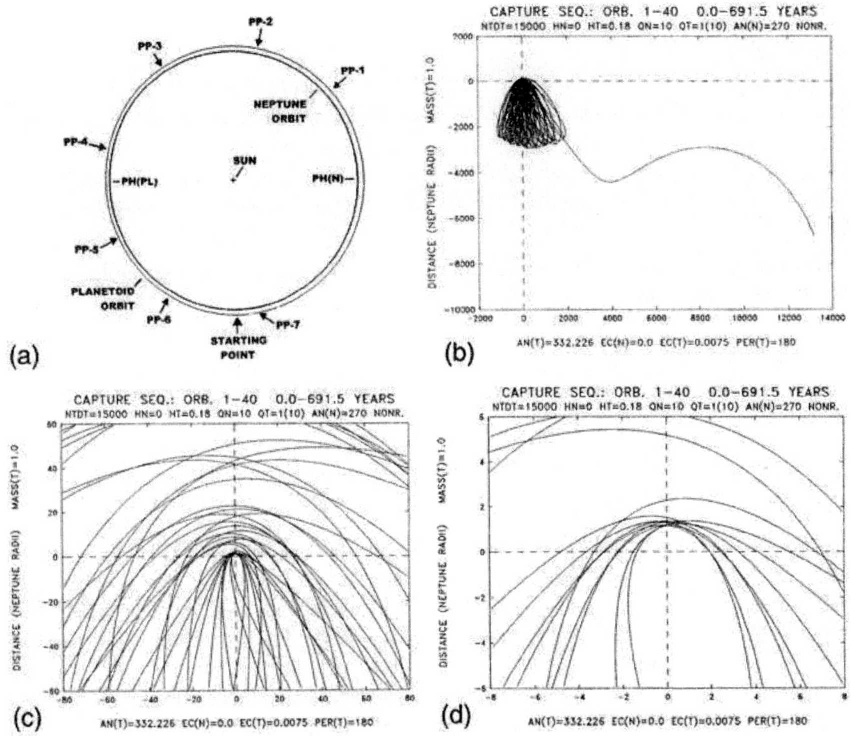

Fig. 8.81 Results of a stable retrograde capture scenario. (**a**) Heliocentric orientation showing the positions of the first seven perinept passages. (**b**) Overview of the orbital pattern of the first 40 orbits (691.5 earth years) are shown on this neptocentric plot. (**c**) A closer view of the orbits for this stable capture scenario. (**d**) Pattern of close encounters relative to the position of Neptune

8.6.6 Summary for Neptune-Triton Simulations

The Neptune-Triton system is located on the outer edge of the planetary portion of the solar system. The source of planetoids for possible capture scenarios are (1) the Kuiper Belt which is somewhat distant from Neptune and (2) planetoids in heliocentric orbits that are considered as "left overs" from the accretion process (i.e., planetoids that escape the accretion process for some reason or another). These "left-overs" would be in quasi-stable orbits in the vicinity of the orbits of Neptune and Uranus. The prograde SCZs are small to non-existent for Neptune. The inside SCZ is limited but stable capture within this SCZ is possible if the encountering planetoid has a high displacement Love number and a sufficiently low specific dissipation factor (Q). The SCZ for prograde encounters from an orbit that is somewhat larger than that of Neptune (an outside orbit) is undefined (i.e., no stable captures can occur).

Thus, when exploring exoplanets in a similar orbital position to Neptune relative to the central star, we could expect only retrograde satellites if the satellites originated via capture. Prograde satellites can form in other ways.

Now for a question: How probable are terrestrial exoplanets relative to icy-gassy exoplanets?

Extensive calculations by Laskar (1989, 1994) and Laskar et al. (1992) suggest that the orbits of the outer planets are very stable but that the orbits of the inner (terrestrial) planets of our solar system are on the verge of instability. The orbits of Earth and Venus are the most stable of the terrestrial planet group. But with very little change in the arrangement of the orbits of Mercury and Mars even Earth and Venus can become unstable (Baszo et al. 2010). This instability associated with the terrestrial planet group may be part of the reason that we have trouble finding terrestrial exoplanets!

8.7 Comparative Geometry of the Stable Capture Zones for Mercury, Venus, Earth, Mars, and Neptune

Figure 8.82 shows the geometric plots for the prograde SCZs for the five planets and associated planetoids: Mercury, Venus, Earth, Mars, and Neptune. Note that there is a full circle (360 °) of planet anomaly positions associated with each plot. The eccentricity axis values for the terrestrial plots is 4% but the eccentricity value for the Neptune plot is 10%.

Figure 8.83 shows the geometric plots for the retrograde SCZs for the five planets and associated planetoids: Mercury, Venus, Earth, Mars, and Neptune. Note that there is a full circle (360°) of planet anomaly positions associated with each plot. The eccentricity axis values for the terrestrial planet plots are in the range of 4–5% but the eccentricity value for the Neptune plot is 10%.

8.8 Discussion of the Implications in the Search for Habitable Exoplanets

Judging from the viewpoint of our set of terrestrial planets, the chances of habitable conditions are not very good. Earth is a very habitable planet! It has supported primitive life forms for at least 3.5 billion years and higher life forms, like plants and animals, for about 0.5 billion years.

There is very little doubt that the Moon is and was involved in helping Earth to become habitable and continued to maintain habitable conditions on the planet over geologic time. The process of lunar capture helped to destroy (via recycling) a portion of the primitive crust into the mantle. Without this destruction of the primitive crust, plate tectonics could never have occurred. And, in my and others opinions, lunar tides are still helping the process along (Nelson and Temple 1972; Moore 1973; Bostrom 2000; Malcuit 2015; Scoppola et al. 2006; Doglioni and

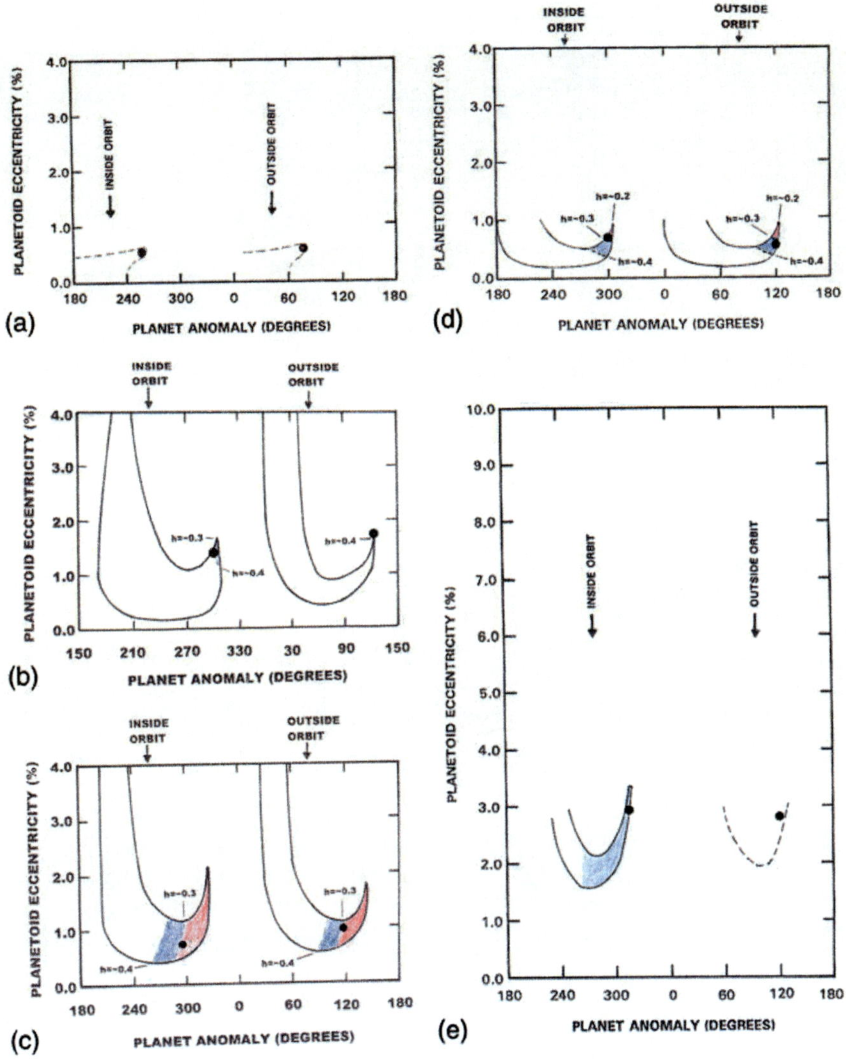

Fig. 8.82 Comparative geometry of the Prograde Stable Capture Zones for Mercury, Venus, Earth, Mars, and Neptune. Note that the x-axis is 360 degrees in all five plots but that the circumference of the planetary orbits increases considerably from Mercury to Neptune. The small black dots mark positions of the encounter scenarios illustrated in the text

Anderson 2015; Doglioni and Panza 2015). Also, without a primitive rotation rate of 10–14 hr/day, evolution of our planet would have been a good bit different.

Our twin sister planet, Venus is in the current era a real environmental "hell-hole". In my view, the story of planet Venus is a result of capture of a large satellite in the retrograde direction mainly because of the lack of effective prograde SCZs for

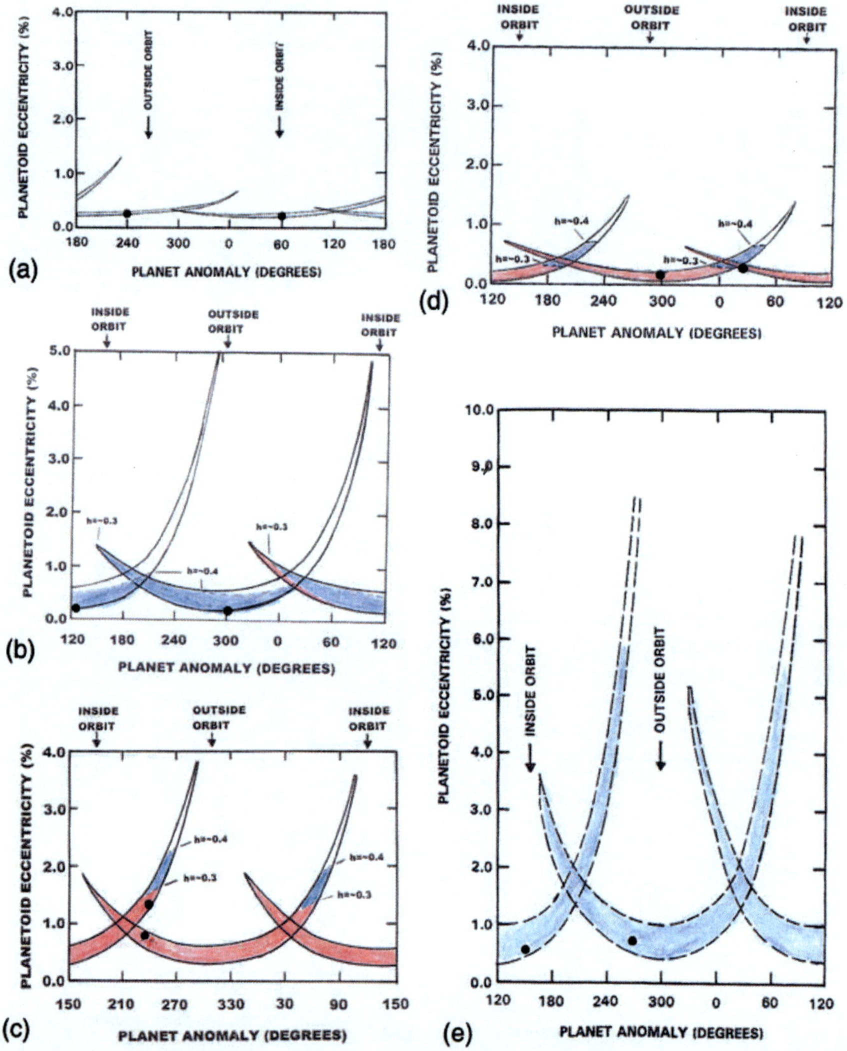

Fig. 8.83 Comparative geometry of the Retrograde Stable Capture Zones for Mercury, Venus, Earth, Mars, and Neptune. Note that the x-axis is 360 degrees in all five plots but that the circumference of the planetary orbits increases considerably from Mercury to Neptune. The small black dots mark the positions of the encounter scenarios illustrated in the text

Venus. Planet Mercury is simply too close to the Sun for habitability and any terrestrial planet in a similar orbital situation can be passed up for habitability. Planet Mars has a transparent atmosphere and an assortment of rocks and sediments on its surface and Mars is considered by many planetologists to be a typical terrestrial planet (Laskar 1994; Nimmo and Tanaka 2005). Mars appears to have large sections of its original crust, volcanoes, rift zones, etc., and a rotation rate similar to that of Earth today. Mars lacks a sizeable satellite. If it had a 0.1 moon-mass satellite (+ or – some mass) in a prograde orbit like that of the Earth and its moon, then the rotation rate would be much slower at the present time (~88 hr/day) and the planet would be even less attractive for habitability than it is today. (Note: A more favorable scenario will be discussed in Chap. 9.)

So if we are searching for potentially habitable planets, we should search for a terrestrial planet with a large satellite, a reasonable rotation rate, a reasonable distance from the central star, oceans of water, and some continental crust above the level of the global ocean. The large satellite does not need to be in a prograde orbit. The satellite can be in a retrograde orbit if the planet is rotating in the retrograde direction.

The point of this chapter is that the geometry of SCZs for a planet-satellite pair can give us a good bit of information on the evolution of the planet as well as on its potential for habitability.

8.9 Summary

The concept of a STABLE CAPTURE ZONE is very important for assessing the probability of capture for a planetoid that is in the proper viscosity range to store the energy of tidal distortion and to subsequently dissipate most of this stored energy during one close encounter to cause its own capture. The planet is mainly a passive bystander in the capture process furnishing a gravitational field for tidally distorting the encountering planetoid. The concept of the SCZ can be applied to the process of collisional capture (Goldreich et al. 1989). In a collisional capture process the orbit of the planetoid after the collision with the smaller body must be within the boundaries of a Stable Capture Zone or the encounter will result in an orbital collision with the planet even if sufficient energy is taken from the orbit of the planetoid for capture.

When attempting to determine the origin of exosatellites orbiting about exoplanets, all of the models of satellite origin that have been proposed for the origin of satellites in the solar system must be considered (i.e., fission, co-formation, capture, giant impact). If a capture origin is suspected, then the concept of the SCZ may be helpful.

Appendix

Some illustrations that demonstrate the unstable nature of encounters that are outside the stable capture zones of the system. (NOTE: These illustrations are for the Earth-Moon system but they apply to all encounter simulations of planetoids that have the deformation parameters that could result in stable capture.)

There are many encounter simulations with initial conditions outside the boundaries of the stable capture zones as shown in Figs. 8.82 and 8.83. For other simulations the SCZ is undefined such as the case for triton-like bodies encountering a Neptune-like planet from an orbit that is slightly larger than the orbit of Neptune (see Figs. 8.69, 8.70, 8.71, and 8.72). In this appendix I am presenting two encounter scenarios with very similar, but not identical, initial conditions. Figure 8.84 shows both prograde and retrograde stable capture zones for the Earth-Moon system. Figure 8.85 shows a prograde encounter that is near the edge of the prograde stable capture zone shown in Fig. 8.84a. If sufficient energy is dissipated, a lunar-like planetoid can be inserted into a stable prograde orbit as shown in Fig. 8.84a (as well as in Figs. 8.42, 8.43, 8.44, 8.45, 8.46, 8.47, 8.48, 8.49, 8.50, and 8.51 in the section

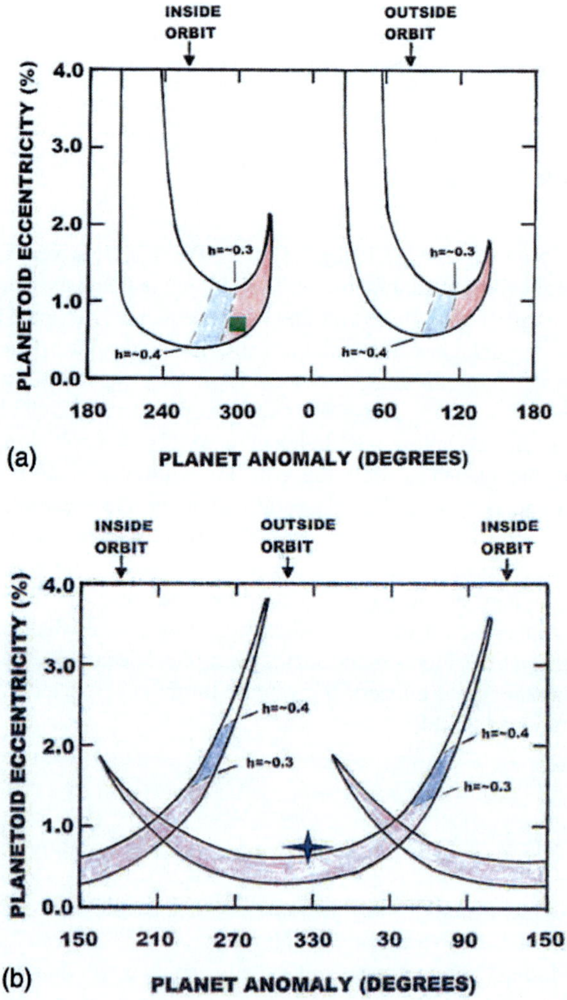

Fig. 8.84 Prograde and retrograde SCZs showing the position of the planet anomaly (320) for the following series of encounters

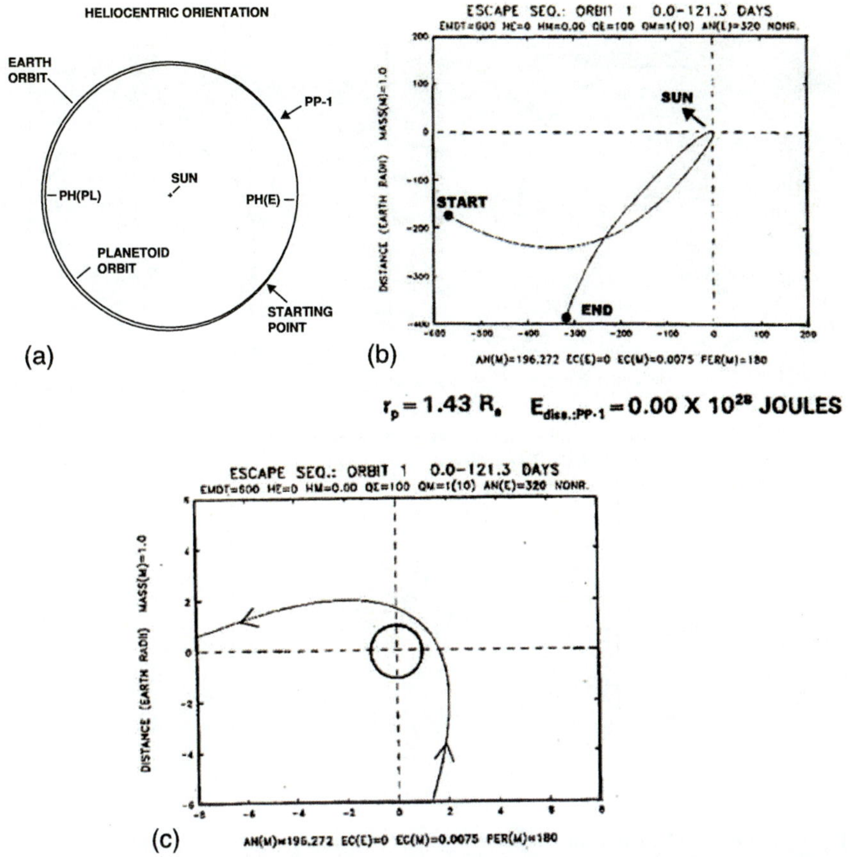

Fig. 8.85 Results of an prograde encounter with no energy dissipation. (**a**) Heliocentric orientation. (**b**) Geocentric orientation of the encounter. (**c**) Close-up of the encounter. View is from the north pole of the solar system

of this chapter featuring encounter simulations for the Earth-Moon system). In Fig. 8.86 the position on an SCZ diagram is the same but the planetoid is encountering the planet in the retrograde direction. The only parameter that is changed is that the planetoid anomaly (the position of the planetoid relative to the planet at the beginning of the simulation) is changed from 196.272 degrees to 191.529 degrees. This causes the planetoid to encounter the planet in the retrograde direction at 1.43 earth radii. The position of the starting point for this encounter is shown on the retrograde SCZ as a 4-pointed star in Fig. 8.84b. The encounter parameters for a retrograde encounter are outside the limits of the retrograde SCZ A prediction for this encounter scenario is that the planetoid will either escape back into a heliocentric

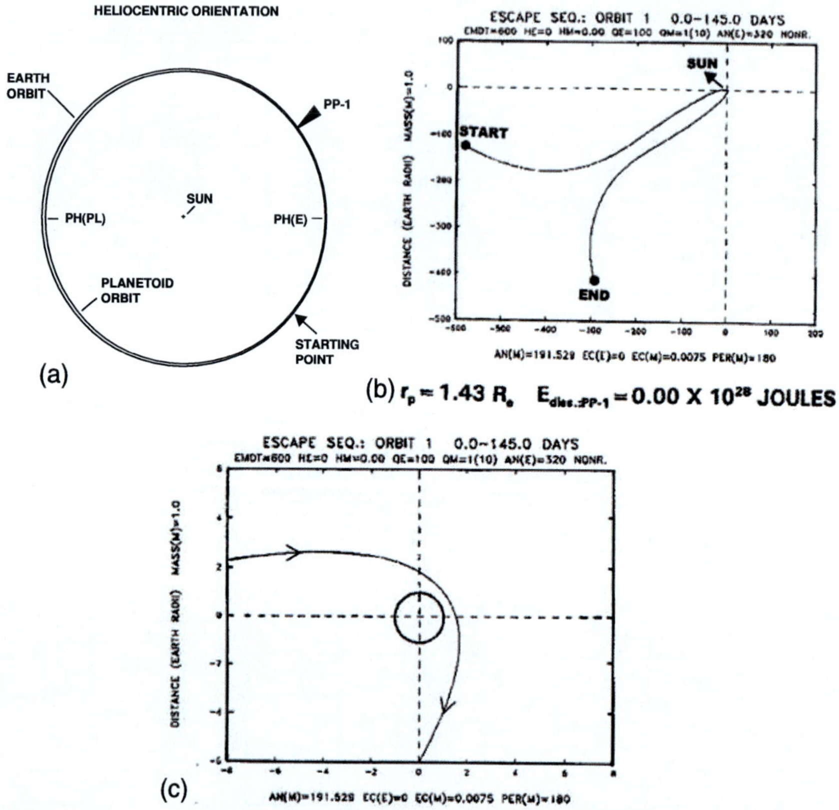

Fig. 8.86 Results of a retrograde encounter with the same starting position as the previous set of figures in Fig. 8.86 but the encounter is in the retrograde direction. (**a**) Heliocentric orientation. (**b**) Geocentric orientation for the encounter. (**c**) Close-up of the encounter. View is from the north pole of the solar system

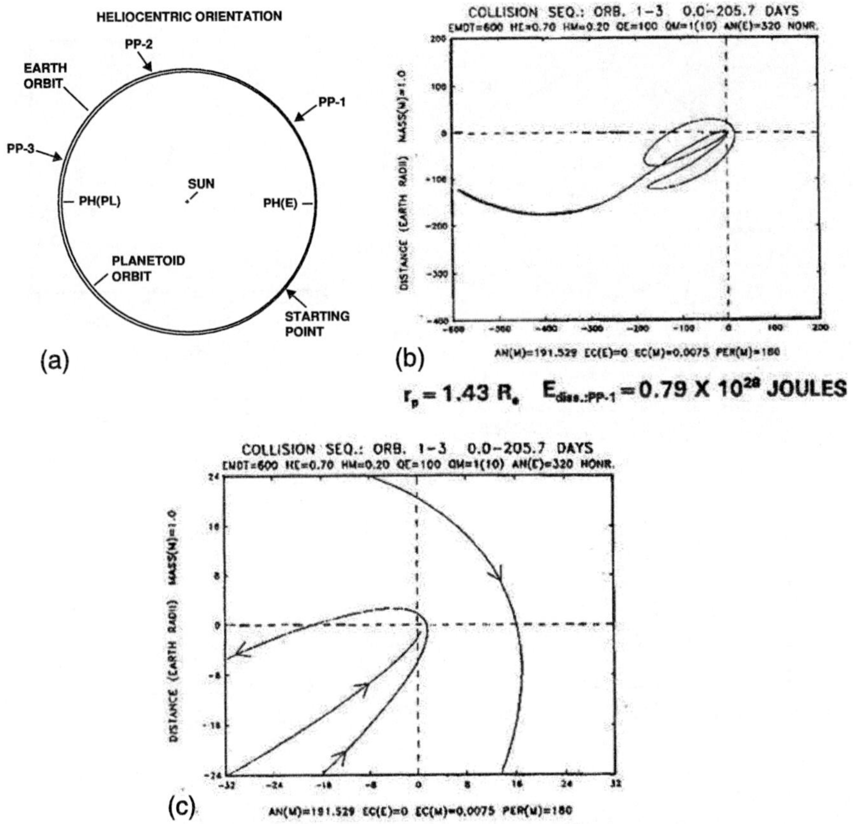

Fig. 8.87 Results of an encounter which is outside the boundaries of a retrograde SCZ. This is for an encounter from an orbit which is larger than the orbit of the planet. (**a**) Heliocentric orientation. (**b**) Geocentric orientation for the encounter sequence. (**c**) Close-up of the encounter. Note that sufficient energy is dissipated for capture but the orbit is unstable and the orbital scenario results in collision on the third encounter. The encounter sequence after the initial retrograde encounter is PP-2 (prograde) and PP-3 is a prograde collision. View is from the north pole of the solar system

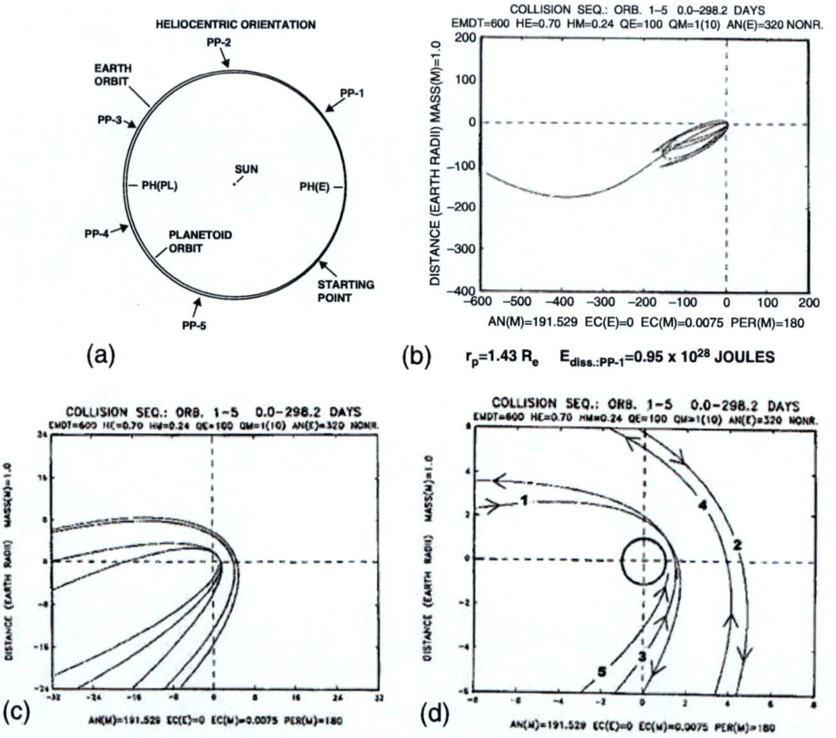

Fig. 8.88 Results of an encounter with the same parameters as in the previous figure but with a higher displacement Love number for the encountering planetoid. (**a**) Heliocentric orientation. (**b**) Geocentric orientation for the encounter. (**c**) Closer view of the encounters. (**d**) Close-up view of the perigee passages. Note that although the planetoid does not collide with the planet on orbit 3 it does collide on orbit 5. The encounter sequence after the initial retrograde encounter is PP-2 (retrograde), PP-3 (prograde), PP-4 (prograde), and then a prograde collision on PP-5. View is from the north pole of the solar system

orbit or, if sufficient energy is dissipated to keep the planetoid in orbits within the Hill sphere, the planetoid will eventually collide with the planet.

Figure 8.86 shows the situation in which no energy is dissipated and the planetoid escapes back into a heliocentric orbit. In Fig. 8.87 there is sufficient energy dissipation for capture but the scenario results in a collision on the third perigee passage. The initial encounter is in the retrograde direction and the following two encounters are in the prograde direction, the last encounter results in collision with the planet. In Fig. 8.87 and Fig. 8.88 there is sufficient energy dissipated for capture and the orbit alternates between prograde and retrograde directions. The encounter scenario in Fig. 8.87 results in a collision in the prograde direction on the third perigee passage and on the fifth perigee passage in Fig. 8.88.

The bottom line is that if there is an encounter with initial conditions that are outside the boundaries of the SCZ, the results are either escape back into heliocentric orbit or collision (either grazing or solid) with the planet after a finite number of perigee passages.

References Cited

Baszo, A., Dvorak, R., Pilat-Lohinger, E., Eybl, V., & Lhotka, C. (2010). A survey of near-mean-motion resonances between Venus and Earth. *Celestial Mechanics and Dynamical Astronomy, 107*, 63–76.

Bostrom, R. C. (2000). *Tectonics consequences of earth's rotation* (p. 266). New York/Oxford: Oxford University Press.

Burns, J. A. (1973). Where are the satellites of the inner planets? *Nature Physical Science, 242*, 23–25.

de Pater, I., & Lissauer, J. J. (2001). *Planetary sciences* (p. 528). Cambridge: Cambridge University Press.

Doglioni, C., & Anderson, D. L. (2015). Top driven asymmetric mantle convection. *Geological Society of America Special Paper, 514*, 51–63.

Doglioni, C., & Panza, G. (2015). Polarized plate tectonics. *Advances in Geophysics, 56*, 1–167.

Goldreich, P., Murray, N., Longaretti, P. Y., & Banfield, D. (1989). Neptune's story. *Science, 245*, 500–504.

Laskar, J. (1989). A numerical experiment on the chaotic behaviour of the Solar System. *Nature, 338*, 237–238.

Laskar, J. (1994). Large-scale chaos in the Solar System. *Astronomy and Astrophysics, 287*, L9–L12.

Laskar, J., Quinn, T., & Tremaine, S. (1992). Confirmation of resonant structure in the Solar System. *Icarus, 95*, 148–152.

Malcuit, R. J. (2015). *The twin sister planets, Venus and Earth: Why are they so different?* (p. 401). Cham: Springer International Publishers.

Moore, G. W. (1973). Westward tidal lag as a driving force of plate tectonics. *Geology, 1*, 99–100.

Nelson, T. H., & Temple, P. G. (1972). Mainstream mantle convection: A geological analysis of plate motion. *American Association of Petroleum Geologists Bulletin, 56*, 226–246.

Nimmo, F., & Tanaka, K. (2005). Early crustal evolution of Mars. *Annual Reviews of the Earth and Planetary Sciences, 33*, 133.

Piro, A. L. (2018). Exoplanets torqued by the combined tides of a moon and parent star. *The Astronomical Journal, 156*(54), 10.

Roy, A. E. (1965). *The foundations of astrodynamics* (p. 385). New York: The Macmillan Company.

Scoppola, B., Boccaletti, D., Bevis, M., Carminati, E., & Doglioni, C. (2006). The westward drift of the lithosphere: A rotational drag? *Geological Society of America Bulletin, 118*, 199–209.

Shakar, A., Driscoll, P., Weinberger, A., & Cody, G. (2019). What makes a planet habitable? *Science, 364*, 434–435.

Ward, P. D., & Brownlee, D. (2000). *Rare earth; Why complex life is uncommon in the Universe* (p. 333). New York: Copernicus (An imprint of Springer).

Chapter 9
Discussion of Some Real and/or Theoretical Effects of Captured Satellites on Both Terrestrial Planets and Gaseous Planets

> The habitable zone in a planetary system is generally defined as a region in which a terrestrial-type planet may have liquid water at its surface during part of the year (the period of revolution around its parent star) . . . The notion of the habitable zone should not be considered as a very rigid one.—From Ollivier et al. (2009a)

In this chapter we are exploring the conditions which help to make a terrestrial planet habitable. We will explore the possibility of development of habitable conditions on a rocky planet like Mars that has no sizeable satellite. Then we will consider the advantages of capturing a 0.1 moon-mass satellite in a prograde orbit. Does such a satellite have any significant effect on the hydrospheric and lithospheric evolution of the planet? Then we will consider the advantages and disadvantages of capturing a 0.1 moon-mass satellite into a retrograde orbit. Does this mode of capture have any significant effect on the hydrospheric and lithospheric evolution of the planet? Then we will consider the possible effects of increasing the mass of the captured body to 0.2 moon-mass for both the prograde capture scenario as well as for the retrograde capture scenario.

9.1 Implications for Finding Habitable Terrestrial Planets for Algae and Bacteria and Possibly Higher Forms of Life

The theme of this book is to demonstrate pathways that nature could have taken to make Mars habitable for algae and bacteria for some segment of geologic time. We have considered five of these pathways. (1) Evolution of Mars to its present condition (this was nature's pathway – a maroformed planet). (2) Gravitational capture of a 0.1 moon-mass body in prograde orbit and the subsequent evolution of the system (an attempt at geoforming Mars), (3) Gravitational capture of a 0.1 moon-mass body in retrograde orbit and the subsequent evolution of the system

R. Malcuit, *Geoforming Mars*, https://doi.org/10.1007/978-3-030-58876-2_9

(an attempt at venoforming Mars). (4) Gravitational capture of a 0.2 moon-mass body in prograde orbit and the subsequent evolution of the system (another attempt at geoforming Mars). (5) Gravitational capture of a 0.2 moon-mass body in a retrograde orbit and the subsequent evolution of the system (another attempt at venoforming Mars).

Now we will examine pictograms of these five evolutionary pathways for Mars and then compare and contrast these evolutionary pathways with those of the other terrestrial planets (i.e., Mercury, Venus, and Earth). Following that we will discuss some of the implications for finding terrestrial exoplanets that may have habitable conditions.

9.1.1 *CASE I: The Mars that We Think We Know Something About*

> The rotational momenta of the major planets and Mars are consistent with the hypothesis that there is a simple relationship between the rotational momentum and mass of a planet.—From MacDonald (1964, p. 522)

> Mars is far from the Sun, and its satellites Phobos and Deimos have masses far too small to slow its rotation, so that its present rotational period of 24 hours 37 minutes is likely to be close to its primordial rotation period.—From Laskar (1995, p. 104)

> Mars is in some ways the type example of a terrestrial planet. It is neither geologically stillborn, like Mercury or the Moon, nor so active that most of the geological record has been destroyed, like Venus or the Earth. The prolonged geological evolution of Mars is recorded in the physical and chemical characteristics of its crust.—From Nimmo and Tanaka (2005)

Some notable features of planet Mars are:

- A dichotomy of the features of the northern and southern hemispheres (Willhelms and Squyres 1984; Elkin-Tanton et al. 2003; Nimmo and Tanaka 2005; Scheinberg et al. 2014; Rosenblatt et al. 2016).
- Largest volcanoes (conic forms) in the solar system (Baker 2005; Nimmo and Tanaka 2005; Bell 2008)
- Abundance of channels many of which appear to have been formed by flowing water (Baker et al. 1991; Carr 1996; Phillips et al. 2001; Salese et al. 2019)
- A strong internal magnetic field early in its history (100 Ma or so) as manifested by remanant magnetic strips imprinted in the ancient crust of Mars (Arkani-Hamed 2009; Kobayashi and Sprenke 2010)
- Abundance of solid body impact craters representing a variety of impactor sizes and directions of impact (Nimmo and Tanaka 2005; Frey 2008)
- Evidence of abundant liquid water (oceans of water) early in martian history (Hurowitz et al. 2017; Citron et al. 2018)
- Absence of a sizeable satellite
- Absence of a long-lived magnetosphere to prevent volatile escape from Mars

Figure 9.1 is a pictogram of my interpretation of some features of martian history that may lead to perhaps 1 Ga of habitability for algae and bacteria.

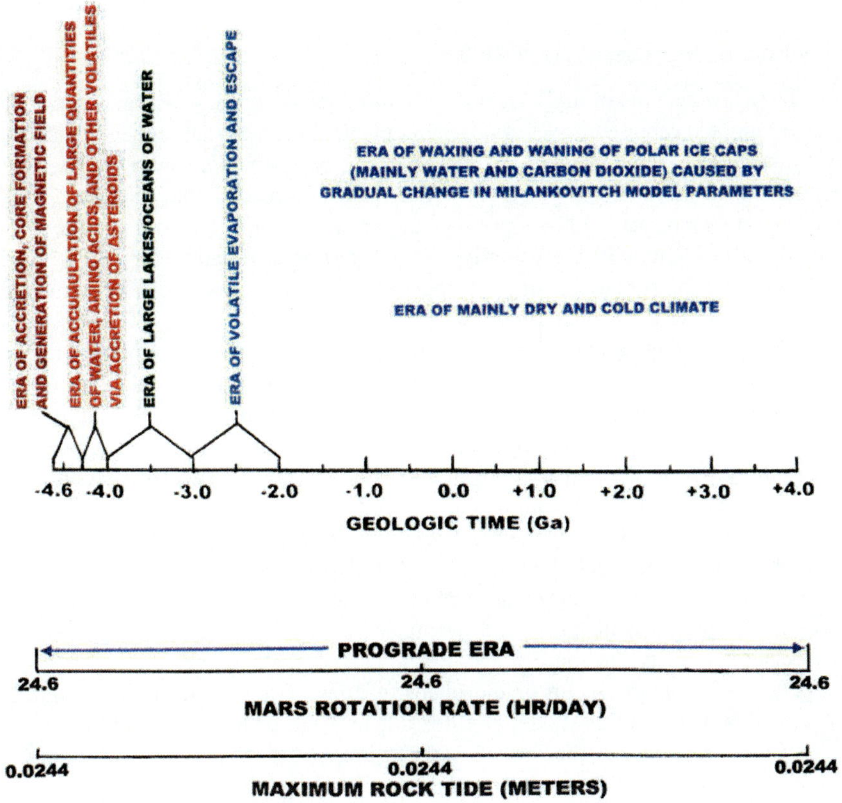

Fig. 9.1 The case of no stable capture of a sizeable satellite. The major events are accretion, core formation, magma ocean formation, separation of a planetary crust from the magma ocean, generation of a core dynamo while a portion of the planetary crust is cooling through the Curie temperature, and accumulation of a large volume of water and other volatiles (including amino acids) via impact of Themis-type asteroids as well as Jupiter-captured comets (comets from the Oort Cloud that are inserted into asteroidal orbits after one, or a few, favorable encounters with Jupiter). Because of the distance from the Sun, the unidirectional solar gravitational tide would be very low (~2.4 cm for a circular martian orbit of about 1.52 AU; $h_{ma} = 0.5$). Because of this very weak tidal influence, the rotation rate of the planet would change very little over geological time. This is the type example of a MAROFORMED PLANET

9.1.2 CASE II. Theoretical Development of a Model in which a 0.1 Moon-Mass Planetoid Is Captured by a Mars-Mass Planet in a Prograde Marocentric Orbit

When tides raised by a satellite dissipate energy in a spinning planet, the planet-satellite system evolves dynamically toward one of three possible final conditions. The satellite orbit may decay inward until the satellite is destroyed; the orbit may decay outward toward escape, although at an ever decreasing rate; or the satellite's orbital and planet's spin periods may approach stable synchronism. Which of these outcomes of tidal evolution is approached depends on the satellite-planet mass ratio and on the initial spin and orbital angular velocities.—From Counselman III (1973)

The major features of this model are all of the features of CASE I in addition to stable prograde capture of a 0.1 moon-mass planetoid and the subsequent evolution of the system (covered in Chap. 4). The early eras of this model would not be much different than those shown in Fig. 9.1. The main difference is the ocean and rock tidal activity associated with the post-capture orbit evolution. The rock tidal activity would cause sections of the primitive crust, along with ocean water and oceanic sediments, to be subducted and incorporated into the mantle rocks. The mantle rocks, as well as the included gaseous constituents (i.e., H_2O, CO_2, CO, N_2, etc.) would be reentered into the martian atmosphere via ocean rise volcanism and subsequent subduction processes. The activity of this global recirculation system decreases as the marocentric orbit circularizes and the associated rock tidal amplitudes decrease. Since the primordial rotation rate of mars, in this model, is about 24.6 hr/day, the total angular momentum of the system is established soon after capture. Since angular momentum is transferred to the orbit of the satellite via tidal action, the system slowly approaches a condition in which the period of the satellite is nearly synchronous with the rotation rate of planet. The surface condition would then return to a "stagnant-lid" condition but the planetary rotation rate would be very slow in the prograde direction and eventually the ocean water would evaporate, dissociate, and escape to space.

Figure 9.2 is a pictogram of the major features of this exercise in theoretical planetology. Perhaps this model could result in 2 Ga of habitable conditions for bacteria and algae before the surface water begins to disappear via evaporation and molecular dissociation. Depending on the h and Q values of the planet and planetoid, the time scale for this orbit evolution can be short or as long at 4 Ga.

9.1.3 CASE III: Theoretical Development of a Model in which a 0.1 Moon-Mass Planetoid Is Gravitationally Captured into a Stable Retrograde Orbit Early in Martian History

When tides raised by a satellite dissipate energy in a spinning planet, the planet-satellite system evolves dynamically toward one of three possible final conditions. The satellite orbit may decay inward until the satellite is destroyed; the orbit may decay outward toward escape, although at an ever decreasing rate; or the satellite's orbital and planet's spin periods may approach stable synchronism. Which of these outcomes of tidal evolution is approached depends on the satellite-planet mass ratio and on the initial spin and orbital angular velocities.—From Counselman III (1973)

Figure 9.3 is a pictogram of the major features for this exercise in theoretical planetology. This could lead to habitable conditions for bacteria and algae for 4.0+ Ga although there would be a few very "trying" surface conditions for organisms.

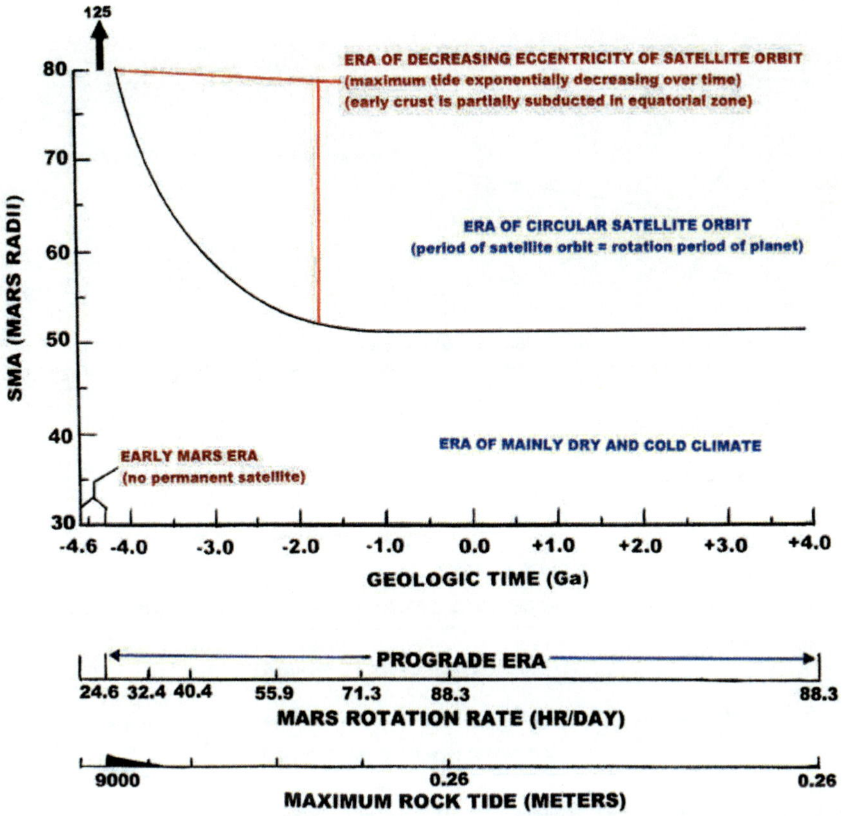

Fig. 9.2 Semimajor axis of a successful prograde capture of a 0.1 moon-mass body vs. geologic time. The timescale is a result of a two body numerical simulation which features a mars-like body tidally interacting with a 0.1 moon-mass planetoid. The circularization era is shortened considerably (~50%) when the effects of solar gravitational perturbations are considered (see Malcuit 2015, Chap. 4 for a more detailed explanation of the orbit circularization process). Note that in this scenario, the prograde rotation rate of the mars-like body decreases exponentially from the primordial rate of ~24.6 hr/day to about ~88.3 hr/day. Then there is very little change because most of the prograde angular momentum of planet rotation has already been transferred to the satellite orbit. Eventually the period of the satellite orbit will equal the period of rotation of the planet. This is an attempt to develop a GEOFORMED PLANET

The major features of this model include all of the features of the model scenario shown in Fig. 9.1 in addition to all the activity associated with the capture episode, and then the subsequent circularization sequence for the retrograde orbit followed by the evolution of the circular orbit of the satellite. The main difference from the model presented in Fig. 9.2 is that, following capture and the subsequent orbit evolution, the orbit gets progressively smaller and the ocean and rock tides get progressively higher. Eventually the rock tides will reinvigorate the process of crustal subduction

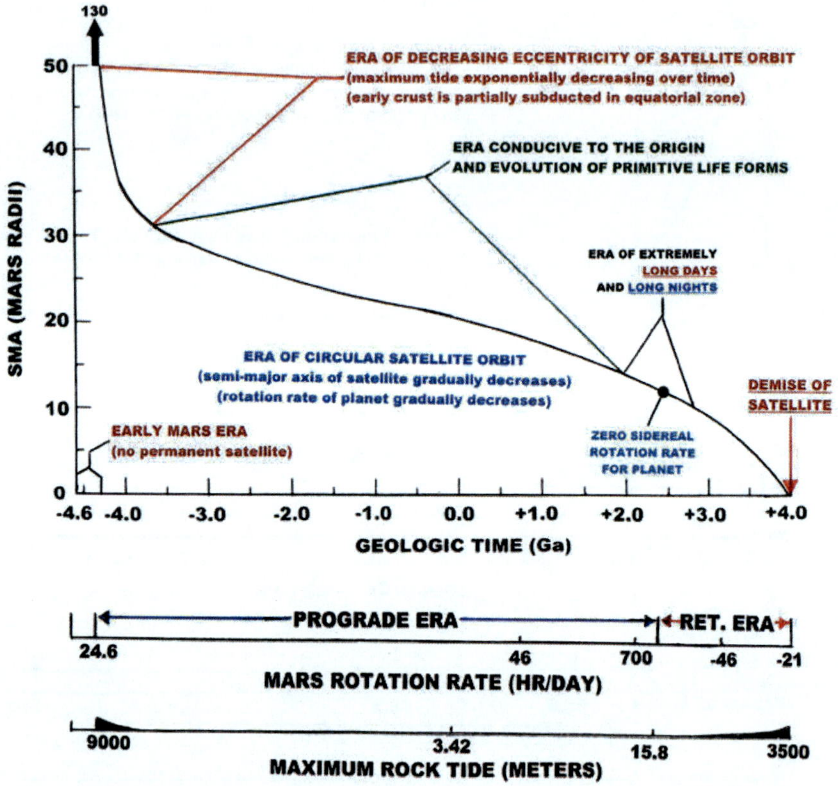

Fig. 9.3 Semi-major axis vs. geologic time for this stable retrograde capture of a 0.1 moon-mass body. During the capture episode and the subsequent orbit circularization sequence, significant portions of the original crust would undergo disruption and subduction because of the high rock tides (up to 9 km) associated with the capture and subsequent orbit circularization process. After orbit circularization the rock and ocean tidal amplitudes would be moderate but gradually increasing in amplitude and frequency. There would be a lengthy period (perhaps 5 Ga long) that could be conducive to the development of life forms. The prograde rotation rate of the planet gradually decreases to zero and then changes to the retrograde direction. The rock tides during the retrograde orbit era increase from ~15.8 to 3500 m as the large satellite coalesces with the mars-like planet [i.e., the satellite is destroyed in the terms of the Counselman (1973) quote above] as the mars-like planet is transformed into a venus-like "Hades" scene with lava flows covering the surface, an accumulation of a dense carbon dioxide rich atmosphere, and a planetary retrograde rotation rate of about 21 hr/day. This is a good example of a VENOFORMED mars-mass planet

in the equatorial zone which leads to subduction of ocean crust and associated sedimentary rocks with their water and other volatiles. These volatiles, then, get circulated through the mantle to be re-introduced into the hydrosphere-atmosphere system via volcanism at ocean spreading centers and subduction zones. This reprocessing system will end when the satellite breaks up in orbit, falls to the surface in parts to be incorporated within the mass of the planet.

9.1.4 CASE IV. Theoretical Development of a Model in which a 0.2 Moon-Mass Planetoid Is Captured by Planet Mars in a Prograde Orientation

When tides raised by a satellite dissipate energy in a spinning planet, the planet-satellite system evolves dynamically toward one of three possible final conditions. The satellite orbit may decay inward until the satellite is destroyed; the orbit may decay outward toward escape, although at an ever decreasing rate; or the satellite's orbital and planet's spin periods may approach stable synchronism. Which of these outcomes of tidal evolution is approached depends on the satellite-planet mass ratio and on the initial spin and orbital angular velocities.—From Counselman III (1973)

Figure 9.4 is a pictogram of the major features of this exercise in theoretical planetology. Perhaps this model could result in 2 Ga of habitable conditions for bacteria and algae before the surface water begins to disappear via evaporation and molecular dissociation. Depending on the h and Q of the planet and planetoid, the time scale for this orbit evolution can be short or as long at 4 Ga.

The main effects of capturing a 0.2 moon-mass planetoid is that the ocean and rock tides on the planet will be twice as high. The down-side of the process of capturing a more massive body is that the orbit circularization process occurs more rapidly and the satellite will approach a near stationary orbit over a shorter period of time. A biological system may attain greater abundance and diversity because of the higher amplitude tidal activity but the bio-system would struggle to survive on a very slowly rotating planet featuring alternating long days and long nights in the broadly defined equatorial zone. Perhaps the polar zones could harbor the last living organisms of such a biological system, the details depending on the tilt angle of the planet.

9.1.5 CASE V: Theoretical Development of a Model in which a 0.2 Moon-Mass Planetoid Is Stably Captured by Planet Mars in a Retrograde Direction. This Case Is the Best Chance of Developing Habitable Conditions on Mars for Significant Portions of Geologic Time

When tides raised by a satellite dissipate energy in a spinning planet, the planet-satellite system evolves dynamically toward one of three possible final conditions. The satellite orbit may decay inward until the satellite is destroyed; the orbit may decay outward toward escape, although at an ever decreasing rate; or the satellite's orbital and planet's spin periods may approach stable synchronism. Which of these outcomes of tidal evolution is approached depends on the satellite-planet mass ratio and on the initial spin and orbital angular velocities.—From Counselman III (1973)

Figure 9.5 is a pictogram of the major features of this scenario. The major features of this model for the first 300 Ma are the same as those of CASE I and then a 0.2 moon-mass satellite comes onto the scene and gets gravitationally captured in the

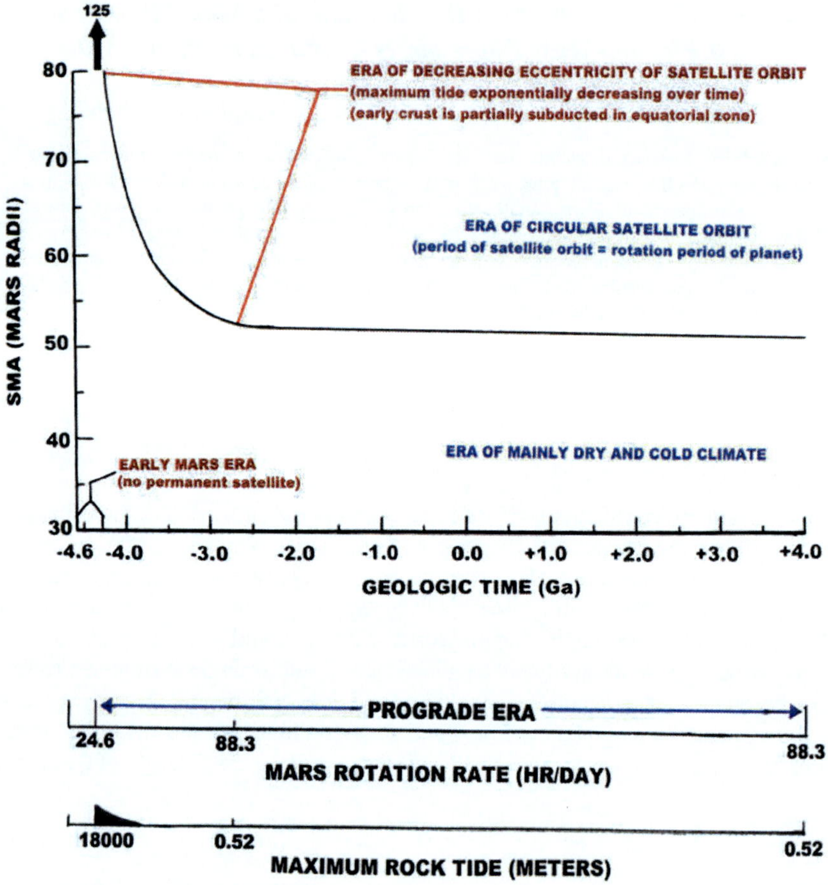

Fig. 9.4 Semi-major axis of a successful prograde capture of a 0.2 moon-mass body vs. geologic time. The 0.2 moon-mass satellite is captured into a highly elliptical orbit. The orbit circularization sequence occurs much more rapidly than in CASE II and the rock tidal amplitudes are twice as high (up to 18 km). The rotation rate after circularization is about 88.3 hr/day and this rotation rate does not change much over geologic time because most of the angular momentum of the rotation of the planet has already been transferred to the orbit of the satellite. After orbit circularization the prograde rotation rate is probably too slow to support organic life (the day-night cycle would be too long). The habitable era for this scenario is about 1.5 Ga long. In time, the period of the prograde satellite would be equal to the prograde rotation of the planet. This is another example of a GEO FORMED PLANET

retrograde orientation. The scenario after retrograde capture of a 0.2 moon-mass satellite is very similar to CASE III but the circularization of the orbit occurs much more rapidly and the tidal amplitudes are about twice as high. The changeover from prograde to retrograde rotation happens much earlier in time and the ocean and rock tidal activity has about twice the amplitude relative to the model in Fig. 9.3. In the case featuring a 0.2 moon-mass satellite there could be two eras of biological activity in the broadly defined equatorial zone possibly bridged by biological survival in the polar zones of the planet. Eventually, as in the model shown in Fig. 9.3, the biological system would probably cease to exist with the demise of the planet-satellite system (i.e., when the planetoid breaks up in orbit and eventually the debris of the planetoid coalesces with the planet).

9.1.6 Comparison of Models of Martian Evolution to the Evolution of Other Terrestrial Planets as well as Terrestrial Exoplanets

A major problem for this analysis is that terrestrial exoplanets appear to be rare in our inventory of exoplanets (Piro 2018; Shahar et al. 2019). We will discuss the five models in **terms of the** terrestrial planets in our solar system.

9.1.6.1 MODEL I: No Stable Capture but Perhaps Several Attempts at Capture

This model may relate to the situation of planet Mercury, but there are significance differences between Mars and Mercury.

- Mercury is in a spin-orbit resonance with the Sun. The 3:2 resonance was established many eons ago; this keeps the rotation rate constant at ~58.6 days in the prograde direction.
- Mercury has been "despun" by the action of solar rock tides and the prograde rotation was stabilized by the spin-orbit resonance with the Sun.
- Mercury has a much higher body density than Mars. Thus Mercury probably contains much more iron than Mars.

And there are similarities between Mars and Mercury:

- Mercury and Mars have no large satellites at the present time.
- Mercury and Mars have crater density patterns that suggest that they have large areas of ancient surfaces.
- Mercury and Mars have remanant magnetic rock patterns consistent with the presence of an internal magnetic field: Mars in the distant past and Mercury perhaps continuing to the present time.
- Solar Gravitational Rock tides:
  - For Mercury the rock tides are associated with alternate high and low rock tides associated with perimerc and apomerc passages; especially during periods of increased orbital eccentricity.

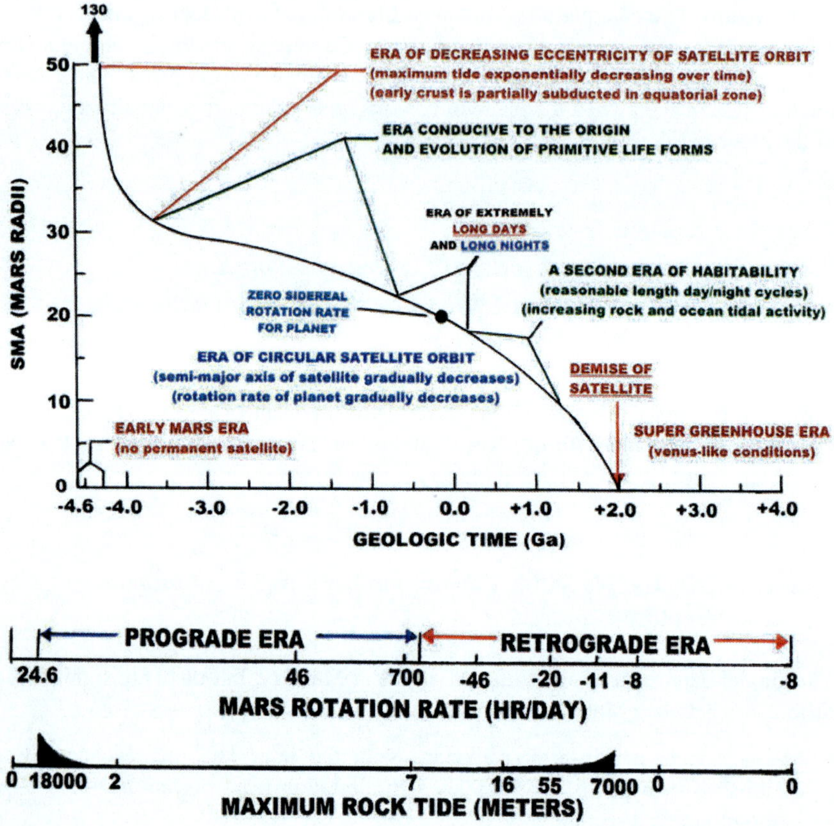

Fig. 9.5 Semi-major axis vs. geologic time for this stable retrograde capture scenario of a 0.2 moon mass planetoid. This retrograde capture scenario can result in about 2 Ga of habitability. The major advantages of capturing a 0.2 moon-mass body is that the rock and tidal amplitudes (and ranges) are twice as high as the 0.1 moon-mass case. The main disadvantage is that the timescale of habitable conditions is considerably shorter. The circular marocentric orbit at the time of zero rotation rate for this scenario is at about 22 mars radii. If organisms can survive the era of extremely long days and nights, then there is a possible second era of habitability before the ultimate fate of planet-satellite coalescence and the transformation of the planet into a venus-like condition of volcanic flows covering the surface, accumulation of a dense carbon dioxide rich atmosphere, and a retrograde rotation rate of about 8 hr/day. This is another example of a VENOFORMED mars-mass planet

– For Mars the solar-induced rock tides are much lower amplitude than on Mercury but the eccentricity of the heliocentric orbit is much more variable causing an alternation of perihelion and aphelion tidal amplitudes.

9.1.6.2 MODEL II: Stable Prograde Capture of a 0.1 Moon-Mass Planetoid

This model relates fairly directly to the situation of the Earth-Moon system as presented by Malcuit et al. (1989, 1992a, b) and Malcuit (2015). Planet Mars acquires water and other volatiles after accretion from a combination of Jupiter-captured comets and asteroids from the Outer Asteroid Belt. Then during the stable capture and subsequent orbit circularization, some of the water and other volatiles get subducted into the mantle and subsequently get emitted to the surface and recycled several times over geologic time.

A major difference between such a Mars-planetoid system and the Earth-Moon system is that the initial prograde rotation rate of planet Mars is very low (~24 hr/day). Thus the angular momentum of prograde planetary rotation is transferred to the satellite early in the history of the planet (probably within about 2 Ga). The satellite then acquires a near stationary orbit about the planet so that the length of the month is about equal to the length of the day. Since Mars is farther from the Sun than is the Earth, the solar component for the Martian rock tides is essentially ineffective for "braking" the prograde rotation of Mars and the evolution of the Mars-planetoid system will remain unchanged for the limited history of the solar system.

9.1.6.3 Model III: Stable Retrograde Capture of a 0.1 Moon-Mass Planetoid

This model is similar to the evolutional history of planet Venus and a 0.5 moon-mass planetoid as presented by Malcuit and Winters (1996, 2009), and Malcuit (2015). Once the captured planetoid goes through its orbit circularization sequence, then cancellation of angular momentum leads to an ever decreasing radius of the lunar orbit. In the case of Venus, this coalescence scene (which in my view, is the cause of the global resurfacing event) occurs somewhere between 1.0 and 0.5 Ga ago (Herrick 1994). For Mars and a 0.1 moon-mass planetoid, this coalescence will occur about 4.0 Ga in the future. For the Venus-Adonis case, the angular momentum of the planet is cancelled at about the time of the coalescence. In the case of the Mars-planetoid system, the angular momentum associated with the prograde planetary rotation is cancelled when the planetoid orbit is at about 12 Mars radii. At this point in time the sidereal planetary rotation goes to zero and the planet will slowly begin to rotate in the retrograde direction (clock-wise direction as view from the north pole of the solar system). The zero rotation situation may be critical for a biological system. If the biological entities survive this zero rotation crisis, then there will be plenty of rock and ocean tidal action for their evolution in their future. This model is by far the most favorable scenario for development of a biological system on a small terrestrial planet like Mars.

The main difference between Model III and the Venus model is that Venus has a much more stable heliocentric orbit than Mars. The maximum variation in heliocentric eccentricity for the orbit of Venus is about the same as that of Earth [the highest value is about 6% (Laskar 1994)]. In contrast, variation in heliocentric orbital eccentricity for Mars goes from a low of 8% to a high of 20% over geologic time (Laskar 1994). Likewise, the variation in heliocentric orbital inclination for Mars is significantly higher than for Venus and Earth. The variation in inclination for Mars goes from a low of 4 degrees to a high value of 10 degrees. The variation for both Earth and Venus is within a band between 3.5 and 5.0 degrees (Laskar 1994).

9.1.6.4 MODEL IV: Stable Prograde Capture of a 0.2 Moon-Mass Planetoid

This model is very similar to MODEL II in that it is a PROGRADE CAPTURE MODEL but in this case the mass of the captured planetoid is twice as large. The time-scale for circularization of the lunar orbit is somewhat shorter but the rotation of the planet decreases much more rapidly and eventually the rotation period of the planet and the orbital period of the satellite are about the same. This model is much more conducive to the origin and evolution of a biological system than the case where there is no satellite. However, the length of the martian day decreases rapidly during and after orbit circularization. The water would eventually evaporate and undergo molecular dissociation in the upper atmosphere of the mars-like planet. The hydrogen molecules would eventually escape from the planet and the oxygen would combine with iron and other reduced metals on the surface of the planet. Thus we are left with a cold and dry planet that is essentially despun. Astronauts exploring such a terrestrial planet would have some deformed crust to study along with former subduction zones and ocean rises and the "smile of the Cheshire Cat" in the form of a satellite in a stationary orbit above the planet.

9.1.6.5 MODEL V: Stable Retrograde Capture of a 0.2 Moon-Mass Planetoid

This model is very similar to MODEL III in that it is a RETROGRADE CAPTURE MODEL but, again, the captured satellite is twice as large.

Future astronauts would have an interesting planet to explore. If near the present time line (about 0.0 Ga), then the Mars-like planet would have a very slow rotation, ~50 hr/day in the retrograde direction, and the satellite would be in a circular retrograde orbit at about 20 mars radii. There would be long days and long nights for most of the surface of the planet, the details depending on the obliquity (tilt-angle) of the planet. The polar zones could be interesting and some biological entities could be present in these polar zones.

Side Bar on a Prograde Gravitational Scenario that Has the Best Chance of Resulting in a Habitable Mars-like Planet (a Geoformed Mars)

This capture scenario is very similar to the one presented in Chap. 4. The main difference is that the pre-capture rotation rate for this scenario is 10 hr/day rather than 24.6 hr/day. This initial rotation rate makes all the difference for developing habitable conditions on a mars-like planet in a mars-like heliocentric orbit.

A quote from Counselman III (1973) is appropriate here:

> When tides raised by a satellite dissipate energy in a spinning planet, the planet-satellite system evolves dynamically toward one of three possible final conditions. The satellite orbit may decay inward until the satellite is destroyed; the orbit may decay outward toward escape, although at an ever decreasing rate; or the satellite's orbital and planet's spin periods may approach stable synchronism. Which of these outcomes of tidal evolution is approached depends on the satellite-planet mass ratio and on the initial spin and orbital angular velocities.—From Counselman III (1973)

Note that the capture conditions are the same as the scenario in Chap. 4 and only the "initial spin" or, in this case, the pre-capture rotation of the mars-like planet is 10 hr/day. Without that extra angular momentum the planet despins rapidly after capture and the day-night cycle is too long for habitable conditions in the equatorial zone of the planet. For example, in Table 4.6 we see that after 0.5 Ga the rotation rate for Mars has decreased from the initial rate of 24.6 to 32.4 hr/day. Then at 1 Ga the rotation rate for Mars is 40.4 hr/day. The long night and long day syndrome would be firmly established.

A question is: How could nature increase the pre-capture rotation rate for a mars-like planet from 24.6 to 10 hr/day? A few well-placed prograde tangential collisions of moderate-sized planetoids could be invoked to accomplish the task. The probability of this happening in our solar system is not favorable because as explained in Chap. 4, retrograde orbital collisions are much more probable than prograde orbital collisions. For this planet-satellite scenario, however, we will assume that it happened.

It is probably becoming clear to the reader at this time that I am trying to build a mini earth-moon system at the heliocentric distance of planet Mars. Mars is about 0.5 the diameter of Earth and about 0.1 of the mass of Earth. The martian satellite is about 0.01 mars-mass and the mass ratio for this scenario is 1:100 (that of the earth-moon system is 1:81). The post-capture satellite orbit with no angular momentum transfer from the rotating planet is about 30 mars radii. The obliquity (tilt) angle for this model is 30 degrees. The year for the mars-like planet is 780 earth days.

The gravitational capture can be accomplished as illustrated in Chap. 4, Figs. 4.8, 4.9, 4.10, and 4.11. After orbit circularization we end up with the orbital geometry and yearly cycle as shown in the cartoon in Fig. 9.6. When considering a reasonable version of angular momentum exchange between a

(continued)

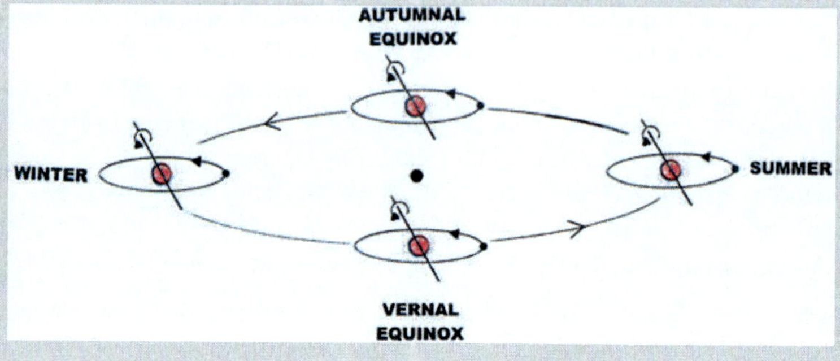

Fig. 9.6 Schematic pictogram of orbital cycle for a mars-large satellite system for one martian year (780 earth days; each quarter of the year = 195 days). With a rotation rate of 24.6 hr/day, there would be 761 days/martian year. With a rotation rate of 10 hr/day, there would be about 1872 mars days/martial year. With a large satellite, the obliquity would remain fairly constant over millions of years. Without a large satellite, the obliquity would vary considerably over thousands of years. Seasons are for the northern hemisphere

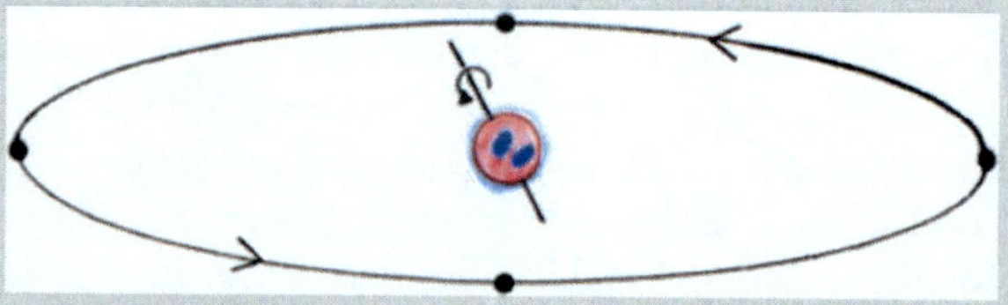

Fig. 9.7 Schematic pictogram of a monthly lunar cycle for a large martian moon. This diagram shows the prograde satellite, Concordia, at four different positions along a circular satellite orbit. The obliquity (tilt) angle would remain nearly constant at 30 degrees for millions of years. Because of the rock tide action during capture, much of the primitive crust would be recycled into the mantle and water and other volatiles would be released to the atmosphere of the mars-like planet. There would be ocean basins (dark blue), land including volcanoes (pink), and atmosphere (light blue)

mars-like planet rotating at 10 hr/day and the captured large satellite, the orbit, after about 1 Ga of orbital evolution, would be at about 45 mars radii and the rotation rate of the planet would be about 15 hr/day. Thus with reasonably high rock and ocean tides and a fairly short day/night cycle, the conditions for habitability for a primitive biota would be favorable. Figure 9.7 illustrates a typical monthly lunar cycle for a Mars-Concordia system.

This planet-satellite system would have a chance of being habitable (i.e., the planet would be "GEOFORMED"). The message is that when we search for habitable terrestrial exoplanets, I suggest that we also search for terrestrial exoplanets with a large satellite relative to the mass of the planet and a reasonably rapid rotation rate. The satellite must be large enough to cause the desired tidal activity but not so large as to despin the planet too rapidly (i.e., the Goldilocks Principle).

9.1.6.6 Discussion of the Models

The five basic models for the evolution of Mars and/or a Mars-planetoid system over geologic time (one real and four theoretical) do relate to the other three planets and/or planet-planetoid systems that are present in the terrestrial planet realm in our solar system. BUT, the special angular momentum exchange relationship (Grinspoon 1997; Laskar 1994, 1995, 1996; Baszo et al. 2010) between the orbits of Earth and Venus which induces orbital stability to these terrestrial planets may be very important for the development of a biological system with significant longevity on planet Earth (Fig. 9.7).

9.2 Some Notes on the Situation of Earth, Venus, and Mercury Without a Satellite

A question that we need to explore in this section is: WHAT WOULD THE EARTH BE LIKE WITHOUT THE MOON??? My first guess is that it might be somewhat similar to the condition of planet Mars at the present time in solar system history, but the Earth is ~10 times more massive and twice as large in diameter relative to Mars. A similar statement works for the possible condition of Venus without the influence of a large phantom satellite. And the situation of planet Mercury would be as it is at present because it never did have a large satellite. The pictogram in Fig. 9.8 shows my concept of what the earth would be like <u>without</u> our large, gravitationally captured, Vulcanoid planetoid satellite in prograde orbit that I call LUNA. First of all let us consider some of the outstanding features of an earth-like planet in an earth-like heliocentric orbit but without a large satellite.

1. The early history and the development of a Bedard (2006) type crust would occur.
2. Themis-type asteroids would bring water and other volatiles, including amino acids, to the earth-like planet.
3. Oceans of water would cover the primitive crust and a substantial atmosphere would blanket the planet.
4. The rotation rate would be about 10 hr/day (perhaps 12 hr/day) because of the "breaking" influence of solar gravitational tides) and there would be considerable dynamic activity in the atmosphere and oceans because of the Coriolis force and associated gyre activity on a planet with a rotation rate about twice what we have today.
5. There would probably be no land above sea level with the exception of raised rims of impact craters. These raised rims would be eroded to sea level in a brief period of geologic time.
6. There would probably be a magnetic field and associated magnetosphere. The strength and characteristics of the magnetic field are difficult to assess.

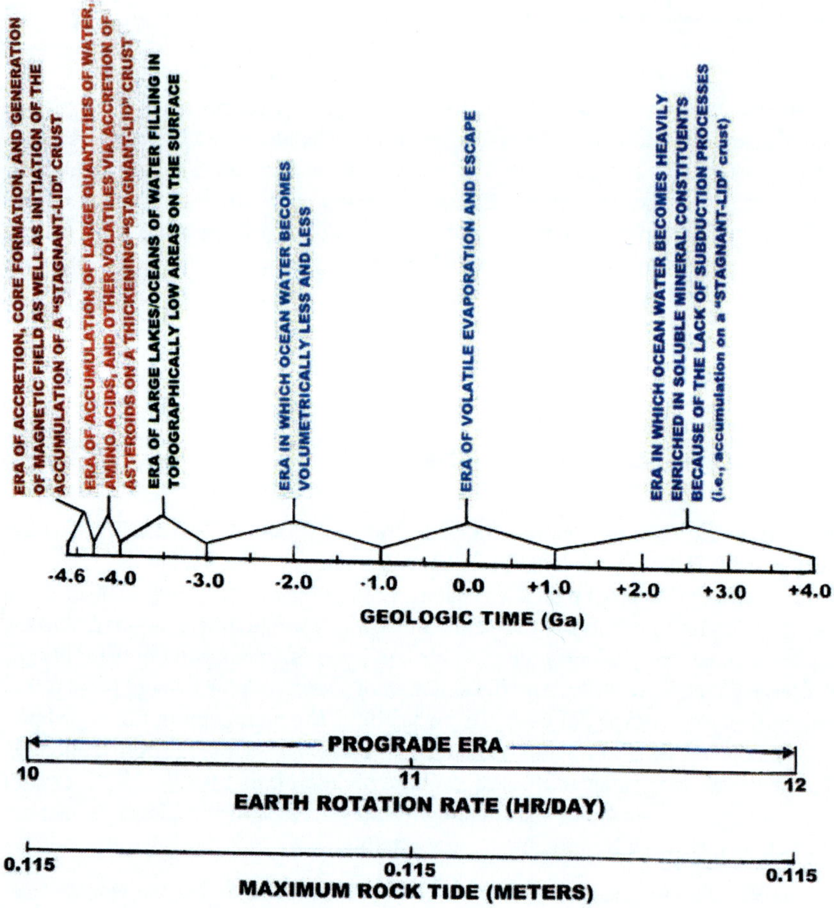

Fig. 9.8 Pictogram of the major eras for the evolution of an earth-like planet without association with a large satellite. Since I propose that gravitational capture of the Earth's moon occurred about 3.95 Ga ago, the events until the time of capture are the same for both models because in neither model is there a large satellite. For the case of a moon-less Earth there would be abundant quantities of water and other volatiles and these materials would remain in or near the surface of the planet for a long period of geologic time. There would be plenty of activity in the realm of the oceans and the atmosphere because of the rapid rotation of the planet and the associated Coriolis force. Because of the consistent but very, very low amplitude solar gravitational tide (about 11.5 cm at 1.0 AU), there would be no plate tectonics and probably no (or very little) crust of the planet above sea level. Eventually, large conic form volcanoes would form on a stagnant-lid crust above mantle hot spots (similar in nature to the large conic volcanic forms on Mars). The earth-like planet would tend to hold its water and volatiles for a long period of geological time, because of its large size and gravity relative to Mars, but eventually the surface may look something like the surface of Mars. The water and other atmospheric volatiles would evaporate and dissociate without replenishment due to a lack of a water storage and recycling mechanism because of the lack of slab-pull crustal subduction and associated tidally assisted mantle convection

9.2.1 Earth

Figure 9.8 is a pictogram illustrating my concept of the major features of earth evolution over geologic time without a large satellite (*i.e.*, a maroformed earth-mass planet).

The following is a synopsis of my rendition of the history of an earth-like planet in an essentially circular orbit at 1.0 AU from the sun:

4.6 to 4.3 Ga

Accretion, core formation, and generation of a magnetic field and establishment of a magnetosphere. Earth also accumulates an inventory of water, other volatiles, and amino acids via Themis-type asteroids (i.e., via Albarede 2009, model). The quantity of water would probably be about 2× to 4× the quantity of water in the oceans today.

4.3 to 4.0 Ga

Accumulation of primitive crust of Earth via vertical serial processing of basalt via the Bedard (2006) model. A small volume of rocks of felsic composition (including zircon crystals) would be included in this essentially stagnant-lid crustal complex. However, the bulk of this crustal complex would be of basalt/gabbro composition.

4.0 to 3.0 Ga

Continued vertical processing of crust featuring the accumulation of very large basaltic-andesitic conic-form volcanoes. The hot-spot heat source would be in the mantle and the crust-mantle complex would be rotating together (no significant differential rotation). A thickening stagnant-lid crust is developing on the planet. The large volcanic structure would depress the crustal complex locally but hydrologic cycle activity would be continually eroding and redistributing the volcanic debris to near sedimentary base level (i.e., sea level) or below.

3.0 to 1.0 Ga

Large quantities of water would be present but no deep ocean basins. Rotation rate of the planet remains at about 10 hr/day. Rapid erosion of volcanic forms continues. The atmosphere is very active because of rapid rotation rate and much water vapor is in the atmosphere. Rapid movement of atmospheric and oceanic gyres continues. Perhaps nine Hadley cells would form from pole to pole on this type of terrestrial planet. There is no moon so the obliquity and precession rates would be very irregular (Laskar 1994, 1995, 1996) but the heliocentric eccentricity cycles would continue to operate as usual because they are controlled by the gravitational perturbations of Jupiter and Saturn. There is significant atmospheric escape during this era via (1) dissociation of carbon dioxide to carbon monoxide plus free oxygen and (2) water to hydrogen (escape) and free oxygen. There is never enough oxygen to oxidize all surface rocks so there is no accumulation for free oxygen in the atmosphere and NO OZONE layer development. Biological entities could continue to form in hydrothermal vents on the ocean floor but there appears to be very little impetus for biological evolution to higher forms of life.

1.0 to +1.0 Ga

Much the same surface processes and conditions but surface water becomes volumetrically less and less. If there is no subduction then there is no mechanism for getting the surface water into the mantle so that it can be re-emitted by volcanic processes. Thus, as time goes on, the volume of ocean water gets less and less and the depth of the ocean basins gets less and less.

+1.0 to +4.0 Ga

During this era ocean water becomes heavily enriched in minerals because there is no subduction or subsidence mechanism for getting these mineral enriched segments of the crust into the mantle. This sedimentary rock material then remains at the surface. The result is a stagnant-lid crust with local down-warps and up- warps. Occasional bolide impacts disrupt the crust but fairly rapid sedimentation "heals" the impact features but this bolide impact mechanism is one mechanism for moving small sections of the crust into the earth-like planet's mantle. Bacteria and algae can be present from about 4.0 Ga but the stresses for organic evolution would be limited to (1) rapid rotation rate and associated atmospheric and oceanic gyres and (2) the irregular "wind-tide" fluctuations caused by atmospheric and oceanic currents operating on the limited shorelines of the planet.

COMMENT: Needless to say, this theoretical construct of an earth-like planet in an earth-like heliocentric orbit contrasts greatly with the global environmental conditions of our planet today as well as what we know about the geological evolution of our planet.

9.2.2 *Venus (Without a Large Satellite in its History as a Planet)*

What would a venus-like planet in a venus-like heliocentric orbit look like if it had no interaction with a large satellite? In my view, the features would be very similar to its TWIN SISTER planet Earth. The timescale for development is a bit shorter because of the planet's shorter distance to the Sun, a slightly different primordial rotation rate, and an increase in solar gravitational tidal influence. The heliocentric orbital resonance between Earth and Venus as discussed by Baszo et al. (2010) would be operational throughout the histories of both planets. Although this orbital resonance maintains the heliocentric orbit eccentricities within narrow limits, it still would not, in my view, make either planet very habitable for a biological system as we understand it. I think that regular tidal action caused by having a large satellite in orbit is very important for development of a biological system beyond the bacterial and/or algal stage.

Figure 9.9 is a pictogram illustrating my concept of the major features of the evolution of Venus without a large satellite.

The following is a synopsis of my rendition of the history of a venus-like planet in an essentially circular orbit at about 0.7 AU from the Sun (i.e., a maroformed venus-mass planet):

4.6 to 4.3 Ga

Accretion, core formation, and generation of a magnetic field and associated magnetosphere. The early crust forms as a silicate "slag" over a more mafic mantle; the "slag" than collects and thickens to form a global silicate layer. The planet then accumulates an inventory of water, other volatiles, and some amino acids via the Albarede (2009) model. Venus probably accumulated less of this water and volatile rich asteroidal material than Earth but enough for a global ocean. The primitive crust continues to accumulate, mainly via vertical accretion, under a partial or complete cover of ocean water.

4.3 to 4.0 Ga

Continued accumulation and development of primitive crust via a process of vertical serial processing via the Bedard (2006) model which includes delamination of higher specific gravity material at base of crust which then is assimilated by the underlying mantle. Like Earth and Mars, late impacts in the retrograde direction would predominate over prograde impactors. Nearly all prograde impactors would need to be direct hits or heavily grazing encounters because prograde encounters in which there is insufficient energy dissipation for capture result in escape back into heliocentric orbit rather than collision. In contrast, most retrograde encounters in which there is insufficient energy dissipation for capture result in a collision after a few orbits. Some of these failed retrograde encounters result in prograde collisions but the vast majority of these collisions are in the retrograde direction. Thus, most failed capture scenarios result in retrograde collisions and each of these would cancel some of the angular momentum of the prograde rotation of the venus-like planet. Thus the primordial rotation rate of Venus [somewhere between 13.5 and 16 hr/day, depending on the placement of the line on the MacDonald (1963) diagram] decreases over geologic time. This impact slowing of rotation rate operates in conjunction with a very slow process of decreasing the rotation of the venus-like planet by solar gravitational rock and ocean tides. As on the earth-like planet, large conic-form volcanoes of andesitic and basaltic andesitic composition would tend to accumulate over mantle hot spots because of the lack of differential rotation between the crustal complex and the mantle.

4.0 to 3.0 Ga

This is a lengthy era during which the prograde rotation rate of Venus is very slowly decreased via rock and ocean tides caused by the constant and persistent solar gravitational tides. With a constant and persistent rock tidal amplitude of about 30 cm (the ocean tidal amplitude is about 1.4 times higher at 42 cm), the rotation rate of the planet would slowly decrease by a few hours per day over geologic time. Since solar energy is somewhat greater at the distance of venus-like planet vs. the earth-like planet, the evaporation and molecular dissociation processes causing escape of some of the planet's volatiles would occur more rapidly

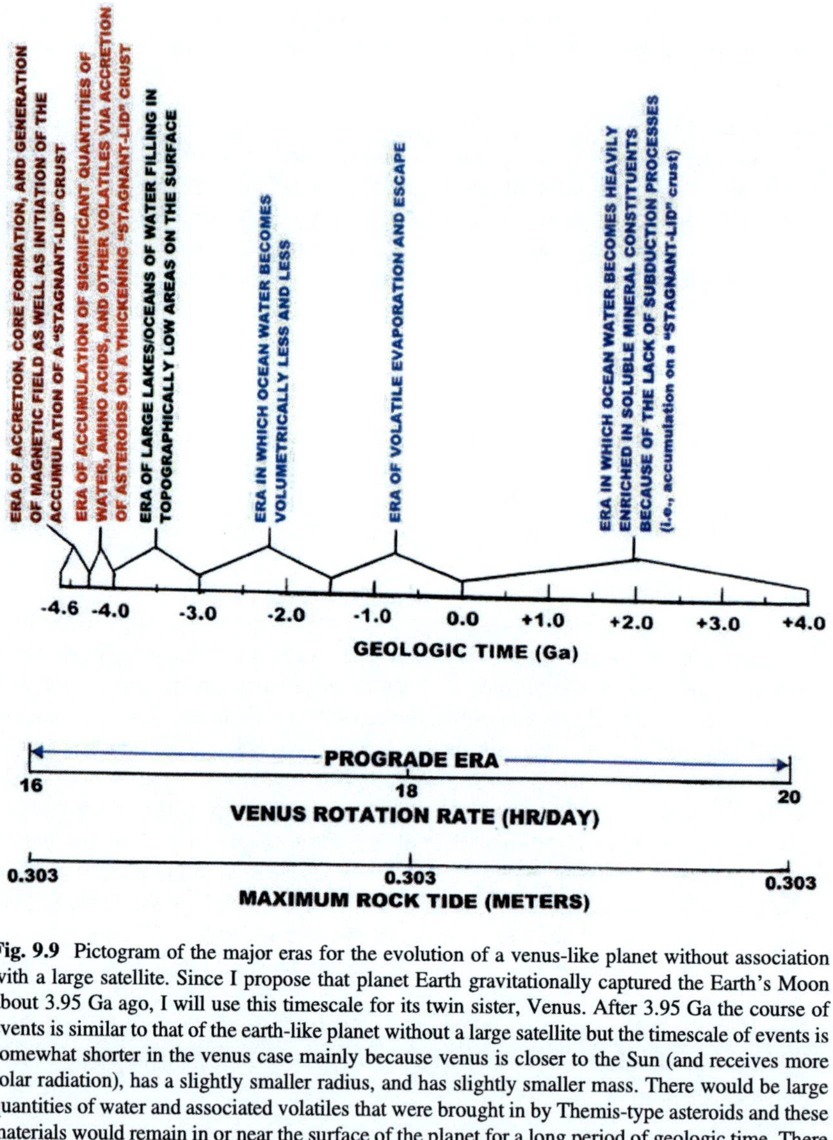

Fig. 9.9 Pictogram of the major eras for the evolution of a venus-like planet without association with a large satellite. Since I propose that planet Earth gravitationally captured the Earth's Moon about 3.95 Ga ago, I will use this timescale for its twin sister, Venus. After 3.95 Ga the course of events is similar to that of the earth-like planet without a large satellite but the timescale of events is somewhat shorter in the venus case mainly because venus is closer to the Sun (and receives more solar radiation), has a slightly smaller radius, and has slightly smaller mass. There would be large quantities of water and associated volatiles that were brought in by Themis-type asteroids and these materials would remain in or near the surface of the planet for a long period of geologic time. There would be significant activity in the oceanic and atmospheric systems because of the rapid rotation rate of the planet and the associated Coriolis force. The primordial rotation rate for this model is between 13.5 and 16 hr/day (MacDonald 1964). The solar gravitational tidal action (rock and ocean tides) would slow the rotation rate from 16 hr/day to perhaps 20 hr/day over the 8 Ga of time illustrated on the diagram. Because of the weak solar rock tidal action, there would be no plate tectonics and probably no or little crust above sea level. Eventually, large conic-form volcanoes would form on a stagnant-lid crust above mantle hot spots (similar in nature to the large conic volcanic forms on planet Mars at the present time) and some impact craters with raised rims above sea level. This venus-like planet would tend to lose water and volatiles more rapidly than the earth-mass planet discussed earlier because of its closer position to the sun and lower gravitational field strength but eventually the surface of this venus-like planet may look something like the surface of

on Venus. As long as there is a "healthy" hydrologic cycle operating on the planet, conic volcanic forms are weathered and eroded to form immature sedimentary rocks. In time the ocean water slowly evaporates and some conic volcanic features continue to accrete and dominate the landscape of the venus-like planet.

3.0 to 1.5 Ga

This era is characterized by continuing water and other volatile loss via evaporation of molecular dissociation in the upper atmosphere of the planet. Like Earth without a moon, bacteria and algae may form and thrive but the big question is: Is there much of an impetus in this venus-like planet system for organic evolution? There is no doubt that there would be much surface action via vigorous atmospheric and oceanic circulation patterns with an 18 hr/day prograde rotation rate and some continuing conic form volcanism but there would be very little rhythmic activity along the shorelines from the low amplitude solar rock and oceanic gravitational tides. Wind-driven air and ocean currents would be the main incentive for organic evolution during this era. In addition, hydrothermal vent activity on the ocean floor could harbor some life forms. The Milankovitch parameters would be operating from the end of the accretion process. The heliocentric orbital eccentricity is controlled by gravitational interaction with the Sun and neighboring planets. The obliquity and precession activity, however, is very irregular and this irregularity can have an effect on any life forms on this version of a venus-like planet without a large satellite.

1.5 to 0.0 Ga

This era is more of the same but the internal heat engine powered by decay of radioactive elements is decreasing in intensity resulting in a decrease in mantle convection activity and surface volcanism. The planet continues to very slowly lose angular momentum via the solar tidal breaking mechanism during this era and water and atmospheric volatiles continue to escape from the upper atmosphere of the planet.

0.0 to +4.0 Ga

This era is characterized by continued precipitation of soluble ions from ocean water because of the continuing processes of evaporation of ocean water and dissociation of gaseous molecules via solar ultraviolet rays. The ocean water becomes highly enriched in various soluble constituents that were derived from the surface rocks. These dissolved constituents then get precipitated on the ocean floor and remain there because of the lack of subduction processes. This results in extensive layers of precipitated sedimentary rocks on a stagnant-lid continental crust of regional or global extent.

Fig. 9.9 (continued) Mars. The water and other atmospheric volatiles would evaporate and dissociate in the upper atmosphere. Without a replenishment mechanism for water and other volatiles due to the lack of crustal subduction and mantle convection, large accumulations of sedimentary precipitates would accumulate on the shallow ocean floors of a stagnant-lid crustal complex

COMMENT: Needless to say, this theoretical construct of a venus-like planet in a venus-like heliocentric orbit contrasts greatly with the global environmental conditions of planet Venus today (i.e., a good example of a "hades-like" condition). This contrast in surface and atmospheric conditions is reasonable evidence that planet Venus was interacting with a large satellite that was captured into a retrograde orbit early in the history of Venus. In general, the condition of planets Earth and Venus without the influence of captured satellites would be very similar (i.e., something like "TWIN SISTERS").

9.2.3 Mercury (Without a Large Satellite)

What would planet Mercury be like without a large satellite? Reasonable answer: JUST LIKE IT IS TODAY! That is, planet Mercury without a large satellite is the Mercury that we have today (the prototype of a mercoformed planet).

Figure 9.10 is a pictogram illustrating my concept of the major features of the evolution of a mercury-like planet in a mercury-like heliocentric orbit. A mercury-like planet in a mercury-like heliocentric orbit is the planet Mercury that we have today.

The following is a synopsis of my rendition of the history of a mercury-like planet in a mercury-like heliocentric orbit (i.e., about 0.4 AU from the Sun):

This mercury-like planet accreted from the materials that were in or near its accretion torus. Since the Iron Line (Malcuit 2015) is within the accretion torus of the mercury-like planet, the resulting planet has a much higher content of iron and nickel than would be expected for a planet at the distance of Mercury from the Sun without the Iron Line. Much of the iron and nickel would settle to make a metallic core for this planet and differential rotation caused by the accretion process and core formation would cause the generation of a global magnetic field very early in the history of the planet.

A primitive crust forms from the lower density silicate material associate with the planet and a serial processing of this primitive crust begins via the Bedard (2006) model. The main difference between this mercury-like planet and the other terrestrial planets is that this mercury-like planet is very rich in iron and nickel. Although most of the surface and near surface material is composed mainly of silicates, the composition of these silicates is more iron-rich than those associated with the other terrestrial planets. The higher metal content of this magma is characterized by a lower viscosity than normal basalts (Taylor 2001). Metal-rich, high specific gravity, and low viscosity lavas are expected to accumulate on the surface of the planet. The large conic forms associated with andesitic compositions that are characteristic of the model venus-like planet and the earth-like planet are much less common or absent.

The primordial rotation rate following the accretion process as estimated from the chart in MacDonald (1963) is about 20 hr/day. This rotation rate is decreased over a brief period of geologic time initially by the persistent 70 cm solar gravity induced

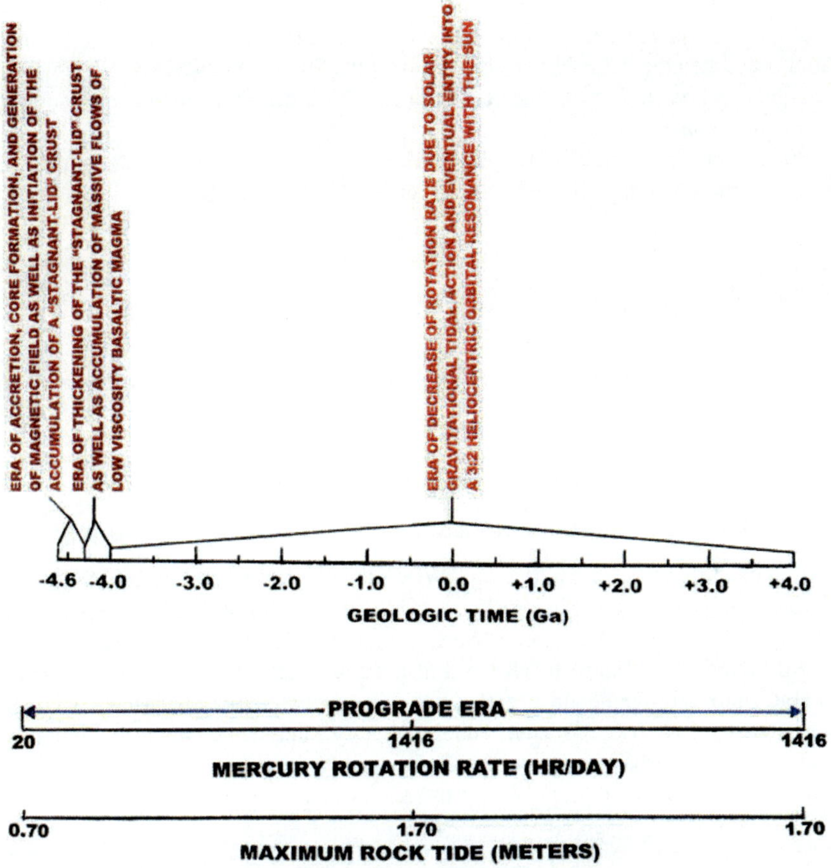

Fig. 9.10 Pictogram of the major eras for the evolution of a mercury-like planet without association with a large satellite. For the early history of the mercury-like planet I will use the same timescale as for the venus-like planet as well as the earth-like planet. The remainder of the history of the mercury-like planet is roughly congruent with what we know about planet Mercury today. This history will be presented in the following section

rock tide and eventually by the 170 cm perigean solar gravitational tide associated with the 3:2 heliocentric spin-orbit resonance with the Sun. This 3:2 resonance takes control of the heliocentric orbital eccentricity as well as the rotation rate of a mercury-like planet.

A few Themis-type asteroids can be deflected by planet Venus and eventually impact on the mercury-like planet but any water and associated volatiles and amino acids would have a short lifetime on the surface of Mercury because of the intense solar radiation. These sensitive volatile and pre-biotic materials, however, could survive much longer in the polar zones of the planet than at lower latitudes.

The Milankovitch parameters may affect planet Mercury for a short period of geological time. Any early control of the heliocentric orbital eccentricity would disappear when the 3:2 orbital resonance with the Sun commences. Any changes in obliquity and precession are strongly controlled by the nearby Sun and the action of the 3:2 resonance.

In general, the condition of planet Mercury today is very similar to what is expected for a mercury-like planet at about 0.4 AU from the Sun.

9.2.4 *Discussion for this Section*

Let us look again at a quote from Nimmo and Tanaka (2005) that occurred earlier in this chapter:

> Mars is in some ways the type example of a terrestrial planet. It is neither geologically stillborn, like Mercury or the Moon, nor so active that most of the geological record has been destroyed, like Venus or the Earth. The prolonged geological evolution of Mars is recorded in the physical and chemical characteristics of its crust.—From Nimmo and Tanaka (2005)

I contend that the reason that the TWIN SISTER planets, Venus and Earth, are devoid of most of their surface rocks is because they had prolonged interaction with gravitationally captured satellites. Venus captured a 0.5 moon-mass satellite into a retrograde orbit early in its history as a planet and Earth captured a 1.0 moon-mass satellite into a prograde orbit early in its history. Mercury and Mars, although separated by over 1.0 AU have endured a very similar geological evolution to each other mainly because they did not have any interaction with stably captured large planetoids relative to the size of these planets.

9.3 Implications for the Analysis of Gaseous Exoplanets in Light of the Models for the Evolution of Planets Neptune and Uranus via Retrograde Capture of Fairly Large "Icy" Satellites for these Planets

When we examine the current data set for exoplanets we find very few examples of terrestrial planets (e.g., Ollivier et al. 2009b; Shahar et al. 2019). We may ask the questions: Are terrestrial planets rare? Or are most exoplanets similar in nature to our gaseous planets? I think that most scientists that are interested in habitability issues would agree that we should concentrate on rocky planets but, in some cases, gaseous planets with satellites may also give us insight into the origin of planet-satellite systems.

The major point of this section is that the gravitational capture process works just as well for gaseous planets as for terrestrial planets mainly because the planet is mainly a passive participant simply furnishing the gravitational field for dissipating the tidal energy for capture within the encountering planetoid. But when we examine gaseous exoplanets we may find satellites in unusual angular configurations relative to the planet or unusual orbital relationships relative to what we may or may not find in our solar system. For example, we may find a gaseous planet that has an obliquity of 98 degrees relative to the common plane of the planets of that star-planet system. We have an example of this in our Solar System in the Uranus-satellite system but we do not have agreement on a good explanation. Another example is the Pluto-Charon system. Pluto has an unusually high heliocentric eccentricity with an inclination angle of about 15 degrees. There are very few models that have been proposed for the origin of Pluto and its major satellite, Charon, and the other very small satellites of Pluto. It is my view that these two unusual planet-satellite relationships may both involve the process of retrograde gravitational capture as illustrated in Chap. 8.

9.3.1 Analysis of the Situation of Planet Uranus

The currently popular model for explaining the 98 degree obliquity of planet Uranus along with the belt of uranian satellites in the equatorial plane of Uranus is that a body of somewhere between <u>one and three earth masses</u> impacted obliquely and nearly tangentially on planet Uranus. The orientation of the collision and the impact parameter of the collision were within a range to instantaneously cause the high (~98 degrees) obliquity (Slattery et al. 1992).

Some problems with this model are (1) the source region in the solar system for such a large body and (2) the mechanism for reorienting the natural satellite system of Uranus to the equatorial plane of the planet after the planet was essentially instantaneously tilted to 98 degrees.

An alternative model, described in some detail in Sect. 6.4.5.2 of this book, was presented by Kobo-Oka and Nakazawa (1995). Their model features retrograde capture of a ~ 20 moon-mass (~2 mars masses; 0.2 earth masses) planetoid into a urocentric orbit near the common plane of the planets of the solar system. After orbit circularization, the retrograde orbit of the captured planetoid gets smaller because of tidal interaction with Uranus. As a result, the obliquity of Uranus and its natural satellite system are gradually increased to 20 degrees, 50 degrees, 80 degrees and finally to 98 degrees. Concomitantly, the orbital radius of the large retrograde satellite decreases as the obliquity of the planet and associated satellite system increases. The obliquity increase <u>ends</u> as the large retrograde satellite breaks up in orbit and eventually coalesces with the body of Uranus.

A question is: How does this model work with a larger or smaller captured satellite in retrograde orbit about a Uranus-type? To answer this question, let us assume that the post-capture orbit after circularization is about the same dimension as in the previous model (about 10 Uranus radii).

9.3.1.1 Case A: A 4.0 Mars-Mass Planetoid as a Captured Retrograde Satellite

In this case the process of obliquity increase continues and would result in an obliquity of about 180 degrees. The large satellite, which now would be orbiting in the same direction as the planet is rotating and in the same direction of the natural satellites of Uranus are revolving, will eventually enter the Roche limit for a solid body (about 1.5 Uranus radii), break up in orbit, and the pieces, large and small, would coalesce with the body of Uranus. With an obliquity of 180 degrees, Uranus would be rotating in the retrograde direction with a rotation rate of about 10 hr/day and the satellite system would be revolving in the retrograde direction also. Depending on the timing of the events of the orbital evolution, the large retrograde satellite could still be intact in a retrograde orbit relative to the solar system but in the same direction as Uranus is rotating. Eventually the body of the captured satellite would break up in orbit and coalesce with the body of Uranus.

9.3.1.2 Case B: A 1.0 Mars-Mass Planetoid as a Captured Retrograde Satellite

In this case the obliquity increase of Uranus and its satellite system is limited to about 45 degrees. At this stage of orbital evolution the mars-mass planetoid is entering the Roche limit, breaking up in orbit, and the pieces are about to coalesce with the planet. Thus Uranus, along with its satellite system are simply undergoing an obliquity increase relative to the normal state.

What are the implications for exoplanets in the size range of Uranus and Neptune? Any planet in this size range can end up with a variety of obliquity states via this process of retrograde capture of sizeable satellites. We must keep in mind that impact of icy planetoids, large and small, can also occur from time to time. If they are in the mars-mass range, the effects on a Uranus or Neptune mass planet are negligible.

9.3.1.3 Discussion of the Uranus-Large Satellite Model and Application to Gaseous Exoplanets

Any stable gravitational capture needs to be near the plane of the planets. Prograde capture probabilities are very low for planets like Neptune and Uranus but the stable retrograde capture zones are much larger (see Chap. 8 for details). Thus, any capture model for these two planets needs to be in the retrograde direction.

9.3.2 *Analysis of the Neptune-Triton System*

Neptune is the only planet with an extant retrograde satellite. There are two models that have been proposed for the capture of Triton. Goldriech et al. (1989) suggest a collision assisted capture; Malcuit et al. (1992a, b) suggest a normal gravitational capture via tidal energy dissipation in Triton during a close encounter with Neptune. Neptune also has several satellites in prograde orbits; these bodies are considered to be from the original "swarm" of natural satellites. Nearly everyone agrees that given enough time, the now nearly circular orbit of Triton will decrease in radius and eventually, enter the Roche limit for a solid body, break up in parts, and coalesce with the body of planet Neptune. Note that the small mass of Triton is not sufficient to "tilt" planet Neptune.

What lessons are to be learned from analysis of the Neptune-Triton System that may be useful for analyzing icy or gaseous exoplanets? First of all, the geometry of stable capture zones for Neptune and a Triton-mass planetoid strongly suggest that prograde capture in this region of space is highly improbable (see Sect. 8.7 for the geometry of stable capture zones for planet Neptune and a Triton-mass planetoid). In contrast, the fairly large extent of retrograde stable capture zones make retrograde capture a much more probable scenario. We must remember that the fate of most retrograde satellites is eventual coalescence with the planet. The exception is when a retrograde satellite is large enough to despin the planet and then torque up the rotation rate to match the orbital motion of the satellite. For low-mass satellites that somehow get captured in retrograde orbit, the time-scale for eventually coalescence with the planet can be very long.

9.4 Summary

This chapter features mainly the analysis of the geological effects of the process of satellite capture by planets and the subsequent evolution of the planet-satellite systems. Planet Mars, the most "normal" of the terrestrial planets, is discussed in detail. It is suggested that any biological system on Mars as a stand-alone planet with a 24.6 hr/day rotation rate could possibly exist for about 2 billion years. Prograde capture of a 0.1 or 0.2 moon-mass satellite could harbor life for about 2 or 2.5 billion year. The major detriment to biological systems on Mars for the prograde capture model is the fairly rapid decrease in rotation rate, and concomitant increase in length of day, because of the fairly slow primordial rotation rate for Mars. If somehow a mars-like planet can be endowed with a pre-capture prograde rotation rate of about 10 hr/day, then prograde capture of a sizeable satellite could result in a mars-like planet with conditions favorable for habitability. Capture of either a 0.1 or 0.2 moon-mass satellite in retrograde orbit for Mars looks favorable for development of a biological system on the planet for a period of 3 billion years and perhaps up to

4 billion years. But eventually, given enough time, most retrograde capture scenarios result in a "Hades-like" planet.

Analysis of planets Mercury, Venus, and Earth without a large satellite illustrates the value of having a large satellite relative to the size of the planet. Mercury is similar to the present day condition of Mars in that it is a one-plate planet; Mercury is different in that is has lost its original rotation rate via tidal interaction with the Sun.

Without interactions with large moons, planets Venus and Earth would probably look like "TWIN SISTERS". They would probably be one-plate planets with rotation rates very close to their primordial rates as presented in the MacDonald (1963, 1964) plot. Venus and Earth would enter into a heliocentric orbital resonance described by Baszo et al. (2010). A biological system would probably form and may survive in some form for 4.6 billion years (i.e., as long as liquid water is at the surface) but because of the lack of significant tidal influence, the impetus for causing organic evolution, in my opinion, would be lacking.

The bottom line is that large satellites can be very important for making terrestrial planets habitable. In some cases retrograde capture may yield the best results for explaining the condition of the planet (i.e., planet Venus, the "HADES" planet). In other cases, prograde capture yields a favorable outcome (i.e., planet Earth, the "PARADISE" planet).

Although satellite capture processes may be important for the outer gassy-icy planets of the Solar System, such capture processes have little to do with development of habitable conditions on either the planet or the satellite but they do suggest solutions to some Solar System problems. The 98 degree tilt (obliquity) angle of Uranus may be explained by gravitational capture of a 20 moon-mass planetoid in retrograde orbit early in the history of Uranus. The retrograde orbit of Triton (a major satellite of Neptune) may be explained by gravitational capture of a Kuiper Belt planetoid in retrograde orbit around Neptune early in the history of Neptune. The orbit of Triton has evolved via tidal dissipation processes mainly in Triton into a circular retrograde orbit with the rotation of Triton synchronous with its orbital motion around Neptune. The orbit of Triton has an inclination of about 20 degrees relative to the common planet of the solar system.

References Cited

Albarede, F. (2009). Volatile accretion history of the terrestrial planets and dynamic implications. *Nature, 461*, 1127–1233.

Arkani-Hamed, J. (2009). Did tidal deformation power the core dynamo of Mars? *Icarus, 201*, 31–43.

Baker, V. R. (2005). Picturing a recently active Mars. *Nature, 434*, 280–283.

Baker, V. R., Strom, R. G., Gulick, V. C., Kargel, J. S., Komatsu, G., & Kale, V. S. (1991). Ancient oceans, ice sheets and the hydrologic cycle on Mars. *Nature, 352*, 589–594.

Baszo, A., Dvorak, R., Pilat-Lohinger, E., Eybl, v., & Lhotka, C. (2010). A survey of near-mean-motion resonances between Venus and Earth. *Celestial Mechanics and Dynamical Astronomy, 107*, 63–76.

Bedard, J. H. (2006). A catalytic delamination-driven model for coupled genesis of Archaean crust and sub-continental lithospheric mantle. *Geochimica et Cosmochimica Acta, 70*, 1188–1214.

Bell, J. (2008). *The martian surface: Composition, mineralogy, and physical properties* (p. 636). Cambridge: Cambridge University Press.

Carr, M. (1996). *Water on mars* (p. 229). New York: Oxford University Press.

Citron, R. I., Manga, M., & Hemingway, D. J. (2018). Timing of oceans on Mars from shoreline deformation. *Nature, 555*, 643–646.

Counselman, C. C., III. (1973). Outcomes of tidal evolution. *The Astrophysical Journal, 180*, 307–314.

Elkin-Tanton, L. T., Parmentier, E. M., & Hess, P. C. (2003). Magma ocean fractional crystallization and cumulate overturn in terrestrial planets: Implications for Mars. *Meteoritics and Planetary Science, 38*, 1753–1771.

Frey, H. (2008). Ages of very large impact basins on Mars: Implications for the late heavy bombardment in the inner solar system. *Geophysical Research Letters, 35*, L13203. https://doi.org/10.1029/2008GL033515.

Goldriech, P., Murray, N., Longaretti, P. Y., & Banfield, D. (1989). Neptune's story. *Science, 245*, 500–504.

Grinspoon, D. H. (1997). *Venus revealed: A new look below the clouds of our mysterious twin planet* (p. 335). Reading: Addison-Wesley Publishing Co.

Herrick, R. R. (1994). Resurfacing history of Venus. *Geology, 22*, 703–706.

Hurowitz, J. A., Grotzinger, J. P., Fischer, W. W., & (and 20 additional co-authors). (2017). Redox stratification of an ancient lake in Gale crater, Mars. *Science, 356*, (10 p.), eaah6849. https://doi.org/10.1126/science.aah6849.

Kobayashi, D., & Sprenke, K. F. (2010). Lithospheric drift on early Mars: Evidence in the magnetic field. *Icarus, 210*, 37–42.

Kobo-Oka, T., & Nakazawa, K. (1995). Gradual increase in the obliquity of Uranus due to tidal interaction with a hypothetical retrograde satellite. *Icarus, 114*, 21–32.

Laskar, J. (1994). Large-scale chaos in the Solar System. *Astronomy and Astrophysics, 287*, L9–L12.

Laskar, J. (1995). Large scale chaos and marginal stability in the Solar System. In *XIth International Congress of Mathematical Physics* (p. 120). International Press.

Laskar, J. (1996). Large scale chaos and marginal stability in the solar system. *Celestial Mechanics and Dynamical Astronomy, 64*, 115–162.

MacDonald, J. G. F. (1963). The internal constitution of the inner planets and the Moon. *Space Science Reviews, 2*, 473–557.

MacDonald, J. G. F. (1964). Tidal friction. *Reviews of Geophysics, 2*, 467–541.

Malcuit, R. J. (2015). *The twin sister planets, Venus and earth: Why are they so different?* (p. 401). Cham: Springer International Publishing.

Malcuit, R. J., & Winters, R. R. (1996). Numerical simulations of gravitational capture scenarios for planets Venus and Earth: Implications for the contrasting thermal histories of these planets. In *Abstracts Volume, 30th International Geological Congress* (Vol. 3, p. 506). Beijing.

Malcuit, R. J., & Winters, R. R. (2009). A retrograde planetoid capture model for planet Venus: Implications for the Venus oceans problem, an era of habitability for Venus, and a global resurfacing event about 1.0-0.5 Ga ago. *Geological Society of America: Abstracts with Programs, 41*(7), 266.

Malcuit, R. J., Mehringer, D. M., & Winters, R. R. (1989). Numerical simulation of gravitational capture of a lunar-like body by earth. In *Proceedings, 19th Lunar and Planetary Science Conference* (pp. 581–591). Houston: Cambridge University Press/Lunar and Planetary Institute.

Malcuit, R. J., Mehringer, D. M., & Winters, R. R. (1992a). *Numerical simulation of retrograde tidal capture of a triton-like planetoid by a neptune-like planet: Two-dimensional limits of a stable capture zone: Abstracts, Lunar and Planetary Science Conference XXIII* (pp. 827–828). Houston: Lunar and Planetary Institute.

Malcuit, R. J., Mehringer, D. M., & Winters, R. R. (1992b). A gravitational capture origin for the Earth-Moon system: Implications for the early history of the Earth and Moon. In J. E. Glover & S. E. Ho (Eds.), *Proceedings Volume, 3rd International Archaean Symposium* (Vol. 22, pp. 223–235). Crawley: The University of Western Australia.

Nimmo, F., & Tanaka, K. (2005). Early crustal evolution of Mars. *Annual Reviews of the Earth and Planetary Sciences, 33*, 133–161.

Ollivier, M., et al. (2009a). Chapter 9: The search for life in planetary systems. In *Planetary systems: Detection, formation and habitability of extrasolar planets* (pp. 319–320). Berlin: Springer-Verlag.

Ollivier, M., Encrenaz, T., Roques, F., Selsis, F., & Casoli, R. (2009b). *Planetary systems: Detection, formation and habitability of extrasolar planets* (p. 340). Berlin: Springer-Verlag.

Phillips, R. J., Zuber, M. T., Solomon, S. C., Golombek, M. P., Jakosky, B. M., Banerdt, W. B., Smith, D. E., Williams, R. M. E., Hynek, B. M., Aharonson, O., & Hauck, S. A. I. I. (2001). Ancient geodynamics and global-scale hydrology on Mars. *Science, 291*, 2587–2591.

Piro, A. L. (2018). Exoplanets torqued by the combined tides of a moon and parent star. *The Astronomical Journal, 156*, 54. (10 p.).

Rosenblatt, P., Charnoz, S., Dunseath, K. M., Terao-Dunseath, M., Trinh, A., Hyodo, R., Glena, H., & Toupin, S. (2016). Accretion of Phobos and Deimos in an extended debris disc stirred by transient moons. *Nature Geoscience, 9*, 581–583.

Salese, F., Pondrelli, M., Neeseman, A., Schmidt, G., & Ori, G. G. (2019). Geological evidence of planet-wide groundwater system on Mars. *Journal of Geophysical Research, 124*, 374–395. https://doi.org/10.1029/2018JE005802.

Scheinberg, A., Elkins-Tanton, L. T., & Zhong, S. J. (2014). Timescale and morphology of martian mantle overturn immediately following magma ocean solidification. *Journal of Geophysical Research: Planets, 119*, 454–467.

Shahar, A., Driscoll, P., Weinberger, A., & Cody, G. (2019). What makes a planet habitable? *Science, 364*, 434–435.

Slattery, W. L., Benz, W., & Cameron, A. G. W. (1992). Giant impacts on a primitive Uranus. *Icarus, 99*, 167–174.

Taylor, S. R. (2001). *Solar System evolution: A new perspective* (2nd ed., p. 460). Cambridge: Cambridge University Press.

Willhelms, D. E., & Squyres, S. W. (1984). The hemispheric dichotomy may be due to a giant impact. *Nature, 309*, 138–140.

Chapter 10
A Discussion of Three Major Paradigms in the Earth and Planetary Sciences

> With humans' initial exploration of the Moon at the dawn of the space age came the birth of modern planetary science, which seeks to understand the complex history of our solar system in enough detail to piece together the story of how we got to where we are. To what extent was Earth an inevitable outcome of planetary evolution? To what extent was it a quirk of fate?—From Denevi (2017).

The dynamic atmospheric, hydrospheric, and lithospheric subsystems of planet Earth are indeed complex operations and are difficult for we humans to comprehend and understand in detail. The same can be said for subsystems of planet Mars. Although the Mars operation is smaller in size, the planet must be observed from a distance which should make it a bit easier to study, but really somewhat more difficult to understand in detail. For both planets, the present conditions are much easier to assess than the long-term evolution of the two planets.

In this chapter I am presenting information on the two leading paradigms in the Earth Sciences, the paradigm of PLATE TECTONICS and the paradigm of the MI LANKOVITCH CYCLES. Then I am going to demonstrate how another major concept may relate to these two beacons of research. This third idea brings our large satellite, the MOON, into focus as part of the Earth-Moon system. We could call it the " EVOLUTION OF THE EARTH-MOON SYSTEM"paradigm, or the "LUNAR" paradigm or the "EARTH TIDE"paradigm. I will admit here that there is a good bit of literature on the ocean tides and how they operate on Earth at present and how they operated throughout earth history but there is very little information in the literature on the operation and importance of the rock tides on our planet. Nevertheless, it is the rock tides (i.e., the solid earth tides) that appear to be an ancillary force in the operation of both PLATE TECTONICS and the MILANKOVITCH CYCLES.

Upon completion of our discussion of these three related major concepts for operation of planet Earth, then we will apply all three of them to the dynamics of planet Mars.

R. Malcuit, *Geoforming Mars*, https://doi.org/10.1007/978-3-030-58876-2_10

10.1 The Earth Tide "Paradigm" for Planet Earth

The earth tides (the rock tides are a constant and persistent unidirectional process operating on Planet Earth and are thought by some scientists to be an ancillary driving force for the operation of Plate Tectonics (e.g, Bostrom 2000; Scoppola et al. 2006; Doglioni and Anderson 2015; Doglioni and Panza 2015, and several others).

10.1.1 A Brief History of the Concept

In the last chapter of the 3rd edition (which was published in German in 1922 and translated to English in 1924), Wegener (1924) addressed what he called geographic features that may relate to an apparent drift of the continents that might be explained by a westward "drag" on the Earth's surface by the weak, unidirectional, continuously operating rock tides. Wegener noted that some of these features were previously reported by Kreichgauer (1902) and Taylor (1910). Wegener (1924) proposed that the earth tide mechanism may be a major cause of the apparent westward movements of the continents over geologic time.

In an article titled "Westward displacement of the lithosphere," Bostrom (1971) rejuvenated the controversy by emphasizing the evidence for this westward drift from paleomagnetic studies. Then two petroleum geologists (Nelson and Temple 1972) published an article titled "Mainstream mantle convection: A geologic analysis of plate motion." This lengthy and well-illustrated paper emphasized the asymmetry of subduction zones on the planet. The main pattern they reported is that westward-dipping subducted plates tend to have a steep angle of descent and that the eastward-dipping subducted plates tend to have a low angle of descent. Another important pattern they noted was that marginal seas were mainly associated with westward-dipping subduction zones. Their explanation for these patterns was that the surface plates were somehow retarded by the earth tide mechanism and that there was a "mainstream" in the asthenosphere that moved the portion of the mantle below the asthenosphere at a slightly more rapid rate than that of the surface plates. These features are well-illustrated by the diagram from Ricard et al. (1991) reproduced in Fig. 10.1. There are several articles and/or sections in books published in subsequent decades that supported the concepts presented by Nelson and Temple (1972) (e.g., Moore 1973; Uyeda and Kanamori 1979; Hagar and O'Connell 1979; Cross and Pilger 1982; Doglioni 1994; Windley 1995, and several others).

An article by Jordan (1974) slowed progress on the Earth Tide Mechanism as a driving force for plate motion. Jordan (1974) demonstrated that if he used the estimated viscosity of the upper mantle, then the retarding force of the earth tides via the continuously operating tidal torque mechanism was several orders of magnitude too small to have much effect on plate tectonics. The article by Jordan (1974) did not negate any of the geographic surface patterns that had been noted by previous authors but it did cause investigators to search for a suitable physical mechanism within the context of the operation of the EARTH TIDES.

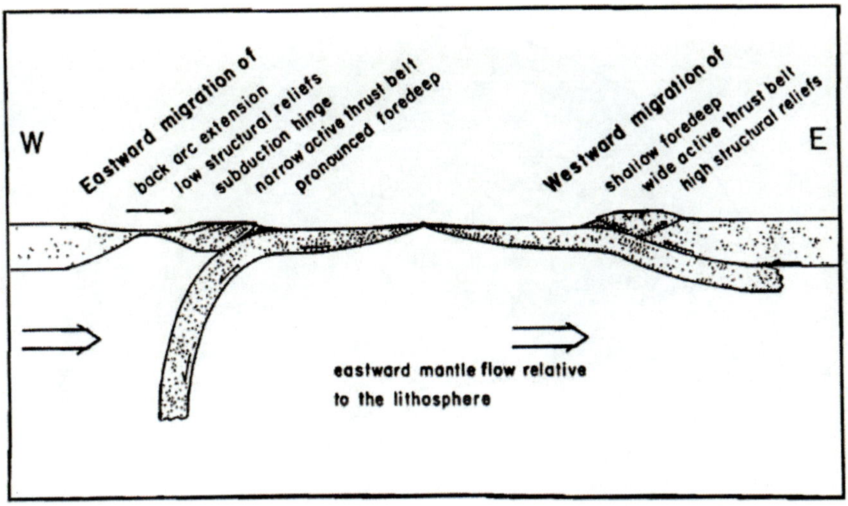

Fig. 10.1 Diagrammatic cross-section of two subduction zones showing contrasting angles of subduction. The large arrows indicate the direction of the mainstream of mantle flow. The mainstream of flow in the mantle is thought to cause the high subduction angle for the westward dipping Wadati-Benioff zone as well as the low subduction angle for the eastward dipping Wadati-Benioff zone. (Diagram from Ricard et al. (1991, Fig. 3), with permission of the American Geophysical Union)

A long-term student of earth tides, Robert C. Bostrom, published a book in 2000 titled "TECTONIC CONSEQUENCES OF EARTH ROTATION". In this book he gave a name to a phenomena that he referred to in a 1976 article titled "WESTWANDERUNG AND THE LUNAR TIDAL COUPLE: MODULATION OF CONVECTION BY TIDAL STRESS". He named the process "TIDAL VORTICITY INDUCTION" (TVI). In his book Bostrom suggests that the continuous operation of the unidirectional, low range (currently about 0.5 m in the equatorial zone), semi-diurnal earth tides could cause a rotation of the strain ellipsoid that, over time, may produce convection cells in the asthenosphere and cause a slow westward drift (or creep) of the lithosphere relative to the upper mantle. This is illustrated in Fig. 10.2.

Then in a summary article by Scoppola et al. (2006) titled "The Westward Drift of the Lithosphere: A Rotational Drag?" the authors challenged the basic tenants of the article by Jordan (1974). Scoppola et al. (2006) suggested that the viscosity of the asthenosphere may be orders of magnitude lower than the value used by Jordan (1974). Scoppola et al. (2006) suggested that a significantly lower value of viscosity could be attained by having a few channels of hydrous mantle material mixed between more viscous layers and that these channels of very low viscosity material would permit the rhythmic vertical and horizontal action of the earth-tides to power a mainstream mantle convection system. Figure 10.3 illustrates the major concepts of Scoppola et al. (2006).

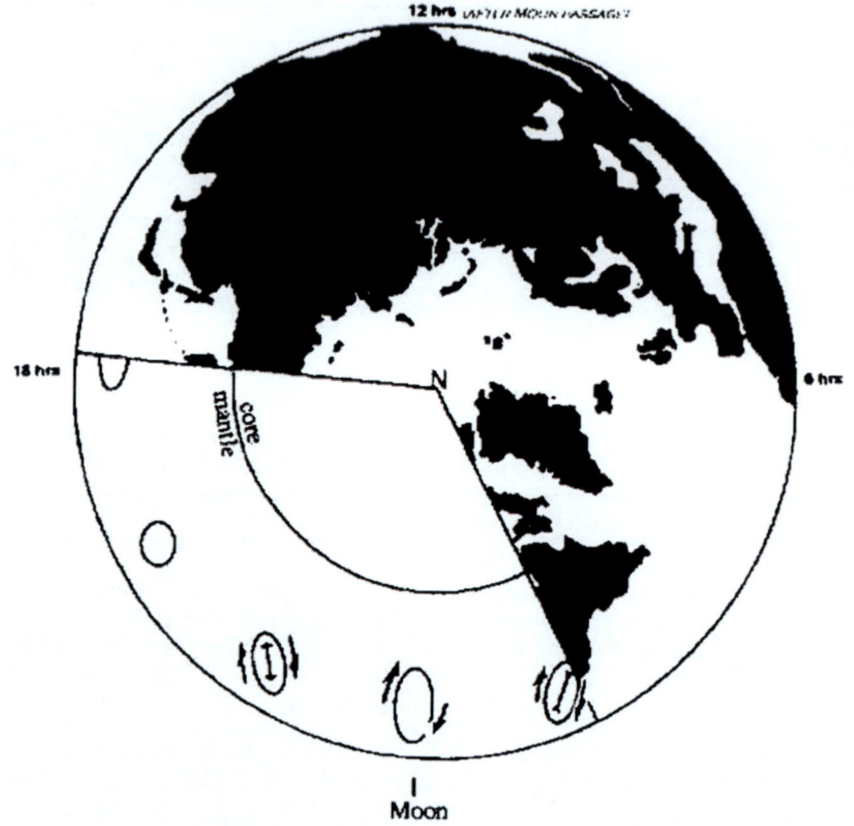

Fig. 10.2 Diagram illustrating the concept of the rotation of the strain ellipsoid in the equatorial zone of Earth caused by the action of the lunar gravitational rock tide. Note that the rotation of the planet is prograde (counterclockwise) and the strain ellipsoid is rotating in the retrograde (clockwise) direction. This diagram illustrates the concept of "tidal vorticity induction". View is from the north pole. (Diagram from Bostrom (2000, Fig. 3.8), with permission from Oxford University Press)

In their article on an asymmetric earth, Doglioni and Panza (2015) explained many of the unusual features of Planet Earth (e.g., asymmetry of the arc-trench complexes of the planet, the generation of earthquakes in specific geotectonic environments, the general absence of earthquake activity on the polar regions of the planet, and others.) A major necessary feature to facilitate westward drift of the lithosphere relative to the mantle in the models of both Scoppola et al. (2006) and Doglioni and Panza (2015) is the presence of one or a few layers of very low viscosity material in the upper part of the upper mantle. Doglioni and Panza (2015) propose some features of an automobile engine (see quote below) as an analogue for the low viscosity layers which appear to be necessary to facilitate the earth-tide mechanism for westward drift of the lithosphere.

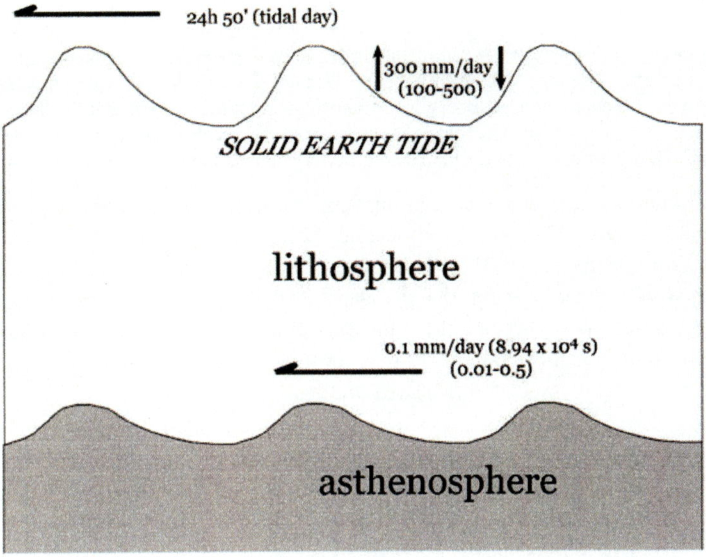

Fig. 10.3 Diagram illustrating how the rhythmically pulsating earth-tide mechanism can cause a unidirectional migration of the lithosphere relative to the mantle of the planet. (Diagram from Scoppola et al. (2006, Fig. 5), courtesy of the Geological Society of America). Doglioni and Panza (2015, Fig. 89, p. 133) suggested that this unidirectional motion is analogous to the progressive "arching" motion of a segmented biological worm

> The Earth's rotation contributes to plate tectonics, and if a very-low-viscosity layer is present in the upper asthenosphere, this layer enables the horizontal component of the tidal oscillation to generate a torque able to slowly shift the lithosphere relative to the mantle. This is true even in the case the dimensions of the very-low viscosity layer are so small to be beyond the reach of tomography. For example, in a car engine, shear friction between piston and sleeve is minimized by an extremely thin film of lubricating fluid that is practically incomprehensible in the direction normal to the piston motion. If the extremely thin film is absent the engine cannot work.—From Doglioni and Panza (2015, p. 147).

Perhaps such a low-viscosity layer has already been located:

> Underneath the resistive oceanic lithosphere, the magnetotelluric data reveal a high-conductivity layer confined to depth of 45–70 km. Because partial melts are stable at these depths in a warm damp mantle,[8] we interpret the conductor to be a partially molten layer capped by an impermeable frozen lid that is the base of the lithosphere. A conductivity anisotropy parallel to plate motion indicates that this melt has been sheared into flow-aligned tube-like structures.[9] We infer that the LAB beneath young plates consists of a thin, partially molten, channel of low viscosity that acts to decouple the overlying brittle lithosphere from the deeper convecting mantle. Because this boundary layer has the potential to behave as a lubricant to plate motion, its proximity to the trench may have implications for subduction dynamics.—From Naif et al. (2013).

Many mineralogical and seismological studies have resulted in interpretations of a hydrous transition zone (~410 to 660 km below surface) because of the physical properties of Ringwoodite. A quote form Schulze et al. (2018, p. 9) is appropriate here:

> Ringwoodite, the dominant mineral at depths between 520 km and 660 km, can store up to 2–3 wt.% of water in its crystal structure, making the Earth's transition zone a plausible water reservoir that plays a central role in Earth's deep water cycle. Experiments show that hydration of ringwoodite significantly reduces elastic wave velocities at room pressure, but the effect of pressure remains poorly constrained. . . . . Our data show that the hydration-induced reduction of seismic velocities almost vanishes at conditions of the transition. Seismic data thus agree with a wide range of water contents in the transition zone.

Other authors suggesting a hydrous upper mantle and/or transition zone are Thompson (1992), Bercovici and Karato (2003), Liu et al. (2016), Fei et al. (2017), Gruninger et al. (2017), Buchan et al. (2018), and Sobolev et al. (2019). Wang et al. (2015) also suggest that the water saturated condition of the basalts in the transition zone may be important for the extrusion of continental flood basalts. The massive outpourings of flood basalts may, in turn, be involved in episodes of global climate change and associated mass extinctions.

10.1.2 How Does the Lunar Capture Model Relate to the Earth Tide Paradigm?

The Lunar Capture Model (Malcuit 2015) gives a natural starting point for the Earth Tide Paradigm. Gravitational Capture of the Earth's Moon at 3.95 Ga is a super important event in the history of planet Earth (Malcuit 2015) and Fig. 10.4. The capture event places a well-defined time-line between the operation of a Cool Early Earth which is characterized by very placid, low amplitude, solar-induced earth and water tides. In contrast, the time immediately following capture is a very turbulent era characterized by very high amplitude rock and ocean tides. The equilibrium rock tides on Earth are up to 20 km at the time of capture and decrease exponentially as the lunar orbit circularizes. The lunar orbit circularizes mainly by tidal energy dissipation within the body of the Moon which causes a re-melting of a portion of the earlier formed lunar magma ocean. This tidal deformation and energy dissipation within the Moon causes a rejuvenation of an internally generated lunar magnetic field for about 300 Ma of time. As the orbit circularization progresses, the ratio of energy dissipated in planet Earth relative to that dissipated in the Moon increases.

The result of the rock tide activity on Earth associated with the capture process is the nearly complete destruction of the primitive crust of the Earth especially in the broadly defined equatorial zone at about 3.95 Ga via subduction of the primitive crust into earth's mantle. Contrary to the commonly held concept that the planet would be severely heated by the capture process (Singer 1968, 1970, 1977, 1986), in reality the thermal energy dissipated in the earth is sufficient to raise the near surface rocks only a few degrees Celsius. The displacement Love number (h) (an index of the deformability) is high for the primitive earth (i.e., 0.6 or 0.7) and the Q (specific dissipation factor) is high (100 or higher). The energy dissipation per deformation cycle is related to the inverse of the value of Q. Thus with a Q value of 100, only about 1% of the energy stored due to tidal deformation is subsequently dissipated as

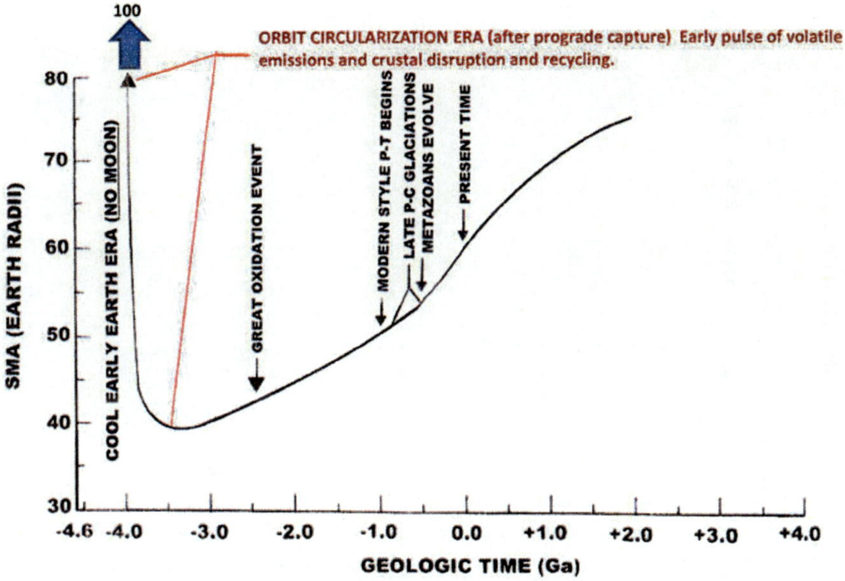

Fig. 10.4 Plot of semi-major axis of the lunar orbit vs. geological time for a “standard” version of the evolution of the lunar orbit. The assumption for this plot is that once the lunar orbit circularizes after the capture episode (a period of about 600 Ma), the lunar orbit does not deviate much from a circular orbit and maintains an orbital eccentricity near the present value of 0.0550

thermal energy during an encounter cycle. On the other hand, the Q value for a body like the earth’s moon can be very low (i.e., 1, 2, or 3) for intermediate values of body viscosity (Ross and Schubert 1986). Thus with a low h value between 0.2 and 0.3, and a Q value of about 1, sufficient energy for capture can be stored and dissipated during one tidal cycle during a very close encounter between the earth and moon. Only a few percent of the energy for capture is dissipated in the planet but the encountering body is heated severely during the initial encounter of a capture sequence of events. Malcuit (2015, Chap. 4, p. 72) lists four paradoxes associated with the gravitational capture process. Paradox # 1 is: THE ENCOUNTERING PLANETOID (the smaller body) MUST ABSORB MOST OF THE ENERGY FOR ITS OWN CAPTURE. This may seem counterintuitive but such is the nature of a paradox. (Note: All four paradoxes are discussed in detail in Malcuit (2015, Chap. 4).)

After post-capture orbit circularization, rock (and ocean) tidal processes are acting on the surface and near-surface regions of the planet Earth in a constant and persistent unidirectional process throughout the history of the Earth and will persist into the distant future of the planet. It is this rock tide activity that is considered by many scientists to be a major cause of unidirectional movement of the crustal complex relative to the upper mantle in earlier eras. It is this constant and persistent operation of the rock (and ocean) tidal cycles that have caused some sort of unidirectional crust-mantle action since the lunar capture event at about 3.95 Ga.

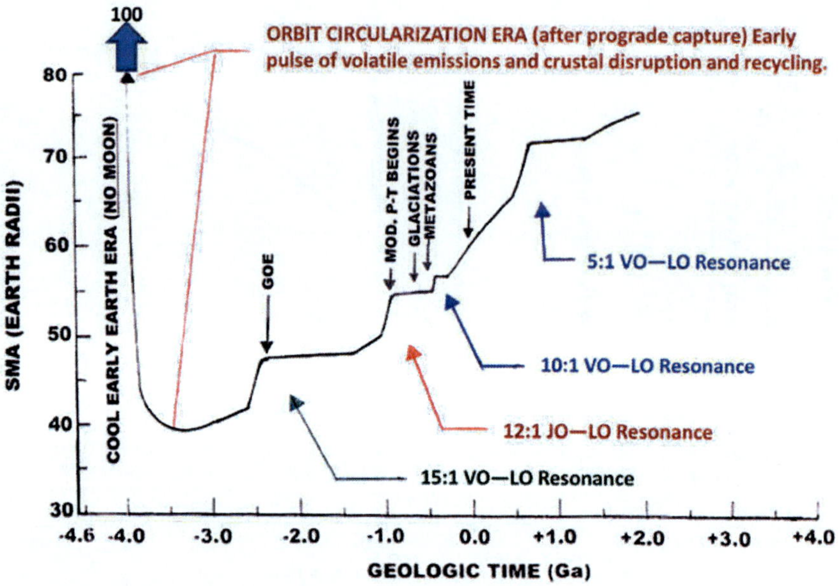

Fig. 10.5 Plot of semi-major axis vs. geologic time for the history or the lunar orbit in which the PO-LO resonance episodes are represented. Note that the lunar orbit eccentricity increases rapidly when each of the resonances is initiated and then simply fades away as the lunar orbit decreases in eccentricity as angular momentum is transferred from the rotating Earth to the lunar orbit. Such rock and ocean tidal episodes could cause, and apparently did cause, significant "punctuation marks" in the rock records of both the Earth and Moon. (From Malcuit (2015, Fig. 9.2) with permission from Springer)

A question is: Has the lunar orbit always been in a circular mode after the post-capture orbit circularization as shown in Fig. 10.4? Or has the lunar orbit experienced excursions of increased eccentricity which would cause a significant increase in rock and ocean tidal amplitudes and ranges on both the earth and the moon as depicted in Fig. 10.5?

A dynamical sequence of events for causing episodes of increased lunar orbit eccentricity is a Planet Orbit-Lunar Orbit (PO-LO) resonance (Malcuit 2015, Chap. 8). Peale and Cassen (1978) mentioned the possible importance of a Jupiter Orbit-Lunar Orbit resonance in which the period of the perigean cycle (prograde) of the lunar orbit is synchronous with the heliocentric orbital period of Jupiter. In this case the lunar orbit is locked into a long-term alignment with Jupiter (i.e., the apogee of the lunar orbit is always pointing toward Jupiter) and for a few months each earth year the gravitational "tugs" of Jupiter increase the eccentricity of the lunar orbit. This increased lunar orbit eccentricity results in an increase in the amplitude and range of the rock and ocean tide during perigee passages and can result in melting lunar mantle rocks which, in turn, can cause fissure lava eruptions on the lunar surface.

In this model the lead-in phase of this resonance episode is characterized by fairly rapid increases in the major axis and eccentricity of the lunar orbit but the angular momentum of the lunar orbit remains nearly constant. Once the lunar orbital period for the resonance is attained then the eccentricity of the lunar orbit is controlled by the energy dissipation properties of the moon. Early in the resonance episode the lunar energy dissipation is low but as the lunar body absorbs more tidal energy, then the energy dissipated per perigee passage increases and eventually the gravitational "tugs" due to the resonance are matched by the energy dissipation per unit time and the lunar orbital eccentricity reaches a peak value. Concomitantly, the tangential tidal mechanism on earth imparts angular momentum to the lunar orbit. This angular momentum transfer results in the earth's rotation slowly decreasing and the minor axis of the lunar orbit increasing (i.e., the angular momentum of the lunar orbit is increasing) but the major axis (i.e., the period of the lunar orbit) remains constant and the orbital resonance continues. The JO-LO resonance becomes ineffective as the major axis increases beyond the resonance limit (~53.4 earth radii) because of the transfer of angular momentum from the rotating planet to the lunar orbit.

Because of the apogee-perigee oscillation associated with the increased orbit eccentricity during the establishment of the resonance, there would be a significant increase in rock and ocean tidal activity during perigee passages and, concomitantly, a diminution of tidal activity during the apogee phase of the lunar orbit. This rhythmic perigee-apogee tidal activity continues month after month, year after year, and millennia after millennia, until the resonance becomes ineffective.

This JO-LO resonance can occur only once in the history of the Earth-Moon system but other similar resonance scenarios can occur with planet Venus [i.e., Venus Orbit—Lunar Orbit (VO-LO) resonances; also mentioned by Cuk (2007)]. VO-LO resonances can occur at three different times in the history of the Earth-Moon system (see Fig. 10.5). The first resonance episode occurs at about 2.5 Ga, the second episode at about 500 Ma, and the third episode at about 500 Ma in the future. For these VO-LO resonances planet Venus is aligned with the major axis of the lunar orbit during inferior conjunctions with Earth. For the 15:1 VO-LO resonance, this alignment occurs once every 15 earth years; for the 10:1 VO-LO resonance, the alignment occurs once every 10 years; and for the 5:1 VO-LO resonance, the alignment occurs once every 5 earth years. The features of these resonance states are well illustrated in Malcuit (2015, Chap. 8). For these VO-LO resonances as well as for the JO-LO resonance, the times of perigee passage are characterized by higher than normal rock and ocean tidal amplitudes and ranges and the apogee phases of the lunar orbit are characterized by lower than normal tidal amplitudes and ranges. Thus, these times of PO-LO resonances should generate a signature in the nearshore sedimentary rock record in the form of tidal rhythmite sequences. Indeed, several tidal rhythmite sequences have been reported from sedimentary sequences of these eras (e.g., the Huronian Complex of Ontario, Canada; several Late Precambrian sedimentary sequences, and Early to Middle Ordovician sedimentary sequences). These models of PO-LO resonances are very testable via the tidal rhythmite sequences of these eras. Zanazzi and Triaud (2019) suggest that excessive tidal stress may be a factor in the initiation of plate tectonic activity. Perhaps the tidal stress of PO-LO resonances may be directly related to an increase in plate tectonic activity in those eras.

The 15:1 VO-LO resonance may be related to the Great Oxidation Event (GOE) starting at about 2.45 Ga which, in turn, may be associated with the onset of an aerobic nitrogen cycle as described by Zerkle et al. (2017). An extensive episode of rapid movement of continental plates may be associated with the increased rock tidal activity associated with such an orbital resonance scenario (Malcuit 2015, Chap. 8). The 12:1 JO-LO resonance, and associated rock tidal activity, may be the initiating cause of slab-pull plate tectonics and the formation of associated blueschist complexes (Stern et al. 2016).

10.2 The Plate Tectonics Paradigm and the Earth Tide Paradigm

The normal version (i.e., the textbook version) of the Plate Tectonics Model postulates that the main driver of plate movements is mantle convection and supplemental activity is in the form of mantle plumes rising from the core-mantle boundary (Burke 2011). Burke's argument is that plate tectonics in some form or other has been happening throughout geologic time in about the same manner but, he admits that it is probably a bit slower now than in the past (this is essentially a uniformitarian approach to plate tectonics) (see Fig. 10.6 for an example). If we accept this argument, then there is no need for the earth tides to do any work. In this model, the moon is orbiting the earth but the moon is treated like a "night lantern" and the moon is postulated to have a negligible effect on plate tectonics.

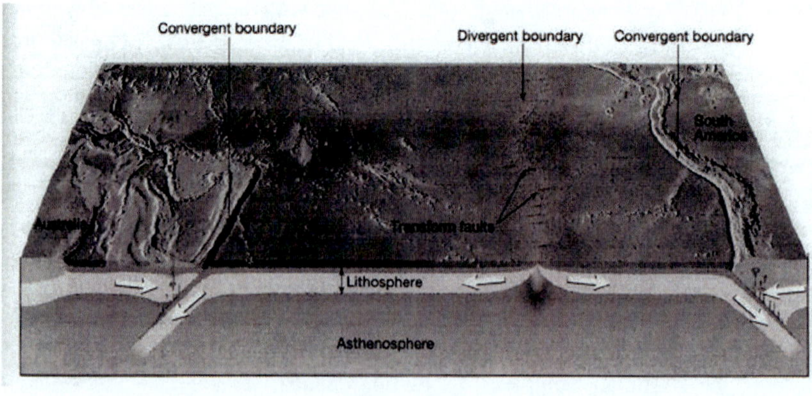

Fig. 10.6 Diagram illustrating the normal "uniformitarian" model of plate tectonics. Note that the eastward and westward dipping subduction zones have about the same angle of descent into the mantle. The only driver in this version of plate tectonics is mantle convection. (From Condie and Sloan (1998, Fig. 2.2), courtesy of Prentice-Hall)

If we consider the directional properties associated with Plate Tectonics as suggested by Scoppola et al. (2006), Doglioni et al. (2011), Doglioni and Anderson (2015), Doglioni and Panza (2015), and others, then we can establish a definite relationship with the Earth Tide paradigm. In this case it is the constant and persistent unidirectional mechanism of the earth tides that cause the directionality and polarity of plate movement. In this combined (composite) model the primary movement of the plates is still caused by mantle convection but the ancillary force to cause the directionality and polarity of plate tectonics is the earth tide mechanism. The probable earth tide mechanism for this ancillary force to move the plates is the low amplitude (and range) tidal action proposed by Scoppola et al. (2006), Doglioni and Anderson (2015), and Doglioni and Panza (2015) illustrated in Fig. 10.3.

The Planet Orbit-Lunar Orbit resonance concept illustrated in Fig. 10.5 may also relate to the origin and emplacement of diamond deposits. It is well known that the age of formation of diamond crystals is much older than the age of emplacement of the kimberlite deposit that contains the diamonds. For example, a kimberlite rock mass emplaced at about 60 Ma before present may have diamond crystals as old as 3.5 or 3.0 Ga. The crystallization age of the diamond crystals is determined by dating inclusions within the crystals (Meyer 1985; Nestola et al. 2017). The general concept is that diamonds and most of the included minerals must be formed in a pressure environment equivalent to 200 km or greater below the surface of earth. Then when conditions are right, the mass of kimberlite with its enclosed diamonds is emplaced rapidly to the surface within an envelope of carbon dioxide gas, a process first proposed by Kennedy and Nordlie (1968). Thus these kimberlite complexes are emplaced as "rocky torpedoes" from the mantle. The carbon dioxide gas expands and cools during its brief (merely hours of time) trip to the surface and the kimberlite reaches the surface in a "cool" condition (Kennedy and Nordlie 1968).

Stern et al. (2016) show that the vast majority of diamond-bearing kimberlite deposits have been emplaced within the last 600 Ma and that is why they propose that the "modern" regime of slab-pull subduction started no earlier than about 1.0 Ga ago (see Fig. 10.7). Their rationale is that before that time the mantle would have been too warm to permit slab-pull subduction. There are, however, some kimberlites that were emplaced as far back as Archean time (i.e., before 2.5 Ga). See Fig. 10.8 for the distribution of kimberlites and related mineral and rock complexes over 3 billion years of earth history.

The question is whether or not the ages of emplacement can be explained in terms of the PO-LO resonance models. Examination of Fig. 10.5 suggests that perhaps the 15:1 VO-LO resonance could cause much higher than normal earth tides during that limited time (about 2.5–1.5 Ga) to cause a series of forced subductions of oceanic lithosphere that could result in an episode of plate tectonics activity that might help explain some kimberlite deposits at about 1800 Ma.

Likewise, the elevated rock tides caused by a JO-LO resonance beginning at about 1.0 Ga could help to explain some early Neoproterozoic emplacements of kimberlites. Then from the beginning of the JO-LO resonance to the present time we have "modern-style" slab-pull plate tectonics and an irregular periodic emplacement of kimberlites and the included diamond deposits.

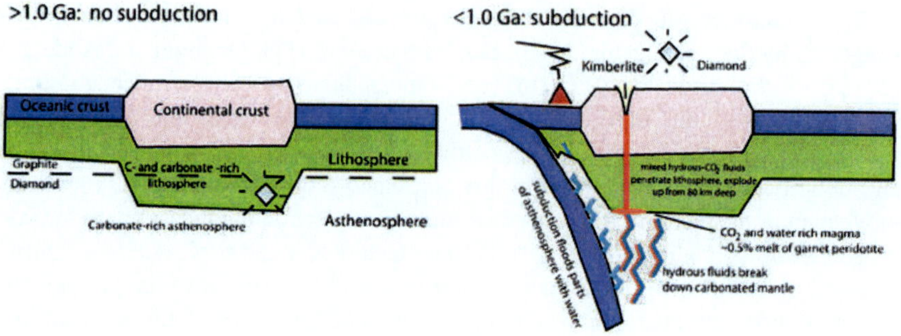

Fig. 10.7 Diagram showing the "slab-pull" subduction (right) in contrast to what was happening to the lithosphere before about 1.0 Ga. Left: Before about 1.0 Ga, very little if any deep subduction occurred. The flux of water into the mantle is low and the fluid pressure at the top of the asthenosphere is low. Right: After 1.0 Ga deep subduction associated with "slab-pull" style of plate tectonics occurs and water is delivered to the deeper part of the upper mantle. Water moving upwards interacts with carbonated mantle peridotite generating water-carbonate fluids and increasing fluid pressure near the top of the asthenosphere. Eventually the fluid pressure results in the near surface emplacement of a mass of kimberlite which may have associated diamonds. In the context of the mainstream convection model of plate tectonics, WEST is to the right of the diagram (i.e., the steep angle of descent is in a westward direction). (From Stern et al. (2016, Fig. 2) courtesy of the Geological Society of America)

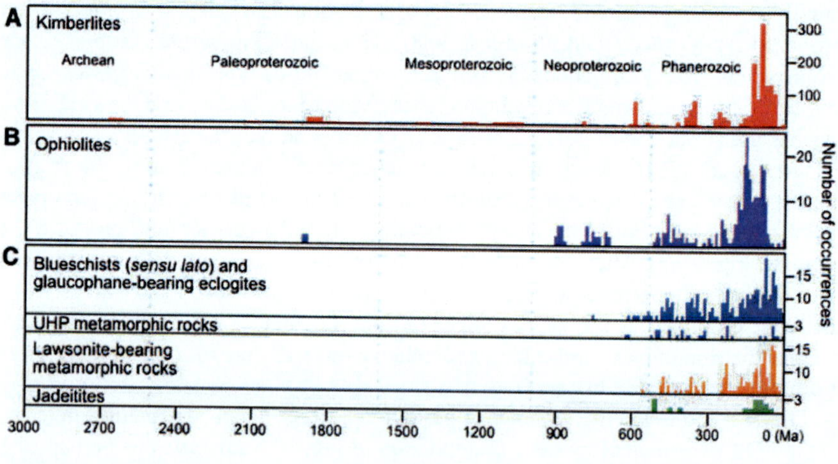

Fig. 10.8 Diagram showing age of emplacement of (**a**) kimberlites, (**b**) ophiolites, and (**c**) blueschists, as well as ultra-high pressure metamorphic rocks, lawsonite-bearing metamorphic rocks, and jadeitites over the past 3 billion years of earth history. (From Stern et al. (2016, Fig. 3) courtesy of the Geological Society of America)

This adventure into the problem of the origin of kimberlite "pipes" and associated diamond crystals is another example of the potential usefulness of the EARTH TIDE (EVOLUTION OF THE EARTH-MOON SYSTEM) paradigm in association with PLATE TECTONICS paradigm to solve some outstanding geological problems.

10.3 The Milankovitch Cycles Paradigm and the Earth-Tide Paradigm

The relationship between these two major concepts for the operation of planet Earth is a bit more cryptic but nonetheless there is a relationship. First of all the main astronomical influences for both models/paradigms are the Sun and the Moon and secondary gravitational influences are by Jupiter, Venus, and Saturn. When we add the influence of Planet Orbit-Lunar Orbit resonances, then both Jupiter and Venus are involved as primary drivers. Thus we do have a chaste of characters to work with to relate these two models/paradigms to each other and some characters do double duty.

The main goal of the calculations by Milutin Milankovitch was to explain the fairly recent episodes of glaciation on Earth (i.e., the Pleistocene glaciations). Major publications of the results of his calculations are Milankovitch (1920, 1941, 1952, 1957). When he passed away at the age of 79 (in 1958) it certainly looked like the idea would pass away with him. Then in the 1950–1970 era, geoscientists at a number of oceanographic institutes and universities found evidence of hot and cold cycles recorded in the oxygen isotope ratios of microfossils in deep ocean deposits (see Imbrie and Imbrie (1979), for more details of this very important collaborative work). The seminal report of their synthesis of information is by Hayes et al. (1976) and is titled "Variations in the earth's orbit: Pacemaker of the Ice Ages".

The critical information on temperature variations is recorded in microfossils in deep ocean sediments but these "hot" and "cold" cycles recorded in the oceanic "muds" did relate to a record of glacial advances and retreats on the continental crust in the northern hemisphere. Computer modelling of the three major parameters identified by Milankovitch were matched with the information from the ocean sediments. The three parameters are (1) heliocentric orbital eccentricity which varies in value from near 0% to a high of near 7% with a major periodicity of about 100 Ka and a supplemental period of about 400 Ka; (2) the obliquity cycle (tilt angle cycle) which varies in value from about 21 degrees to about 25 degrees over a period of about 40 Ka; and (3) the precession cycle with a period of about 20 Ka. See Fig. 10.9 for pictograms of the operation of these three parameters and Fig. 10.10 for a summary of the probable variations of these parameters over the past 250 Ka. Figure 10.11 is a composite diagram that may help the reader to better comprehend this fairly complex model.

The heliocentric orbital eccentricity cycles are controlled mainly by the gravitational influence of the Sun, Jupiter, and Saturn. Both the obliquity and precession cycles are controlled mainly by gravitational interactions with the Sun and Moon. In general, cold temperatures on Earth in the northern hemisphere result from low heliocentric orbital eccentricity (near zero), high obliquity (~25 degrees), and when the longitude of perihelion of the precession cycle is associated with the autumn season. Warm conditions in the northern hemisphere would occur when the orbital parameters are reversed: i.e., high heliocentric orbital eccentricity (~6%), low obliquity (~21 degrees), and when the longitude of the perihelion is associated with the spring season.

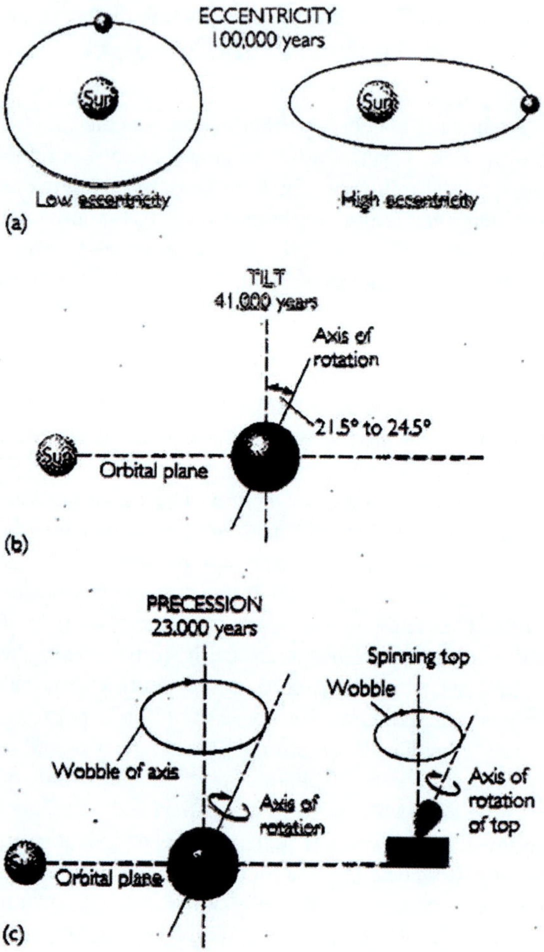

Fig. 10.9 Pictograms of the three parameters associated with the Milankovitch model. (**a**) Eccentricity is the quantity of ellipticity of the earth's orbit around the sun (i. e., heliocentric orbit eccentricity). (**b**) Obliquity (tilt) is the angle between the rotational axis of earth and the normal to the orbital plane of the planet. (**c**) Precession is the path of migration of the north pole, in the retrograde (clockwise) direction, relative to a line perpendicular to the orbital plane. (An analogue for precession is the motion of the rotation axis of a spinning top.) (Diagram from Press and Siever (2001, Fig. 15.34) courtesy of MacMillan Publishers)

So how does the operation of the Milankovitch cycles relate to the Earth Tide (Evolution of the Earth-Moon System) "paradigm"? Most of the controlling characters are the same. The Sun and Moon modulate the spring-neap earth tide cycles (as well as the spring-neap ocean tidal cycles). I will admit that the operation of the earth tides is easier to understand than the operation of the oceanic tides and, furthermore, the earth tide mechanism is much more relevant to the operation of the Milankovitch cycles.

Now let us consider the effect of Planet Orbit-Lunar Orbit resonances illustrated in Fig. 10.5. The strongest of these resonances is the JO-LO resonance. The Sun, Moon, and planet Jupiter are involved in this operation. Although the Milankovitch cycles would be operating during a PO-LO resonance sequence of events, some of the effects of these cycles would be overwhelmed by the tidal activity of the

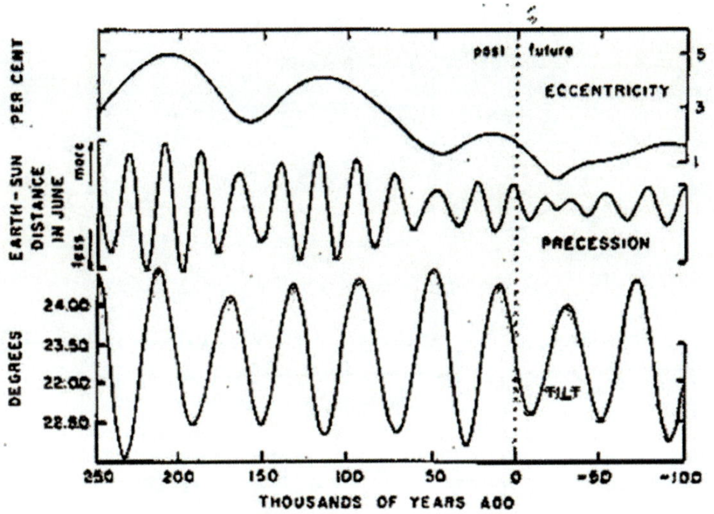

Fig. 10.10 Diagram showing changes in the parameters of heliocentric orbit eccentricity, precession of the polar axis about a line perpendicular to the orbital plane of the planet, and obliquity (tilt angle) relative to a line perpendicular to the plane of the orbit of the planet. The time scale of the diagram is from 250 Ka in the past to 100 Ka in the future. (Diagram from Imbrie and Imbrie (1979, Fig. 41) courtesy of Enslow Press)

resonance. To be more specific the heliocentric orbital eccentricity would be operating near normal but the obliquity and precession operations would be very irregular because of the elevated eccentricity of the lunar orbit. The same can be said for the Venus Orbit-Lunar Orbit resonances. In this case the main players are the Sun, Moon, and planet Venus. The resonance is not as energetic as the JO-LO resonance so the obliquity and precession cycles would not be as greatly affected by the VO-LO resonances as they are by the JO-LO resonance.

The other important feature of the Earth Tide/Evolution of the Earth-Moon System "paradigm" is the gradual transfer of angular momentum from the rotation of the Earth to the lunar orbit via the agency of the rock and ocean tides. Going backward in time, earth rotation increases and the major axis of the lunar orbit decreases. As a consequence, the earth tides become more effective because of the increased gravitational force. The heliocentric orbital eccentricity cycle is not affected by this angular momentum exchange but both the obliquity and precession cycles would change significantly over time. Estimated values for the obliquity and precession parameters for the past 450 Ma are presented in Berger et al. (1989) and reproduced here as Fig. 10.12. The results of that set of calculations suggest (1) that the present obliquity cycle value of about 41 Ka decreases to about 30 Ka at 450 Ma and that the present precession cycle value of about 19 Ka decreases to about 16 Ka at about 450 Ma.

Now that researchers are recognizing the signatures of the Milankovitch model parameters for the Pleistocene Epoch, sedimentologists and stratigraphers are

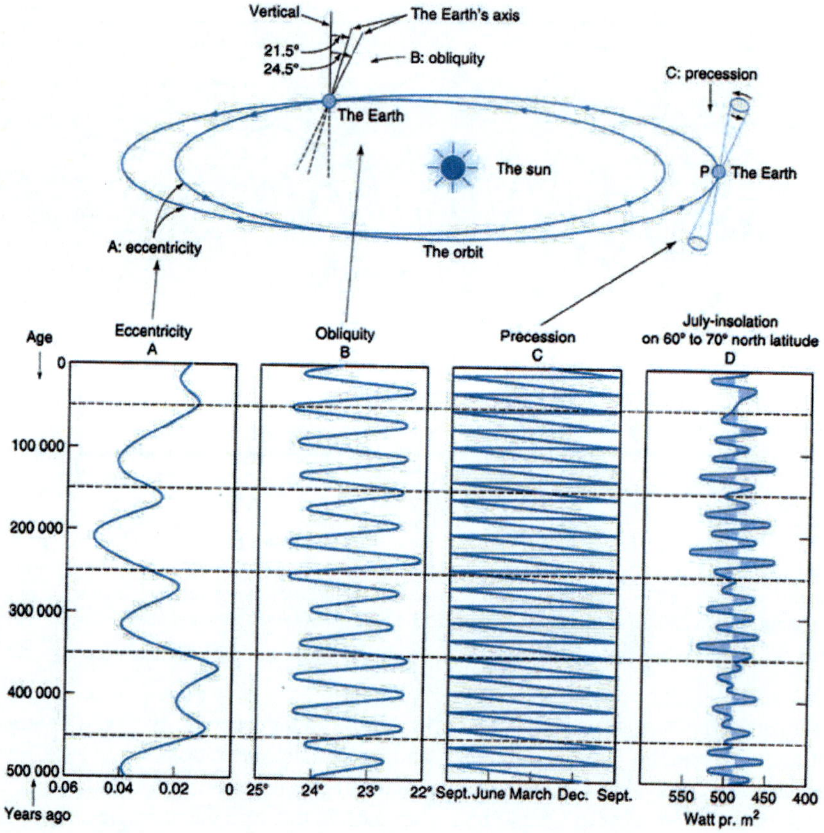

Fig. 10.11 A complex pictogram illustrating the regular variations in the earth's orbit for the Milankovitch model of climate oscillations for 500 Ka of the Quaternary Epoch. (**a**) The eccentricity of the earth's orbit varies from about 0% to 6% with a dominant period of about 100 Ka and a subordinate period of about 400 Ka. (**b**) The obliquity (tilt angle) varies from about 21 to 25 degrees with a period of about 41 Ka. (**c**) The position of the North Pole precesses in a retrograde, clockwise, direction with a period centered on 20 Ka. [Note that the diagram illustrating precession has arrows going in the wrong direction; the direction should be clockwise (retrograde) in the northern hemisphere.] (**d**) The July insolation for 60–70 degrees north latitude in watts per square meter. (Cool periods of potential glaciation are to the right and warm periods of potential interglacial periods are to the left on this diagram.) P is the position of perihelion (nearest approach to the Sun for an elliptical orbit. (Diagram from Rothschild and Lister (2003, Fig. 18.1) with permission of Academic Press)

examining rock sequences and cycles in the geologic record that might relate to the changing parameters of the Milankovitch model. This is the study of Cyclostratigraphy which can be defined for our purposes as the study of sedimentological response to the eustatic fluctuations of global sea level caused by the continually changing parameters of the Milankovitch model which are, in turn,

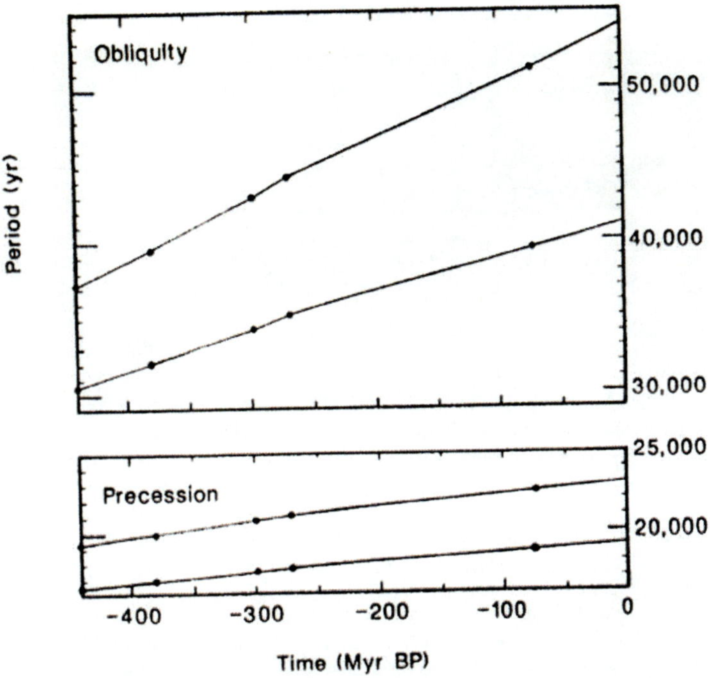

Fig. 10.12 Plot of probable values of planetary obliquity and precession of planet Earth for the past 450 Ma that could be useful in cyclostratigraphic studies. Estimates of earth-moon distance, earth rotation rate, and polar flattening of the figure of the planet with a faster rotation rate are involved in the estimate. (Diagram from Berger et al. (1989) courtesy of Springer-Nature)

due to the evolution of the Earth-Moon system. Figure 10.13 is from an article concerning the formation of the abundant coal deposits in the late Pennsylvanian and early Permian Peroids [about 300–310 Ma ago (Feulner 2017)]. In this article the stratigraphy of the coal deposits is explained in terms of the three Milankovitch model parameters (i.e., heliocentric orbital eccentricity, obliquity of the polar axes, and precession of the planetary axis). Eustatic sea level fluctuations are explained in terms of the waxing and waning of the polar ice regimes during this coal-forming era. The fluctuations in the values of molecular oxygen and carbon dioxide in the atmosphere at that time are also explained by their computer-based modelling. Variations in obliquity and precession motion are directly related to the succession of polar glaciations/deglaciations and the timing of sea level fluctuations associated with the development of these Paleozoic coal deposits and associated strata (i.e., the sequence stratigraphy).

Going further back in time, Mingxiand and Tucker (2013) reported on a 3300 m thick carbonate rock complex located about 50 km east of the Beijing area of China. This carbonate complex is interpreted to record a succession of sea level rises and falls caused by variation of the Milankovitch parameters. The carbonate rock

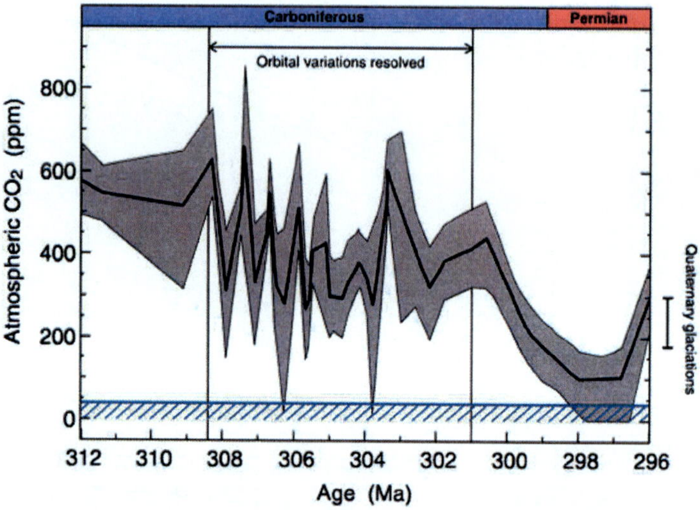

Fig. 10.13 Plot showing estimates of atmospheric carbon dioxide and the global glaciation thrushold about 300 Ma ago compared to the carbon dioxide level during the glaciations of the last million years (marked as "Quaternary Glaciations" on the right side of the diagram). A global climate model was used to calculate the variables of the Milankovitch model in the interval labelled as "orbital variations resolved". Below the blue line the planet could undergo a "global glaciation". Note that the Permian glaciations on Gondwana, centered on the south polar region of that era, were very extensive. (Diagram from Feulner (2017, Fig. 3) with permission of National Academy of Sciences)

complex was deposited between about 1310 and 1207 Ma before present. Sedimentary units (facies) are interpreted to represent repeating sedimentary sequences going from subtidal to intertidal to supratidal to lagoonal supratidal to paleosols. Figure 10.14 shows a generalized model for this well-preserved Middle Proterozoic rock complex.

10.4 The "Astronomical Metronome" and Planet Orbit – Lunar Orbit Resonances

More recently researchers are focusing on the 405 Ka heliocentric orbit eccentricity cycle as a potential "astronomical metronome" for the evolution of the Earth as well as the evolution of the Earth-Moon system. The short story version is that the astronomers tell us that the 405 Ka eccentricity cycle is controlled by the orbital motions of a combination of the large planet Jupiter and our "sister" planet Venus. Since the orbits of these two planets are very consistent over long periods of geologic time (Laskar et al. 2011), their interaction with planet Earth, as recorded in the sedimentary rock record of Earth should be very consistent (Kent et al. 2017; Hinnov

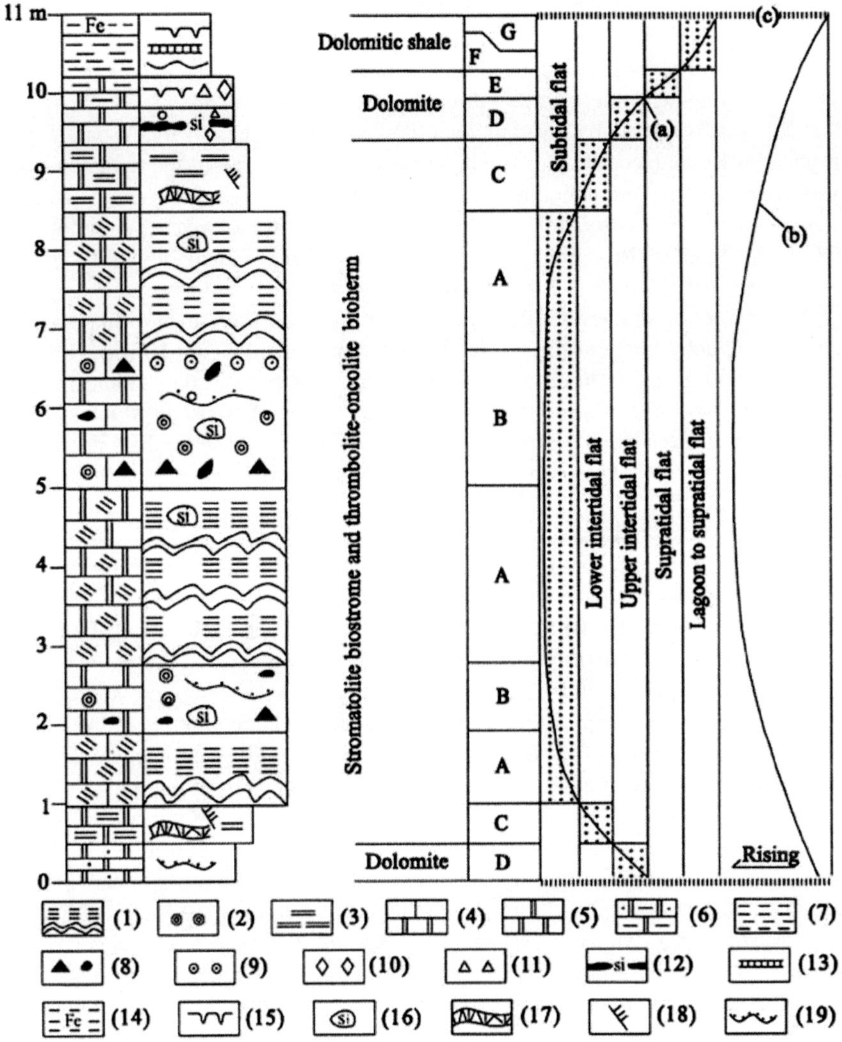

Fig. 10.14 A generalized model of a symmetrical peritidal sedimentary rock sequence from the Wumishan Formation deposited between 1310 and 1207 Ma. The model sequence represents 11 m of sedimentary rocks (a typical complete sequence in this 3.3 km thick complex) which is bounded by short-lived exposure surfaces at top and bottom of each sequence. Explanations of **(a)**–**(c)** on the right-hand portion of the diagram are as follows: **(a)** the changing curve of water depth represented in the features of the rock units (sea level rising to the left and sea level falling to the right); **(b)** the curve of high-frequency relative sea-level changes representing some combination of the Milankovitch parameters of heliocentric eccentricity and precession of the polar axis; **(c)** short-exposure (erosional) surfaces at top and bottom of the sequence. Interpretations of rock symbols 1–19 are as follows: (1) stromatolite, (2) ooncolite, (3) microbial laminate, (4) dolomitic limestone, (5) dolomite, (6) sandy and muddy dolomite, (7) dolomitic shale, (8) thrombolite, (9) oolite, (10) halite pseudomorphs, (11) karst breccia, (12) silcrete, (13) calcrete, (14) paleosols, (15) mud cracks, (16) chert nodule, (17) tabular rudite, (18) cross-bedding, (19) scoured surface. (Diagram from Mingxiang and Tucker (2013, Fig. 5); note that the figure has not been modified but the caption has been altered for clarity) courtesy of Springer)

2018). This concept has led to the idea that the 405 Ka eccentricity cycle can be used as an "astronomical metronome" throughout the history of the Earth. Since the 405 Ka cycle is essentially invariant over time, the slightly more variable parameters of the 100 Ka heliocentric eccentricity cycle of Earth, as well as the monotonically varying precession and obliquity cycles (as illustrated in Fig. 10.12) can be fit on a framework of the "astronomical metronome" of the 405 Ka cycle. The work by Olsen and Kent (1996), Hinnov (2013), Wu et al. (2013), Zhang et al. (2015), and Kent et al. (2017) are good examples of the use of this new framework.

Meyers and Malinverno (2018) relate their work on Proterozoic Milankovitch cyclostratigraphy to the history of the Earth-Moon system. Figure 10.15a is a time-evolution diagrams from their paper showing where their data point for the Xiamaling Formation fits into the story of the evolution of the Earth-Moon system. Figure 10.15b has a green curved line showing my interpretation for the trajectory of the lunar paleo-orbit based on the concept of Planet Orbit – Lunar Orbit resonances as demonstrated in Malcuit (2015, Chap. 8). Although both curves have similar trends, there are significant differences. Future research on both rock sequences and theoretical calculations should help to refine the trend for the Earth-Moon distance over geologic time.

The diagrams in Fig. 10.16 show a broader view of the evolution of the Earth-Moon system over geologic time and into the future. Figure 10.16a shows the normal smooth line rendition with the green curved line now representing the Meyers and Malinverno (2018) interpretation of the trajectory for the evolution of the Earth-Moon system consistent with their interpretation of the data recovered from Xiamaling Formation. Figure 10.16b shows that same green line segment plotted on my analysis of the effects caused by Planet Orbit – Lunar Orbit resonances.

Perhaps the Jupiter Orbit – Lunar Orbit (JO-LO) resonance as well as the Venus Orbit – Lunar Orbit (VO-LO) resonances illustrated in Malcuit (2015, Chap. 8) may serve as mile-post markers along the path of the evolution of the Earth-Moon system. The main drivers of these resonances are the same planets as those involved in the 405 Ka heliocentric orbit eccentricity cycle (i.e., Jupiter and Venus). The processes involved in the PO-LO and VO-LO resonances are presented in detailed graphic form in Malcuit (2015, Chap. 8). Each of the Planet Orbit—Lunar Orbit (PO-LO) resonances is characterized by a fairly rapid increase in the eccentricity of the lunar orbit to a resonance value. The maximum value of lunar orbit eccentricity in these models is determined by the capacity for tidal energy dissipation in the lunar body. When the energy dissipation in the lunar body equals the energy pumped into the lunar orbit via resonance pumping, then the maximum value of the major axis as well as that of the lunar orbit eccentricity is attained. The period of the lunar orbit (which is directly related to the length of the major axis) then remains constant as the rotating planet (the Earth in this case) imparts angular momentum from its rotation to the lunar orbit via the agency of a combination of the ocean and rock tidal activity. Since the period of the lunar orbit is directly related to the major axis of the orbit, the period of the lunar orbit remains constant as the earth rotation slowly decreases because of angular momentum transfer from the rotation of the planet to the lunar orbit.

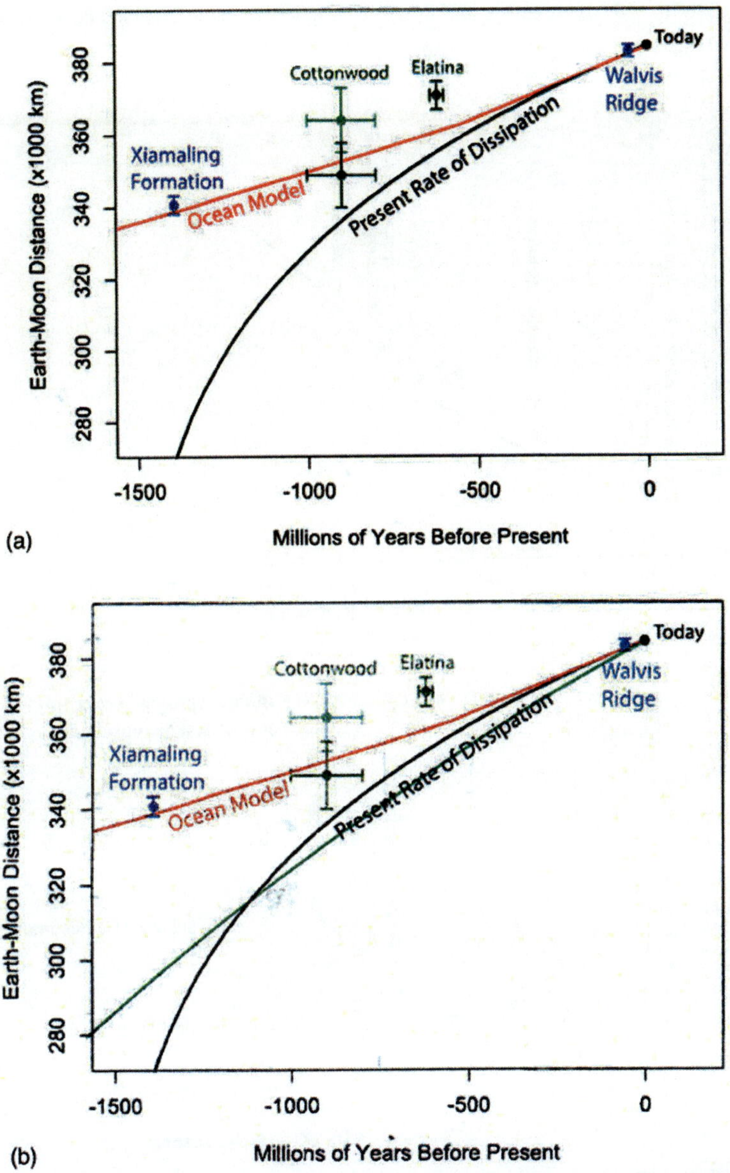

Fig. 10.15 Two diagrams, one original and the second one modified with an additional curve, illustrating two different pathways for the evolution of the Earth-Moon system over geologic time. The x-axis is geologic time in millions of years before present and the y-axis is the earth-moon distance in kilometers ($\times 10^3$). (**a** This is the original diagram from Myers and Malinverno (2018, Fig. 3), with permission of the National Academy of Sciences). The black line shows the evolution of the Earth-Moon system (from Bills and Ray 1999) using the estimated present rate of energy dissipation in the earth which controls the rate of angular momentum transfer to the lunar orbit. The red line is a trace-back trajectory for the earth-moon distance that is consistent with an interpretation of the rhythmite record of the Xiamaling Formation by Myers and Malinverno (2018). (**b**) Same diagram with a green line superimposed. This green line in this diagram is my estimate of the trajectory of the earth-moon distance which is consistent with the Planet Orbit—Lunar Orbit resonance models outlined in Malcuit (2015, Chap. 8)

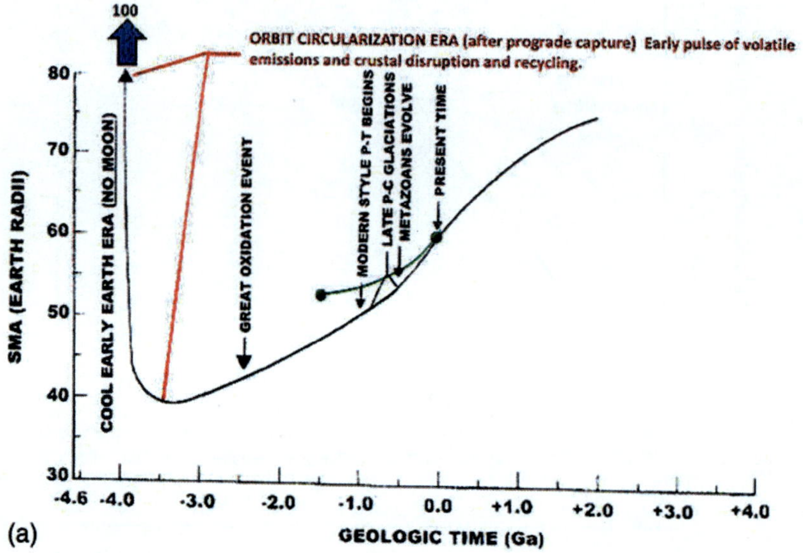

(a)

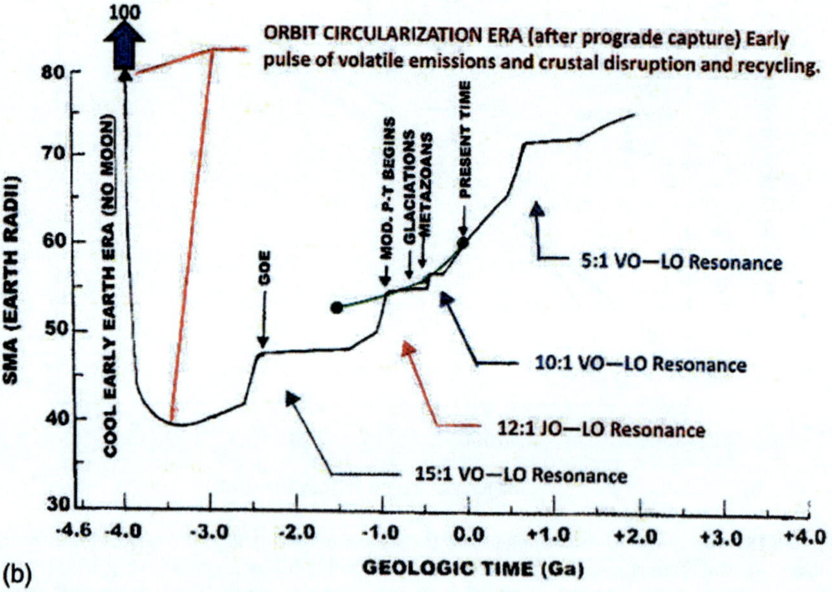

(b)

Fig. 10.16 Two diagrams (from Malcuit 2015, Chap. 8, Figs. 9.1 and 9.2) for the evolution of the Earth-Moon system over geologic time. The x-axis is geologic time and the y-axis is distance of separation of the Earth and Moon in earth radii. (**a**) A normal (smooth curve) version of the evolution of the Earth-Moon system (Malcuit 2015, Fig. 9.1) with a green line showing the trace-back trajectory for the earth-moon system (from Meyers and Malinverno 2018, Fig. 3 and shown in Fig. 10.15b). (**b**) A version of the evolution of the Earth-Moon system (Malcuit 2015, Fig. 9.2) featuring PO-LO resonance episodes with the green line segment representing the interpretation of Meyers and Malinverno (2018, Fig. 3)

The PO-LO resonances that can occur in the geologic past are the 15:1 VO-LO resonance at about 2.5 Ga which may relate to the Great Oxidation Event as well as an episode of Plate Tectonic activity. A JO-LO resonance (and there is only one of these) occurring at about 1.1 Ga may relate to a number of tidal rhythmite sequences as well as the beginning of the modern era of slab-pull plate tectonics (Stern 2005, 2007; Stern et al. 2016) as well as deep ocean oxygenation and an increase in sulfate levels in ocean water about that time (Stolper and Bucholz 2019). The 10:1 VO-LO resonance may relate to the Elatina Formation rhythmites and other tidal rhythmites of that era as well as the origin of metazoans.

In general, these PO-LO resonance episodes may be considered as special punctuation marks in the geologic record where the major axis of the lunar orbit as well as the rotation rate of the earth may be obtained from the tidal rhythmite record. Then the "astronomical metronome" of the 405 Ka heliocentric eccentricity cycle can be "draped" over the PO-LO resonance punctuation marks. These two astronomical mechanisms are not only compatible with each other but are supportive of each other. After all, the main drivers for each of these indices for marking geologic time are "big brother" JUPITER and "twin sister" VENUS.

10.5 Discussion of the Compatibility of these Three Models (Paradigms)

The Plate Tectonics Model can be related very directly to the Earth Tide Model. The operation of plate tectonics has not operated uniformly over geologic time but in this combined model there is some continuous unidirectional relative movements between the lithosphere and the upper mantle since the Lunar Capture Episode about 3.95 Ga ago. The model predicts that there should be a major change in the process of subduction because of the action of a JO – LO resonance about 1.0 Ga ago. Davies (1992), Stern (2005, 2007), and Stern et al. (2016) make the argument that the thermal regime of the upper mantle would not permit a "slab-pull" style of plate tectonics before about 1.0 Ga ago. I am suggesting that perhaps this change in subduction style may have been initiated by the high amplitude tidal activity associated with the JO-LO resonance as presented in Malcuit (2015, Chaps. 8 and 9).

The Orbital Parameter Variation Model (the Milankovitch model) appears to relate fairly directly to the Earth Tide Model in that the chaste of characters is nearly the same (i.e., Sun, Earth, Moon, Venus, Jupiter, Saturn). Also, some combination of earth and ocean tidal action over geologic time causes a gradual, very predictable change in the Milankovitch parameters of obliquity of the polar axis orientation relative to the earth's orbital plane as well as the retrograde precession of the polar axis relative to the earth's orbital plane as illustrated in Fig. 10.12.

In summary, many of the complexities of the rock record of both the Earth and the Moon may be explained by a combination of the PLATE TECTONICS paradigm, the MILANKOVITCH MODEL paradigm, and the EARTH TIDE "paradigm".

10.6 Application of these Three Major Concepts to the Operation of Planet Mars

We must keep in mind that planet Mars appears to be the most normal of our group of terrestrial planets (Nimmo and Tanaka 2005). Plate Tectonics, if it ever operated on planet Mars, persisted for a very brief time when a magma ocean operated in the martian mantle or during the formation of the lowlands of the north polar region of Mars (Sleep 1994). The origin of the Borealis lowlands is still unresolved. Is it the result of a magmatic event associated with a bolide impact (Andrews-Hanna et al. 2008; Rosenblatt et al. 2016), or the result of subsidence of a more dense primitive crust into the martian upper mantle of Mars and subsequent replacement with a new (younger) surface rock complex (Sleep 1994)? Future seismic experiments in this Borealis region will help to resolve this controversy.

In contrast, the Milankovitch model of Heliocentric Orbit and Planetary Parameter Variations has apparently had a real presence in martian climate throughout martian history (Schorghofer 2008). All three components of the Milankovitch model are represented. According to the calculations of Laskar et al. (2004), the heliocentric orbit variations are large (averaged eccentricity over 5 Ga = 0.069 with standard deviation of 0.0299: total variation from 0.00 to 0.12) relative to the 0–7% variation of Earth's heliocentric orbit. The obliquity cycle variation is great over 5 Ga (averaged value is 37.62 degrees with a standard deviation of 13.82 degrees and a maximal value of 82.035 degrees). Furthermore, the obliquity of Mars is "strongly chaotic" (Laskar et al. 2004). The retrograde precession cycle is long and highly variable and is dependent on the obliquity of that era. The present precession rate is −7.83 + and – 0.30 arc sec/yr (Yoder and Standish 1997). The martian moons have no significance effect on the precession rate and only solar gravity is involved. In general, the Milankovitch cycles are the main cause of temperature variations on Mars and the main cause of many surface phenomena during most of the history of the planet.

Although we do not yet have evidence from the martian sedimentary rock record of orbital phenomena caused by interaction with the Earth, we do have evidence from the sedimentary rock record of Earth that is interpreted as recording a secular resonance between the orbits of Earth and Mars (Paillard 2017). Wu et al. (2018) did a detailed analysis of a rock sequence of Carboniferous Age in South China. They interpreted a sequence of lamina as relating to a chaotic motion between Earth and Mars that occurs when a key resonance occurs between the motions of orbital perihelia and nodes of Mars and Earth that result from a transition from one orbital state to another. This chaotic orbital motion has also been recorded by Ma et al. (2017) for the Cretaceous Western Interior Basin of North America and by Fang et al. (2018) for yet another sequence of marine rocks of Carboniferous-age in South China. These secular resonance transitions were predicted by Laskar (1990).

The rock tides on Mars are miniscule because the natural satellites, Phobos and Diemos, are very small and the solar gravitational tide, although in continuous operation, is very small. Thus the rotation rate of Mars has changed very little over the 4.6 Ga of martian history (Laskar 1995; Laskar et al. 2004). Thus, if decrease in rotation rate has anything to do with causing or maintaining habitable conditions on a terrestrial planet, Mars has missed out on this factor also. Nonetheless, Mars does have a solid surface for humans from Earth to establish a settlement and has limited resources necessary for existence on the planet.

The main event that could have changed the surface and upper mantle of Mars to make it more habitable is to have captured a sizeable satellite in orbit about the planet for a segment of geologic time. There is much evidence for water and related volatiles (oceans of water) on Mars early in its history (Cirton et al. 2018; Carr and Head 2019; Salese et al. 2019). Capture of a sizeable satellite in either prograde orbit (leading to a geoformed Mars) or retrograde orbit (leading to a venoformed condition) would have been a major event in the history of Mars to make it much more habitable. The gigantic rock and ocean tides associated with the capture episode would have forced subduction of a portion of the older crust, along with water and associated volatiles, into the deep mantle. As the orbit circularization progressed, the rock and ocean tides would diminish in amplitude but the process of crustal subduction, associated volcanism, and rehydration of the martian mantle would have continued. Retrograde gravitational capture would have continued. Retrograde gravitational capture would have been more beneficial for developing and perpetuating a biological system mainly because of the longer duration of a favorable tidal regime (Lathe 2004). These various capture scenarios, although very beneficial for life forms, would not have resulted in the environmental conditions of planet Earth. But it would have resulted in a nearby planet that would have many of the atmospheric, hydrospheric, and lithospheric features of our planet but on a much smaller scale. And it could have been a place that could be more easily engineered (i.e., terraformed) for limited human habitation.

10.7 Summary

Earth is a complex planet mainly because of the continuous interaction with a large satellite over ~85% of geologic time. For the past billion years, and perhaps longer, it has had a mantle thermal regime conducive to slab-pull plate tectonics (Davies 1992; Stern 2005, 2007; Stern et al. 2016). The celestial mechanisms of the Milankovitch cycles have been operating throughout the planet's history but the pre-capture regime would have been much different than the post-capture operation because the moon has a great influence on both the precession and obliquity cycles.

Likewise, rock tides have been operating throughout the planet's history. Before 3.95 Ga only solar gravitational tides were operating. During the capture episode as well as during the early part of the orbit circularization sequence the rock and ocean tides were severe. However, the polar zones would have been sheltered from much of the severe tidal action. After lunar orbit circularization, the ocean and rock tidal amplitudes and ranges are significantly lower during circular orbit eras than during periods of elevated orbital eccentricity. In general, the rock tidal amplitudes and ranges (as well as those of the ocean water tides) have been decreasing over the past 3.0 billion years of geologic time. As the lunar orbit expands, the earth rotation rate decreases and we humans and associated biota are fairly "content" with our existence on the planet.

In general, it is the interactions with a large satellite in prograde orbit that is the LIFE BLOOD of our habitable planet. The low-probability event of planetoid capture appears to be a major cause of our existence on this THIRD PLANET FROM THE SUN.

Planet Mars is a rocky planet with a transparent atmosphere. It has many of the broad features of the Earth. At present it has an atmosphere, a lithosphere, and some features of a hydrosphere. In the past Mars had a much more massive hydrosphere and atmosphere but after about 3 Ga ago, the atmosphere-hydrosphere complex was greatly diminished. For a brief time in its early history, Mars had an active magnetic field, probably driven by an impact mechanism that featured bodies in retrograde orbit (Arkani-Hamed 2009). But within about 0.6 billion years, the global magnetic field was gone. It is the author's opinion that the history of planet Mars could have been much different if, by chance, it had captured a sizable satellite in either prograde or retrograde orbit. Such a satellite in a prograde orbit could have been favorable for short term habitability and a large satellite in retrograde orbit could have fostered long term habitability. A large satellite would also cause much greater climate stability and a much greater predictability for weather and climate than there is on present day MARS.

Now let us return to a quote by Denevi from the beginning of this chapter.

> With humans' initial exploration of the Moon at the dawn of the space age came the birth of modern planetary science, which seeks to understand the complex history of our solar system in enough detail to piece together the story of how we got to where we are.
>
> To what extent was Earth an inevitable outcome of planetary evolution? To what extent was it a quirk of fate?—From Brett Denevi (2017, p. 39)

The reader can ponder an answer to this question!!!

References Cited

Andrews-Hanna, J. C., Zuber, M. T., & Banerdt, W. B. (2008). The Borealis basin and the origin of the martian crustal dichotomy. *Nature, 453*, 1212–1215.

Arkani-Hamed, J. (2009). Did tidal deformation power the core dynamo of Mars? *Icarus, 201*, 31–43.

Bercovici, D., & Karato, S. (2003). Whole mantle convection and the transition zone. *Nature, 425*, 39–44. https://doi.org/10.1038/nature01918.

Berger, A., Loutre, M. F., & Dehant, V. (1989). Pre-quaternary Milankovitch frequencies. *Nature, 342*, 133.

Bills, B. G., & Ray, R. D. (1999). Lunar orbital evolution: A synthesis of recent results. *Geophysical Research Letters, 26*, 3045–3048.

Bostrom, R. C. (1971). Westward displacement of the lithosphere. *Nature, 234*, 536–538.

Bostrom, R. C. (1976). Westwanderung and the lunar tidal couple: Modulation of convection by tidal stress. *The Moon, 15*, 109–117.

Bostrom, R. C. (2000). *Tectonic consequences of earth's rotation* (266 p). Oxford: Oxford University Press.

Buchan, J., Marquardt, H., Speziale, S., Kawazoe, T., Boffa-Ballaran, T., & Kurnosov, A. (2018). High-pressure single-crystal elasticity of wadsleyite and the seismic signature of water in the shallow transition zone. *Earth and Planetary Science Letters, 498*, 77–87. https://doi.org/10/1016/j.espl.2018.06.027.

Burke, K. (2011). Plate tectonics, the Wilson cycle, and mantle plumes: Geodynamics from the top. *Annual Reviews of the Earth and Planetary Sciences, 39*, 1–29.

Carr, M., & Head, J. (2019). Mars: Formation and fate of a frozen Hesperian ocean. *Icarus, 319*, 433–434.

Cirton, R. I., Manga, M., & Hemingway, D. J. (2018). Timing of oceans on Mars from shoreline deformation. *Nature, 555*, 643–646.

Condie, K. C., & Sloan, R. E. (1998). *Origin and evolution of earth: Principles of historical geology* (p. 498). Upper Saddle River: Prentice-Hall.

Cross, T. A., & Pilger, R. H. (1982). Controls of subduction geometry location of magmatic arcs, and tectonics of arc and back-arc regions. *Geological Society of America Bulletin, 93*, 545–562.

Cuk, M. (2007). Excitation of lunar eccentricity by planetary resonances. *Science, 318*, 244.

Davies, G. F. (1992). On the emergence of plate tectonics. *Geology, 20*, 663–966.

Denevi, B. (2017). The new moon. *Physics Today, 70*(June), 38–44.

Doglioni, C. (1994). Foredeeps versus subduction zones. *Geology, 22*, 271–274.

Doglioni, C., & Anderson, D. L. (2015). Top driven asymmetric mantle convection. *Geological Society of America Special Paper, 514*, 51–63.

Doglioni, C., & Panza, G. (2015). Polarized plate tectonics. *Advances in Geophysics, 56*, 1–167.

Doglioni, C., Ismail-Zadeh, A., Panza, G., & Riguzzi, F. (2011). Lithosphere-asthenosphere viscosity contrast and decoupling. *Physics of the Earth and Planetary Interiors, 189*, 1–8.

Fang, Q., Wu, H., Wang, X., Yang, T., Li, H., & Zhang, S. (2018). Astronomical cycles in the Serpukhovian-Moscovian (Carboniferous) maring sequence, South China and their implications for geoshronology and icehouse dynamics. *Journal of Asian Earth Sciences, 156*, 302–315. https://doi.org/10.1016/j.jseaes.2018.02.001.

Fei, H., Yamazaki, D., Sakurai, M., Miyajima, N., Ohfuji, H., Katsura, T., & Yamamoto, T. (2017). A nearly water-saturated mantle transition zone inferred from mineral viscosity. *Science Advances, 3*(6), e1603024. https://doi.org/10.1126/sciadv.1603024.

Feulner, G. (2017). Formation of most of our coal brought earth close to global glaciation. *Proceedings of the National Academy of Sciences, 114*, 11333–11337.

Gruninger, H., Armstrong, K., Greim, D., Boffa-Ballaran, T., Frost, D. J., & Senker, J. (2017). Hidden oceans? Unraveling the structure of hydrous defects in the Earth's deep interior. *Journal of the America Chemical Society, 139*(30), 10499–10505. https://doi.org/10.1021/jacs.7b05432.

Hagar, B. H., & O'Connell, R. J. (1979). Kinematic models of large-scale flow in the earth's mantle. *Journal of Geophysical Research, 84*, 1031–1048.

Hayes, J. D., Imbrie, J., & Shackleton, N. J. (1976). Variations in the earth's orbit: Pacemaker of the ice ages. *Science, 194*, 1121–1132.

Hinnov, L. A. (2013). Cyclostratigraphy and its revolutionizing applications in the earth and planetary sciences. *Geological Society of America Bulletin, 125*, 1703–1724. https://doi.org/10.1130/B30934.1.

Hinnov, L. A. (2018). Astromomical metronome of geological consequence. *Proceedings of the National Academy of Sciences, 115*, 6104–6106.

Hirchmann, M. M. (2010). Partial melt in the oceanic low velocity zone. *Physics of the Earth and Planetary Interiors, 179*, 60–71.

Hoink, T., Jellinek, A. M., & Lenardic, A. (2011). A viscous coupling at the lithosphere-asthenosphere boundary. *Geochemistry, Geophysics, Geosystems, 12*, QOAK02.

Imbrie, J., & Imbrie, C. P. (1979). *Ice ages: Solving the mystery* (p. 224). Short Hills: Enslow Publishers.

Jordan, T. H. (1974). Some comments on tidal drag as a mechanism for driving plate motion. *Journal of Geophysical Research, 79*, 2141–2142.

Kennedy, G. C., & Nordlie, B. E. (1968). The genesis of diamond deposits. *Economic Geology, 63*, 495–503.

Kent, D. V., Olsen, P. E., & Muttoni, G. (2017). Astrochronostratigraphic polarity time scale (APTS) for the Late Triassic and Early Jurassic form continental sediments nand correlation with standard marine stages. *Earth Science Review, 166*, 153–180.

Kreichgauer, D. (1902). *Die aquatorfrange in der geologie* (p. 248). Munich: Steyl.

Laskar, J. (1990). The chaotic motion of the solar system: A numerical estimate of the size of the chaotic zones. *Icarus, 88*, 266–291.

Laskar, J. (1995). Large scale chaos and marginal stability in the solar system. In D. Lagolnitzer (Ed.), *Proceedings volume, XI International Congress of Mathematical Physics* (pp. 75–120).

Laskar, J., Correia, A. C. M., Gastineau, M., Joutel, F., Lavard, B., & Robutel, P. (2004). Long term evolution and chaotic deffusion of the insolation quantities of Mars. *Icarus, 170*, 343–364.

Laskar, J., Fienga, A., Gastineau, M., & Manche, H. (2011). La2010: A new orbital solution for the ling-term motion of the earth. *Astronomy and Astrophysics, 532*, A89. https://doi.org/10.1051/0004-6361/201116836.

Lathe, R. (2004). Fast tidal cycling and the origin of life. *Icarus, 168*, 18–22.

Liu, J., Li, J., Hrubiak, R., & Smith, J. S. (2016). Origin of ultralow velocity zones through slab-derived metallic melt. *Proceedings of the National Academy of Sciences, 113*, 5547–5551.

Ma, C., Meyers, S. R., & Sageman, B. B. (2017). Theory of chaotic orbital variations confirmed by Cretaceous geological evidence. *Nature, 542*, 468–470. https://doi.org/10.1038/nature21402.

Malcuit, R. J. (2015). *The twin sister planets, Venus and Earth: Why are they so different?* (p. 401). Cham: Springer International Publishers.

Meyer, H. O. A. (1985). Genesis of diamond: A mantle saga. *American Mineralogist, 70*, 344–355.

Meyers, S. R., & Malinverno, A. (2018). Proterozoic Milankovitch cycles and the history of the solar system. *Proceedings of the National Academy of Sciences, 115*, 6363–6368.

Milankovitch, M. (1920). *Theorie mathematique des phenomenenes thermiques produits per la radiation solaire*. Paris: Gauthier-Villars.

Milankovitch, M. (1941). *Kanon der erdbestrahlung und seine andwendung auf das eiszeitenproblem* (Vol. 133, pp. 1–633). Belgrade: Royal Serbian Academy Special Publication. (English translation published in 1969 by Israel Program for Scientific Translations available from U. S. Department of Commerce.).

Milankovitch, M. (1952). Memories, experiences and perceptions from the years 1909-1944, Belgrade: Serbian Academy of Sciences, CXCV, p. 1–322. (in Serbo-Croatian).

Milankovitch, M. (1957). *Astronomische theorie der klimaschwankungen ihr werdegang und widerhall* (Vol. 280, p. 58). Belgrade: Serbian Academy of Sciences, Monograph.

Mingxiand, M., & Tucker, M. E. (2013). Milankovitch-driven cycles in the Precambrian of China: The Wumishan formation. *Journal of Palaeogeography, 2*, 369–389.

Moore, G. W. (1973). Westward tidal lag as the driving force of plate tectonics. *Geology, 1*, 99–100.

Naif, S., Key, K., Constable, S., & Evans, R. L. (2013). Melt-rich channel observed at the lithosphere-asthenosphere boundary. *Nature, 495*, 356–359. https://doi.org/10.1038/nature11939.

Nelson, T. H., & Temple, P. G. (1972). Mainstream mantle convection: A geological analysis of plate motion. *American Association of Petroleum Geologists Bulletin, 56*, 226–246.

Nestola, F., Jung, H., & Taylor, L. A. (2017). Mineral inclusions in diamonds may be synchronous but not syngenetic. *Nature Communications, 8*, 14168. https://doi.org/10.1038/ncomms14168 (2017).

Nimmo, F., & Tanaka, K. (2005). Early crustal evolution of Mars. *Annual Reviews, Earth and Planetary Sciences, 33*, 133–161.

Olsen, P. E., & Kent, D. V. (1996). Milankovitch climate forcing in the tropics of Pangea during the Late Triassic. *Palaeogeography, Palaeoclimatology, Palaeontology, 122*, 1–26.

Paillard, D. (2017). Predictable ice ages on a chaotic planet. *Nature, 542*, 419–420.

Peale, S. J., & Cassen, P. (1978). Contributions of tidal dissipation to lunar thermal history. *Icarus, 36*, 245–269.

Press, F., & Siever, R. (2001). *Understanding earth* (3rd ed., p. 620). New York: W. H. Freeman Co.

Ricard, Y., Doglioni, C., & Sabadina, R. (1991). Differential rotation between lithosphere and mantle: a consequence of lateral mantle viscosity variations. *Journal of Geophysical Research, 96*, 8407–8415.

Rosenblatt, P., Charnoz, S., Dunseath, K. M., Terao-Dunseath, M., Trinh, A., Hyodo, R., Glenda, H., & Toupin, S. (2016). Accretion of Phobos and Deimos in an extended debris disc stirred by transient moons. *Nature Geoscience, 9*, 581–583.

Ross, M., & Schubert, G. (1986). Tidal dissipation in a viscoelastic planet: Proceedings of the 16th Lunar and Planetary Science Conference. *Journal of Geophysical Research, 91*(p), D447–D452.

Rothschild, L. J., & Lister, A. M. (2003). *Evolution on planet Earth: The impact of the physical environment* (p. 438). London: Academic Press.

Salese, F., Pondrelli, M., Neeseman, A., Schmidt, G., & Ori, G. G. (2019). Geological evidence of planet-wide groundwater system on Mars. *Journal of Geophysical Research: Planets, 124*, 374–395. https://doi.org/10.1029/2018JE005802.

Schorghofer, N. (2008). Temperature response of Mars to Milankovitch cycles. *Geophysical Research Letters, 35*, L18201. https://doi.org/10.1029/2008GL034954.

Schulze, K., Marquardt, H., Kawazoe, T., Ballaran, T. B., McCammon, C., Koch-Muller, M., Kurnosov, A., & Marquardt, K. (2018). Seismically invisible water in Earth's transition zone? *Earth and Planetary Science Letters, 498*, 9–16. https://doi.org/10.1016/j.epsl.2018.06.021.

Scoppola, B., Boccaletti, D., Bevis, M., Carminati, E., & Doglioni, C. (2006). The westward drift of the lithosphere: A rotational drag? *Geological Society of America Bulletin, 118*, 199–209.

Singer, S. F. (1968). The origin of the moon and geophysical consequences. *Geophysical Journal, Royal Astronomical Society, 15*, 205–226.

Singer, S. F. (1970). The origin of the moon and its consequences. *Transactions, American Geophysical Union, 51*, 637–641.

Singer, S. F. (1977). The early history of the Earth-Moon system. *Earth Science Reviews, 13*, 171–189.

Singer, S. F. (1986). Origin of the moon by capture. In W. K. Hartmann, R. J. Phillips, & J. G. Taylor (Eds.), *Origin of the moon* (pp. 471–485). Houston: Lunar and Planetary Institute.

Sleep, N. H. (1994). Martian plate tectonics. *Journal of Geophysical Research, 99*, 5639–5655.

Sobolev, A. V., Asafov, E. V., Gurenko, A. A., Arndt, N. T., Batanova, V. G., Portnyagin, M. V., Garbe-Schonberg, D., Wilson, A. H., & Byerly, G. R. (2019). Deep hydrous mantle reservoir provides evidence for crustal recycling before 3.3 billion years ago. *Nature, 571*, 555–559.

Stern, R. J. (2005). Evidence from ophiolites, blueschists, and ultrahigh-pressure metamorphic terranes that the modern episode of subduction tectonics began in Neoproterozoic time. *Geology, 33*, 556–560.

Stern, R. J. (2007). When and how did plate tectonics begin? Theoretical and empirical considerations. *Chinese Science Bulletin, 52*, 578–591.

Stern, R. J., Leybourne, M. I., & Tsujimori, T. (2016). Kimberlites and the start of plate tectonics. *Geology, 44*, 799–802.

Stolper, D. A., & Bucholz, C. E. (2019). Neoproterozoic to early Phanerozoic rise in island arc redux state due to deep ocean oxygenation and increased marine sulfate levels. *Proceedings of the National Academy of Sciences, 116*, 8746–8755.

Taylor, F. B. (1910). Bearing of the tertiary mountain belt in the origin of the earth's plan. *Geological Society of America Bulletin, 21*, 179–226.

Thompson, A. B. (1992). Water in the Earth's upper mantle. *Nature, 358*(6384), 295–302. https://doi.org/10.1038/358295a0.

Uyeda, S., & Kanamori, H. (1979). Back-arc opening and the mode of subduction. *Journal of Geophysical Research, 84*, 1049–1061.

Wang, X. C., Wilde, S. A., Li, Q. L., & Yang, Y. N. (2015). Continental flood basalts derived from the hydrous mantle transition Zone. *Nature Communications, 6*, 7700. https://doi.org/10.1038/ncomms8700.

Wegener, A. (1924). *The origin of continents and oceans* (p. 212). London: Methuen.

Windley, B. F. (1995). *The evolving continents* (3rd ed., p. 526). Chichester: Wiley.

Wu, H. C., Zhang, S. H., Hinnov, L. A., Jiang, G. Q., Feng, Q. L., Li, H. Y., & Yang, T. S. (2013). Time-calibrated Milankovitch cycles for the Late Permian. *Nature Communications, 4*, 2452. https://doi.org/10.1038/ncomms3452.

Wu, H., Fang, Q., Wang, X., Hinnov, L. A., Qi, Y., Shen, S., Yang, T., Li, H., Chen, J., & Zhang, S. (2018). An ~34 m.y. astronomical time scale for the uppermost Mississippian through Pennsylvanian of the Carboniferous System of the Paleo-Tethyan realm. *Geology, 47*, 83–86. https://doi.org/10.1130/G45461.1.

Yoder, C. F., & Standish, E. M. (1997). Martian precession and rotation from Viking lander range data. *Journal of Geophysical Research, 102*(E2), 4065–4080.

Zanazzi, J. J., & Triaud, A. H. M. J. (2019). The ability of significant tidal stress to initiate plate tectonics. *Icarus, 325*, 55–66.

Zerkle, A. L., Poulton, S. W., Newton, R. J., Mettam, C., Claire, M. W., Bekker, A., & Junium, C. K. (2017). Onset of the aerobic nitrogen cycle during the great oxidation event. *Nature, 542*, 465–467. https://doi.org/10.1038/nature20826.

Zhang, S., Wang, X., Hammarlund, E. U., Wang, H., Costa, M. M., Bjerrum, C. J., Connelly, J. N., Zhang, B., Bian, L., & Canfield, D. E. (2015). Orbital forcing of climate 1.4 billion years ago. *Proceedings of the National Academy of Science, 112*, E1406–E1413.

Glossary

Absolute age A position of the age of a substance on a numerically calibrated time scale (also see relative age).

Accretion process Accumulation of particles from a cloud of debris to form a planet or planetoid.

Accretion torus A heliocentric doughnut-shaped geometric form from which smaller bodies are gravitationally attracted to participate in the planet-building process.

Achondritic meteorite A meteorite with no chondrules.

Adonis Proposed name for a phantom satellite of planet Venus.

Aeolis Dorsa area A broad depression located along the highland/lowland boundary of Mars. This area has many erosional features.

Algae Primitive aquatic single-cell or colonial organisms capable of photosynthesis.

Allende meteorite A carbonaceous chondrite meteorite that fell to Earth in 1969 and contained the first intensely studies calcium-aluminum inclusions (CAIs).

Andesite A fine-grained igneous rock characterized consisting mainly of intermediate plagioclase and hornblende. The coarser-grained equivalent is called diorite.

Andesitic crust A section of the surface of a planet or planetoid consisting of the chemistry and mineralogy of andesite and diorite.

Angra (Angra dos Reis) Phantom parent body for the Angrite group of meteorites.

Angrite meteorites A group of achondritic meteorites that have very similar characteristics and are thought to be fragments of a parent body (Angra) that has not yet been identified. The chemical composition of the angrite meteorites is very similar to that of the Earth's Moon.

Angular momentum A physical property of a rotating body. Similar to linear momentum but is momentum along an arcuate trajectory.

Angular momentum density A term used in the analysis of the rotational properties of planets and planetoids in units of radians cm^2/sec.

Anorthosite A rock composed mainly of the mineral Plagioclase. This is the main rock type in the primitive crust of the Moon.

R. Malcuit, *Geoforming Mars*, https://doi.org/10.1007/978-3-030-58876-2

Aphelion The farthest distance from the Sun for a body in a heliocentric orbit.

Argyre Basin A martian impact basin ~1200 km in diameter and ~ 3 km deep.

Argon-38/Argon-36 ratio An index used in the analysis of planetary atmospheres.

Asteroid Small planetoid in a heliocentric orbit commonly located in the Asteroid Belt.

Asteroid belt A zone of heliocentric space between the orbits of Mars and Jupiter which contains most of the orbits of the asteroids.

Asthenosphere A zone within the Earth's upper mantle that deforms plastically.

Astronomical metronome A proposal that the 400 Ka eccentricity cycle of the Milankovitch Model is essentially unchanged over geologic time and can be used as a calibrated time scale for geological purposes. The planets causing this very regular eccentricity cycle to occur are Jupiter and Venus.

Astronomical unit (AU) The average distance between the center of the Sun and the orbit of Earth.

Atmosphere The gaseous envelope surrounding a planetary body.

Atmospheric drag A process in which a celestial body interacts with a planetary atmosphere resulting in a decrease in the velocity of the encountering body.

A-type asteroid Small olivine-rich bodies that are thought to be fragments from a differentiated asteroid.

Bacteria Single-celled microscopic biota that are building blocks for more complex life forms.

Basal magma ocean dynamo A model proposed by Scheinberg et al. (2018) as a possibility for generation of an internal lunar magnetic field. This model requires complete melting of the body of the Moon. The driving mechanism is not specified.

Basalt Fine-grained igneous rocks that are the main rock type on Earth's ocean floors as well as the main rock type on the surface of planet Venus. The major minerals in basalt are plagioclase, pyroxene, and olivine.

Basaltic asteroid An asteroid with the composition of basalt.

Basaltic crust A surface or section of the surface of a planet or planetoid with the general composition of basalt.

Basaltic-gabbroic complex Gabbro is the coarse-grained equivalent of basalt. Thus, a rock complex composed of a combination of basalt and gabbro is a basaltic-gabbroic complex.

Big Brother Jupiter Planet Jupiter has a significant gravitational effect on most bodies in the inner part of the Solar System. The effects are greatest in the asteroid zone but Jupiter is also the main cause of the heliocentric eccentricity cycles of the Milankovitch model as well as for a purported Jupiter Orbit–Lunar Orbit resonance. This situation does seem analogous to a Big Brother doing kind positive deeds for a litter sister, Earth.

Biological system Organized group of several organisms on a planet or planetoid.

Biosphere The totality of locations on a planet that is characterized by biological entities.

Bipolar outflow Massive outflow of material (dust and gas) in the polar regions of a nascent sun-like star. The outflow appears to be guided by magnetic flux lines. Much of this material escapes from the star.

Body viscosity and Q factor According to Ross and Schubert (1986) the Q factor is a function of body viscosity. Low Q values are associated with intermediate values of planetary viscosity for lunar-sized bodies.

Bolide A small to intermediate-sized body in a heliocentric orbit.

Bolide impact Impact of a bolide onto a planet. The result is normally a crater on a rocky surface of a planet or planetoid.

Borealis Basin The north polar lowlands of Mars. The origin of this lowland is debatable.

Bradbury crater Landing site within Gale Crater on the martian surface and the location of Curiosoty Rover activities.

Calcium-aluminum inclusions (CAIs) These are inclusions rich in calcium and aluminum that are found in some chondritic meteorites. These small particles formed under very high temperature conditions. A composite of all types of CAIs yields a chemical composition very similar to that of the Earth's Moon (Gast 1972).

Capture model A proposal that a planet can acquire a satellite from a heliocentric orbit.

Capture encounter An encounter of a planetoid with a planet in which sufficient energy is dissipated within the interaction bodies to insert the planetoid into a stable planetocentric orbit.

Carbonate rocks Rocks composed of calcium carbonate and related carbonate minerals.

Carbonaceous chondrites Stony meteorites containing chondrules and abiotic organic matter that have not been altered by heating since their formation in the early history of the Solar System.

Carbon dioxide An oxide of carbon that is an important constituent in the atmospheres of Venus, Earth, and Mars.

Cataclysm's end The concept of a Late Heavy Bombardment (a cataclysm) was proposed in the 1975–1977 era. The major bombardment was thought to take place between 4.0 and 3.8 billion years ago. No one has been successful in identifying the source of the impactors. According to Mann (2018) the concept is about to be assigned to the history of solar system science.

Celestial mechanics The study of the movement and interactions of bodies in space.

Ceres The largest and most massive asteroid (~914 km in diameter).

Ceres-type asteroid Ceres is considered to be the "normal" type of asteroid. It is composed of a high percentage of hydrous minerals.

Charonoid satellites Four small satellites of Pluto that are in "stable" orbits. Charon is a large satellite of Pluto.

Chemical differentiation A process in which chemical units are layered in a planetary body based mainly on density. The heavy elements like iron end up in the core of the planet or planetoid.

Chondritic meteorites Meteorites that contain chondrules.

Chondrules Millimeter-sized spherical particles that were at one time glassy beads in space before they got accreted into chondritic meteorites.

Circular lava lakes Many of the lunar maria areas appear to be solidified basaltic lava lakes.

Co-formation model A proposal that suggests that a planet and its satellite, or satellites, formed at about the same time and from similar material. This model works fairly well for the satellites of some of the gassy/icy planets of the Solar System but not for the Earth-Moon system.

Collisional capture model A proposal suggesting that an encountering planetoid collides with another small body during the timeframe of a close encounter with a planet. This collision of the small body and the planetoid then accounts for much of the energy dissipation necessary for capture of the encountering planetoid into a planetocentric orbit.

Comet A body composed of ice and dust that revolves around the Sun in a highly elliptical orbit.

Cometary source of water Martian rocks have a deuterium to hydrogen ratio very similar to that of comets. A reasonable conclusion is that Mars got some, or most, of its water from either a collision with a comet or a collision with a Jupiter-captured comet (*i.e.*, a comet that was inserted into an asteroidal orbit after a gravitational encounter with Jupiter).

Compressed density Density of a body as it exists in nature.

Concordia Name proposed for a large theoretical prograde satellite of Mars.

Continental drift A concept proposed by Alfred Wegener in the early 1900's that suggested the continents are moving relative to one another over periods of geologic time. This idea was the precursor to the Plate Tectonics model.

Continental shelf A low-lying section of continental crust that is flooded with ocean water.

Core dynamo A mechanism needed for generation of a global magnetic field if the core of the body is metallic.

Crustal uplift A section of planetary crust that is elevated by tectonic processes.

Current sheets A major source of thermal energy for the Disk-Wind model for chondrule formation.

Cyclostratigraphy A study of astronomically forced climate cycles recorded in sedimentary sequences. Astronomical cycles are periodic variations of the Earth's heliocentric orbital eccentricity, planetary obliquity, and planetary precession motion due to gravitational interaction with the Sun, Earth's Moon, and neighboring planets.

Default model An explanation that remains as a reasonable solution after all other models have been considered and discarded. This is the "bottom of the barrel" of ideas on the subject! But, really, this is the way most controversial subjects in science are settled.

Deimos A small natural satellite of Mars located in a nearly circular orbit at about 6.92 mars radii.

Deuterium A hydrogen atom with a proton and a neutron in its nucleus (i.e., heavy hydrogen).

Deuterium to hydrogen ratio (D/H ratio) A ratio of heavy hydrogen to normal hydrogen (protium) that is characteristic of certain materials.

Diamond-bearing kimberlite deposits Some kimberlite deposits are barren of diamonds, others contain diamonds.

Diamond deposits Diamonds can only form with a confining pressure relating to over 100 km depth within the Earth. They then get emplaced to the surface of Earth as "torpedos" of rock masses (kimberlite rock type) within an envelop of carbon dioxide gas. The carbon dioxide gas cools as it is emplaced to low pressure levels. Thus very little thermal metamorphism is associated with the emplacement of diamond deposits.

Differential rotation A body characterized by some inner concentric zone, or zones, rotating at a different rate than the outer part of the body. This implies that there is a zone of low viscosity within the body.

Diogenite Achondritic meteorite consisting mainly of magnesium-rich orthopyroxene. Related to eucrites and howardites (the HED group) and are thought to be from asteroid 4 Vesta.

Directionality and polarity of plate tectonics The mainstream mantle convection model suggests that the lithosphere of the Earth rotates at a different rate then the mantle of the planet. The proposed "slippage" zone is the asthenosphere and the retarding force is the unidirectional lunar rock tide. Thus the lithosphere migrates in a westerly direction and the maximum displacement would be in the equatorial zone with minimal displacement in the polar areas.

Discordia A name proposed for a phantom satellite of Uranus that could be the cause of the 98 degrees obliquity of Uranus. See Chap. 6 for details.

Disc Wind model A proposal that electromagnetic current sheets caused "flash melting" episodes that, in turn, caused the formation of most of the chondrules in chondritic meteorites. The Disc Wind would succeed the X-Wind in the early history of solar system development.

Displacement Love number A deformation parameter for a planetary body that is related to its deformability due to tidal stresses.

Dynamics of the capture episode A sequence of body deformation events associated with the capture of a sizeable planetoid from a heliocentric orbit as well as the subsequent planetocentric orbit evolution. Once stable capture occurs, the tidal activity on the planet and satellite is long-lived.

Earth accretion torus A theoretical three-dimentional, doughnut-shaped heliocentric zone representing the space from which the planet accumulated meteoritic debris during the accretion process.

Earth-Moon-Sun system The system consists of three bodies: Earth, Moon, and Sun. The three bodies interact in a multitude of ways which affect the obliquity (tilt angle) of the planet, the precession of the equinoxes, and the eccentricity of Earth's orbit. This complex interaction over geologic time has resulted in a habitable planet for a multitude of biota.

Earth radius Distance from the center of the planet to the surface (~6371 km).

Earth tide model Theoretical development for analyzing the effect of earth (rock) tides.

Earth tide paradigm Concept that earth (rock) tides are important for the operation of planet Earth.

Earth tides Systematic deformation of the lithosphere of the planet due mainly to gravitational interaction with the Moon and Sun.

Eccentricity of the orbit of the planet Degree to which a planetary orbit departs from a circular orbit.

Eccentricity of the orbit of the planetoid Degree to which a satellite orbit departs from a circular orbit.

Elliptical orbit A section of a cone at an angle to its base in the form of a closed geometric form having an eccentricity between 0 (perfect circle) to 1 (a hyperbole if over 1).

Energy necessary for capture Energy dissipation by some process or combination of processes that results in causing the orbit of a planetoid to change from a heliocentric mode to a planetocentric mode.

Energy stored Energy of deformation of a body caused by tidal stretching. A fraction of this energy is subsequently dissipated as the tidal stress is relaxed.

Enstatite chondrite Stony meteorite containing chondrules and composed primarily of the mineral enstatite, a magnesium-rich pyroxene. A composite enstatite chondrite appears to be a good choice for representing the composition of planet Earth.

Equatorial zone A zone roughly plus and minus 23.5 degrees latitude from the equator of Earth.

Equilibrium chemical condensation models A calculation for determining the chemical composition of minerals that would crystallize in various pressure-temperature conditions from the primitive solar nebula. The results relate fairly well with the chemistry of the planets of the solar system.

Equilibrium rock tides Deformation of a planetary surface caused by tidal stress.

Equilibrium tidal amplitude The deviation of the tidal bulge form the normal shape of the body.

Eucrite parent body Probably the asteroid 4 Vesta. However, there could be other asteroids that could yield this composition.

Eucrite A basaltic achondrite meteorite that may have originated from asteroid 4 Vesta (a member of the HED group).

Exoplanets Planets associated with stars other than our Sun.

Exosatellites Satellites associated with exoplanets.

Fission model A proposal suggesting that the material that formed the Moon fissioned from a rapidly rotating Earth early in earth history to form the Earth-Moon system.

Flinders Range A mountain range in South Australia.

Fluvial-deltaic sequence A sequence of sedimentary rocks (or sediments) caused by deposition of the bottom load of a stream as the stream enters a standing body of water. Repeated cycle of erosion and deposition can result in a complex stratigraphy sequence.

Fourth-order Runge-Kutta A numerical integration process for precise calculation of planetary and satellite orbits.

Frost line The distance from the Sun at which ice begins to freeze in the evolution of the solar nebula. This frost line is commonly thought to have fluctuated in the vicinity of the orbit of Jupiter causing Jupiter to be the most massive planet in the system.

FU-Orionis stage A stage in the development of the solar nebula closely following the Disc Wind action. The FU-Ori stage is characterized by electromagnetic radiation in the microwave range. This radiation could cause significant surface and near-surface melting of planetoids in orbits between the Sun and Mercury and significant metamorphism and hydrothermal metamorphism at greater distances from the nascent Sun.

Ga One billion years.

Gabbro A coarse-grained chemical equivalent of basalt. The major minerals are calcium plagioclase, pyroxene, and olivine.

Gas-drag capture model A proposal that the energy for planetoid capture can be dissipated by one or more encounters through a dense planetary atmosphere. This capture converts the path of the planetoid from a heliocentric orbit to a planetocentric orbit.

Gaseous planet A planet with a planetary surface of a mixture of gases (*i.e.*, there is no obvious solid surface). Good examples are planets Jupiter and Saturn.

Geocentric orbit A planetoid orbit of whatever eccentricity with Earth as the focus.

Geocentric orientation A computer plot in which the orbit of a planetoid is shown relative to Earth at the origin of the plot.

Geoforming A theoretical planetary development process resulting in conditions on a planet or exoplanet that are similar to those on Earth at the present time or at some time in the past or future evolution of planet Earth

Geologic time A time scale relating to the history of Earth.

Geometry of stable capture zones A geometric plot of Planet Anomaly vs. Planetoid Eccentricity that can be used to predict which values can result in stable capture scenarios if sufficient energy is dissipated within the interacting bodies.

Giant impact hypothesis A proposal that the Earth's Moon was formed by the collision of one or more planetoids coming from heliocentric orbits. A portion of the debris cloud caused by the collision(s) then accretes to form the body of the Moon.

Giant impact model A calculated simulation of the origin and evolution of the Earth-Moon system that features a Mars- mass body colliding with Earth and the Moon forms from the debris of the collision.

Giant impact paradigm A generally accepted model for the origin of the Earth-Moon system. This model, however, does not relate well to the size and chemistry of the lunar body. In addition, this model has no reasonable explanation of lunar rock magnetization patterns.

Global equivalent layer of water (GEL) A theoretical model for explaining the volume of water on a planet. For example, Mars may have had a global equivalent layer of liquid water of 100 meters in the past but it is not there now.

Global magnetic field A magnetic field that has a flux covering most or all of the surface of the planet. The implication is that this magnetic field is generated by an internal dynamo.

Global resurfacing event A model proposed to explain the apparent youthful surface of planet Venus. The main evidence for the interpretation is a fairly uniform distribution of impact craters on the surface of the planet.

Gravitational capture model A proposal that the orbit of the Earth's Moon was changed from a heliocentric orbit to a geocentric orbit by gravitational interactions of the two bodies. Since only the Moon can have sufficient deformation and energy dissipation properties to store and dissipate the energy for capture, the sink for energy must be within the lunar body.

Gravitational constant A constant used in all equations involving mass attraction calculations.

Gravitational infall stage A stage in the evolution of the Sun (and other stars) in which a mass of dust and gas from previous stellar explosions forms a cloud of debris that accumulates to form a proto- Sun (or proto-star).

Gravitational perturbations Systematic small gravitational tugs on a planetary body.

Great circle pattern Any pattern of features along a great circle. A number of lunar maria are located very near to a lunar great circle. A number of circular/elliptical impact structures on the martian surface are located along a great circle.

Grazing impact An impact of a small body onto a larger body that has some overlap of the normal surfaces of the bodies. There are many grades of grazing impacts from slightly grazing to massive grazing.

Habitable zone A heliocentric zone for planets which is suitable for a biological system to exist. The possibility of liquid water to exist on the surface of the planet for at least part of the annual cycle appears to be very important for this determination.

Habitable exoplanets Exoplanets that are located within the habitable zone of their central star.

HED meteorites A group of meteorites that are thought to be derived from the asteroid 4 Vesta. These are the howardites, eucrites, and diogenite meteorites.

Heliocentric orbit The path of a planet or planetoid around the Sun, the orbit can be circular or elliptical.

Heliocentric orientation diagram A plot used in computer simulations that shows the position of the planet relative to the Sun in angular measure.

Hellas Basin A large impact basin in the southern hemisphere of Mars and the largest recognizable impact feature on the planet.

Hemispheric dichotomy A condition in which one hemisphere of a body has notably different characteristics than the other. An example is planet Mars where the northern hemisphere has polar lowlands and the southern hemisphere has polar highlands.

Hesperian age The Hesperian period had a duration of about 800 Ma (~3.7 to~2.9 Ga).

Hesperian-age ocean This period is characterized by the presence of significant quantities of surface water on Mars especially in the north polar region.

Hill sphere A nearly spherical boundary around a planet that defines the gravitational sphere of influence of the body.

Howardite meteorite Brecciated basaltic achondite consisting of fragments of eucritic or diogenitic compositions.

Hydrated minerals Minerals with water of hydration in their structure usually in the form of (OH) or H_2O.

Hydrologic cycle The water cycle on a planet or planetoid. This included some combination of ground water, surface water, and atmospheric water.

Hydrosphere A term that included all water on or near the surface of the planet (i.e., surface, near surface, and atmosphere).

Hydrothermal metamorphism A process of recrystallization in which some combination of heat, pressure, and hot water, steam, or some other volatile substances are involved.

Hypothesis An essentially untested statement for explanation of a natural phenomenon.

Ice cap A mass of glacial ice that commonly forms over a pole of a planet during the winter season. If cold winter seasons prevail, then the mass of ice can grow year after year.

Ice sheets Large accumulations of ice on a planetary surface. When the weight of the ice exceeds the bearing strength of ice, then the ice and spread over the land surface to form a sheet of ice.

Icy planet or planetoid A planet or planetoid that is completely covered with ice. Europa is a good example of an icy satellite; the "snowball earth" episode in earth history is a good planetary example.

Impact velocity (or speed) The velocity (or speed) of the impactor at the time of impact on the target body.

Inclination The angle of deviation of an orbit from its orbital plane.

Inferior conjunction A situation in which a planet in an orbit nearer to the Sun moves between an outer planet and the Sun (e.g., Venus passes between Earth and the Sun).

Inorganic processes A process or reaction in which no organisms are involved.

Intact capture model A model in which the encountering planetoid maintains its integrity during the capture process.

Io The innermost Galilean satellite of Jupiter that is volcanically active because of tidal heating.

Ionized cloud A mass of vaporized material with charged particles and thus can interact with the magnetic flux of the planet or planetoid.

Iron Line A heliocentric zone within the proto-solar midplane that gets enriched in metallic minerals because of repeated episodes of vaporization of metallic ions due to heat pulses and condensation of these vaporized elements into metallic mineral structures upon cooling.

Joules heating Energy that results in raising the temperature of a substance.

Jupiter The largest planet in the Solar System located at about 5.2 AU

Jupiter-captured comet Comets that have been captured into heliocentric asteroid-like orbits after a close encounter with planet Jupiter.

Jupiter Orbit – Lunar Orbit (JO-LO) resonance A resonance caused by a lengthy series of gravitational tugs by Jupiter on the lunar body that cause the major axis and eccentricity of the lunar orbit to gradually increase.

Ka Thousand years.

Kona Conference A conference on the "Origin of the Moon" that was held in Kona (HI) in October 1984. The Giant Impact Model emerged from this meeting as the favorite model of lunar origin.

KREEP A term for some lunar basalts, especially in the Oceanus Procellarum region, that are enriched in potassium (K), rare earth elements (REE), and phosphorous (P). These lunar basalts are interpreted to be the final crystallization products in the primitive lunar magma ocean.

Kuiper Belt A heliocentric belt that extents from the orbit of Pluto out to at least 500 AU and is populated by icy bodies of various masses.

Kuiper belt objects Icy bodies (planetoids) in irregular orbits within the Kuiper belt.

Late Heavy Bombardment A purported massive bombardment of the Moon by external bodies between about 4.0 and 3.8 Ga. The source of these impactors is still unknown. The concept of the LHB is now under review and may be gradually abandoned according to Mann (2018).

Late Precambrian "global" glacial deposits During this era of geologic time there are many sedimentary rock sequences that have glacial deposition signatures and paleomagnetic data that suggest many of these deposits formed in very low paleolatitudes settings. Analysis of this information has led to the development of the concept of a "snowball earth" at that time.

Lineation patterns A pattern featuring straight lines on a planetary surface.

Lithosphere The outer part of the surface of a rocky planet. The lithosphere consists of the crust of the planet as well as the solid mantle material beneath it.

Lithospheric drift Movement of the lithosphere of a planet relative to the layers of rock beneath it. This type of movement of a planetary crustal complex is very important for paleomagnetic studies.

Luna A name for the Moon before it was captured by Earth.

Lunar achondrite meteorites Meteorites found on Earth that have very similar compositions to those in the Apollo rock collection. Thus, they are interpreted as meteorites from the Moon.

Lunar capture model A model that features gravitational capture of Luna from a heliocentric orbit into a geocentric orbit.

Lunar KREEP basalts Basalts rich in potassium, rare earth elements, and phosphorous. This type of basalt is thought to be the last magma to crystallize from a lunar magma ocean and also the first to melt when that region of lunar magma ocean is reheated to above its melting point.

Lunar magma ocean A zone about 600 km thick that was apparently melted at the time of lunar formation or remelted soon after lunar formation. Differentiation of this zone resulted in the development of an all-enclosing anorthositic crust.

Lunar mare Large areas of lunar basalt on the lunar surface that have either circular, elliptical, or irregular outlines.

Lunar rock magnetization There are two eras of magnetization recorded in the crust of the Moon. The older magnetized rocks are associated with the primitive anorthositic crust of the Moon and these rocks accumulated a magnetic pattern as the primitive crust was cooling through the Curie temperature. The younger magnetized rocks are associated with mare basalts and this era of magnetization appears to record an exponentially decreasing strength of the field during this time.

Ma Million years.

Magma ocean A zone of molten rock that can form in the outer portion of a planetoid or planet via joule heating.

Magnetic amplifier A mechanism by which a weak (stray) magnetic field can be amplified by wrapping the magnetic flux lines in a systematic way.

Magnetic anomaly pattern A pattern on a planetoid or planet surface that has many irregularities. In many cases, these irregularities yield valuable information about the generative mechanism.

Magnetic field Magnetic flux lines caused by a magnetic dynamo or amplifier.

Magnetic field flux lines Lines of a magnetic force field. If a magnetic dynamo is located in the core of the body, then the magnetic flux lines will emerge from one polar area and reenter the body at the other polar area.

Magnetic rock patterns These can be patterns in surface rocks or patterns in rock or mineral samples.

Magnetized chrondrules Many chrondrules record a magnetic field. The strength of the magnetism recorded in the chrondrules may relate to the strength of a solar magnetic field at the time of the formation of the chrondrules.

Main asteroid belt (MAB) A zone of orbits of small bodies located between the orbits of Jupiter and Mars. This is the source of most meteorites that are found on Earth.

Mainstream mantle convection model A model developed by Nelson and Temple (1972) to explain the directional features and asymmetries associated with subduction zones as well as other directional features associated with the plate tectonic model

Major axis of an orbit The maximum length of an elliptical orbit.

Mantle A zone of the Earth's interior that is beneath the crust and above a metallic core. This concept of a mantle can be applied to any terrestrial planet or planetoid.

Marocentric orbit An orbit of a body around planet Mars.

Maroforming A theoretical planetary development process resulting in conditions on a planet or exoplanet that are similar to those on Mars at the present time or at some time in the past or future evolution of planet Mars.

Marologic time A time scale for recording the events on planet Mars over billions of years.

Mars rotation rate The rotation rate is about 24.6 hr/day. This rate is thought by many investigators to be close to the primordial rotation for the planet.

Martian core There is a metallic core of some sort and a major constituent is iron.

Martian core dynamo There is a magnetic signature from rocks on the martian surface and there is general agreement that it was generated by movement within a central metallic core.

Martian mantle The zone between the crust of Mars and the core. Many details are lacking on the zonation of this zone.

Martian meteorite There are several meteorites found on Earth that are thought be from planet Mars. Many of these meteorites have been found in the Dry Valley of Antarctica.

Martian satellites There are two very small satellites of Mars, Phobos and Deimos. They are of such low mass that they have very little dynamical influence on Mars.

Mercocentric orbit An orbit of a body around planet Mercury.

Mercocentric orientation Computer plots that have Mercury at the origin of the plot.

Mercoforming A theoretical planetary development process resulting in conditions on a planet or exoplanet that are similar to those on Mercury at the present time or at some time in the past or future evolution of planet Mercury

Mercury Planet Mercury is the terrestrial planet nearest to the sun with a mean orbital radius of ~0.4 AU.

Metallic core When terrestrial bodies undergo gravitational differentiation in their early histories, the metals generally settle to a metallic central area (a core). If a magnetic field is generated, it is usually associated with differential movement of the core relative to the outer zone of the body.

Metallic meteorites Meteorites composed mainly of metallic minerals.

Metal-sulfide core A composition that has been proposed for a core in the Moon, but there is no credible physical evidence for such a lunar core.

Messenger Mission to Mercury Mission completed on April 30, 2015. The goal was to explore from orbit the surface and near surface features of the planet.

Microwaving process for melting rocks A process of joule heating by way of intense exposure to electromagnetic waves in the microwave range. This type of melting has been done in microwave ovens in laboratories and it could be done to the exterior of planetoids between the Sun and the orbit of Mercury in the early history of the Solar System.

Microwave wind A "solar wind" in the microwave range of the electromagnetic spectrum.

Milankovitch cycles Cycles of heliocentric orbital eccentricity, planetary obliquity (tilt), and planetary precession as proposed and calculated by Mulitin Milankovitch.

Milankovitch model A very complex model to explain climate cycles over geologic time. The Sun, Earth, and Moon as well as Venus, Jupiter, and Saturn are involved in the model.

Modern style of plate tectonics Stern (2005) and Davies (1992) and others think that the Earth's mantle was too warm to permit slab-pull subduction before about 1.0 Ga ago. The slab-pull subduction mechanism is thought to be a major driving force for the modern style of plate tectonics.

Milky Way Galaxy The galaxy that includes the Sun and all parts of our Solar System.

Moment of inertia A property related to the distribution of mass within a body. The coefficient of the moment of inertia is an important factor for this characterization. For example, a homogeneous body has a coefficient of the moment of inertia of 0.4. It is interesting to note that the coefficient of the moment of inertia of the Moon is very near 0.4.

Nebular dust and gas stage Material from previous stellar explosions are gathered together by gravity to form the proto-Sun.

Neptocentric orbit A planetoid orbit of whatever eccentricity that has Neptune at the focus.

Neptocentric plot A computer-generated plot of planetoid encounters or orbits which has Neptune at the origin of the plot.

Neptune Eighth planet from the Sun.

Nice model A complex model for the early history of the evolution of planetary orbits which features a major rearrangement of the orbits of the outer planets which, in turn, causes major instability in the region of the Asteroid Belt. The model was designed to help explain the Late Heavy Bombardment episode that was thought to affect the Moon and other bodies in the inner Solar System from about 4.0 to 3.8 Ga ago. The model was named for the city of Nice (FR). An early article on the model was by Gomes et al. (2005). According to Mann (2018) the promoters of the model are in the process of abandoning the idea. (An added note is that the Prograde Gravitational Capture model, described in Chap. 7, explains nearly all of the features associated with the Late Heavy Bombardment.)

Noachian period The Noachian is a geological period on Mars during which the oldest surface features were formed. It is characterized by high rates of impact of space debris of many different sizes. There is also evidence of abundant surface water. Age estimates are from ~4.1 to 3.7 Ga.

Non-capture encounter An encounter between a planetoid and a planet in which insufficient energy is dissipated for capture into a planetocentric orbit.

Non-rotating coordinate system A plotting procedure for planetocentric orbits in which the plot does not rotate as it theoretically revolves around the Sun.

Oblique impact An impact onto the surface of an object that is <u>not</u> vertical to the surface (i.e., impact at an angle).

Obliquity of a planet The numerical value of the tilt angle of the body's axis of rotation.

Obliquity cycle The tilt angle cycle for Earth at the present time is about 41 Ka. The variation is about plus and minus one degree from the present value of 23.5 degrees.

Ocean basin A crustal area of high density that causes the crust in the area to "sink" to a lower level. The resulting depression is filled with water (perhaps ocean water) if surface water is available.

Ocean tidal amplitude Tidal distortions of ocean water on a planet or planetoid.

Ocean tidal range Tidal distortions of ocean water on a planet or planetoid as the planet rotates under a sizeable satellite. The total tidal range in an area of open ocean is about 1.5 times the tidal amplitude.

Ocean tides Ocean tides on planet Earth are simple in principle but complex in detail. The ocean tidal bulge (amplitude) is seldom directly associated with the position of the satellite in a given geographical region. Each ocean basin has a system for the rotation of tidal waters that is controlled by the Coriolis Force, (Check a website on ocean tidal phenomena for more details.)

Olympus Mons Largest volcano in the Tharsis area of Mars as well as the largest known volcano in the Solar System at the present time.

Oort cloud A spherical zone surrounding the Solar System beyond the Kuiper Belt. This Oort Cloud is thought to be the place of origin of most Comets.

One-plate planet A planet on which the lithosphere moves as one unit. However, there can still be a zone of plastic deformation below the lithosphere to permit movement of the entire surface.

Orbit circularization sequence Process of the evolution of a highly elliptical orbit to an orbit of low eccentricity. The energy sink can be rock tidal action or gas drag in an extensive atmosphere of a planet.

Orderly component of planetary rotation This model features the accumulation of small bodies from orbits of low eccentricity by a favored planetary embryo on a circular orbit. Small planetoids on elliptical orbits from the outer regions of the accretion zone of the planet generally induce a more rapid prograde rotation.

Origin of lunar maria The question is whether they were formed over a short period of time or over an extensive period of time. The Gravitational Capture model suggests that they were formed in about 2 hours of time about 3.95 Ga ago via a close, tidally disruptive, encounter. Solid body impact models suggest at least 200 Ma for the impact basins to be flooded with mare basalt. More age data are needed from rock samples via sample-return missions to the Moon to choose between the models.

Origin of satellites The main models are co-formation models and capture models. Fission models have not been demonstrated to work under reasonable conditions and impact, mainly giant impact models, are not solving many problems in solar system science (see Chap. 6 for more details.)

Outcomes of tidal evolution The outcomes of tidal evolution of a planet-satellite system depend on (1) the rotation rate and mass of the planet at the initiation of the scenario, (2) the direction of movement of the satellite relative to the planet, and (3) the radial distance and mass of the satellite. The time frame of the orbital evolution scenario is determined by the deformation parameters of the interacting bodies. See Counselman (1973) for more details.

Outer asteroid belt (OAB) The part of the asteroid zone that is nearest to planet Jupiter. The orbits of these asteroids are strongly affected by gravitational perturbations by Jupiter.

Outgassing A process of removing gaseous material from the body of a planet or planetoid. The most common process is some sort of volcanic activity.

Oxygen A very reactive gas that is a major constituent of the Earth's atmosphere; it is ~21 volume percent of the atmosphere at the present time.

Oxygen isotope ratios Ratio of O-16 (the light isotope) relative to either O-18 or O-17 or both; these isotope ratios are sensitive to the temperature of formation of certain materials.

Paleo-shorelines Reconstructions of ocean or lake shorelines using physical evidence of shoreline erosion and deposition. This is essentially the procedure by mappers to identify the shorelines of the North Polar Sea on Mars (Citron et al. 2018).

Pallas Second largest asteroid by diameter (~522 km); third most massive asteroid.

Paradigm A commonly accepted concept or model for the explanation of natural phenomena.

Paradox A statement that seems contradictory, unbelievable, or absurd but may be actually true in fact.

Pebble accretion A process of using an accumulation of small planetesimals to grow planets.

Perihelion The nearest approach distance to the Sun for a body in heliocentric orbit.

Perigee The nearest approach distance to the Earth for a body in geocentric orbit.

Perimar The nearest approach distance to Mars for a body in marocentric orbit.

Perinept The nearest approach distance to Neptune for a body in neptocentric orbit.

Periven The nearest approach distance to Venus for a body in venocentric orbit.

Planet anomaly The position of the planet in degrees, at the beginning of a three-body numerical simulation calculation.

Planetary differentiation Internal separation of planetary materials mainly based on density.

Planetary precession Retrograde motion (clockwise as viewed from the N-pole) of the polar axis around the celestial pole of the Solar System.

Planetesimals Small planetoids in heliocentric orbit.

Planetocentric orbit An orbit that has a planet at the focus.

Planetological evolution The geological development of the entire planet over geologic time.

Planet Orbit – Lunar Orbit resonance A resonance caused by a lengthy series of repeated and systematic gravitational tugs by a planet in an unrelated heliocentric orbit which can increase the eccentricity and major axis of the orbit of the satellite of a planet.

Planet-satellite system The orbital configuration of a satellite revolving around a planet.

Plate tectonics Formation and movement of a system of rocky plates on a planet; the main processes that result in movement are the formation of new ocean floor at ocean rises and the subduction of plates of ocean floor at subduction zones.

Phantom satellite A satellite that no longer exists but one that may have had a notable influence on the history of a planet.

Phobos The innermost small satellite of Mars located in a nearly circular orbit at about 2.76 mars radii.

Pluto-Charon system Pluto has a large satellite (Charon) relative to the size of the planet; Pluto also has several small satellites that can be referred to a Charonoids. The origin of this unusual planet- satellite system is still unresolved. See Chap. 6 for more details.

Polar wandering A result of movement of the crust relative to the magnetic poles of a planet or planetoid. The magnetic poles are considered to be fairly stable. This implies that the crust is moving relative to the magnetic poles.

Potassium vs. Uranium (K/U) ratio The K/U fraction is the basis for the Potassium Index for solid bodies in the Solar System; this Potassium Index appears to be related to the place of origin of planets, planetoids, and meteorites in the Solar System.

Precession cycle This cycle features the retrograde movement of the north pole of a planet relative to the celestial pole; this cycle at the present time in geological history is ~20 Ka.

Post-capture energy dissipation Major quantities of energy must be dissipated for gravitational capture. Then less energy is dissipated per encounter during the post capture orbit circularization sequence because the average perigee distance keeps increasing and the value of Q increases as the body viscosity of the planetoid decreases. These changes in the energy dissipation parameters and distance of separation result in a lengthy time scale for the post-capture evolution of a planetoid orbit.

Primordial rotation rate for Earth According to the plot of MacDonald (1963) the primordial rotation rate for Earth should be about 10 hr/day.

Probability of capture of planetoids by planets A numerical value can be obtained from the geometry of an appropriate Stable Capture Zone if the energy dissipation characteristics of the interacting planet is within a range to store and dissipate the energy for capture.

Prograde Movement in a counter-clock-wise direction relative to the north pole of the system.

Prograde Gravitational Capture Model (PGCM) Depiction of a process in which the captured planetoid is inserted into a planetocentric orbit in the prograde direction.

Progressive devolatilization X-wind action on the solar nebular disc causes progressive devolatilization of materials from near the proto-Sun outward to the vicinity of planet Jupiter.

Protium A normal atom of hydrogen which has only one proton in the nucleus.

Proto-planetary cloud A stage characterized by material from previous stellar explosions falling into a central region by gravity to form a somewhat spherical mass called a proto-Sun.

Proto-planetary disk After a solar magnetic field is established the angular momentum associated with the rapidly rotating proto-Sun is gradually transferred to the surrounding cloud transforming it into a disc-shaped mass. The proto-Sun decreases in rotation rate as the rotation rate of the surrounding disc is increased.

Proto-Pluto This is one possible story for the origin of proto-Pluto and this is a slightly modified scheme proposed by Lyttleton (1936). A prograde satellite of Neptune has a close encounter with a gravitationally captured satellite of Neptune in retrograde orbit, namely Triton. During a very close, slightly grazing, encounter between proto-Pluto (the smaller body) and Triton, a mass of ice is gravitationally extracted from proto-Pluto and goes into orbit around Pluto. The combination of the newly formed Pluto and its satellite, Charon, accompanied by several other small satellites, Charonoids, gravitationally escape from Neptune and enter into an unusual heliocentric orbit. (There are some graphic illustrations of this physically possible celestial activity in Chap. 6.)

Q (specific dissipation factor) 1/Q is the fraction of the tidally stored energy that is dissipated during a close encounter between a planet and planetoid. The smaller the Q value, the greater the quantity of energy that is dissipated. A very low Q value for an encountering planetoid is necessary for successful gravitational capture.

Quasi-three-body calculation of post-capture orbital evolution This is a procedure for making a two- body calculation of orbit evolution more realistic by including the effects of close encounters caused by solar gravitational perturbations. These adjustments to the two-body calculation shorten the time scale of the post-capture orbital evolution by about 30 percent.

Reconnection of magnetic flux lines A major source of energy in the X-Wind model; this "reconnection" energy is caused by the breaking and reconnection of magnetic flux lines and is thought to be a major source of energy for the formation of calcium-aluminum inclusions that are found in chondritic meteorites.

Reconnection Ring A shifting belt of intense thermal activity in the proto-solar mid-plane of the X-wind model. This is the place of formation of CAIs and the belt has a central value of ~0.06 AU.

Relative age The age of rock unit relative to another rock unit as determined by some combination of position in sequence, crosscutting relations, overlapping ejecta aprons, crater density patterns, etc.

Remanent magnetism The direction and intensity of a magnetic field as recorded by certain iron- bearing rock-forming minerals. This magnetic information is recorded in the rock as it cools through the Curie Point.

Retrograde Movement of a body in the clock-wise direction when viewed from the north pole of the system.

Retrograde capture Capture of a planetoid into a planetocentric orbit from a heliocentric orbit in the retrograde direction. The energy dissipation mechanism is not specified.

Retrograde gravitational capture model (RGCM) Depiction of a process in which the captured planetoid is inserted into a planetocentric orbit in the retrograde direction.

Rift system A fracture on a planetary surface along which movement has taken place. Examples are the East African Rift Zone on Earth and the Valles Marinaris Zone on Mars.

Roche limit for a planet The radial distance at which a liquid body becomes unstable and begins to undergo a necking process at the sub-planet point as well as at the anti-sub-planet point. The limit is at ~2.89 planet radii.

Roche limit for a solid body A planetocentric position at which a "solid" but deformable body becomes unstable and begins to undergo a disintegration or necking process at the sub-planet point as well as at the anti-sub-planet point. This limit is at ~1.6 planet radii.

Rock tides Tidal action on the lithosphere of a planet that mimics the ocean tidal action. The rock tides are of lower amplitude and range but are operating somewhere at all times.

Rotating coordinate system A plotting procedure for planetocentric orbits in which the plot rotates as it theoretically revolves about the Sun.

Runge-Kutta numerical integration method A fourth-order Runge-Kutta numerical integration scheme is used to step position and velocity coordinates of the bodies forward in time. The precession of the calculations presented in this book are a few parts in 10E5 for energy and a few parts in 10E8 for angular momentum.

Satellite A body in a planetocentric orbit.

Satellite magnetometer survey A survey of the magnetic patterns recorded in the surface rocks of a body with a magnetic detection device in an artificial satellite in orbit around the body.

Scale of credibility Wilson (1998) suggested a scale of credibility for judging new ideas in science. The scale goes from Interesting to Suggestive to Persuasive to Compelling and finally to OBVIOUS. See Chap. 7 for more details.

Scientific method A procedure for solving problems in the natural sciences.

Shallow-shell magnetic amplifier A mechanism that was first proposed for generation of a magnetic field in the Moon by Roman Smoluchowski in 1973. The general principle is that convection cells in the lunar magma ocean region under certain conditions can amplify a weak "seed" magnetic field of Earth origin.

Single-impact hypothesis A term that relates to the giant impact model. In this case the debris field from which the Moon formed resulted from the impact of one large body.

Slab-pull plate tectonics A model for a subduction zone on Earth in which a cooler (denser) part of the lithosphere is subducted into a warmer (less dense) mantle. The denser slab aids in the subduction process by "sinking" into the mantle.

SNC meteorites These are the shergottite, nakhlite, and chassignite group of meteorites. They are all thought to be impact debris from Mars. For more details see McSween (1999).

Solar composition Composition of the Sun as determined by spectroscopic examination of the Sun.

Solar composition devolatilized The composition of chondrite meteorites is very similar to that of solar composition without volatiles. Chondritic meteorites also have very similar abundance ratios of elements to those of the Sun.

Solar gravitational perturbations These are gravitational tugs or influences by the Sun on a planet or planetoid. In simulations of gravitational capture, solar gravitations perturbations, under certain conditions, can reduce the energy dissipation necessary for capture by 50%.

Solar nebula A cloud of dust and gas that was inherited from previous stellar explosions in our part of the galaxy. This, then, was the starting composition for the Solar System.

Solar system A system of planets, planetoids, satellites, Kuiper Belt objects, and comets that are associated gravitationally with the Sun.

Sphere of influence A region of space in which the gravity of the planet is predominate. This is called the Hill Sphere in planetary dynamics.

Spin-orbit coupling A condition in which the orbit of a satellite controls the spin rate of the planet.

Stable capture zone (SCZ) A region of parameter space (planetoid anomaly vs. planetoid eccentricity) in which the orientation of the orbit of the encountering planetoid is favorable for a stable post- capture orbit if sufficient energy is dissipated for its capture into a planetocentric orbit.

Stagnant-lid planet A one-plate planet. However, there can be movement relative to the planetary interior.

Stochastic component of planetary rotation This model suggests that any systematic (i.e., orderly) component of planetary rotation is overwhelmed by the random process of accumulation of bodies of various sizes impacting onto the planet.

Stochastic impact model A impact model dominated by large bodies impacting on a body in essentially a random pattern in space and time.

Sub-mars point A point in a circum-martian orbit that is closest to Mars.

Synestia model The most recent version of a GIM. This model features much more kinetic energy and much more angular momentum than the standard giant impact model.

Tangential impact Impact of a planetoid at a low angle in order to impart maximum rotational angular momentum to the planet.

Tectonic consequences of earth rotation A book by Bostrom (2000) that explains the potential impact of Earth's rotation on the operation of planet Earth.

Terraforming A combination of processes by which a planet can be processed to be more like planet Earth.

Terrestrial planet A planet something like Earth (i.e., a rocky planet).

Terrestrial-type exoplanets Rocky exoplanets.

Tharsis Plateau An elevated region in the equatorial zone of Mars that has several conic-form volcanoes.

Tharsis Rise Equivalent to Tharsus Plateau.

The eighteenth-pass scenario A capture scenario in which the 18th close encounter of a capture sequence is close enough to cause tidal gravitational extraction of mare basalts from the lunar magma ocean. See Chap. 7, Figs. 7.15, 7.16, 7.18, and 7.19 for details.

Theia A name for a mars-mass planetoid that is the theoretical impactor for many scenarios of the Giant Impact Model.

Themis-type asteroid An asteroid that is very rich in water-bearing minerals. This type of asteroid is thought by some investigators to be the main source of water for the terrestrial planets.

Theoretical planetology A branch of planetary science devoted to the theoretical development of star- planet systems, planets, and planet-satellite systems.

Theory A well-tested explanation of a natural phenomena.

Tidal amplitudes on a planet Equilibrium tidal distortions of a planet.

Tidal amplitudes on a planetoid Equilibrium tidal distortions of a planetoid.

Tidal disruption encounter For tidal disruption to occur on a body like the Earth's Moon, the encounter must be well within the Weightlessness Limit and there must be a zone of molten rock material in the lunar magma ocean prior to the very close encounter.

Tidal energy dissipation The quantity of tidal energy that is dissipated during a close encounter depends on h (displacement Love number) and Q (specific dissipation factor) of the body.

Tidal friction This term relates to energy that gets dissipated by tidal activity. At the present time in geological history most of the tidal energy is dissipated by the action of ocean water. In the earlier history of the planet the rock tides may have been an important source for dissipation of tidal energy.

Tidal range The equilibrium tidal range is 1.5 times the value of the equilibrium tidal amplitude.

Tidal rhythmite sequences Sedimentary rock sequences that have organized packets of rhythmically- layered sediments. The accepted interpretation is that the couplets (light and dark and/or coarse and fine) represent diurnal or semi-diurnal cycles and that the larger packets represented fortnightly cycles. Fortnightly cycles can be packaged into annual cycles of deposition.

Tidal vorticity induction This results from a process by which rock tidal action causes convection cells to form due to differential rotation between the core and crust of a planet or planetoid. This process is described in more detail in Bostrom (2000)

Thin liquid basaltic shell dynamo A term used by Smoluchowski (1973) in his proposal that the lunar magnetic field may have been generated by tidal action with the Earth.

Traceback calculation An calculation which starts with the present condition of the Earth-Moon system and attempts to trace the system back dynamically to earlier eras in Earth-Moon history. Since there are many variables to consider, the results of such calculation can vary greatly depending on how the variables are treated.

Triton A large retrograde satellite of planet Neptune.

T-Tauri stars This term is used to describe star formation but the meaning of the term has changed somewhat over the past few years. T-Tauri stars were studies by observational astronomers because they were thought to be analogs for the evolution of the Sun. In more recent terminology, the T-Tauri phase of the sun-like star is the fourth in line for energy output. The sequence is: X-Wind, Disk-Wind, FU Orionis, and then T-Tauri phenomena. In Chap. 2, it is the FU Orionis joule heating phase that is proposed for melting either the outer portions of Vulcanoid planetoids or, in the case of small bodies, the entire Vulcanoid planetoid.

Type of encounters There are non-capture encounters, collision encounters, and capture encounters.

Ubertas A name proposed by the author for a large retrograde satellite for Mars that could have severely altered the course of the evolution of planet Mars. Ubertas is a Roman Goddess of Agriculture and Abundance.

Uncompressed density The uncompressed density of a planetary body is estimated by theoretically decompressing the mineral structures of successive layers within the body. The density, then, should relate to the density of the meteoritic bodies that accreted to form the body.

Uranus The seventh planet from the Sun.

Uranus radii The distance from the center of Uranus to the surface (~25,500 km).

Valles Marineris The Grand Canyon of Mars.

Velocity at infinity Velocity of a body at the edge of the Hill Sphere (i.e., before is comes under the influence of the planet).

Venocentric orientation The orientation of a computer plot with Venus at the origin.

Venoforming A theoretical development process resulting in conditions on a planet or exoplanet that are similar to those on Venus at the present time or at some time in the past or future evolution of planet Venus.

Venus The second planet from the Sun.

Venus orbit – lunar orbit resonance (VO-LO) model There are three possible occurrences in the history and future of the Solar System. Two of them have already left their signatures in the geological record and one is scheduled for the future. All three are caused by a synchronization of the perigean cycle of the Earth-Moon system and the heliocentric orbital motion of Venus.

Viscosity A term relating to the resistance to flow or deformation.

Volatiles Material in a gaseous state at a stated temperature and pressure.

Volatile siderophile elements (VSEs) This is a group of elements that are depleted in the composition of the Moon, as well as in Vesta, Angra, and basaltic meteorites. The abundance patterns of the VSEs may eventually help to "solve" the lunar origin problem.

Volcano A volcanic structure usually in a cone-shape.

Volcanic-type (V-Type) asteroids These are achondritic meteorites that have all undergone significant heating and melting so that basalt is (or was) present on their surfaces. Basalt is a common volcanoic rock on Earth.

Vulcanoid zone A zone in the Solar System located between the Sun and the Orbit of Mercury. This is interpreted by the author to be a source region for Luna, Adonis, Angra, Vesta, and many other volcanic asteroids.

Wadati-Benioff zone The name given for a subduction zone on Earth.

Weightlessness limit (W-limit) This is the center-to-center distance at which a particle at the sub-planet point will become weightless. That is, the particle would move toward the planet if it is within the W-limit. See Chap. 7 for more details.

Wumishan Formation A late Precambrian rock formation in central China that contains many shallow water sedimentary rock units that can be interpreted in terms of the Milankovitch model of climate variation.

Xiamaling Formation A late Precambrian rock formation in central China that contains many sedimentary rock units that are being related to cyclostratigraphy, an astronomical metronome, and the history of the Earth-Moon system. See Chap. 10 for more details.

X-Wind The earliest phase and the most energetic phase of the X-wind Model of Shu et al. (2000).

X-Wind model This model was proposed to explain the origin of the very high temperature components found in chondritic meteorites (the CAIs). The very high temperature and very dynamic processes associated with the X-Wind model have been duplicated in part in laboratories but have not been observed in nature because of the opacity of the observable features of nascent stars of roughly solar mass.

X-wind stage Earliest era of a solar evolution model and the one that relates to the formation of calcium-aluminum inclusions that are found in chondritic meteorites.

Yamato achondrite A meteorite that has the composition and texture of a typical lunar basalt. Thus, the Moon is thought be the parent body for this meteorite. This specimen was collected from the Yamato Mountain Range in Antarctica.

Zircon crystals Crystals composed of zirconium silicate and crystals that are very refractory. Zircon crystals commonly contain inclusions of uranium and thorium bearing minerals. These mineral inclusions contain the age-dating elements for the U-Th-Pb system of radiometric dating.

Author Index

R. Malcuit, *Geoforming Mars*, https://doi.org/10.1007/978-3-030-58876-2

Subject Index

R. Malcuit, *Geoforming Mars*, https://doi.org/10.1007/978-3-030-58876-2